SURFACTANTS

Surfactants

In Solution, at Interfaces and in Colloidal Dispersions

Bob Aveyard

OXFORD

UNIVERSITY PRESS

Great Clarendon Street, Oxford, OX2 6DP,
United Kingdom

Oxford University Press is a department of the University of Oxford.
It furthers the University's objective of excellence in research, scholarship,
and education by publishing worldwide. Oxford is a registered trade mark of
Oxford University Press in the UK and in certain other countries

Published in the United States of America by Oxford University Press
198 Madison Avenue, New York, NY 10016, United States of America

British Library Cataloguing in Publication Data
Data available

Library of Congress Control Number: 2019937511

ISBN 978–0–19–882860–0

DOI: 10.1093/oso/9780198828600.001.0001

Printed and bound by
CPI Group (UK) Ltd, Croydon, CR0 4YY

Preface

Surfactants are ubiquitous in nature and have been known, made and used by man for millennia. They adsorb from aqueous solution to various kinds of interfaces, and if sufficiently concentrated in solution, they simultaneously aggregate into sub-microscopic units (*micelles*). This ability to adsorb and to aggregate makes surfactants enormously useful in many industrial processes and in a wide range of everyday products.

The behaviour of surfactants in solution and at interfaces (i.e. *surfactant science*) comes under the broad umbrella of colloid science which, as whole, underwent a renaissance in the second half of last century. It is now very much an accepted academic subject, although not as widely taught as it might be given its importance to the understanding and control of many technological and biological processes. It underpins and is part of the relatively new nanoscience and nanotechnology. Insights into basic surfactant science have been exploited extensively in industry, and what was once empirical 'know how' (of say formulators) has become increasingly informed by the underlying science.

In the main, colloid science has been served well by academic texts, although the basic aspects of surfactant science have been less extensively covered. As mentioned, surfactant behaviour has direct relevance to a range of chemical technologies, for example, those involved in the production of foodstuffs, personal care and household products, pharmaceuticals, agrochemicals, and so on, and much of what has been written is directed primarily to the practical and industrial application of surfactants. Consequently there is room for a pedagogical text covering some of the more fundamental aspects of the physical science of surfactant behaviour, at a level suitable for senior undergraduates and postgraduates in chemistry and the allied disciplines of physics, chemical engineering, food science, pharmacy and biological and medical sciences. It is hoped that this book will help towards meeting such a need, and at the same time be of use to researchers in industry faced with the need to familiarize themselves with some background aspects of surfactant behaviour.

In writing the book, it has been assumed that the reader has a knowledge of basic physical chemistry, particularly chemical thermodynamics, and of simple physics (e.g. electrostatics). Mathematics (elementary algebra and calculus) is kept at a simple level, consistent with the straightforward derivation of many of the equations presented. Where an equation is simply presented, a reference to an original research paper is usually given. There are some well-established textbooks which are referred to throughout the book, but the majority of the references made are to original research papers in the scientific literature. These relate to the examples discussed in the text, and in this way the reader will be able to gain a wider appreciation of the subject matter.

A novel feature of the book is a section entitled 'Themes and Connections'. The way in which the material has been assembled and broken into chapters is to some extent arbitrary since topics can often be grouped in a variety of useful ways, depending on the interests of the reader. In part this is catered for in the text by providing cross references, and of course connections can be made by referring to the book index. However, in order to help the reader a little more explicitly, a series of major topics has been selected and relevant material to be found throughout the book is referred to under the topic headings. A student of surface chemistry or a research worker embarking on an industrial project involving the use of surfactants, or indeed someone assigned to produce a university or industrial course in surface chemistry or surfactant science, may wish to understand in a general way something about, for example, interfacial tension, capillarity, the curvature properties of surfactant monolayers or wetting, spreading and thin liquid films. The material referred to under a given topic (which is illustrative rather than exhaustive) is what might have been included in a chapter devoted to that topic alone. It is hoped that the presentation of some of the material in this way will help to give the reader a broader and more coherent understanding of the subject matter.

Increasingly, small and colloidal particles are being used as (in place of) surfactants, for example in the stabilization of emulsions. As background to the use of particles as quasi surfactants, chapters are included covering the adsorption of single particles and the formation of particle monolayers at fluid interfaces (Chapter 17) and aspects of the use of particles in the stabilization of (Pickering) emulsions (Chapter 18).

Although colloid and surfactant science gained 'respectability' rather late in the day, their scientific foundations have had contributions from a number of great scientists over the centuries. To reflect this, short biographical notes are given throughout the text in recognition of the contributions of such scientists as Newton (intermolecular forces), Franklin (early experiments on insoluble monolayers), Faraday (early preparation and recognition of colloidal dispersions), Plateau (behaviour of collections of soap films in foams), and Gibbs (thermodynamics of surfaces).

I am most grateful to John Clint, Paul Fletcher, and Brian Vincent for reading the manuscript of the book and making valuable and constructive suggestions. Of course the responsibility for errors, misconceptions and omissions lies entirely with me. I have learned much of the surfactant science I know alongside, and often from my friends and co-research workers at Hull University over the years, particularly Bernie Binks, John Clint, Paul Fletcher, Tommy Horozov, and Vesko Paunov. Thanks to them, as well as to all the Ph.D. students and postdoctoral research workers I have had the pleasure of working with. Last and by no means least, my thanks go to my wife for her great forbearance during the preparation of the book.

Bob Aveyard
Beverley, December 2018

Contents

Section IV Surface forces and thin liquid films

Themes and connections   523

Index   535

Principal symbols

The equation or section numbers refer to the place where the symbol first appears or is defined.

$\mathcal{A}$	surface area §3.2
A	Helmholtz free energy §4.2.2; area per mol or per molecule §5.3
A_o	molecular co-area §5.3.2
A_σ	Helmholtz free energy per unit area of surface (4.2.16)
A	Debye–Hückel constant §4.3.3
A_H	Hamaker constant (11.2.3) (subscript H is omitted if phase subscripts used)
a	activity (4.3.6); (subscript) denotes air §3.2; area available to head group at micelle surface §9.4
aw	(subscript) denotes air/water interface
a_o	optimal head group area §9.5.1
$a_\pm$	mean ionic activity §4.3.3
ads	(subscript) denoting adsorption
b	neutron scattering length density of adsorbed molecule §5.5.2
c	molar concentration; curvature §10.2.2
c_K	critical coagulation concentration §13.2
c_o	spontaneous curvature §10.2.2
D	diffusion coefficient §6.2.2
d	(superscript) denoting from dispersion forces §3.3
d	(subscript) denoting liquid drop §3.4.2
d	distance of closest approach of centres of molecules §5.3.2; surfactant layer thickness §5.5.2; crystal spacing (9.11.3)
d_o	equilibrium separation of centres of molecules §5.3.2
E	electric field §A10.1; entry coefficient (12.3.1)
E_c	curvature energy of surface, area $\mathcal{A}$ (10.2.9)

E^g	generalized entry coefficient (12.3.3)
$\mathbf{e}$	mathematical constant
e^-	electronic charge
F	force
F_u	free energy of forming unit area plane surface from droplets §10.2.2
f	activity coefficient
$f(x)$	normal Gaussian) distribution §9.10
f_{inter}	force of interaction between 2 spherical particles (17.4.2)
f^s	activity coefficient in surface (5.5.10)
$f_{\pm}$	mean ionic activity coefficient §4.3.3
$\mathcal{G}$	a Gibbs free energy (4.2.9)
G	Gibbs free energy §2.1; shear modulus §6.2.1
G^*	complex modulus (15.1.5)
G'	storage (elastic) modulus (15.1.6)
G''	loss (viscous) modulus (15.1.6)
g	acceleration due to gravity
g_c	curvature energy per unit area (10.2.1)
$g(\tau)$	correlation function in dynamic light scattering (A10.8)
H	enthalpy §2.1; mean curvature (10.2.11)
h	Planck constant; separation between surfaces (11.2.3)
I	intensity (light, scattered neutrons) §A10.1
I	ionic strength §4.3.3
$I(Q)$	intensity of scattered neutrons (A10.11)
i	(subscript) generic component label
$\mathcal{J}(t)$	shear creep compliance (15.1.4); Janus balance §17.3.2
j	flux §6.3.1
K	equilibrium constant §9.2.1
k	Boltzmann constant
k^-	rate constant for monomer dissociation from micelle (9.10.1)
k_H	Huggins coefficient (15.3.5)

l	(subscript) denoting liquid
m	molality
mic	(subscript) denoting micelle formation §9.2.1
N	number of molecules/ions
$N(\infty)$	number density in bulk solution (5.3.28)
n	number of moles §2.1; neutron refractive index (8.1.2); refractive index §8.3; aggregation number §9.2.1; diffraction order (9.11.3); number molecules per unit volume (11.2.4)
n^o	ion number concentration in bulk solution (11.2.15)
n^+	number concentration of cations (11.2.15)
n^-	number concentration of anions (11.2.15)
$\tilde{n}$	complex refractive index (8.3.6)
n_{agg}	aggregation number defined in (9.5.2)
N_A	Avogadro number
o	(superscript) denotes standard quantity §2.1
oa	(subscript) denotes oil/air interface
ow	(subscript) denotes oil/water interface
P	wetting perimeter §A3.1; packing factor (9.5.4)
$P(Q)$	form factor (A10.2)
p	(superscript) denoting from polar forces §3.3
p	pressure; vapour pressure §3.4.2; polydispersity index (10.2.8)
p_c	capillary pressure §12.2.2
p^o	saturated vapour pressure §3.4.2
$p(n)$	micellar size distribution (9.4.10)
Q	momentum transfer (modulus of the wave vector) (8.1.1) and (A10.4)
q	heat absorbed by system (4.2.1); charge
q_{im}	image charge (17.4.1)
R	gas constant
R_θ	Rayleigh ratio §A10.1
R_G	radius of gyration of a particle (A10.3)
$R(Q)$	reflectivity §5.2.2

R_H	hydrodynamic radius (A10.7)
$\mathcal{R}$	molar ratio water: aggregated surfactant in droplet §10.1.2
r_g	radius of gyration of polymer molecule §11.3.3
r^*	radius of critical nucleus §3.4.2
S	classical spreading coefficient §(16.2.1); entropy (4.2.2)
S^g	generalized spreading coefficient (16.2.8)
s	(superscript) denoting use of surface phase model §4.1.1
sa	(subscript) denotes solid/air interface
so	(subscript) denotes solid/oil interface
sw	(subscript) denotes solid/water interface
T	temperature
T_m	average micellar lifetime (9.10.7)
t	time
t_r	relaxation time in Maxwell model (15.2.6)
$trans$	(subscript) denoting transfer between phases §5.3.1
U	internal energy (4.2.1)
V	volume; interaction energy between 2 molecules (A5.3)
$V(h)$	interaction energy per unit area between surfaces, separation h (11.2.3)
V_A	attractive interaction energy per unit area between surfaces §11.2.1
V_R	repulsive interaction energy per unit area between surfaces §11.2.3
V_T	$V_A + V_R$ §11.2.4
V_S	steric interaction energy per unit area §11.3.1
V_{Sm}	osmotic contribution to steric interaction energy §11.3.1
V_{dep}	attraction energy from depletion forces §11.3.3
v_p	volume of particle (A10.11)
v_m	molar volume of liquid §3.4.2
v	subscript denoting vapour §3.4.2; molecular volume §9.4; neutron velocity (A10.10); drop velocity (14.6.3)
vol	(subscript) referring to Volmer model (5.3.23)
$W_A; W^A$	work of adhesion §3.3—W^A is used when phase subscripts are present
W^C	work of cohesion §16.2.1

W	hydration interaction energy (11.3.8)
w	work done on system (4.2.1); interaction energy between molecules Box 11.7
x	mol fraction; fraction of bulk phase concentrations §5.5.1; number of nearest neighbour sites per site (A5.3)
z	valence of an ion
	Greek symbols
α	a phase label §2.1; van der Waals interaction parameter (5.3.25); ratio of concentrations §5.5.3 and §9.7; compound excluded area §5.5.4; micellar degree of dissociation (9.6.2); polarizability of molecule (11.2.2); angle indicating extent of apolar cap on spherical Janus particle §17.3.1
α_c	micellar degree of dissociation (9.6.4)
α_p	degree of surfactant dissociation at plane surface (9.6.5)
β	a phase label §2.1; interaction parameter in regular solution theory (5.5.17); interaction parameter (5.5.18)
Γ	surface concentration
Γ^s	surface concentration §4.1.1.
Γ^σ	surface excess concentration §4.1.2
$\Gamma_2^{(1)}$	relative adsorption of component 2 (4.3.5)
γ	surface or interfacial tension (3.2.1)
γ_c	interfacial tension at the *cmc* §10.2.4; critical surface tension (16.3.9)
γ^d	part of interfacial tension arising from dispersion forces §3.3
γ_f	film tension (12.1.1)
γ^p	part of interfacial tension arising from polar forces §3.3
$\dot{\gamma}$	shear or strain rate §6.2.1
γ	shear strain §6.2.1
$\dot{\gamma}^*$	complex shear strain rate (15.2.1)
Δ	change in thermodynamic quantity §2.1; ellipsometric parameter (8.3.5)
$\Delta\chi$	surface potential Box 5.5
δ	amount of nonionic surfactant §10.4.2; adsorbed polymer layer thickness §11.3.1; phase angle §15.1; lens angle §16.2.3
δ_{rs}, δ_{rp}	phase shift on reflection of s and p-polarized components (8.3.2)

ε	permittivity; Gibbs elasticity Box 12.2
ε_o	permittivity of free space
ε_r	relative permittivity
ε^*	dilational modulus (6.2.4); complex dilational viscoelastic modulus (6.2.8)
ε'	storage (elastic) modulus (6.2.8)
ε''	loss modulus (6.2.8)
η	viscosity
η^*	complex viscosity (15.1.9)
η_d	surface dilational viscosity (6.2.5)
η_s	surface shear viscosity (6.2.2)
η_r	relative viscosity (15.3.1)
η_P	plastic viscosity §15.4.3
η_{sp}	specific viscosity (15.3.3)
$[\eta]$	intrinsic viscosity (15.3.4)
θ	angle; contact angle §3.3; fractional surface coverage §7.4
θ_A	advancing contact angle §16.3.2
θ_{CB}	Cassie–Baxter contact angle (16.3.22)
θ_R	receding contact angle §16.3.2
θ_W	Wenzel contact angle (16.3.21)
θ_a	contact angle of apolar part of Janus particle with oil/water interface §17.3.1
θ_c	contact angle at $\tau = \tau_c$ §17.2.2
θ_m	contact angle at $\tau = \tau_m$ §17.2.2
θ_p	contact angle of polar part of Janus particle with oil/water interface §17.3.1
$(\theta_p)_t$	value of θ_p at which particle transfers from water to oil (17.3.3)
κ	bending elastic modulus §10.2.2; reciprocal of double layer thickness (11.2.20)
$\overline{\kappa}$	Gaussian elastic modulus §10.2.2
λ	wavelength §5.5.2
λ_c	critical wavelength (12.4.1)
μ	chemical potential §2.1; apparent viscosity §15.1
ν	frequency of oscillation §6.2.2
ν_e	absorption frequency in UV (11.2.10)

ξ	persistence length (10.2.12); ratio of adsorptions §16.3.4
π	spreading or surface pressure §4.4
Π	osmotic pressure (11.2.26)
Π_D	disjoining pressure (11.3.7)
Π_{el}	electrical contribution to disjoining pressure (11.6.8)
ρ	density of a bulk phase §3.4.1; neutron scattering length density §5.5.2; complex amplitude reflection ratio (8.3.1); volume charge density (11.2.16); number density of particles §11.3.4
Σ	specific surface area (of a solid) §7.1
σ	surface charge density §5.3.2;(superscript) denoting use of Gibbs model for surface §4.1.2; (subscript) denoting per unit area of surface §4.2.2; standard deviation of micelle size §9.10.1
τ	surface thickness §4.1.1; shear stress §15.1; line tension §16.2.3
τ_c	relaxation time §6.2.2
τ_1	fast micelle relaxation time (9.10.1)
τ_2	slow micelle relaxation time §9.10.2
τ_{12}	slow micelle relaxation time (9.10.5)
τ_B	Bingham yield stress (15.4.3)
τ_c	critical line tension §17.2.2
τ_m	maximum line tensions §17.2.2
$\bar{\tau}$	reduced line tension (17.2.2)
ϕ	correction factor in drop volume method §A3.3; phase §10.4.1
φ	correction factor in du Noüy ring method §A3.1; phase shift (6.2.7); volume fraction
χ	Flory-Huggins parameter §11.3.1
ψ	electrical potential; ellipsometric parameter (8.3.4); meniscus slope angle §12.4.3
ω	angular velocity §A3.5; angular frequency §6.2.2; partial molar surface area (4.2.18)
ω_c	characteristic frequency §6.2.2

Section I

Background

Section I comprises the first three chapters of the book. The chemical nature of surfactants, what they can do, and why they are important are described in the first chapter. A characteristic of surfactant molecules is that, as a result of their amphipathic nature, they adsorb from water to a variety of interfaces and aggregate in aqueous solution. These key processes are often energetically driven by the removal of the hydrophobic groups in the molecule from water. In Chapter 2 the antipathy between nonpolar molecules and groups and water is explored. Capillarity, wetting, and interfacial tension are the subjects covered in Chapter 3. These are relevant to many aspects of behaviour in systems containing surfactants as well as to the varied applications of surfactants.

1

What are surfactants?

1.1 The nature of surfactants

Surfactants . . . today you have probably eaten some or rubbed others on your bodies in one form or another. Plants, animals (including you), and microorganisms make them. Many everyday products contain surfactants or surfactants have been involved during their manufacture. So what are they, how do they behave in solution, at interfaces, and in colloidal dispersions, and why are they useful? This is the subject of the book.

The term surfactant is a contraction of *surf*ace *active agent*, and surfactants are so called because they adsorb at surfaces (interfaces). The tendency to adsorb arises from the dual (amphiphilic) nature of surfactant molecules, which contain a hydrophobic moiety (the 'tail' group, which is often an alkyl chain) attached to a hydrophilic ('head') group. A surfactant molecule can be represented for present purposes, as shown in Fig. 1.1a. In Fig. 1.1c a single layer (*monolayer*) of surfactant molecules is shown adsorbed at an air/water or oil/water interface. The driving force for the adsorption arises either from the removal of the surfactant tails from water (for surfactant originally in water) or the removal of the heads from (nonpolar) oil if the molecules were originally in the oil. In either case, in the monolayer the heads are in aqueous surroundings and the tails are out of contact with bulk water. Since, for example, long chain alkanes are virtually insoluble in water, the water-solubility of surfactant molecules arises from the very favourable interaction of the head groups with water (through dipole–dipole and/or ion–dipole interactions). The chemical nature of some typical head groups is covered in §1.3.

Adsorption of molecules at a surface or interface (Box 1.1) always lowers the surface or interfacial tension (Chapter 4). If the concentration of surfactant in an aqueous phase is sufficiently high, and when the adsorbed monolayer is effectively full (i.e. *close-packed*), some of the molecules in solution aggregate into *micelles* (or other aggregates), which remain in equilibrium with adsorbed molecules and un-aggregated molecules (in the context usually termed 'monomers') in the aqueous phase, as discussed in Chapter 9. The concentration at which micelles just begin to form is termed the *critical micelle concentration* (*cmc*) (§9.1). A cross-section of, for example, a simple spherical micelle is represented in Fig. 1.1b; the hydrophobic tails form the core of the micelle, which is liquid-like, while the heads are at the surface of the micelle in contact with water. The

Surfactants: In Solution, at Interfaces and in Colloidal Dispersions. Bob Aveyard. © Bob Aveyard 2019.
Published in 2019 by Oxford University Press. DOI: 10.1093/oso/9780198828600.001.0001

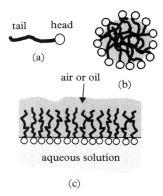

tail   head

(a)

air or oil   (b)

aqueous solution

(c)

Figure 1.1 *(a) Representation of a surfactant molecule; (b) cross-section of a spherical micelle in water; (c) an adsorbed monolayer at an air/water or oil/water interface.*

diameter of such a micelle is approximately (but less than) twice the extended surfactant molecular length (say 3 or 4 nm), and very approximately a spherical micelle might contain of the order of a hundred molecules. Micelles are able to take up ('*solubilize*') water-insoluble materials. At the *cmc*, the adsorbed monolayer is close-packed (Box 1.2) and the lowering of surface tension by the surfactant adsorption is maximum and roughly constant. Because surfactants lower surface and interfacial tension, adsorb at interfaces, and solubilize materials in micelles they are extremely useful in a very wide variety of products, systems, and processes.

Box 1.1   Interface and interfacial tension

Interfacial tension is defined as the work required to form unit area of an equilibrium interface at constant temperature, pressure, and volume, as explained in §4.1. It quantifies the tendency of an interface to contract; small free drops of liquid for example are spherical in shape, the sphere having the minimum area (hence minimum surface free energy) for a given volume of liquid. Adsorption at an interface always lowers interfacial tension. The term *surface tension* is often used for an interface between a condensed and a vapour phase, whereas *interfacial tension* is frequently taken to refer to the region between two condensed phases. In general however, an interface is an area over which two phases (in any physical state, solid, liquid, or gas) are in contact. Physically it has a finite thickness and contains material from each of the contacting phases (Chapter 4).

Box 1.2   What does 'close-packed' mean?

The term *close-packed* simply implies that no further adsorption occurs into the monolayer with increasing bulk concentration of surfactant. The area available per adsorbed molecule in the close-packed state is not however simply related to a molecular dimension as represented

Box 1.2 *Continued*

by say a molecular model (see §5.3.2 and §9.5.1). In the present context close-packing corresponds to the condition where the chemical potential of surfactant is the same for surfactant in the monolayer and in micelles, which are in equilibrium with single surfactant molecules in solution.

1.2 Where are surfactants encountered?

Over the centuries, and probably back to ca. 3000 bc, the most commonly encountered surfactant was undoubtedly soap, which is an alkali metal (usually sodium) salt of a carboxylic acid. Soap can be prepared from animal fats or vegetable oils, which contain mixtures of triglycerides, together with lye (containing sodium hydroxide) obtained by extraction from plant ashes. The process of the alkaline hydrolysis of fats and oils is called saponification; the term possibly arises from the name of the Italian town Savona where soap was manufactured as early as the ninth century ad. The French for soap is savon.

Various plants known to our ancestors produce natural surfactants; for example, saponins (natural glycosides) can be extracted from the foliage of soapwort. Our bodies produce natural surfactants or surfactant-like molecules, perhaps the most obvious being lung surfactants (Box 1.3). Our livers produce bile acids, the salts of which are biological surfactants. Phospholipids (e.g. lecithin) are produced by the liver and are founds in cell membranes. Lecithin is also used in the production of food emulsions (e.g. ice cream), since it is a very effective emulsifying agent. Bacteria produce biosurfactants (§1.3.5).

Box 1.3 Lung surfactants

Lung, or pulmonary, surfactants are present in alveolar surfaces within the lung and consist of a mixture of phospholipids and proteins. The mixture is very surface active (i.e. adsorbs very strongly) at the surfaces of the alveoli and reduces their interfacial tension. Thus the production of lung surfactant makes it possible to expand alveolar surfaces easily on inhalation. In a condition termed respiratory distress syndrome (RDS), which affects 10% of premature infants and which is life-threatening, the lungs do not produce sufficient lung surfactant without which the lungs are unable to inflate.

Soap is a good cleaning agent, but it produces a precipitate or scum when used in hard water because the calcium and magnesium carboxylates are water-insoluble. New synthetic surfactants were introduced in the 1920s which do not suffer in this way and now there are many types of surfactants available for a whole range of applications and formulations. As will be seen they can, for example, act as emulsion stabilizers (Chapter 14), promote wetting of surfaces (Chapter 16), produce foaming (Chapter 14), effect the dissolution and dispersion of 'water-insoluble' material in water

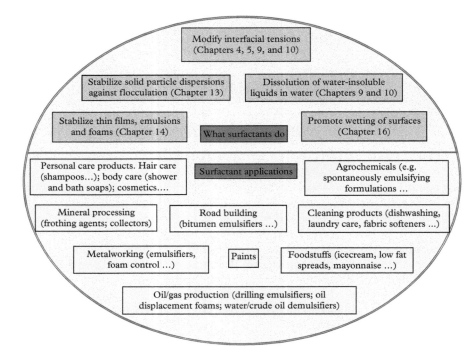

Figure 1.2 *Some of the things surfactants can do and examples of their applications.*

(Chapter 9), and stabilize dispersions of solid particles against flocculation (Chapter13). As a result of this remarkable versatility, surfactants are used in industries ranging from agrochemical through pharmaceutical to crude oil production and processing. The worldwide surfactant industry is a multi-billion pound business. In 2014, for example, the global consumption was 17,500 kilotons, over half of which was used in household detergent formulations. Other significant amounts were used in industrial and institutional cleaning, production of personal care products, oil production, in the textile industry, food manufacture, mineral production, and pulp and paper manufacture (see Fig. 1.2).

1.3   Types of surfactant and their classification

Although soaps (originating from renewable vegetable and animal fats) are still significantly used, the present market is dominated by products derived from petrochemicals, notable amongst these being alkylbenzene sulfonates and alkylphenol ethoxylates. Modern detergent powders typically contain linear alkylbenzene sulfonates. Nonetheless, there is now pressure to use renewable plant feedstocks more and reduce the use of non-renewable petrochemicals. In this connection, the use of 'sugar surfactants' is becoming important; these are made from carbohydrates and plant oils and find use as emulsifiers in foodstuffs and cosmetics. Alkyl polyglucosides are used in detergent products.

Surfactants are usefully classified in terms of the nature of their hydrophilic head group, which can be nonionic, ionic, or zwitterionic (amphoteric) (Tadros, 2005). Ionic surfactants can of course be anionic or cationic. Polymeric surfactants form another class. In addition fluorocarbon surfactants constitute an important group of speciality surfactants which are particularly effective in lowering the surface tension of water. Perfluoroalkyl chains form the tail of fluorocarbon surfactants, which can have ionic, nonionic or amphoteric head groups.

In what follows just a few examples of the various types of surfactants are given. Authoritative and fuller accounts can be found, for example, in books by Porter (1994), Holmberg et al. (2003), Holmberg (2003), Rosen and Kunjappu (2012), and Tadros (2005).

1.3.1   Nonionic surfactants

The hydrophilic moieties in nonionic surfactants are commonly polyoxyethylene groups, which if attached to a straight alkyl chain hydrophobe gives rise to so-called alcohol ethoxylates: $C_nH_{2n+1}(CH_2CH_2O)_mOH$. The formula is often contracted to C_nE_m, where n and m represent the tail and head group lengths, respectively. If n is relatively small and m is relatively large, it is clear for these surfactants that the head group can be considerably larger that the tail group. Alcohol ethoxylates have often been used in academic investigations of the behaviour of nonionic surfactant solutions since pure monodisperse samples can be prepared with a range of both n and m values. Alkylphenol ethoxylates, in which the hydrophobic groups are alkylphenol residues rather than straight alkyl chains, are widely used as heavy duty degreasing detergents.

Mono-, di-, and triesters of sorbitan with carboxylic acids (e.g. octadecanoic acid, oleic acid) form an important group of nonionic surfactants, often referred to as Spans (an Atlas trade name). These materials can be ethoxylated giving so-called Tweens; Spans are insoluble in water whereas Tweens are usually water-soluble. The surfactants (see Fig. 1.3) are used as cosmetics emulsifiers and in pharmaceutical preparations.

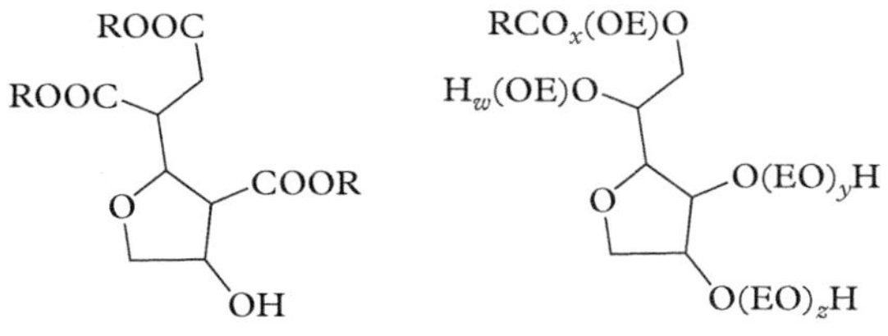

(a)   (b)

Figure 1.3 *(a) A sorbitan ester, where R is H or an alkyl chain. If, for example all $R = C_{18}H_{37}$, the structure is that of Span 65. (b) An example of an ethoxylated sorbitan ester, termed a Tween.*

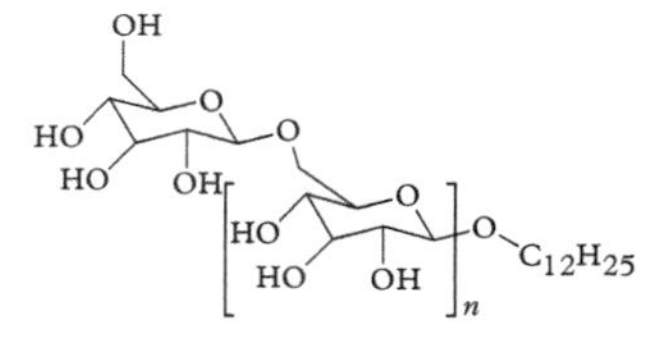

Figure 1.4 *A dodecyl polyglycoside.*

As mentioned, surfactants with head groups based on carbohydrates are becoming of more importance. In addition to sorbitan, glucose and sucrose for example can be esterified giving respectively alkyl polyglycosides (APGs, used in detergents) (Fig. 1.4) and sucrose esters (used in food manufacture). Surfactants of this type are often referred to generically as 'sugar surfactants'.

1.3.2   Ionic surfactants

Anionic surfactants

The alkali metal carboxylates (soaps) have already been referred to and have long been produced. Soaps are cheap, easily made and biodegradable, but their alkaline earth metal salts are insoluble in water and produce a scum. Animal fats and plant oils are used in their production, giving alkyl chain lengths above 10 and below 20. Shorter chains render the soaps too soluble in water and longer chains result in insolubility in water, although the long chain homologues become oil-soluble. The alkali metal soaps are used (particularly in developing countries), for example, in toilet soap and other personal care products, in cutting oils, ore flotation, and emulsion polymerization.

Sulfonates and sulfates are the most commonly encountered anionic surfactants (Fig. 1.5). The sodium salts of linear alkylbenzene sulfonates are very widely employed in industrial and household detergent formulations. Other cations can be used and surfactants based on amine salts (e.g. of triethanolamine, isopropylamine) have found uses in dry-cleaning and production of agricultural emulsions. A very widely encountered alkyl sulfate is sodium dodecylsulfate (SDS), which is both industrially important and widely used in academic studies. SDS is used, for example, in low temperature detergents, in toothpaste, and in food and cosmetic emulsions.

To make the sulfates and sulfonates more water-soluble and increase their resistance to precipitation in electrolyte solutions, ethyleneoxy (EO) groups can be incorporated giving, for example, an alcohol ether sulfate $R(CH_2CH_2O)_xSO_4^-M^+$ (x is say 2 or 3). Addition of EO groups to carboxylates (producing $RO(CH_2CH_2O)_xCH_2COO^-$) can also render the Ca^{2+} and Mg^{2+} salts more water-soluble.

An anionic surfactant worth mentioning here, which has been used widely in academic studies (see Chapter 10), is Aerosol OT (sodium di(2-ethylhexyl)sulfosuccinate) (Fig. 1.6).

Cationic surfactants

Cationic surfactants are widely encountered in everyday life. Since the head group carries a positive charge, cationics tend to adsorb at solid/water interfaces which usually

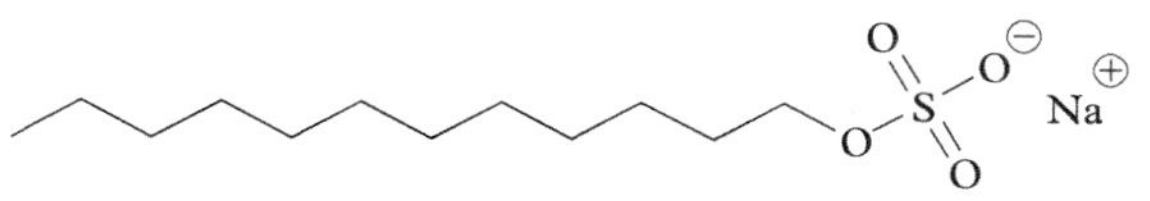

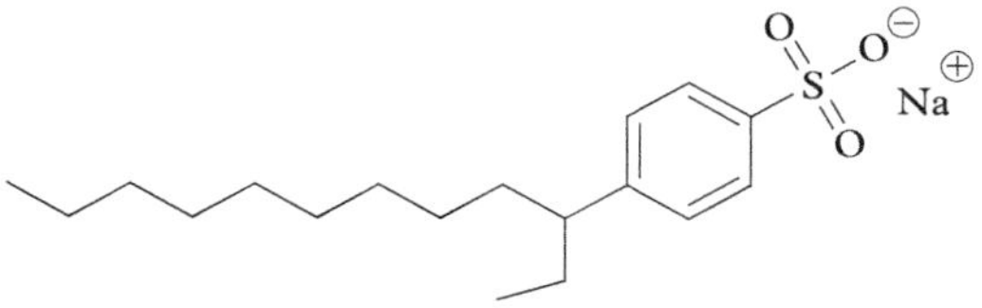

Figure 1.5 *Top: Sodium dodecylsulfate. Bottom: An isomer of sodium p-dodecylbenzene sulfonate.*

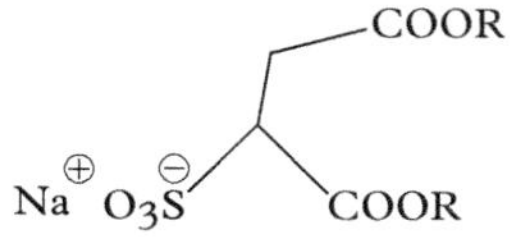

Figure 1.6 *Structure of Aerosol OT; R represents the 2-ethylhexyl group.*

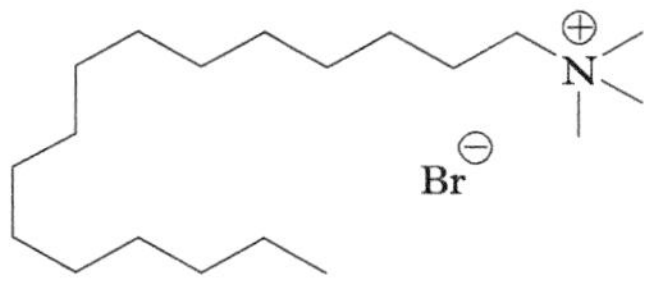

Figure 1.7 *Hexadecyltrimethylammonium bromide (CTAB).*

carry a negative charge (but see Chapter 7). The most common examples of cationic surfactants are the quaternary ammonium salts mono-, di-, and trimethylammonium halides. Cetyltrimethylammoniun bromide (usually abbreviated to CTAB) has been widely used in fundamental studies of surfactant behaviour (Fig. 1.7). Cetyl is the common name for the hexadecyl group; the single-chain surfactants usually have chain lengths between 8 and 18. The dialkyl chain surfactants are of course less water-soluble than the single chain surfactants for a given chain length. Other cationic head groups are possible, as for example in dialkylbenzylammonium chlorides, used in disinfectants. Cationics are also employed *inter alia* in hair conditioners and as antistatic agents.

1.3.3 Amphoteric surfactants

Amphoteric surfactants carry a pH-dependent charge; the charge is positive at low pH and negative at high pH. At some intermediate pH, the so-called *isoelectric point (i.e.p.)*, the head group is net neutral. At high pH amphoteric surfactants behave as anionic surfactants, at low pH like cationics and at the *i.e.p.* like nonionics. The aqueous-phase solubility is minimum at the *i.e.p.* Amphoteric surfactants are widely used in non-irritant

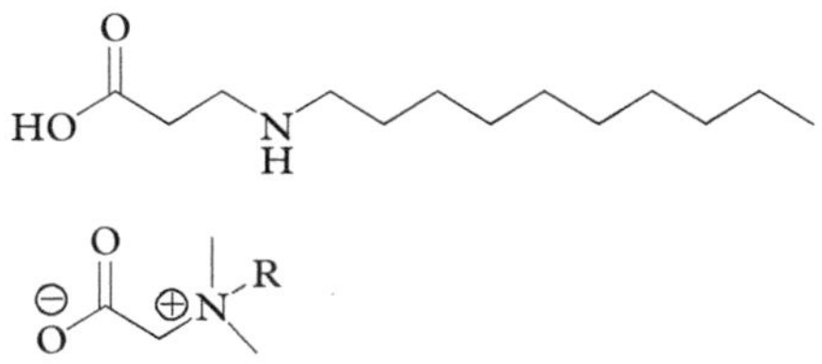

Figure 1.8 *Top: β-N-decylaminopropionic acid (isoelectric point pH = 4). Bottom: N-alkylbetaines are zwitterionic above the i.e.p. and cationic below, but are never anionic.*

mild detergent products such as shampoos and bath products. Some examples of amphoteric surfactant types are shown in Fig. 1.8. Some zwitterionic surfactants, for example, sulfobetaines, $RN^+(CH_3)_2(CH_2)_xSO_3^-$ (where x is the number of methylene groups in the molecule), are not sensitive to pH.

1.3.4 Polymeric surfactants

So-called polymeric surfactants do not always have very high molecular weights, and some might be better termed oligomeric surfactants. Block copolymers of ethylene oxide (EO) and propylene oxide (PO) are important examples of polymeric surfactants, for example, the simple ABA type $(EO)_x(PO)_y(EO)_x$. The EO and PO moieties are, respectively, hydrophilic and hydrophobic in nature. Relative molar masses range from about 1000 to 30,000, and these materials can be tailor made with different amounts of EO and PO for different applications. Uses include pigment dispersants in latex paints (high molar mass members with high EO content), and petroleum demulsifiers and foam control agents (low molar mass and low EO content) (see Chapter 14).

Graft copolymers, in which side chains are grafted onto a polymeric backbone, form an important group of both natural and synthetic surfactants. The backbone can be hydrophobic with hydrophilic side chains or vice versa. The backbone can also contain both hydrophilic and hydrophobic groups (so the backbone might be considered a block copolymer) and both hydrophilic and hydrophobic side chains can be attached simultaneously to the backbone. As can be appreciated many variations are possible (Holmberg et al., 2003), but a common feature in polymeric surfactants is that the molecules can be so oriented as to present hydrophilic groups to aqueous surroundings and hydrophobic groups to a nonpolar region (e.g. the oil at an oil/water interface). An example of a polymeric surfactant with a hydrophilic backbone and hydrophobic side chains is hydrophobized starch (an emulsion stabilizer and 'associative thickener'); there are many surfactants with a hydrophobic backbone and hydrophilic side chains, including ethoxylated alkyd resins and the important class of silicone surfactants. The siloxane backbone in silicone surfactants is the hydrophobe; silicone polyethers (Fig. 1.9.) are commonly used surfactants which can contain one or more polyoxyalkylene chains (e.g. polyoxyethylene or polyoxypropylene) attached to the backbone.

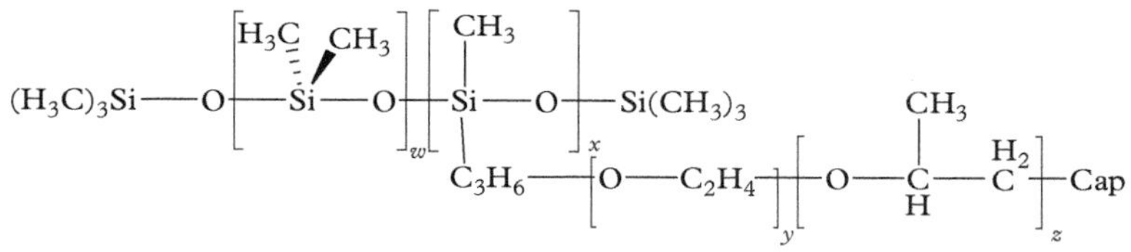

Figure 1.9 *A silicone polyether surfactant.*

From what has been said so far it is clear that the use and importance of surfactants is very wide indeed. Improvements in performance are always being sought, as are novel structures for traditional as well as for emerging commercial and research uses. Given this, it is a little difficult to fit, in an explicitly useful way at least, all surfactant types of interest into the above simple classification. For this reason some surfactant types which have been of more recent interest, namely biosurfactants and *gemini* and *bolaform* surfactants, are mentioned below. In addition, small solid particles are known to adsorb, often very strongly, at fluid/fluid interfaces (Chapter 17). It has long been known that particles can stabilize emulsions in the absence of surfactants (Chapter 18), and there is now a growing interest in using particles in place of surfactants for some uses.

1.3.5 Biosurfactants

Many microorganisms produce amphiphilic materials termed biosurfactants. The surfactants are formed on cell surfaces or excreted into the surroundings. There is a wide variety of structures of biosurfactants, often complex, with either high or low molar mass. Many have been purified and identified, whilst others have been detected by the effect they have on the surface tension of water. High molar mass biosurfactants are usually polymeric hetero-saccharides containing proteins whilst those with low molar masses are commonly glycol-lipids, the hydrophilic groups being based on trehalose, sophorose, or rhamnose. Examples, together with space filling models, are given in Fig. 1.10.

A widely studied example of a lipopeptide biosurfactant is surfactin (from *Bacillus subtilis*). It has the structure shown in Fig. 1.11, the hydrophilic group being made up of seven amino acids. The versatile behaviour illustrates the considerable potential of biosurfactants in a number of areas. Surfactin is an antibiotic and has antifungal and antiviral properties; it also has potential in crude oil recovery (enhanced oil recovery, EOR) and environmental applications . Biosurfactants are probably best known for their potential in bioremediation of polluted soil. The surfactants, which can be added to soil or created *in situ*, increase the dispersion (and/or solubility) of hydrocarbon based contaminants and hence their bioavailability to bacteria which break down the pollutants to less harmful products. Biosurfactants are less toxic and more readily biodegradable than conventional synthetic surfactants and are likely to find many uses (Banat et al., 2000), for example in the production of pharmaceuticals, toiletries, and paints.

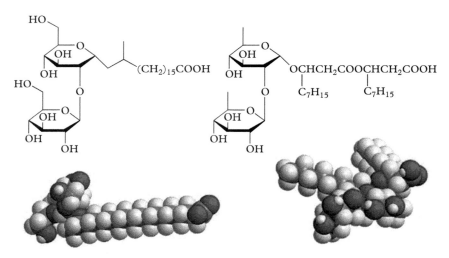

Figure 1.10 *Examples of glycolipid biosurfactants based on (left) sophorose and (right) rhamnose. In space-filling models the red spheres represent the O atoms. In the model of the rhamnolipid surfactant (bottom right) the two heptyl chains are at the top/back of the molecule.*

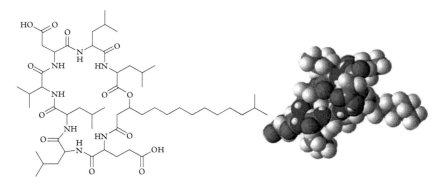

Figure 1.11 *The lipopeptide biosurfactant, surfactin. In the molecular model the alkyl chain is on the middle right. Red atoms are oxygen and blue are nitrogen atoms.*

1.3.6   Gemini and bolaform surfactants

Gemini[1] surfactants form a relatively new class of surfactants (Zana, 2002) and are essentially surfactant dimers in which two monomers are joined by a spacer group at or close to the head groups. The most commonly studied Gemini surfactants have been the di-cationic quaternary ammonium materials of the kind shown in Fig. 1.12. The example is a dimer of decyltrimethylammonium bromide (C_{10}TAB) joined by a

---

[1]   Gemini is the constellation containing the stars Castor and Pollux, the so-called 'heavenly twins'. A bola is a missile used in South America consisting of two heavy balls one at each end of a rope.

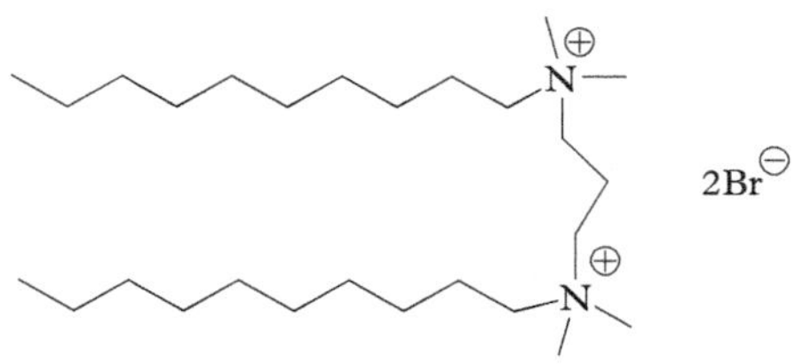

Figure 1.12 *An example of a di-cationic Gemini surfactant, designated 10-3-10 (see text).*

trimethylene spacer. The general designation of this kind of Gemini surfactant is *n-s-n*, where *n* represents the hydrophobic chain length and *s* the spacer group chain length. The hydrophilic moiety in Gemini surfactants can be anionic or nonionic as well as cationic and it is possible also to have asymmetric surfactants where the two hydrophobic chains are of different lengths.

Gemini surfactants are of particular interest for several reasons. The *cmc* (§1.1) of a simple Gemini surfactant can be 1–2 orders of magnitude less than that of the constituent monomer. For example, for the 12-3-12 surfactant of the type shown in Fig. 1.12 the *cmc* is 0.055 wt per cent whereas that for the monomer dodecyltrimethylammonium bromide is 0.5 wt per cent. Since the surface tension of an aqueous solution at the *cmc* is at its lowest, for a given concentration of a Gemini surfactant just above its *cmc* the lowering of surface tension of water caused by the Gemini surfactant is very much greater than that for the constituent monomer. In addition, aqueous solutions of Gemini surfactants with short spacer groups at concentrations above the *cmc* can give solutions with high viscosity and viscoelasticity (Chapter 15). This arises from the shape of the micelles which can be worm-like and become mutually entangled in solution. Gemini surfactants are produced commercially.

Bolaform surfactants are related structurally to Gemini surfactants in that they have two hydrophilic head groups, one at each end of a spacer, but the spacer is much longer than in Gemini surfactants. Thus Gemini surfactants are monomers joined at (or close to) the head groups, whereas bolaform surfactants can be considered to be monomers covalently joined through their chains. The head groups can also have hydrophobic tails attached but the spacer group is significantly longer than the tail length. The head groups can be the same or different. Whereas Gemini surfactants are synthetic, bolaform surfactants can be produced biologically in the membranes of bacteria. The surfactants are able to span and confer stability on the membranes. Examples of bolaform surfactants are given in Fig. 1.13.

1.3.7   Small solid particles as surfactants

Dispersions of very small particles in water do not exhibit all the types of behaviour associated with surfactants; for example, they do not aggregate in solution in the same way that surfactant molecules do, although they may flocculate. What they usually do very well however is adsorb at the surface of water or at oil/water interfaces (Chapter 17). A given particle may, for example, be more wetted by water than by oil. It is observed

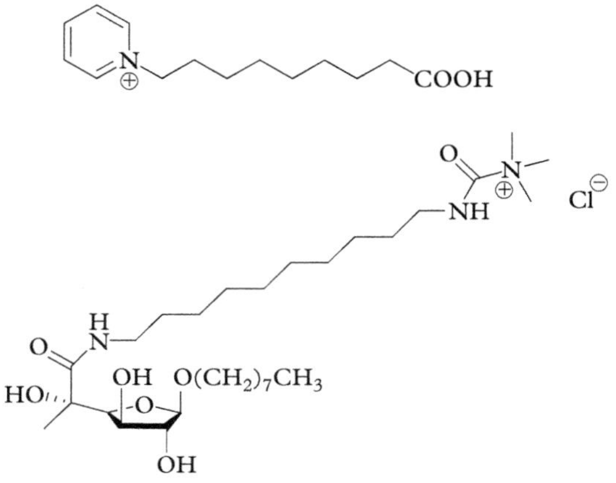

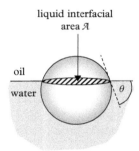

Figure 1.13 *Bolaform surfactants. Both examples are of unsymmetrical surfactants and both contain a cationic head group. The example below has one head group based on a sugar moiety.*

Figure 1.14 *A spherical particle resting at an oil/water interface. The particle is more wetted by water than by oil so that the angle θ (the contact angle) is less than 90°.*

nevertheless that it can be strongly adsorbed from water to the oil/water interface, where part of the particle obviously resides in oil rather than water. It turns out that the driving force for adsorption is the removal of liquid interface (circular area $\mathcal{A}$ in Fig. 1.14). The interfacial tension of a surface is the work required to *form* unit area of that interface; this work is regained if the interface is *removed*.

It has long been known that solid particles can stabilize emulsions (Chapter 18), but it is only relatively recently that particle adsorption and monolayer properties have been sufficiently well-understood to tailor make particles with the required properties (e.g. surface wettability) for the production of emulsions. It is likely that particles will be increasingly used rather than surfactants for the stabilization of certain kinds of emulsions in the future. Silica particles (with diameters of only a few nm) can be prepared with the required wettability for various uses, and of course silica is chemically inert and hence environmentally friendly.

· ·

GENERAL BIBLIOGRAPHY

A selection of supplementary resources, that cover some of the same subject matter as the present book, is given below. There are a number of texts that, unlike the present book, are directed primarily to more applied aspects of surfactant behaviour.

S. Abbott, *Surfactant Science Principles & Practice*. DEStech Publications, Inc., 2018.
This is a stimulating presentation and technical resource aimed in the main at industrial formulators; it is available free online. A nice interactive feature is that formulae have links to apps so you can see how graphs change with input values.

T. Cosgrove (ed.), *Colloid Science: Principles, Methods and Applications*. Blackwell Publishing, Oxford, 2005.
A useful book (288 pages) of contributed chapters based on a spring school on colloid science given by the Bristol Colloid Centre in the UK. It covers colloids, surfactants, and polymers.

E. Dickinson, *An Introduction to Food Colloids*. Oxford University Press, Oxford, 1992.
A short (207 pages) introductory text. The book covers some areas in common with the present book, at a little lower level but with great insight. There is of course a focus on areas appropriate for food scientists.

D. F. Evans and H. Wennerström, *The Colloidal Domain Where Physics, Chemistry and Biology Meet*. 2nd edition. Wiley-VCH, Weinheim, 1999.
This book of 630 pages is a classic treatment of colloid science theory, methods, and applications for advanced students and researchers in industry. The level is higher and more detailed, and the approach more mathematical than that in the present book.

J. Goodwin, *Colloids and Interfaces with Surfactants and Polymers*, 2nd edition. Wiley, Chichester, 2009.
An introduction (280 pages) to fundamental colloid science with less emphasis on surfactant systems than in the present text.

K. Holmberg, B. Jönsson, B. Kronberg, and B. Lindman, *Surfactant and Polymers in Aqueous Solution*, 2nd edition. John Wiley and Sons Ltd, Chichester, 2003.
Surfactants are often used together with polymers to impart required properties to a formulation, and this valuable book considers polymer+surfactant systems from a practical rather than theoretical standpoint.

J. N. Israelachvili, *Intermolecular and Surface Forces*, 3rd edition. Academic Press, 2011.
The behaviour of colloidal systems is intimately associated with the operation of surface forces, and this monograph covers the theories and concepts of such forces. It is the standard work on the subject and is presented at a level suitable for students and research workers. Jacob Israelachvili, who died in September 2018, contributed immeasurably to our understanding of surface forces and the ways in which they operate.

D. Myers, *Surfactant Science and Technology*, 3rd edition. Wiley-Interscience, New York, 2006.
A well-regarded introduction (400 pages) to surfactants, surface activity, and surfactant applications. It is aimed primarily at beginners and non-specialists who seek a practical knowledge of surfactant systems.

M.J. Rosen and J.T. Kunjappu, *Surfactants and Interfacial Phenomena*, 4th edition. Wiley, 2012.
A standard text (616 pages), now in its 4th edition, on the properties and applications of
surfactants. Intended for industrial chemists as well as for teaching purposes, it has a minimum
of mathematical treatment and contains very useful tables of collected data on surfactants.

A number of books by T.F. Tadros cover various aspects of surfactant and colloid behaviour in a
range of systems:

(a) Books published by Walter de Gruyter GmbH & Co KG:

Basic Theory of Interfacial Phenomena and Colloid Stability, 2018; Basic Principles of Interface
Science and Colloid Stability, 2017; Basic Principles of Formulation Types, 2018; Industrial
Applications I: Pharmaceuticals, Cosmetics and Personal Care, 2017; Pharmaceutical, Cos-
metic and Personal Care Formulations, 2018; Polymeric Surfactants: Dispersion Stability and
Industrial Applications, 2017; Emulsions: Formation, Stability, Industrial Applications, 2016;
An Introduction to Surfactants, 2014.

(b) Books published by John Wiley and Sons:

Colloids in Agrochemicals, 2009; Formulation of Disperse Systems, 2014; Emulsion Formation
and Stability, 2013; Dispersion of Powders in Liquids and Stabilization of Suspensions, 2012;
Colloids in Paints, 2011; Self-Organized Surfactant Structures, 2011.

(c) Published by Wiley-VCH Verlag:

Colloid Stability and Application in Pharmacy, 2007.

· ·

REFERENCES

I.M. Banat, R.S. Makkar, and S.S. Cameotra, Potential commercial applications of microbial
surfactants. *Appl. Microbiol. Biotechnol.*, 2000, **53**, 495–508.

K. Holmberg (ed.), *Novel Surfactants. Synthesis, Applications and Biodegradability*, 2nd edition.
Marcel Dekker, New York, 2003.

K. Holmberg, B. Jönsson, B. Kronberg and B. Lindman, *Surfactants and Polymers in Aqueous
Solution*, 2nd edition. John Wiley and Sons, New York, 2003.

M.R. Porter, *Handbook of Surfactants*, 2nd edition. Chapman and Hall, London, 1994.

M.J. Rosen and J. T. Kunjappu, *Surfactants and Interfacial Phenomena*, 4th edition. Wiley-
Interscience, New York, 2012.

T.F. Tadros, *Applied Surfactants*. Wiley-VCH, Weinheim, 2005.

R. Zana, Dimeric and oligomeric surfactants. Behavior at interfaces and in aqueous solution:
a review. *Adv. Colloid Interface Sci.*, 2002, **97**, 205–253.

2

Oil and water do not mix—hydrophobic hydration

2.1   The thermodynamic parameters of dissolution of a solute in water

The poor solubility of nonpolar groups in water provides the driving force for, amongst other processes, adsorption of surfactants from aqueous solution to air/water and oil/water interfaces (Chapter 5), and for micelle formation (Chapter 9). It is useful therefore to have some insight into the way in which hydrocarbon molecules and moieties interact with water.

Imagine the dissolution, at constant temperature T and pressure p, of one mole of a solute X, originally in the form of an ideal gas (phase α, containing only X) to give an ideal dilute solution in water (phase β). Since both phases are ideal dilute systems, there are no X–X interactions in either phase, and in the aqueous phase the only interactions are those between water and molecules of X. The chemical potentials of X in phases α and β, μ_X^α and μ_X^β respectively, are defined as

$$\mu_X^\alpha = \left(\partial G^\alpha / \partial n_X^\alpha\right)_{T,p}; \quad \mu_X^\beta = \left(\partial G^\beta / \partial n_X^\beta\right)_{T,p,n_B}$$

where G are the Gibbs free energies of the superscripted phases. The n_X are numbers of moles of X in the phases denoted by the superscripts. The solvent (water) is denoted B. Differences in the chemical potentials are responsible for the spontaneous flow of the solute between the two phases until equilibrium is attained. The molar free energy of transfer of X from α to β, $\Delta_{trans}^{\alpha \to \beta} \mu$, is

$$\Delta_{trans}^{\alpha \to \beta} \mu = \mu_X^\beta - \mu_X^\alpha \tag{2.1.1}$$

At distribution equilibrium (i.e. when the aqueous phase is saturated with X under the prevailing conditions) the two chemical potentials are equal and $\Delta_{trans}^{\alpha \to \beta} \mu$ is zero.

Surfactants: In Solution, at Interfaces and in Colloidal Dispersions. Bob Aveyard. © Bob Aveyard 2019.
Published in 2019 by Oxford University Press. DOI: 10.1093/oso/9780198828600.001.0001

There are two contributions to chemical potential μ, one arising from the concentration (number density), c, of the solute and the other, μ^o, which is associated with the interactions of a solute molecule with its local surroundings:

$$\mu_X^\alpha = \mu_X^{o,\alpha} + RT\ln\left(c_X^\alpha/c_X^{s,\alpha}\right); \ \mu_X^\beta = \mu_X^{o,\beta} + RT\ln\left(c_X^\beta/c_X^{s,\beta}\right) \tag{2.1.2}$$

Here, μ^o are *standard* chemical potentials and $c_X^{s,\alpha}$ and $c_X^{s,\beta}$ are the concentrations of X in the α and β phases in their respective *standard states*. Usually the c_X^s are chosen as unity (in the same units used for c_X) and in which case they are omitted from the expressions for μ in (2.1.2). Then, the standard chemical potential of the solute is equal to its chemical potential when its concentration is unity, that is, when it is in its standard state.

From (2.1.1) and (2.1.2), taking standard concentrations of X of unity in both α and β

$$\Delta_{trans}^{\alpha\to\beta}\mu = \Delta_{trans}^{\alpha\to\beta}\mu^o + RT\ln\left(c_X^\beta/c_X^\alpha\right) \tag{2.1.3}$$

where $\Delta_{trans}^{\alpha\to\beta}\mu^o = \mu_X^{o,\beta} - \mu_X^{o,\alpha}$. From what has been said, $\Delta_{trans}^{\alpha\to\beta}\mu^o$ contains that part of $\Delta_{trans}^{\alpha\to\beta}\mu$ that arises from the interactions of the solute with water accompanying dissolution.[1] At equilibrium $\Delta_{trans}^{\alpha\to\beta}\mu = 0$ and so, from (2.1.3)

$$\Delta_{trans}^{\alpha\to\beta}\mu^o = -RT\ln\left(c_X^\beta/c_X^\alpha\right)_{equ} \tag{2.1.4}$$

The ratio $\left(c_X^\beta/c_X^\alpha\right)_{equ}$ is called the Ostwald absorption coefficient. Corresponding changes in standard entropy, enthalpy and heat capacity that accompany dissolution can be obtained from the temperature dependence of $\Delta_{trans}^{\alpha\to\beta}\mu^o$ using the usual thermodynamic relationships.

2.2   Hydrocarbons are only sparingly soluble in water—hydration of CH$_2$ groups

Many surfactants contain hydrocarbon moieties as the hydrophobic groups, and the driving force for the transfer of these groups from water to a nonpolar or vapour environment arises in some way from the antipathy between the hydrocarbon groups and water. A water molecule and a methylene group (the constituent unit of say an alkyl chain) are of similar size and the dispersion forces (see §11.2.1) operating between the pairs H_2O and H_2O, CH_2 and CH_2, and H_2O and CH_2 are expected to be similar,

---

[1] In addition, $\Delta_{trans}^{\alpha\to\beta}\mu^o$ also depends on the choice of standard states in phases α and β, as can be appreciated from the text.

and are attractive in nature. The antipathy arises therefore from effects other than the straightforward operation of intermolecular dispersion forces. Since water has a degree of tetrahedral structure at room temperature, imposed by hydrogen bonding, introduction of a nonpolar solute might be expected to cause a disruption of the local water structure in some way. The transfer of an alkane molecule from the vapour phase to water would therefore be expected to be accompanied by a more positive entropy change than for transfer to a non-structured nonpolar liquid. In fact, the entropy change is found to be large and *negative*, and in addition, the (positive) changes in heat capacity are anomalously high.

The beginnings of a modern understanding of the origins of the hydrophobicity (hence low solubility in water) of hydrocarbons were given in a landmark paper by Frank and Evans in 1945. They wrote: 'When a rare gas atom or nonpolar molecule dissolves in water at room temperature, it modifies the water structure in the direction of greater "crystallinity"; the water, so to speak, builds a microscopic iceberg around it . . . As the temperature is raised these icebergs melt giving rise to the enormous partial molar heat capacity of these gases in water.'

Standard free energies of solution in water, $\Delta_{trans}\mu^o$, of some normal alkane vapours at room temperature (relative to that for ethane) are plotted against the alkane chain length in Fig. 2.1. It is noted that the methylene group increment denoted $\Delta_{hyd}\mu^o$, which is positive, is independent of the choice of standard states (assuming the *same* standard states are chosen for all systems) and represents the hydration free energy of a methylene group. This increment and those in the corresponding enthalpy and entropy, $\Delta_{hyd}H^o$ and $\Delta_{hyd}S^o$, at room temperature, are given in Table 2.1.

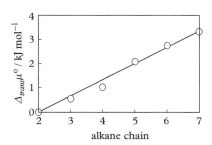

Figure 2.1 *Standard free energies of solution (relative to that for ethane) of linear alkane vapours in water at room temperature. The free energies increase linearly with increasing alkane chain length.*

Table 2.1 *Thermodynamic parameters for the hydration (at 293 K) and adsorption (from vapour to the air/water surface at 285.6 K) of the methylene group. Free energies and enthalpies are given in kJ mol^{-1} and entropies in J mol^{-1} K^{-1}*

$\Delta_{hyd}\mu^o$	$\Delta_{hyd}H^o$	$\Delta_{hyd}S^o$	$\Delta_{ads}\mu^o$	$\Delta_{ads}H^o$	$\Delta_{ads}S^o$
+0.7	−2.8	−11.7	−2.0	−4.4	−8.4

Both the enthalpy and entropy of hydration are temperature dependent, and at room temperature the *positive* increment per CH_2 in $\Delta_{hyd}\mu^o$ ($\approx +0.7$ kJ mol^{-1}) arises from the effect of the entropy of hydration ($-T\Delta_{hyd}S^o \approx +3.5$ kJ mol^{-1} CH_2 whereas the corresponding $\Delta_{hyd}H^o$ is ≈ -2.8 kJ mol^{-1}). At elevated temperatures the contributions of enthalpy and entropy to the free energy of hydration are reversed, with favourable contributions coming now from the entropy of hydration and unfavourable changes being contributed by the enthalpy change. This temperature dependence is reflected in the changes in hydrocarbon solubility with temperature; solubilities pass through a minimum in the region of 310–350 K.

The original picture of microscopic 'icebergs' surrounding nonpolar moieties in water has often been considered to be an over-simplification. Certainly there have been difficulties with the use of direct physical techniques (e.g. neutron scattering and EXAFS) to probe the water structure around dissolved nonpolar molecules because the concentration is, due to the low solubility in water, necessarily low. Nonetheless, nonpolar moieties in molecules containing polar groups (e.g. alkanols) can be probed, although of course the vicinity of a polar group to the nonpolar moieties within the molecule can complicate matters. Some experimental studies have led to the suggestion that the 'structure' induced by nonpolar groups is less ordered than ice and not dissimilar to that in liquid water remote from a dissolved molecule, and that the 'order' may be associated with the orientation (rather than disposition) of water molecules close to nonpolar solute molecules. Recently, however, Grdadolnik et al. (2017), based on IR studies of nonpolar molecules (including methane and ethane) in HDO water, claim that water hydrogen bonding close to the nonpolar solutes is strengthened and that the water displays extensive structural ordering similar to that in clathrates, supporting the original ideas of Frank and Evans (1945). Interested readers should also consult the extensive work of Soper on water structure in water and aqueous solutions (e.g. Soper, 2014).

Whatever the details of the origins of hydrophobic hydration (and the story is probably not yet finished), it is clear that around room temperature hydrophobic moieties perturb water in such a way as to lead to a large negative change in entropy (giving a positive contribution to the free energy) when the groups are introduced into water. Conversely, when hydrophobic groups are removed from aqueous surroundings, in such processes as adsorption of surfactants from aqueous solution and micelle formation in water, there is a negative free energy contribution arising from the dehydration of the nonpolar moiety.

When hydrocarbon groups are in close proximity in aqueous solution, they can become loosely associated by so-called *hydrophobic bonding,* a term introduced by Kauzmann (1959). The total amount of unfavourably perturbed water around two dissolved nonpolar groups is reduced as the two groups mutually approach. This gives rise to a solvent mediated attraction, although 'bonding' might not be a very good term for this. Much of the interest in hydrophobic bonding has arisen out of work on protein folding (Kauzmann, 1959; Tanford, 1997). The hydrophobic side chains of proteins in water undergo mutual hydrophobic bonding which is one of the factors influencing the shape of the protein molecules in aqueous solution.

What has been discussed above applies at molecular length scales to nonpolar groups and molecules in water. It is well known however that there are also strong attractive interactions between *macroscopic* hydrophobic surfaces in water, which can extend over large separations between surfaces. Various possible origins of such forces have been considered and in Chapter 11 these and a variety of other so-called *surface forces* are described.

2.3 How do nonpolar groups interact with water at a water surface?

In a close-packed monolayer of surfactant at the air/water interface it may reasonably be assumed that the surfactant tails are disposed roughly normal to the interface, as illustrated in Fig. 1.1. When surfactant adsorbs to form a very dilute monolayer at the air/water surface however, it is not immediately obvious what the disposition of the alkyl chains might be. Are the chains repelled or attracted by the water *at the surface*? If the surfactant tails can lie along the surface without unfavourably perturbing the local water structure, then it might be expected that the chains would orient in some way parallel to the surface, thereby interacting with the water through van der Waals forces. On the other hand, if the water is unfavourably disrupted by the presence of the nonpolar groups (as it is in *bulk* water) it may be that they coil as much as possible so avoiding contact with the interfacial water. These two extreme possibilities are illustrated schematically in Fig. 2.2.

An idea of the orientation of adsorbed surfactant chains can be obtained by considering the standard thermodynamic parameters of adsorption of a series of normal alkanes from dilute vapour to the air/water interface. The way in which the standard free energy of adsorption, $\Delta_{ads}\mu^o$, is calculated from the lowering of the surface tension with increasing gas pressure (concentration) is discussed in §4.3. The quantity $\Delta_{ads}\mu^o$ per methylene group is analogous to the hydration free energy, $\Delta_{hyd}\mu^o$, discussed in §2.2 and given in Table 2.1, but now the hydrocarbon is present in an infinitely dilute monolayer rather than in an infinitely dilute bulk aqueous solution.

Standard free energies and enthalpies of adsorption, together with $T\Delta_{ads}S^o$ for a series of normal alkanes at 285.6 K are shown in Fig. 2.3.[2] For clarity, values are given relative

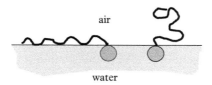

Figure 2.2 *Schematic representation of a surfactant molecule at the air/water surface. (left) Molecule disposed with chain parallel to the interface and (right) coiled and roughly normal to the interface.*

---

[2] The chosen standard state for the alkanes in the surface is a little complicated, but is the same for all the alkanes. The increment per methylene group under these circumstances is independent of the standard state. The standard state for the vapour phase is one atmosphere.

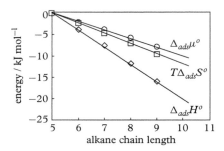

Figure 2.3 *Standard thermodynamic parameters for the adsorption of normal alkanes at 285.6 K from dilute vapour to the air/water surface. Values are given relative to those for pentane.*

to those for pentane. It is evident that the free energies, enthalpies and entropies of adsorption all become more negative with increasing alkane chain length. Incremental values per methylene group are given in Table 2.1, and compared with corresponding values for hydration of the methylene group. It is seen that, although the hydrocarbon moieties interact unfavourably with bulk water, they are attracted by water at an air/water interface, and so it is likely that adsorbed alkane molecules assume orientations that are essentially parallel to the interface.

The (negative) entropy of adsorption per CH_2 group could possibly be accounted for by changes in conformation of the alkanes accompanying adsorption, together with vibrations normal to the surface. The enthalpy change probably arises in the main from the operation of dispersion forces between alkane and water molecules at the interface. This can be appreciated by considering the enthalpy of *cohesion*, $\Delta_{cohesion}H$, of liquid alkanes. As will be discussed in §3.1, the free energy (work) of cohesion is defined as the isothermal work of creating two units of area of liquid surface, that is, $\Delta_{cohesion}G = 2\gamma$ where γ is the surface tension. Enthalpies of cohesion can then be obtained from the temperature variation of the free energies in the usual way. The cohesion parameters relate to unit area; to convert them to molar quantities per CH_2 it is only necessary to obtain a surface area, σ_{CH_2}, occupied by a methylene group at the alkane surface. This can be approximated by the area occupied in the plane of aligned alkane molecules in bulk liquid alkane. The distance between the long axes of the chains is about 0.47 nm and the C–C distance resolved along the chain is 0.126 nm, giving σ_{CH_2} of about 0.06 nm². From surface tensions of alkanes as a function of temperature, and σ_{CH_2}, an enthalpy of cohesion for the methylene group of about 3.8 kJ mol⁻¹ is obtained. This relates to *separating* alkane from alkane; bringing two surfaces together (i.e. destroying the surfaces) is associated with an enthalpy change of −3.8 kJ mol⁻¹. This figure is reasonably close to the molar enthalpy of adsorption of the methylene group, −4.4 kJ mol⁻¹. Since only dispersion forces are operative in alkanes, this implies that the enthalpy of adsorption of alkanes could arise largely from dispersion forces between water molecules and alkanes; it is recalled that the dispersion forces acting between H_2O and CH_2 groups are expected to be similar to those between two CH_2 groups.

In conclusion, it is not necessary to invoke the notion of water 'structure' being perturbed by hydrocarbon groups at the surface of water. This is in contrast to the situation in bulk water, where such a perturbation is implicated in the very low solubility of hydrocarbons in water.

2.4 Concluding remarks

Many surfactants contain alkyl chains as the hydrophobic moiety, and adsorption from aqueous solution to air/water and oil/water interfaces (Chapters 4 and 5), as well as micelle formation (Chapter 9) are driven at least in part by changes in the local environment of the hydrophobic group. The methylene group is a basic unit of an alkyl chain and consideration has therefore been given above to the changes in free energy that result from changes in environment of the CH_2 group. Discussion was centred on the ways in which the methylene group interacts with water, both in bulk aqueous solution and at the air/water surface. The relative free energies of the methylene group in various environments that relate to adsorption and micelle formation are shown schematically in Fig. 2.4.

For the systems represented in Fig. 2.4, the most positive free energy is that for the CH_2 group in ideal dilute aqueous solution. This leads to the low solubility of alkanes in water. The difference in free energy for the group in water and in ideal dilute vapour is the hydration free energy, which as discussed is positive (0.7 kJ mol^{-1}). The difference in free energy of the methylene group in aqueous solution and at the air/water interface, that is, the adsorption free energy, is about -2.7 kJ mol^{-1}, which is less than that when the group is adsorbed to the oil/water interface (ca. -3.2 kJ mol^{-1}) (see §5.3.1). The methylene group free energy is similar for a group at the oil/water interface (where it

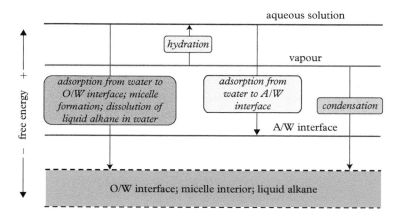

Figure 2.4 *Schematic representation of the free energy of a methylene group in various environments. The micelle interior, bulk liquid alkane, and oil at the oil/water interface are similar, but not equivalent, and the energy in these environments is indicated by a band.*

dips into the oil phase), in a micelle interior (which is liquid-like, §9.3) and in bulk liquid alkane. With reference to the figure, the free energy of micellization is ca. -3.1 kJ mol^{-1} CH$_2$, and that of condensation into bulk liquid alkane is -3.0 kJ mol^{-1} CH$_2$.

....

REFERENCES

H.S. Frank and M.W. Evans, Free volume and entropy in condensed systems III. Entropy in binary liquid mixtures; partial molal entropy in dilute solutions; structure and thermodynamics in aqueous electrolytes. *J. Chem. Phys.*, 1945, **13**, 507–532.

J. Grdadolnik, F. Merzela, and F. Avbelj, Origin of hydrophobicity and enhanced water hydrogen bond strength near purely hydrophobic solutes. *PNAS*, 2017, **114**, 322–327.

W. Kauzmann, Some factors in the interpretation of protein denaturation. *Adv. Protein Chem.*, 1959, **14**, 1–63.

A.K. Soper, Water and ice structure in the range 220–365 K from radiation total scattering experiments. arXiv:1411.1322 [cond-mat.dis-nn], 2014.

C. Tanford, How protein chemists learned about the hydrophobic factor. *Protein Sci.*, 1997, **6**, 1358–1366.

3

Capillarity, wetting, and surface (interfacial) tension

3.1 Capillarity

From everyday experience it is known that, for example, a porous cloth will absorb water and that a watercolour brush will retain paint, in both cases against the action of gravity. It is also observed that a plane surface of a liquid becomes distorted when it comes into contact with, say, a vertical solid surface. Tea in a teacup forms a meniscus where it meets the surface of the cup, and in this case the (wetting) liquid in the concave meniscus[1] is lifted above the level of the plane liquid surface distant from the side of the cup (Fig. 3.1a). In the case of mercury in a glass bottle or capillary tube (e.g. in a thermometer), the meniscus is convex and the (non-wetting) liquid is depressed rather than elevated (Fig. 3.1b).

If a cylindrical capillary tube (i.e. a tube with a hair-like bore) is placed vertically in a liquid, parts of the meniscus on opposite walls within the tube overlap and the liquid either rises (Fig. 3.2) or is depressed within the tube. The term *capillarity* stems from this

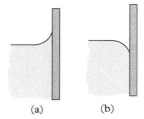

(a)   (b)

Figure 3.1 *(a) A wetting liquid in contact with a vertical solid surface in air; the meniscus is concave and raised above the plane liquid surface. (b) A non-wetting liquid in air in contact with a solid surface. The meniscus is convex and depressed below the plane liquid.*

---

[1] The name *meniscus* derives from the ancient Greek *meniskos* (crescent), a diminutive of *mene*, meaning moon. *Capillus* is the Latin for hair.

Surfactants: In Solution, at Interfaces and in Colloidal Dispersions. Bob Aveyard. © Bob Aveyard 2019.
Published in 2019 by Oxford University Press. DOI: 10.1093/oso/9780198828600.001.0001

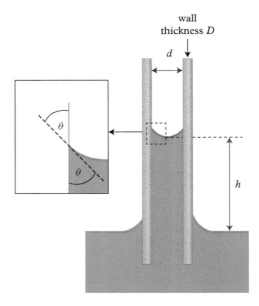

Figure 3.2 *Rise of a wetting liquid, in air, within a capillary tube with internal diameter d and wall thickness D. The angle θ which the liquid makes with the tube wall is termed the contact angle. The curved liquid surfaces are termed menisci.*

phenomenon although it also includes other phenomena associated with liquid menisci. For a given liquid, the extent of the rise (*h* in Fig. 3.2) or depression, depends upon the angle at which the meniscus meets the capillary wall, that is, on the *contact angle θ*. The rise also depends on the internal diameter *d* of the tube but not on the wall thickness *D*. The first accurate observations of capillary rise were made by Francis Hawksbee (Box 3.1) in the early part of the eighteenth century.

Box 3.1 Newton, Hawksbee, capillarity, and intermolecular forces

The study of capillarity has been of great importance in the understanding of intermolecular forces. Francis Hawksbee (The Elder) lived from c.1666 to 1713 and was, from 1703 until his death a 'paid performer of experiments' at the Royal Society in London. Supported by Isaac Newton, he was elected Fellow of the Royal Society in 1705. He made accurate observations on capillary rise, and Newton used Hawksbee's findings to gain insights into what at the time were referred to as 'inter-particle' forces. Newton writes about this in his famous work 'Opticks'. The rise of liquids within, for example, capillary tubes was supposed to involve the attraction between particles (i.e. molecules) of the liquid and of the solid. Since, from Hawksbee's experiments it was known that the extent of elevation of the liquid depended only on the internal diameter of the tube and not on the thickness of the walls, it was concluded that the interparticle forces of attraction were short range, involving only those particles close

Box 3.1 *Continued*

to the solid/liquid interface. At the end of Opticks, Newton wrote a series of Queries. In the last edition of 1730, there are 31 Queries, which are in the form of rhetorical questions, some developing into essays. Query 31 is extensive and begins: 'Have not the small Particles of Bodies certain Powers, Virtues, or Forces, by which they act at a distance...' Later in the Query, having referred to Hawksbee's work on capillary rise, Newton writes concerning forces between particles: 'And therefore where the distance [between particles] is exceeding small, the attraction must be exceeding great... There are therefore Agents in Nature able to make the particles of Bodies stick together by very strong Attractions. And it is the Business of experimental Philosophy to find them out.' This is a profound observation and the problem is one which continues to exercise physical scientists to the present time.

3.2 Surface and interfacial tension

In order to understand the origins of capillary rise it is convenient to introduce the definition and significance of surface/interfacial tension, γ. As mentioned in Chapter 1 (footnote 1), the term *surface* is often used to refer to the interface between a condensed and a vapour phase, and *interfacial* to the interface between condensed phases; however the term *interface* is general and encompasses the area of mutual contact between phases in any state of matter.

It is a common observation that a small free liquid drop tends to take on an effectively spherical shape. For a given volume, a sphere has the minimum surface area and so it can be inferred that molecules in a surface are in a higher energy state than those in the bulk. Molecules in a condensed phase always possess a net attractive interaction energy due to the balance of attractive and repulsive forces acting in three dimensions. This is why condensation occurs. Molecules at a surface however have negligible energy of interaction with the vapour phase and so work must be done to create a surface, which has an *excess* free energy compared to (the same amount of) material in bulk. This is why surfaces tend to contract. At the interface between two immiscible or partially miscible condensed phases, α and β, however, the molecules at each of the surfaces of the contacting phases have attractive interaction energies with the surface molecules of the other phase. Surface/interfacial tension is defined as the isothermal reversible work of forming a unit area of interface, and the interfacial tension $\gamma_{\alpha\beta}$ between two condensed phases is less than (or possibly equal to) the sum of the two surface tensions γ_α and γ_β, as is shown in (§12.3.2). The excess free energy of molecules at a surface gives rise to a lateral tension within the surface, and Thomas Young (in 1805) noted that, mechanically, an interface behaves as if a (fictitious) uniformly stretched membrane of infinitesimally small thickness existed between contacting phases (Box 3.2). In thermodynamic terms, the surface or interfacial tension, γ, of a material at temperature T and pressure p can be expressed as (see also §4.2.2)

$$\gamma = (\partial G/\partial \mathcal{A})_{T,p,n} \qquad (3.2.1)$$

Box 3.2   Thomas Young—a remarkable polymath

Thomas Young, 1773–1829. An English polymath who, amongst other achievements, made important scientific contributions in the fields of light and solid mechanics. World History Archive/Alamy Stock Photo.

Thomas Young (1773–1829) was not only a physicist but also a physician and Egyptologist. He made early discoveries in deciphering Egyptian hieroglyphics such as those on the Rosetta Stone, which was carved in 196 bc (Robinson, 2007). The stone, which is now housed in the British Museum in London, was discovered by French soldiers in Egypt in 1799 in the small town of Rosetta. The French Egyptologist Champollion built upon Young's early work to decipher completely the Rosetta script. As a physicist Young is perhaps most famous for his contributions to the theory of light and of elasticity. In the present context he founded the theory of capillarity in terms of the concept of surface tension (1804). In an article on capillary action appearing in the Encyclopaedia Britannica (1911, volume 5, p. 258), James Clerk Maxwell notes that Young's methods 'though always correct, and often extremely elegant, are sometimes rendered obscure by his scrupulous avoidance of mathematical symbols'. The subject of capillarity was subsequently examined by Pierre Simon Laplace whose results were in many respects identical with those of Young, but were obtained using mathematical methods.

where G is the Gibbs free energy of the system, with surface area $\mathcal{A}$, containing n mol material. From (3.2.1) the units of γ are seen to be those of energy per unit area, and are frequently quoted in mJ m^{-2}. Since the excess surface energy results in a tension acting laterally within the surface (hence interfacial *tension*), γ is equivalently defined as the force acting parallel to the surface and at right angles to a line of unit length anywhere within the surface. The dimensions of energy/area are the same as those of force/length, and 1 mJ m^{-2} is identical to 1 mN m^{-1}. In some contexts it is convenient to consider interfacial tension as an excess free energy per unit area and in others as a force per unit length. Some values of interfacial tensions for liquid/vapour and water/liquid interfaces,

Table 3.1 *Some values of surface tension and interfacial tension (against water) of pure liquids that are immiscible or partially miscible with water, at 293.15 K, and their temperature coefficients. Tensions are given in mN m^{-1}*

	liquid/vapour tension	water/liquid tension	$-d\gamma/dT$ liquid/vapour	$-d\gamma/dT$ water/liquid
water	72.8		0.16	
decane	23.8	52.3	0.09	0.09
hexadecane	27.4	53.8	0.09	0.09
decanol	28.6	9.0	0.08	0.07
methyl decanoate	28.5	22.5	0.10	0.03

Note that the surface tension of water is high relative to that for the other liquids represented. Water has a tetrahedral structure in bulk as a result of hydrogen bonding, and the formation of surface with vapour results in disruption of this structure giving rise to the high surface tension. In contrast, the polar materials decanol and methyl decanoate have surface tensions very similar to that of the alkanes. The polar liquids do not have the extensive bulk structuring that water has and presumably surface molecules can still retain the kind of polar interactions at the surface that are present in bulk. However, as a result of polar interactions with water at liquid/liquid interfaces, the polar materials have lower interfacial tensions with water than do the alkanes.

and their temperature coefficients, are given in Table 3.1. The measurement of interfacial tension for liquid/vapour and liquid/liquid interfaces is described in the Appendix to this chapter.

In the process of capillary rise, an area of solid/air interface is replaced by solid/liquid interface in the capillary tube. Since interfacial tension is the work of forming unit area of interface, it is expected that capillary rise should occur if the interfacial tension between the solid and liquid, γ_{sl}, is less than the tension between solid and air, γ_{sa}. Conversely capillary depression (as with mercury in a glass tube) should occur if $\gamma_{sa} < \gamma_{sl}$. This is considered in a little more detail in the next section, where the topics of wetting and adhesion in relation to the contact angle θ are introduced. For a discussion of the definition of surface/interfacial tension for solid interfaces, see §16.3.3.

3.3 Wetting, adhesion, and the contact angle

Amongst other factors, capillary rise or depression in a capillary, say in air, depends on the contact angle θ that the liquid/air interface makes with the solid/liquid interface. The value of the contact angle is an indication of whether it is energetically favourable for a liquid to contact (wet) the solid (in this case in air) or not. The contact angle shown in Fig. 3.2 for a liquid in a capillary tube is equivalent to the angle that the liquid makes with a smooth flat, horizontal plate of the (same) solid immersed in air, as indicated in Fig. 3.3a. In general it is necessary to stipulate through which phase the angle is measured—the liquid in this case. An (incompletely) wetting liquid has a contact angle less than 90°, whereas a (partially) non-wetting liquid has an angle greater than 90°. Examples of completely wetting liquid-solid pairs ($\theta = 0°$) are common whereas completely non-wetting systems ($\theta = 180°$) are not (see §16.3.5).

Around a drop with circular cross-section resting on a plane horizontal solid in air, there exists a circular contact line where the three phases, solid, liquid and air meet. With reference to Fig. 3.4, at mechanical equilibrium the forces acting horizontally at a point in the contact line can be resolved to give Young's equation:

$$\gamma_{sa} = \gamma_{sl} + \gamma_{la} \cos \theta \tag{3.3.1}$$

The quantity $\gamma_{sa} - \gamma_{sl}$ ($= \gamma_{la} \cos\theta$) is referred to as the *wetting tension*.

The *work of adhesion*, W_A, of solid and liquid (in a third medium, say air) is defined as the work associated with parting unit area of solid/liquid interface to give units of area of solid/air and liquid/air interfaces (Fig. 3.5). Thus, noting the definition of interfacial tension, W_A can be expressed in the *Dupré equation*:

$$W_A = \gamma_{sa} + \gamma_{la} - \gamma_{sl} \tag{3.3.2}$$

The interfacial tension of a liquid/fluid interface is readily measurable (see Appendix), but tensions of interfaces involving solids are not, so the expression for W_A in (3.3.2) is not particularly useful. However, combination of (3.3.1) and (3.3.2) gives the more convenient expression for W_A,

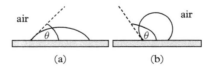

(a)   (b)

Figure 3.3 *Drop of liquid on solid in air with contact angle θ (a) less than 90° and (b) greater than 90°.*

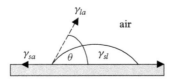

Figure 3.4 *Forces acting at the contact line of a drop of liquid in air resting on a flat smooth solid surface.*

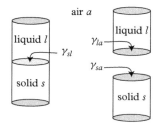

Figure 3.5 *Parting a column of solid from a contacting column of liquid with unit interfacial area of tension γ_{sl} to give unit area of solid/air interface, tension γ_{sa}, and unit area of liquid/air interface, tension, γ_{la}.*

$$W_A = \gamma_{la}\,(1 + \cos\theta) \tag{3.3.3}$$

which is referred to as the *Young–Dupré equation*. Like γ_{la}, θ is readily accessible experimentally and so W_A can easily be obtained. Values of the work of adhesion of water with some solids in air are given in Table 3.2. As is obvious from (3.3.3) and the table, for a given liquid W_A increases as θ falls, that is, as the liquid increasingly wets the solid.

For capillary rise to occur (Fig. 3.2) the contact angle must be less than 90° ($\cos\theta > 0$), and for capillary depression $\theta > 90°$ ($\cos\theta < 0$). It follows therefore from (3.3.2) and (3.3.3) that in the case of capillary rise $W_A > \gamma_{la}$ and hence $\gamma_{sa} > \gamma_{sl}$. For capillary depression $W_A < \gamma_{la}$ and so $\gamma_{sl} > \gamma_{sa}$, as mentioned at the end of §3.2.

It can be useful to suppose that interfacial tensions are made up of additive components arising from the different kinds of interaction between phases at the interface (Fowkes, 1963). For a polar liquid such as water, the surface tension γ can notionally be split into contributions from dispersion forces, γ^d, and 'polar' forces, γ^p, arising from dipole interactions and hydrogen bonds, so that

Table 3.2 *W_A for water with a range of solids in air at room temperature. The interfacial tension of the air/water interface is taken as 72.8 mN m^{-1}.*

solid	θ through water/°	W_A/mJ m^{-2}
PTFE	112	45
hexamethylethane	115	42
C$_{36}$-paraffin	105	54
polystyrene	91	72
cellulose acetate	54	116

$$\gamma = \gamma^d + \gamma^p \tag{3.3.4}$$

For say an alkane (1), only dispersion forces are possible so $\gamma_1 = \gamma_1^d$. When an alkane is in contact with water (2) the interfacial tension γ_{12} is the sum of the two surface tensions, γ_1 and γ_2, *less* a term resulting from the mutual attractive interaction through dispersion forces of the liquids at the interface. Assuming this interaction is the geometric mean, $\sqrt{\gamma_1^d \gamma_2^d}$, the tension in the interfacial region of phase 1 is $\gamma_1 - \sqrt{\gamma_1^d \gamma_2^d}$ and that in 2 $\gamma_2 - \sqrt{\gamma_1^d \gamma_2^d}$ so that γ_{12} is given by

$$\gamma_{12} = \gamma_1 + \gamma_2 - 2\sqrt{\gamma_1^d \gamma_2^d} \tag{3.3.5}$$

Since the work of adhesion (W_A) between 1 and 2 is $\gamma_1 + \gamma_2 - \gamma_{12}$, it follows that

$$W_A = 2\sqrt{\gamma_1^d \gamma_2^d} \tag{3.3.6}$$

The polar contribution, γ_2^p, to γ_2 can be obtained from measured values of γ_1, γ_2 and γ_{12} in the following way. Equation (3.3.5) can be recast to give

$$\gamma_2^d = (\gamma_1 + \gamma_2 - \gamma_{12})^2 / 4\gamma_1^d \tag{3.3.7}$$

and, noting that $\gamma_1 = \gamma_1^d$, (3.3.6) is used to calculate γ_2^d. The polar contribution γ_2^p is then given as ($\gamma_2 - \gamma_2^d$). Values of γ^d and γ^p for many liquids are available in the literature. This approach, and elaborations of it, have also been used in the treatment of solid/liquid interfacial tensions, as explained in §16.3.3.

3.4   Curved liquid interfaces: the Young–Laplace and Kelvin equations

3.4.1   The Young–Laplace equation

Many liquid interfaces encountered are curved, as in the case of liquid in menisci of the type already referred to. Examples of particular interest in surfactant science are the surfaces of emulsion and microemulsion droplets. Often, relevant surfaces are spherically curved, as in the case of small emulsion droplets and liquid menisci in sufficiently narrow capillaries. Consider the rise of liquid in air to an equilibrium height h, in a vertical, cylindrical capillary, as depicted in Fig. 3.6. The pressure in the vapour phase is denoted p'' and the very small difference in p'' with height is ignored. The pressure in the liquid within the capillary at the same height as the plane surface of liquid outside the tube, is also p'' at equilibrium. Since the pressure in the liquid column falls with increasing

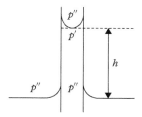

Figure 3.6 *Capillary rise to height h in a vertical cylindrical tube, radius r_c.*

height, it is clear that the pressure p' just below the liquid meniscus must be less than p''. Ignoring the small pressure due to the liquid above the lowest point in the meniscus

$$\Delta p = p'' - p' = h\Delta\rho g \qquad (3.4.1)$$

where $\Delta\rho$ is the difference in density between liquid and vapour phases and g is the acceleration due to gravity. The pressure difference through a curved liquid interface, Δp, is referred to as the *Laplace pressure*.

The Laplace pressure is easily related to the interfacial tension γ by reference to a small spherical vapour bubble in a liquid of the same material. Effects of gravity (causing the bubble to rise) are ignored and it is supposed that the bubble is in equilibrium with the liquid. Imagine an infinitesimal contraction of the bubble giving a change in radius dr There would be a change in surface area of the bubble of $8\pi r dr$, and hence an energy change $8\pi r dr\gamma$. At equilibrium the contraction is just opposed by the excess pressure within the bubble, acting over the surface of the bubble. The work (force x distance) required to expand the bubble back to radius r is $(4\pi r^2 \Delta p)dr$. Thus $8\pi r dr\gamma = 4\pi r^2 \Delta p dr$, and so

$$\Delta p = \frac{2\gamma}{r} \qquad (3.4.2)$$

Equation (3.4.2) is the form of the *Young–Laplace equation* (often simply referred to as the *Laplace* equation) for a *spherically curved* interface. It is fundamental to surface chemistry and shows, for example, that due to the action of interfacial tension a spherically curved liquid interface can maintain a mechanical equilibrium between two fluid phases at different pressures. Note that the pressure is the greater on the concave side of the interface, for example, the inside of a bubble or droplet, irrespective of phase. For example, for a vapour bubble in water ($\gamma = 72.8$ mN m^{-1}), radius $r = 100$ nm, the excess pressure inside the bubble is about 1.5×10^6 Pa (ca. 15 atm) the same as that inside a drop of water (same radius) in water vapour.

For a spherically curved interface, a single radius defines the curvature, that is, the curvature is the same whichever direction is travelled along the surface. To understand how the curvature of *non-spherically curved* liquid interfaces can be described, it is useful first to imagine a *smooth curved line*. At any point on the curve the *curvature c* can be

Figure 3.7 *The curvature of a smooth curve (full line) at a point is described by the radius r of a circular arc.*

described by the radius r of a circular arc, as illustrated in Fig. 3.7; the curvature is the reciprocal of the *radius of curvature*, and can be positive or negative, depending on which way the arc points. At a point on a non-spherically curved smooth surface, the surface bends by different amounts in different directions, and the curvature can be described by the radii of two circular arcs. The *principal radii of curvature r_1 and r_2* at a point on the surface are the radius of the biggest circle, r_1, which is always orthogonal to the circle with the smallest radius, r_2. *Gaussian curvature* (referred to in §10.2.2) at a point on the surface is the product of the principal curvatures at that point; it is positive when the principal curvatures have the same sign, negative when the curvatures are of opposite sign and zero if one of the curvatures is zero (i.e. if a circle degenerates into a straight line). The *mean curvature* is half the sum of the principal curvatures, and for constant mean curvature (CMC) surfaces it is the same at all points on the surface; *minimal surfaces* are those for which the mean curvature is everywhere zero. An equilibrium soap bubble (which encloses a volume) has a CMC surface, and there is a (Laplace) pressure drop across its surface. On the other hand there is no pressure drop across a soap film attached say to a twisted wire loop and open to the atmosphere on both sides.[2] Such surfaces at equilibrium are minimal surfaces; Δp in (3.4.3) is zero and the two principal curvatures are everywhere equal and of opposite sign. The surface minimizes its surface area, given prevailing physical constraints (e.g. attachment to the wire loop).

For non-spherically curved surfaces with principal radii of curvature r_1 and r_2, $2/r$ in (3.4.2) is replaced by $(1/r_1 + 1/r_2)$ so that more generally the Young–Laplace equation is expressed:

$$\Delta p = \gamma \left(\frac{1}{r_1} + \frac{1}{r_2} \right) \tag{3.4.3}$$

The equation has been derived above for macroscopic systems, so that for example the surface tension alluded to is that of a bulk liquid. Use will be made of the equation in the discussion, for example, of emulsion behaviour (Chapter 14), where drop diameters are of the order of microns, for which macroscopic behaviour is expected. It is interesting however that Liu and Cao (2016) show, by the used of molecular dynamics simulations, that the Young–Laplace equation appears to be valid even down to the nanoscale.

---

[2] Consider a soap film on a vertical rectangular frame, such as that shown in Fig. 12.1. Imagine holding top and bottom sides of the frame and twisting one of them in a clockwise direction and the other in an anticlockwise direction. The soap film would become curved rather than planar but still be open to the atmosphere on both sides.

Returning to the rise of a liquid in a vertical cylindrical capillary tube (see Fig. 3.6), suppose the tube has a sufficiently small radius for the meniscus within the tube to be spherically curved. The internal radius of the capillary tube is denoted r_c, and the radius of curvature of the meniscus, r, is related to r_c by $r_c = r \cos \theta$, where θ is the contact angle of the liquid with the side of the capillary. Noting this, and equating expressions for Δp in (3.4.1) and (3.4.2) it follows that

$$\frac{2\gamma \cos \theta}{r_c} = h\Delta\rho g \text{ or } \gamma = 0.5 \left(\frac{r_c h \Delta\rho g}{\cos \theta} \right) \tag{3.4.4}$$

Equation (3.4.4) is the capillary rise equation, and allows the surface tension of a liquid to be accurately determined from precise measurement of the height h to which the liquid rises within the capillary. In practice the method is mainly useful for systems in which the contact angle is reproducibly zero, that is, for liquids or solutions that completely wet the glass; it is not appropriate for the measurement of interfacial tensions of surfactant solutions. For example, cationic surfactants adsorb on glass and can render the surface more or less hydrophobic.

To give an idea of the height to which a liquid (in air) can rise in a capillary tube, consider a cylindrical capillary with internal diameter of 1 mm. By using (3.4.4) and assuming zero contact angle, it is calculated that water will rise about 30 mm and decane ($\gamma = 23.8$ mN m^{-1}, density 0.73×10^3 kg m^{-3}) about 13 mm at 293 K.

3.4.2 The Kelvin equation and homogeneous nucleation of vapours

It is well known that a gas or vapour in contact with a porous solid can condense in the pores at pressures, p, less than the saturated vapour pressure p°. This phenomenon is called *capillary condensation* (Box 3.3), and the condensed liquid in the pores has menisci concave to the vapour phase, as illustrated in Fig. 3.8 for a cylindrical pore closed at one

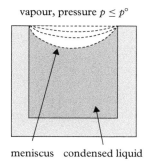

vapour, pressure $p \leq p^\circ$

meniscus   condensed liquid

Figure 3.8 *Condensed liquid in a cylindrical capillary closed at one end. As the vapour pressure increases the meniscus curvature decreases until, when $p = p^\circ$, the liquid surface becomes planar and the capillary is completely filled with liquid.*

Box 3.3   Principle of mercury porosimetry

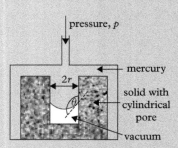

Mercury being forced under pressure into an evacuated cylindrical pore, radius r, within a solid. The contact angle of mercury with the solid surface is θ > 90°.

Capillary condensation occurs if the condensed liquid within pores of a solid wets the pore walls (i.e. the contact angle $\theta < 90°$); p required for condensation to occur is less than p^{o} as discussed. For *non-wetting liquids*, however, $(\theta > 90°)$ liquid has to be *forced into* the pores, and its surface is convex within a pore (as shown in the figure). The forced penetration of liquid mercury into porous solid powders is the basis of the technique of *mercury porosimetry* for probing the pore structure of solids (see §7.2). From a knowledge of the volume of mercury penetrating pores as a function of applied pressure p, a pore size distribution can be obtained. The pressure p is related, by the Laplace equation, to the surface tension γ of mercury and the radius r of (say) a cylindrical closed pore in an evacuated solid by

$$p = -2\gamma\cos\theta/r$$

---

end. Gas condenses in narrow pores at lower pressures than in pores of larger diameter. As the pressure of vapour $(p < p^{o})$ is increased above the pressure at which condensation first occurs, the meniscus curvature decreases until, when $p = p^{o}$, the liquid surface becomes planar, as shown in Fig. 3.8.

The Kelvin equation relates the equilibrium vapour pressure p of a pure liquid to the curvature of its surface. For a planar surface p $(= p^{o})$ is the saturated vapour pressure. In Chapter 14, the Kelvin equation will be used in connection with emulsion behaviour and stability so the equation is derived below by reference to spherical droplets rather than capillary condensed liquid.

If a small amount of material is removed from a bulk liquid and a droplet is formed, the molar free energy of material in the droplet is increased as a result of the formation of the surface. In general the chemical potential μ_{v} of a substance (equal to the molar free energy for a one-component system) in the gas (vapour) phase is related to its pressure p by

$$\mu_v = \mu_v^o + RT \ln p \tag{3.4.5}$$

where μ_v^o is the standard chemical potential in the gas phase (see §2.1 for a discussion of standard states). At equilibrium between liquid and vapour, $\mu_v = \mu_l$, where μ_l is the chemical potential in the liquid phase. Thus, the higher μ_l and hence μ_v, the higher the equilibrium vapour pressure over the drop.

To derive the Kelvin equation the transfer, at constant temperature T, of dn mol of one-component liquid from a bulk phase with plane surface (equilibrium vapour pressure p^o) to a spherical drop, radius r, in equilibrium with vapour at pressure p is considered. The drop radius increases by dr and the change in free energy is $\gamma d\mathcal{A}$ where $\mathcal{A}$ is the surface area of the drop, $4\pi r^2$. The change $d\mathcal{A}$ is $8\pi r dr$ and so the work associated with the transfer is $8\pi r dr \gamma$. The change in volume V of the drop on addition of dn mol liquid, dV, is $4\pi r^2 dr$ which is also equal to $v_m dn$, where v_m is the molar volume of the liquid at temperature T. Thus $d\mathcal{A}/dV = 2/r$ and so

$$\gamma d\mathcal{A} = 2\gamma dV/r = 2\gamma v_m dn/r \tag{3.4.6}$$

In each of the systems with flat and curved surface, the liquid phase is in equilibrium with the corresponding vapour and for a given system chemical potentials in liquid and vapour phases are equal. The chemical potentials of vapour over the plane liquid and the drop surfaces, μ_v^p and μ_v^d respectively, are

$$\mu_v^p = \mu_v^o + RT \ln p^o \text{ and } \mu_v^d = \mu_v^o + RT \ln p \tag{3.4.7}$$

Since in each of the systems vapour and liquid phases are in equilibrium, the transfer of dn mol liquid is energetically equivalent to transferring the same amount of material between the vapour phases, for which the free energy change is

$$\left(\mu_v^d - \mu_v^p\right) dn = dn \, RT \, \ln \left(p/p^o\right) \tag{3.4.8}$$

Equating the expressions for the energy changes in (3.4.6) and (3.4.8) leads to

$$\ln \left(\frac{p}{p^o}\right) = \frac{2\gamma v_m}{rRT} \tag{3.4.9}$$

Equation (3.4.9) is the *Kelvin equation* (Box 3.4) applicable to droplets, that is, convex spherically curved liquid surfaces, and it relates the equilibrium vapour pressure p to r as seen. It will be recalled that for concave menisci $p < p^o$, and the more highly curved the surface, the lower the equilibrium vapour pressure over the surface. From (3.4.9) for droplets however $p > p^o$, and it is seen that the greater the curvature (i.e. the smaller the droplets) the higher p. The sign of the curvature of a convex surface is opposite to that of a concave surface, and so the form of (3.4.9) for, for example, a bubble in a liquid, or capillary condensed liquid, has a negative sign on the right-hand side.

Box 3.4   William Thomson, Lord Kelvin—Professor at 22

William Thomson, later Lord Kelvin, born in Belfast in 1824, died 1907. This is a depiction of Thomson in an abandoned wine cellar converted into a laboratory, lecturing students at Glasgow University, Scotland. Chronicle/Alamy Stock Image.

William Thomson, later (1892) to become Lord Kelvin (named after the River Kelvin flowing past the University in Glasgow, Scotland) studied at Cambridge University and afterwards, in 1846, was appointed Professor of Natural Philosophy at Glasgow University at the age of 22. He held this post for 53 years. He introduced the absolute temperature scale and the term kinetic energy, and he was instrumental in establishing the Laws of Thermodynamics. The discovery of the Joule–Thomson effect arose from his collaboration with James Joule, and his name became widely familiar to the public at large through the Kelvinator fridge. Kelvin was knighted by Queen Victoria in 1866 for his part in the laying of the first transatlantic telegraph cable, and was ennobled in 1892 in recognition of his work on thermodynamics. He had a great love of the sea and of sailing and owned a schooner, the *Lalla Rookh*. In 1874 he sailed to Madeira, and as he approached the harbour he signalled a message to the house of his friend Charles Blandy, asking if his daughter Fanny would marry him. She replied yes and Kelvin and Fanny were married in June 1874, his first wife having died in 1870. One of the great men of nineteenth century physics, his remains lie close to those of Isaac Newton in Westminster Abbey, London.

The Kelvin equation has been verified experimentally in various ways (Israelachvili, 2007). When a solid sphere is placed on a flat solid surface, or when two curved solid surfaces (e.g. of mica) are brought into contact in the presence of a saturated vapour phase, capillary condensation can occur around the point of contact (see §8.2.2). By measuring the effect of such capillary bridges on the forces of adhesion between the solids (as a result of the effects of the Laplace pressure in the condensed liquid) the Kelvin equation has been verified in a number of cases. It appears that, depending on the liquid involved, the equation is valid for meniscus radius of curvature as small as between

0.5 nm (cyclohexane and benzene) and 2 nm (for water). The conclusion appears to be that the surface energy of a very small cluster of molecules can be very similar to that of a bulk liquid (see Israelachvili, 2007, §5.6). The derivation presented above implicitly assumes that the surface tension is independent of surface curvature.

The effects of curvature on vapour pressure only become significant for highly curved surfaces, with radii of curvature of the order of about 10 nm and less. This can be seen in Fig. 3.9 where the variation of the relative pressure, p/p^o, with radius of curvature r is shown for both convex (droplet) and concave (bubble) liquid surfaces for water at room temperature.

Droplets can exist in an 'infinite' vapour phase (e.g. the atmosphere), such that evaporation from the drop does not significantly change the partial vapour pressure. On the other hand, if the vapour space is enclosed and sufficiently small in extent, it is possible for evaporation or condensation from droplets to change the partial pressure of the vapour significantly. Probably the more important of the two cases is the former, which is considered first.

Imagine a drop radius r in contact with its vapour at partial pressure p such that r and p satisfy the Kelvin equation (3.4.9). If some evaporation occurs, the (constant) value of p is less than the equilibrium value for the smaller drop formed and further evaporation would therefore occur. Conversely, if condensation onto the drop were to occur the partial vapour pressure would exceed that for the expanded drop and so further condensation would occur. It is clear then that in such systems the 'equilibrium' represented by the Kelvin equation is an *unstable* equilibrium.

This can be illustrated by considering the free energy change, $\Delta_d\mu$, accompanying the *homogeneous* formation of a liquid droplet (i.e. in absence of foreign particles and wall effects which could cause condensation) containing n_d mol material from vapour at constant pressure p (relative pressure p/p^o). The chemical potential μ_v of a vapour at its saturated vapour pressure p^o is

$$\mu_v = \mu_v^o + RT \ln p^o = \mu_o^o \qquad (3.4.10)$$

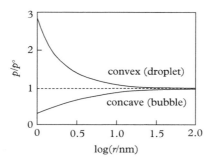

Figure 3.9 *Variation of relative vapour pressure, p/p^o, with radius of curvature, r, for water at 293 K.*

where μ_v^0 is the standard chemical potential in the vapour and μ_0^0 is the chemical potential of the pure liquid in equilibrium with the vapour. Since at *any* pressure p the chemical potential in the vapour is

$$\mu_v = \mu_v^0 + RT \ln p \qquad (3.4.5)$$

the free energy of forming a liquid droplet, radius r containing n_d moles of material from supersaturated vapour at pressure p is

$$\Delta_d\mu = -n_d RT \ln (p/p^o) + 4\pi r^2 \gamma \qquad (3.4.11)$$

where the second term on the right-hand side of (3.4.11) accounts for the formation of the surface of the droplet with surface tension γ. The number of moles n_d in the droplet is $4\pi r^3/3v_m$ where v_m is the molar volume of the condensed material. The expression for $\Delta_d\mu$ therefore has a negative term in r^3 and a positive term in r^2. This leads to a maximum in the relationship between $\Delta_d\mu$ and r, as illustrated in Fig. 3.10, for which parameters appropriate for water have been assumed; p/p^o has been taken to be 1.1 (i.e. 10 per cent supersaturated vapour). The droplet with radius r^* corresponding to the maximum in the curve is termed the *critical nucleus* ($\approx$ 10 nm in the case considered) and any droplet which forms with $r < r^*$ will spontaneously evaporate since $\Delta_d\mu$ always falls as r *decreases*. On the other hand for $r > r^* \Delta_d\mu$ falls continuously with *increasing* r and drops grow spontaneously and indefinitely. For $p/p^o = 1$ the first term on the right-hand side of (3.4.11) is zero and $\Delta_d\mu$ only rises with r so that homogeneous drop formation cannot occur; it requires that the vapour phase be supersaturated.

Stable equilibrium of a drop with its saturated vapour (as opposed to the unstable equilibrium discussed above) can notionally exist in an enclosed space, such as depicted schematically in Fig. 3.11. If the drop is originally at equilibrium, r and p are related through the Kelvin equation. Suppose some evaporation from the drop occurs. This will

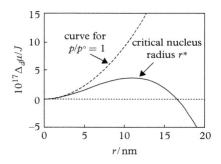

Figure 3.10 *Free energy of droplet formation against droplet radius calculated from (3.4.11) using parameters for water at 298 K: $v_m = 18 \times 10^{-6}$ m^3 mol^{-1}, $\gamma_{lv} = 0.072$ N m^{-1}. The upper dashed curve is for $p/p^o = 1$, and the full curve, showing a maximum at $r = r^*$, is for $p/p^o = 1.1$ (10% supersaturated).*

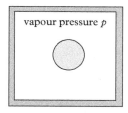

Figure 3.11 *A drop, radius r, in contact with vapour, pressure p, in a closed container.*

reduce r and increase the pressure of the vapour, which if it exceeds the equilibrium vapour pressure for the reduced drop size will cause condensation to occur, increasing drop size and decreasing the pressure. If on the other hand in the original system, some condensation occurs on the drop, the pressure of the vapour falls and r increases. If the reduced pressure is below the equilibrium pressure for the expanded drop, evaporation will occur restoring the original equilibrium. Thus it is possible in principle for a drop to maintain true equilibrium with vapour according (3.4.9) in a limited space since evaporation or condensation gives a significant change in vapour pressure (see Box 3.5).

Box 3.5   An illustrative example

Suppose an enclosed space with volume 1×10^{-19} m^3 contains 1.4573×10^{-19} mol water at 293 K. The molar volume, saturated vapour pressure (p°), and surface tension for water at this temperature are, respectively 1.8×10^{-5} m^3 mol^{-1}, 2340 Pa, and 0.0728 N m^{-1}. Show that in principle in this system, both a stable and an unstable equilibrium between a water droplet and its vapour can exist, and calculate the drop radii for the two equilibria.

   Answer. As a starting point suppose that a droplet is formed containing 10% of the total amount of water present. The drop would have a radius, r, of 3.969 nm, and the vapour pressure that would be in equilibrium with such a drop (here denoted p_{eq}, and calculated using the Kelvin equation) would be 3054 Pa. However the actual vapour pressure, p_{ac}, calculated from the amount of material remaining in the vapour phase, would be 3250 Pa (assuming ideal gas behaviour). Vapour would therefore condense onto the drop, its size would increase and both p_{eq} and p_{ac} would change accordingly. Calculations are made, using a spreadsheet, of p_{eq} and p_{ac} as function of r as material is added to or removed from the drop.

   The results of such calculations are depicted in Fig. 3.12 where the full line shows the variation of p_{ac} with r and the dashed line shows how p_{eq} changes with r. There are two radii for which $p_{ac} = p_{eq}$ ($r = 5.00$ nm at point A and $r = 2.596$ nm at point B in the figure). Noting that if $p_{ac} > p_{eq}$ condensation will occur, whereas if $p_{ac} < p_{eq}$ evaporation ensues, it follows that point A represents a stable equilibrium. Although at point B the Kelvin equation is satisfied, if some evaporation occurs p_{ac} falls below p_{eq} and so further evaporation will occur. If on the other hand some condensation takes place p_{ac} rises above p_{eq} and so further condensation takes place. That is, the equilibrium that exists at B is an unstable equilibrium of the kind already discussed for a drop in an infinite vapour phase.

Box 3.5 *Continued*

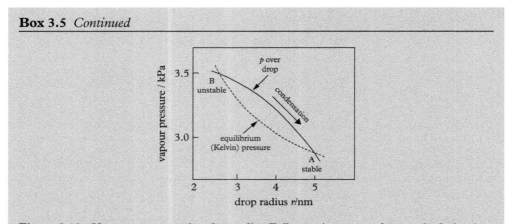

Figure 3.12 *Vapour pressure against drop radius. Full curve is pressure of vapour in the box (see text) and the dashed curve represents the equilibrium (Kelvin) pressure as a function of drop radius. Parameters assumed are given in the text.*

There are analogies between the formation of liquid droplets in vapour and the formation of micelles in solution, as discussed in §9.4. However, condensing droplets (for $r > r^*$ in Fig. 3.10) grow indefinitely and macroscopic phase formation occurs. On the other hand micelle formation results in stable aggregates of a *finite size*; the free energy curve analogous to that in Fig. 3.11 exhibits a minimum at higher radii as a result of repulsive interactions between surfactant head groups at the micelle surface.

The foregoing discussion has been for monodisperse droplets, but in a polydisperse mist of droplets in an infinite vapour phase, it is obviously not possible to have a value of p in (3.4.9) that satisfies more than a single value of r. Since the equilibrium vapour pressure of the small drops is larger than for the large drops, material effectively 'distils' from the small to the large drops. Hence the small drops become smaller and the large drops larger. This process, which happens in various kinds of colloidal dispersions, leading to kinetic instability, is called *Ostwald ripening*. In relation to droplets in emulsions, the Kelvin equation can be applied to the *solubility* of one liquid in another, and this is referred to further in Chapter 14.

Appendix A3:   Measurement of surface and interfacial tensions

Surface and interfacial tensions are fundamental thermodynamic quantities at the heart of the study of interfacial properties and behaviour. In addition to being central to capillary behaviour, as seen in this chapter, they are also directly related to the extent of surfactant adsorption at interfaces (Chapters 4, 5, and 7) and to the wetting properties of solid and liquid surfaces (Chapters 12 and 16). The viscoelastic properties of fluid interfaces (Chapter 15) are determined by the ways in which interfacial tensions vary under surface flow and extension. Various properties of emulsions and foams (and associated thin emulsion and foam films) are closely related to the tensions of the liquid/liquid or liquid/vapour interfaces present (Chapters 12 and 14).

Conveniently, there are many commercial instruments now available for the determination of surface and interfacial tensions, both for routine industrial purposes as well as more elaborate instruments for research use. When measuring surface and interfacial tensions, it is worth bearing in mind that many common contaminants are highly surface active and so can adsorb at surfaces (and *drastically* affect tensions) from very low bulk concentrations in solution. Further, some methods of measurement rely on placing metal elements (e.g. Wilhelmy plate or du Noüy ring—*vide infra*) in the surface, and it is important that the surfaces of the elements are completely wetted by the solutions under investigation. This requires that the metal surfaces are extremely clean and free of surface contamination that can arise for example from contact with the skin. The wettability of a surface can be completely changed by the presence of a single (or partial) monolayer of contaminant.

The phenomenon of capillary rise has been discussed in §3.1 and this furnishes a method for the determination of surface tension. Capillary tubes with precise bores are readily available, but the method requires that the contact angle θ of the liquid/vapour interface with the internal bore of the tube be zero. The method has been used to give very accurate values of surface tensions of pure liquids (which can be used as standards) but it is not appropriate for use with surfactant solutions, for which the contact angle θ (Fig. 3.2) is likely to be non-zero, irreproducible, and not readily measurable. In extreme cases (cationic) surfactant adsorption at the (negatively charged) glass/solution interface can render the surface hydrophobic.

A3.1   Wilhelmy plate and du Noüy ring methods

Two very common approaches to the measurement of surface and interfacial tensions, particularly of surfactant solutions, are referred to by the elements, attached to a microbalance, that are located at the interface under investigation: the du Noüy ring and the Wilhelmy plate methods. In the latter, a thin rectangular plate made from a suitable metal (e.g. platinum–iridium alloy), mica, glass or even filter paper is used. Whatever the material however, it must be completely wetted by the solution under test, or by the aqueous phase in the case of oil/water interfaces. Wettability is often aided by roughening the plate surface. In the case of wetted filter paper, the surface is in effect the solution under test, which forms a thin liquid layer over the paper surface. The wetting perimeter of the plate (i.e. the length of the three-phase contact line when the plate is in contact with the interface) is denoted P, and is usually of the order of 5 cm. The plate attached to a microbalance is initially weighed in air (or oil for an oil/water interface measurement). Then the liquid sample is raised on a suitable stage until the plate just comes into contact with the liquid interface as illustrated in Fig. A3.1. The base of the plate is level with the plane liquid interface remote from the plate so that buoyancy effects are eliminated. A liquid meniscus is formed and the (extra) force F exerted on the plate, due to the action of interfacial tension, is measured. The force is equal to the product of P and the surface tension forces, $\gamma \cos \theta$. Thus

$$\gamma = \frac{F}{P \cos \theta} \tag{A3.1}$$

and since the plate must be completely wetted $\theta = 0$, $\cos \theta = 1$ and hence $\gamma = F/P$.

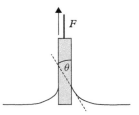

Figure A3.1 *Wilhelmy plate at the surface of a liquid. The bottom of the plate is at the level of the plane liquid surface. The plate is attached to a microbalance and the upward force F on the plate is measured. The contact angle θ of the solution or liquid/liquid interface with the plate should be zero.*

The plate method is absolute and needs no corrections, nor does it require a knowledge of the liquid densities since there are no buoyancy effects. After contact with the interface, the plate does not move in or out of the surface and can be used to monitor changes in interfacial tension with time. The du Noüy ring technique, discussed below, does not share these advantages but often the length of the wetting line around a ring is significantly greater than in the case of the plate and so the ring technique can be more sensitive, meaning that lower interfacial tensions can be accessed. The ring method is also less prone than the plate method to problems arising from incomplete wetting.

The du Noüy ring is circular and made of thin inert metal wire (e.g. platinum–iridium), attached to two uprights of the same material, as illustrated in cross-section in Fig. A3.2. It is attached to a microbalance, the ring being disposed horizontally. The liquid sample in a dish is raised vertically on a platform until the ring contacts the interface. The sample is then lowered again and the meniscus formed around the ring is progressively pulled up above the *plane* liquid interface, as illustrated in Fig. A3.2(a)–(c). For the interface outside the ring to be planar, the liquid container must be sufficiently large so that the meniscus at the container wall does not overlap that on the outside of the ring. As the meniscus changes shape, the direction of the forces due to interfacial tension change, as schematically illustrated by the downward arrows in Fig. A3.2(a)–(c). The outcome of this is that the force F on the ring passes through a maximum (at point (b)) with respect to the distance above the plane interface, as shown in Fig. A3.2(d).

The curvature of the meniscus in contact with the ring is different on the inside and outside of the ring which means that the maximum force (when $\theta = 0$) occurs at different heights of the ring above the planar liquid interface on the two sides. This has to be accounted for by the use of correction factors φ so that unlike the plate technique, the ring method is not absolute. Approximately, F_{max} is given as the product of the total length of the contact line ($4\pi R$) and the interfacial tension, where R is the mean ring radius indicated in Fig. A3.2(e). The meniscus volume is given by

$$V = \frac{4\pi R F_{max}}{\Delta\rho g} \tag{A3.2}$$

where $\Delta\rho$ is the (known) difference between the densities of the two phases and g is the acceleration due to gravity. To obtain the tension from the measured F_{max}, V is first calculated using (A3.2) and the value of $\Delta\rho$. The correction factor φ in the expression for the interfacial tension

$$\gamma = \frac{F_{max}}{4\pi R}\varphi \tag{A3.3}$$

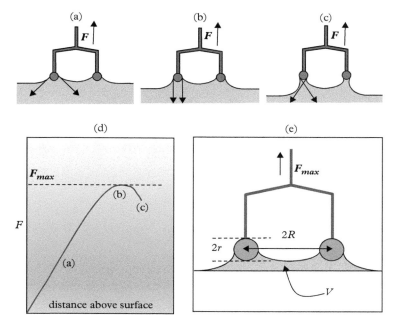

Figure A3.2 *The du Noüy ring method for the determination of surface and interfacial tensions. A vertical force F is applied to the ring which is located horizontally in the interface. The force passes through a maximum, F_{max}, with respect to the distance of the ring above the surface. The positions in (a), (b), and (c) are shown on the plot of F against distance above the surface in (d). The dimensions of the ring and the volume of liquid, V, supported above the plane liquid (shaded) are indicated in (e).*

is obtained by calculating a value of R^3/V (for each determination) and taking φ from tables (e.g. those of Harkins and Jordan, 1930) according to the value of R/r.

A3.2   Methods involving drop shape (optical tensiometry)

In the absence of gravity, a drop of liquid at equilibrium is spherical as a result of the action of surface or interfacial tension. Gravity tends to distort the drop and the drop shape reflects the interplay between gravity and surface tension. If the drop shape can be analysed in some way, the surface tension can be extracted, assuming the densities of the two phases are known. Drops to be analysed can either be pendant or sessile, as illustrated in Fig. A3.3. In the case of the pendant configuration, the drop is formed on a tip of circular cross-section, ensuring the axisymmetric shape of the drop, that is, the vertical cross-section of the drop is the same viewed horizontally from all angles. For the sessile drop (Fig. A3.3(b)), which rests on a smooth horizontal plate, it may be more difficult to ensure such symmetry, depending on the homogeneity of the wetting characteristics of the solid surface used. The drops represented in Fig. A3.3 are denser than the surrounding medium. For the study of liquid/liquid interfaces it is possible to have an inverted

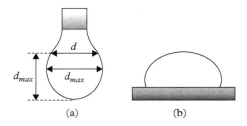

Figure A3.3 *Axisymmetric liquid drops distorted by gravity. (a) Pendant drop hanging on a circular tip. Dimensions d and d_{max} can be used to define the drop shape. (b) Sessile drop on smooth horizontal surface.*

system in which the drop is less dense than the surrounding medium. Equally, bubbles of gas can be studied.

A pendant drop is considered first, although the treatment for a sessile drop is very similar. (For much more detailed discussions of drop and meniscus shapes, see e.g. Padday, 1969, del Rio and Neumann, 1997, Drehlich et al., 2002, and Hartland and Hartley, 1976). The difference in hydrostatic pressure on the two sides of the interface, Δp_{hyd}, changes with vertical height, z, from the apex of the drop and

$$\Delta p_{hyd} = \Delta \rho g z \tag{A3.4}$$

where $\Delta \rho$ is the density difference between the drop and the external medium. The Laplace equation (3.4.3), for any point on the interface, is

$$\Delta p = \gamma \left(\frac{1}{r_1} + \frac{1}{r_2} \right) \tag{3.4.3}$$

and relates the pressure difference, Δp, on the two sides of the interface to the two principal radii of curvature, r_1 and r_2, of the surface. At the apex of the drop ($z = 0$), the two radii are the same, say b, and so here

$$\Delta p = \frac{2\gamma}{b} \tag{A3.5}$$

Noting (A3.4), (3.4.3), and (A3.5) it follows that the drop shape is described by

$$\gamma \left(\frac{1}{r_1} + \frac{1}{r_2} \right) = \frac{2\gamma}{b} - \Delta \rho g z \tag{A3.6}$$

It is convenient to work in terms of the arc length s between the apex and any other point on the drop surface, and the angle θ, defined by $\tan \theta = dz/dx$ as illustrated in Fig. A3.4. It is seen that

$$\frac{dx}{ds} = \cos \theta; \frac{dz}{ds} = \sin \theta \tag{A3.7}$$

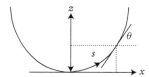

Figure A3.4 *Drop profile showing the coordinates x and z, the slope angle θ, and the arc length s between the drop apex and any point on the drop surface.*

and

$$\frac{d\theta}{ds} = \frac{2}{b} + \frac{\Delta\rho gz}{\gamma} - \frac{\sin\theta}{x} \tag{A3.8}$$

In terms of dimensionless quantities $X = x/b$, $Z = z/b$, and $S = s/b$, (A3.8) can be expressed as

$$(2 + \beta Z) = \frac{d\theta}{dS} + \frac{\sin\theta}{X} \tag{A3.9}$$

where

$$\beta = \frac{\Delta\rho gb^2}{\gamma} \tag{A3.10}$$

The quantity $\Delta\rho g/\gamma$ is a measure of the competing effects of gravity and interfacial tension on drop shape (see §16.2.3). In modern instruments, image analysis software is often included to fit the optically observed drop profile to (A3.9) or equivalent, thus extracting the value of the interfacial tension γ.

As with pendant drops, the shapes of sessile drops are determined by the competing effects of interfacial tension and gravity, but the drop is truncated when θ is equal to the contact angle of the drop with the horizontal solid surface. Thus contact angles can be measured by fitting the drop profile using the above equations, as mentioned in §16.3.2.

An alternative method of extracting the interfacial tension from the shape of a pendant drop involves measuring the drop radius in two selected planes. With reference to Fig. A3.3, one possibility is to measure the maximum diameter, d_{max}, and then the diameter d in the horizontal plane a distance d_{max} from the apex of the drop (Andreas et al., 1938). The interfacial tension is given by

$$\gamma = \frac{\Delta\rho gd_{max}^2}{H} \tag{A3.11}$$

The ratio $d/d_{max} = S$ is termed the shape factor and the parameter H, which depends on drop shape, is obtained from tables of $1/H$ as a function of S (Andreas et al., 1938; Stauffer, 1965; Misak, 1968).

Pendant drop methods are suitable for the study of surfactant solutions both at liquid/vapour and liquid/liquid interfaces. Also, with modern instrumentation and desktop computers, measurements and analysis can be performed very rapidly so that variation of tension with time can be obtained, although obviously not down to very short times (say <10 to 20 s).

A3.3   Drop-volume method

This method involves the extrusion of a drop of (the more dense) liquid from a circular tip of known radius r into a second less dense phase (liquid or vapour) until the drop just becomes detached from the tip (Fig. A3.5). The tip is attached to a syringe and micrometer. A micrometer reading is taken when a drop just detaches, and then a second reading made as the next drop becomes detached, so that the volume V of the *detached* drop is known. Very crudely the interfacial tension is given by $W/2\pi r$ where W is the weight of the detached drop. However, such a simple approach is inadequate because, *inter alia*, not all of the drop becomes detached, a fraction remaining on the tip.

Harkins and Brown (1919) made an empirical calibration of the drop-volume method, using liquids of known surface tension, obtaining a correction factor ϕ such that

$$\gamma = \frac{W}{2\pi r\phi} = \frac{V\Delta\rho g}{2\pi r\phi} \tag{A3.12}$$

where, as before, $\Delta\rho$ is the density difference between the two phases. The correction factor is related to the tip radius and the drop volume, that is,

$$\phi = f\left(\frac{r}{V^{1/3}}\right) \tag{A3.13}$$

The method can be used for both surface and interfacial tensions and can give highly reproducible results. However, during the process of detachment of the drop an expanding neck is formed and it is not clear that adsorption of, say, surfactant can keep pace with this creation of new surface. If the tension at the surface of the neck differs from the equilibrium tension, this could conceivably influence the fraction of the drop that remains on the tip after detachment. Therefore, although it has been widely used in the past for the study of surfactant solutions it is probably best to avoid its use for such systems.

A3.4   Maximum bubble pressure method for surface tension

This approach can be used to measure *dynamic surface tensions* as well as static (equilibrium) values. When gas (air or nitrogen) is blown through a cylindrical capillary tube (radius r_c) immersed in a liquid whose tension is to be determined, bubbles are formed and detached at a measurable rate.

Figure A3.5 *Liquid drop formed on a steel tip with circular cross-section. The tip is attached to a microsyringe and micrometer.*

The tip of the capillary is located a distance h below the surface of the liquid, density ρ, as illustrated in Fig. A3.6. The pressure, p_g, of the gas required to displace the liquid from the capillary is greater than the pressure in the liquid, p_l, as a result of the curvature $(1/r_b)$ of the bubble surface. If the bubble is spherically curved (*i.e.* if r_c is sufficiently small, say <300 μm) the pressure difference, Δp, across the bubble interface is given by the Laplace equation (3.4.2). Thus

$$\Delta p = p_g - p_l = p_g - h\rho g = \frac{2\gamma}{r_b} \tag{A3.14}$$

It follows that the gas pressure varies inversely with r_b. With reference to Fig. A3.6, when the bubble starts to emerge from the tip (surface a) r_b is large and so p_g is relatively small. The bubble radius is minimum and p_g is maximum when $r_b = r_c$, that is, when the bubble is a hemisphere (surface b in Fig. A3.6). As the bubble expands further and ultimately becomes detached from the tip, r_b increases (c) and p_g falls. This sequence of change in gas pressure is illustrated schematically in Fig. A3.7. Clearly, if the maximum gas pressure p_g^{max} is measured and r_c, ρ, and h are known the surface tension can be calculated.

The surface tension obtained for say a surfactant solution depends on the rate of bubble formation. When fresh surface is formed, surfactant adsorption is not immediate since molecules must diffuse to the interface, and surface age is related to bubble frequency. If the bubbles are formed sufficiently slowly the tension obtained is the equilibrium tension. For faster rates, tensions are dynamic tensions (see §6.1). The time interval between one bubble being detached and the next however is not the true surface age. For example, Austin et al. (1967) divided this time interval into two regimes, as shown in Fig. A3.7. In region A, a new bubble begins to form and the gas pressure rises as the bubble approaches a hemispherical shape, when $p_g = p_g^{max}$. Surfactant adsorption occurs during this period which represents the true surface age. In Region B the hemispherical bubble begins to grow very rapidly and then detaches from the tip; this time interval is termed the dead time, t_D. To obtain the age of the surface for which the tension is measured, the dead time must be allowed for giving a corrected bubbling rate, t^{-1}, according to

$$t^{-1} = \left(\frac{1}{f^{-1} - t_D} \right) \tag{A3.15}$$

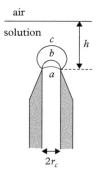

Figure A3.6 *Cylindrical tip under the air/solution interface; a, b, and c represent the menisci of changing radius as a bubble grows. The bubble is hemispherical when its radius is equal to the tip radius, r_c.*

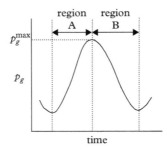

Figure A3.7 *Variation of p_g with time as bubbles are formed and detach from the tip. Region A corresponds to the formation of a bubble up to a hemisphere (maximum p_g) and region B to the rapid growth and detachment of the bubble.*

where f is the measured bubble frequency. Dukhin et al. (1995) give a detailed account of dynamic tensions including their experimental determination. Commercial bubble pressure tensiometers are available and surface ages down to about 5 ms can be accessed for solutions with surface tensions between about 10 and 100 mN m^{-1}.

A3.5   Spinning-drop tensiometry for very low interfacial tensions

Liquid/vapour tensions do not become very low even in the presence of strongly adsorbing surfactant. In contrast however liquid/liquid interfacial tensions can fall to 10^{-3} mN m^{-1} and below (see §10.1.3), and the techniques described above cannot be used for such low tensions. A convenient and widely used method is the spinning-drop technique. A drop of the less dense liquid (density ρ_d) is introduced into the more dense liquid (density ρ_m), which is being spun at high velocity in a horizontal cylindrical tube. At mechanical equilibrium, the shape of the spinning drop is determined by a balance between the forces of interfacial tension and centrifugal forces.

For oil–water systems, the (usually) denser aqueous phase fills a cylindrical tube which is spinning with angular velocity ω around its horizontal axis, as illustrated in Fig. A3.8. At sufficiently high speed of rotation the drop becomes cylindrical with hemispherical caps. In this regime (where the drop length is at least four times the drop diameter $2r$) the interfacial tension is given by the simple expression due to Vonnegut (1942)

$$\gamma = \frac{(\rho_m - \rho_d)\, r^3 \omega}{4} \tag{A3.16}$$

The temperature of the system can be controlled by surrounding the cylindrical tube with an oil thermostat, as illustrated in Fig. A3.8. The interfacial tension is calculated from a knowledge of the density difference between drop and medium, the drop radius and ω (up to rotation frequencies of about 2×10^4 min^{-1}). The drop dimensions are obtained using a lens so the magnification must be known, note being taken that the drop image is viewed through both the thermostat fluid and the more dense (aqueous) phases. As for the other methods discussed above, commercial instruments

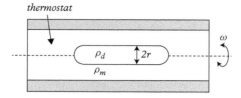

Figure A3.8 *Representation of the spinning-drop technique for the measurement of ultralow interfacial tensions.*

are available, and the ultralow interfacial tensions presented in Chapter 10 have, in the main, been obtained using the spinning drop technique.

Surface laser light scattering has been used to determine ultralow interfacial tensions at oil/water interfaces, although it has not been used as routinely as the spinning-drop method. Fluid/fluid interfaces are deformed by thermal fluctuations in the form of capillary waves (ripplons) with characteristic wavelength and amplitude. The restoring force is the interfacial tension. The ripplons behave as a diffraction grating and scatter incident laser light, which gives information about interface properties. A description of the method is beyond the scope here but the interested reader can gain more information and references from Pouchelon et al. (1980) and Dorshow and Swofford (1990).

....

GENERAL READING

R. Defay, I Prigogine, A. Bellemans, and D. H. Everett, *Surface Tension and Adsorption*. Longmans, London, 1966. The book is devoted to the thermodynamics of surface tension and adsorption.
E.A. Guggenheim, *Thermodynamics*. North Holland, Amsterdam, 1967. This is an authoritative text which contains sections on the thermodynamics of fluid interfaces.
J.S. Rowlinson and B. Widom, *Molecular Theory of Capillarity*. Clarendon Press, Oxford, 1982. This is a classic text on the theory of fluid interfaces.

....

REFERENCES

J.M. Andreas, E.A. Hauser and W.R. Tucker, Boundary tension by pendant drops. *J. Phys. Chem.*, 1938, **42**, 1001–1019.
M. Austin, B.B. Bright, and E.A. Simpson, The measurement of the dynamic surface tension of manoxol OT solutions for freshly formed surfaces. *J. Colloid Interface Sci.*, 1967, **23**, 108–112.
H. Liu and G. Cao, Effectiveness of the Young–Laplace equation at nanoscale. *Sci. Rep.*, 2016, **6**, 23936.
O.I. del Rio and A.W. Neumann, Axisymmetric drop shape analysis: computational methods for the measurement of interfacial properties from the shape and dimensions of pendant and sessile drops. *J. Colloid Interface Sci.*, 1997, **196**, 136–147.

R.B. Dorshow and R.L. Swofford, An adjustable-resolution surface laser-light scattering spectrometer and its application to the measurement of interfacial tension and viscosity in crude oil systems. *Colloids Surf.*, 1990, **43**, 133–149.

J. Drehlich, C. Fang, and C. L. White, in *Encyclopedia of Surface and Colloid Science*. Marcel Dekker Inc., 2002, 3152.

S.S. Dukhin, G. Kretzschmar and R. Miller, *Dynamics of Adsorption at Liquid Interfaces: Theory, Experiment, Application*. Elsevier, Amsterdam, 1995.

F.M. Fowkes, Additivity of intermolecular forces at interfaces. I. Determination of the contribution to surface and interfacial tensions of dispersion forces in various liquids 1. *J. Phys. Chem.*, 1963, **67**, 2538–2541.

W.D. Harkins and F.E. Brown, The determination of surface tension (free surface energy), and the weight of falling drops: the surface tension of water and benzene by the capillary height method. *J. Am. Chem. Soc.*, 1919, **41**, 499–524.

W.D. Harkins and H.F. Jordan, A method for the determination of surface and interfacial tension from the maximum pull on a ring. *J. Am. Chem. Soc.*, 1930, **32**, 1751–1772.

S. Hartland and R.W. Hartley, *Axisymmetric Fluid–Fluid Interfaces*. Elsevier, New York, 1976.

J. Israelachvili, *Intermolecular and Surface Forces*, 3rd edition. Academic Press, Amsterdam, 2011.

M.D. Misak, Equations for determining 1/H versus S values in computer calculations of interfacial tension by the pendent drop method. *J. Colloid Interface Sci.*, 1968, **27**, 141–142.

J.F. Padday, in *Surface and Colloid Science* (ed. E. Matejević), Volume 1. Wiley-Interscience, New York, 1969, 103.

A. Pouchelon, J. Meunier, D. Langevin, and A.M. Cazabat, Light scattering from oil-water interfaces: measurements of low interfacial tensions. *J. Phys. Lett.*, 1980, **41**, 239–242.

A. Robinson, Thomas Young and the Rosetta Stone. *Endeavor*, 2007, **31**, 59–64.

C.E. Stauffer, The measurement of surface tension by the pendant drop technique. *J. Phys. Chem.*, 1965, **69**, 1933–1938.

B. Vonnegut, Rotating bubble method for the determination of surface and interfacial tensions. *Rev. Sci. Instrum.*, 1942, **13**, 6–9.

Section II

Adsorption of surfactants

As a result of their amphiphilic structure, surfactant molecules/ions characteristically adsorb at liquid/fluid and solid/liquid interfaces, as well as forming aggregates in solution. Aggregation processes are described in Section III and adsorption is the subject of the present section. Adsorption lowers interfacial tension and modifies the properties of the interface, which is of crucial importance in wetting behaviour (Section VI) and the stabilization of thin films, particulate dispersions and emulsions (Sections IV and V). Much can be learned about adsorption and adsorbed films from the thermodynamic analysis of the variation of interfacial tensions with the activity of surfactant in solution, as illustrated in Chapter 4. Adsorption at liquid/fluid interfaces, including dynamic aspects, is covered in Chapters 5 and 6, and in Chapter 7 adsorption at solid/fluid interfaces is described. Some methods for the direct characterization of adsorbed surfactant layers are introduced in Chapter 8.

4

Adsorption of surfactants at liquid interfaces: thermodynamics

As a result of their molecular structure, surfactants adsorb at liquid/fluid and solid/fluid interfaces and, above a certain concentration, they also aggregate in solution to form micelles or other assemblies (covered in Section III). The aggregates in solution remain in equilibrium with the adsorbed layers. In this chapter, focus is on the thermodynamics of adsorption of surfactants, particularly at liquid/vapour and liquid/liquid interfaces.

The nature and structure of adsorbed layers can be probed directly using physical techniques, such as neutron reflection and ellipsometry, which are capable of revealing subtle detail, and some such techniques are referred to in Chapter 8. Often however, interest in surfactant systems centres on the way in which surfactants lower interfacial tension, and the processes which are associated with this such as wetting, permeation of liquids through porous media and microemulsion formation. Indeed, a great deal can be gleaned about adsorption and the nature of adsorbed monolayers from an analysis of the dependence of interfacial tension on the concentration (or activity) of surfactant in solution. For this reason a central equation in surfactant science is the Gibbs adsorption equation, which relates the amount of surfactant adsorbed per unit area, to *changes* in interfacial tension with surfactant concentration in solution. In order to derive and apply this equation some thermodynamics relating to surfaces is first introduced. In the next chapter some experimental findings are presented and the nature of adsorbed layers at liquid/fluid interfaces is discussed.

4.1 Conventions for the definition of an interface

The extent (e.g. volume and mass) and composition of a bulk phase are readily ascertained. For a surface however the situation is more problematic. The surface is a region between two contacting bulk phases, say α and β, and the properties, for example, composition and density, change continuously from those of α at one side of the surface to those of β at the other. Generally, however, the thickness, τ, of the surface region is not known. There are two widely used conventions for defining a surface, one of them introduced by Gibbs in his pioneering treatment of surfaces.

Surfactants: In Solution, at Interfaces and in Colloidal Dispersions. Bob Aveyard. © Bob Aveyard 2019.
Published in 2019 by Oxford University Press. DOI: 10.1093/oso/9780198828600.001.0001

4.1.1   The surface phase convention

This is perhaps the simpler of the two conventions to understand physically; it treats the surface as a phase separate from α and β with finite thickness, τ, and volume, V^s. With reference to Fig. 4.1, the planes AA and BB bounding the surface are mutually parallel and placed such that phases α and β are homogeneous up to AA and BB, respectively. All changes in passing from α to β therefore occur in the surface region bounded by these two planes. The concentration profile of an adsorbing species (e.g. surfactant) normal to the interface is shown in Fig. 4.1a. The amount of surfactant present in unit area of the surface, Γ^s, is represented by the shaded area and is termed the *surface concentration*. The superscript s denotes the choice of the surface phase model and is used with other thermodynamic quantities (i.e. free energy, entropy, etc.) as appropriate. As presented in Fig. 4.1a, there is a finite concentration of the surfactant in phase α and phase β, and the ratio of the concentrations is the distribution ratio of the surfactant between the two phases. If one of the phases is vapour then the concentration in that phase is zero.

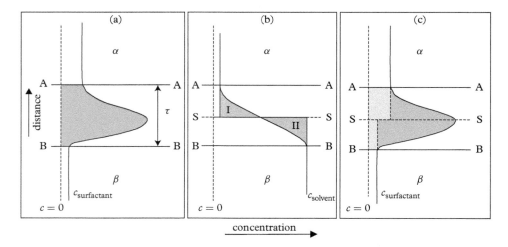

Figure 4.1 *Representation of models for an interface. (a) In the surface phase model planes AA and BB are parallel and placed such that phases α and β are homogeneous up to planes AA and BB, respectively. All changes in concentrations from α to β occur between the planes, which are separated by τ, the surface thickness. (b) This shows the Gibbs dividing surface SS and the profile of solvent concentration through the interface. As SS is moved vertically the relative values of shaded areas I and II change. The usual choice of position for SS is that where the areas I and II are equal, which corresponds to Γ^σ for the solvent of zero. (c) The surface excess concentration of surfactant in the Gibbs model, Γ^σ, is represented by the total shaded areas (which give the total surface concentration as in (a)) less the amount represented by the two yellow shaded rectangular areas. For strong adsorption, as in the case of surfactants, the yellow areas usually become insignificant relative to the total area and in this case Γ^s and Γ^σ are effectively equal.*

4.1.2   The Gibbs model

The model introduced by Gibbs (Box 4.1) is a little more abstract. The surface is taken to be a plane (the *Gibbs dividing surface*), shown as SS in Fig. 4.1b and 4.1c. This plane is placed between and parallel to AA and BB. A quantity called the *surface excess* concentration, Γ^σ, of a component is then defined as the total amount of that component in unit area of interface (i.e. Γ^s), *less* the amount that would be present if phases α and β extended unchanged up to SS (represented by the rectangular yellow areas in Fig. 4.1c). The superscript $^\sigma$ is used to denote the choice of the Gibbs model. From Fig. 4.1c it is evident that for strong adsorption from dilute solution (as for surfactants), the sum of the yellow areas becomes very small in relation to the total shaded area and in this case Γ^s and Γ^σ become effectively equal.

Box 4.1   Josiah Willard Gibbs 1839–1903

Josiah Willard Gibbs (1839–1903), American scientist who made contributions to theoretical physics and chemistry, and to mathematics. Science History Images/Alamy Stock Photo.

Josiah Willard Gibbs (1839–1903), Professor of Mathematical Physics at Yale from 1871 until his death, is most famous to chemists for his theoretical work on thermodynamics (e.g. the Gibbs phase rule and the concepts of free energy and chemical potential) but he also did pioneering work in mathematics and theoretical physics, laying the foundations of statistical mechanics. His work was largely unknown in Europe until translated into French by le Chatelier and into German by Ostwald. J.G. Crowther (1944) wrote 'Gibbs was highly esteemed by his friends, but US science was too preoccupied with practical questions to make much use of his profound theoretical work during his lifetime. He lived out his quiet life at Yale, deeply admired by a few able students but making no immediate impress on US science commensurate with his genius'. One American who was not overly impressed by Gibbs' work however was his cousin who, on taking a look at Gibbs' paper on heterogeneous equilibrium commented: 'It looks full of hard words and signs and numbers, not very entertaining or understandable looking, and I wonder whether it will make people wiser or better.'

The position of SS between AA and BB is arbitrary but a convenient choice, and one that is often used (for a liquid–vapour system), is that where the surface excess concentration of the *solvent* is zero. A solvent concentration profile through an interface is depicted in Fig. 4.1b. In the model (i.e. reference) system (where concentrations in α and β phases remain unchanged up to SS) the concentration profile is a step function. The 'real' system, in which the profile is sigmoid, has an amount of material per unit area more than the reference system, as represented by the green area I. On the other hand the real system has an amount per unit area less than the reference, as represented by area II. If SS is placed so that I and II are equal, then the model and real systems have the same amounts of solvent per unit area so the *excess* amount in the surface is zero, that is, Γ^σ (solvent) $= 0$.

To conclude, the choice of which model to employ when considering surfactant adsorption is a matter of convenience, and in any event usually the surface concentration and surface excess are effectively equal. In the Gibbs model, because the surface is a plane, the surface volume and thickness are zero, that is, $V^\sigma = \tau = 0$. With regard to the placing of the planes AA and BB bounding the surface region, there is a lower limit to their separation since the concentrations of species in α and β must remain unchanged up to AA and BB, respectively. However, although it is perhaps not obvious at this stage, there is no upper limit to their separation, so that the thickness of the surface region need not usually be known.

In systems where adsorption is weak it is obvious from what has been said that there will be a significant difference between Γ^s and Γ^σ for a species. Further, in some systems, for example, aqueous solutions of inorganic electrolytes with strongly solvated cations, solute *desorption* occurs. Here the solute is excluded from the interface and there is a *negative* surface excess concentration. As will be seen, positive adsorption is accompanied by a decrease in interfacial tension, and desorption gives rise to an increase in the tension.

4.2   Some thermodynamic quantities for surfaces

4.2.1   The First Law of Thermodynamics for a system with an interface

Having defined what is meant by an interface, some thermodynamic quantities and expressions for interfaces can be introduced. The starting point is the First Law of Thermodynamics for a closed system (no material leaving or entering) without an interface:

$$dU = dq + dw \tag{4.2.1}$$

Here U, the internal energy of the system, is a state function and so dU is an exact differential. In contrast, the quantities q and w, the heat taken up by the system and work done on the system respectively, are not state functions; dq and dw depend on *how* any change is effected and dq and dw are not therefore exact differentials. For systems at equilibrium however, infinitesimal changes are reversible and the work dw associated with a volume change, dV, at constant pressure p of the system is $-p$dV. Further, from the Second Law of Thermodynamics, for a reversible change at temperature T, d$q = T$dS where S, the entropy of the system is, like U a state function. Noting this, for a reversible change at constant T and p (4.2.1) becomes

$$dU = TdS - pdV \qquad (4.2.2)$$

The terms on the right-hand side of (4.2.2) each consist of an *intensive* quantity (T, p) multiplied by a change in an *extensive* property (S, V).

In general, interest will be centred on open systems where material can enter or leave, so the concentrations of components can be changed, and in this case a term must be added to (4.2.2) to allow for this. The number of moles of a component i in the system is denoted n_i; obviously n_i is an extensive quantity. The intensive quantity associated with n_i is μ_i, the chemical potential of i. For an open system then (4.2.2) becomes

$$dU = TdS - pdV + \sum_i \mu_i dn_i \qquad (4.2.3)$$

Finally, a term must be added for a system with an interface present. The intensive quantity for the interface is the interfacial tension, γ, and the extensive quantity the interfacial area $\mathcal{A}$. Thus for an open system with an interface

$$dU = TdS - pdV + \sum_i \mu_i dn_i + \gamma d\mathcal{A} \qquad (4.2.4)$$

In general, in thermodynamic expressions for systems with an interface, the tern $-p$dV is replaced by $-p$d$V + \gamma$d$\mathcal{A}$. Obviously $\mathcal{A}$ is a two-dimensional analogue of V; interfacial tension is a two-dimensional analogue of p, but surfaces tend to contract rather than expand and the terms in p and γ therefore carry opposite signs.

For a system with two bulk phases α and β, and a surface, s, the internal energy has additive contributions from each of the bulk phases and the interface so $U = U^\alpha + U^\beta + U^s$ and d$U = $ d$U^\alpha + $ d$U^\beta + $ dU^s. Thus, for the α or β phase, it is possible to write

$$dU^\alpha = TdS^\alpha - pdV^\alpha + \sum_i \mu_i dn_i^\alpha$$

$$dU^\beta = TdS^\beta - pdV^\beta + \sum_i \mu_i dn_i^\beta \qquad (4.2.5)$$

and for the surface

$$dU^s = TdS^s - pdV^s + \sum_i \mu_i dn_i^s + \gamma d\mathcal{A} \qquad (4.2.6)$$

Note that the intensive quantities T, p, and μ_i do not carry superscripts since at equilibrium these quantities are equal in all coexisting phases. In (4.2.5) and (4.2.6), the superscript s is used generally to mean surface, not restricting the discussion here to a particular convention for the surface.

4.2.2   Gibbs and Helmholtz free energies for surfaces

The usual definition of the Helmholtz free energy for a surface, A^s, is

$$A^s = U^s - TS^s \tag{4.2.7}$$

and so

$$dA^s = dU^s - TdS^s - S^s dT$$

Noting (4.2.6) therefore, dA^s is given by

$$dA^s = -S^s dT - pdV^s + \sum_i \mu_i dn_i^s + \gamma d\mathcal{A} \tag{4.2.8}$$

There are two definitions of the Gibbs free energy for a surface which have been commonly used, differing by the term $\gamma \mathcal{A}$:

$$G^s = A^s + pV^s \quad \text{and} \quad G^s = A^s + pV^s - \gamma \mathcal{A} \tag{4.2.9}$$

It is easily shown that the differential forms are

$$dG^s = -S^s dT + V^s dp + \gamma d\mathcal{A} + \sum_i \mu_i dn_i^s \tag{4.2.10}$$

$$dG^s = -S^s dT + V^s dp - \mathcal{A} d\gamma + \sum_i \mu_i dn_i^s \tag{4.2.11}$$

Integration of these expressions for constant intensive variables (T, p, γ and μ_i) shows that

$$G^s = \sum_i \mu_i n_i^s + \gamma \mathcal{A} \tag{4.2.12}$$

$$G^s = \sum_i \mu_i n_i^s \tag{4.2.13}$$

The Gibbs free energy of a bulk phase, G, is defined such that $G = \sum_i \mu_i dn_i$ and so it is seen that the definition of G^s, which includes the term γA, is the one analogous to the definition for a bulk phase.

Surface tension, which as seen from (4.2.8) and (4.2.10) is defined by

$$\gamma = \left(\frac{\partial A^s}{\partial \mathcal{A}} \right)_{T,V^s,n_i^s} = \left(\frac{\partial G^s}{\partial \mathcal{A}} \right)_{T,p,n_i^s} \tag{4.2.14}$$

is sometimes referred to as the 'surface free energy', presumably because it is a free energy per unit area of interface. Clearly, however, it is A^s or G^s which is the surface free energy. If for example (4.2.8) is integrated at constant temperature it is seen that, employing the Gibbs model for the surface (now using superscript σ to denote the choice and noting that $V^\sigma = 0$),

$$A^\sigma = \gamma \mathcal{A} + \sum_i \mu_i n_i^\sigma \tag{4.2.15}$$

For unit area of interface (4.2.15) becomes

$$A_\sigma = \gamma + \sum_i \Gamma_i^\sigma \mu_i \tag{4.2.16}$$

where $\Gamma_i^\sigma = n_i^\sigma / \mathcal{A}$ is the surface excess concentration of component i. The surface free energy, A_σ, is the *excess* Helmholtz free energy per unit area of interface and is not in general equal to γ. There is one special case however where the two quantities are equal, that is, where $\sum_i \Gamma_i^\sigma \mu_i = 0$, and that is for a one-component system with the Gibbs dividing surface chosen such that $\Gamma_i^\sigma = 0$. In general, the significance of 'surface free energy' and 'surface concentration' depends on the chosen model for the interface; in contrast, surface tension is a measurable thermodynamic quantity and is independent of any model.

4.2.3 Surface chemical potential

The surface chemical potential of a component i, μ_i^s, is equal to the chemical potential, μ_i^l, of the component in bulk solution at adsorption equilibrium. Two Gibbs free energy functions for a surface, G^s and $\mathcal{G}^s$, were introduced in §4.2.2:

$$G^s = \mathcal{G}^s - \gamma \mathcal{A}. \tag{4.2.9}$$

It follows from (4.2.9) that

$$\left(\frac{\partial G^s}{\partial n_i^s} \right)_{T,p,\gamma,n_j} = \left(\frac{\partial \mathcal{G}^s}{\partial n_i^s} \right)_{T,p,\gamma,n_j} - \gamma \left(\frac{\partial \mathcal{A}}{\partial n_i^s} \right)_{T,p,\gamma,n_j} \tag{4.2.17}$$

From the differential expressions for G^s and $\mathcal{G}^s$, (4.2.10) and (4.2.11) it is seen that $\left(\partial G^s / \partial n_i^s \right)_{T,p,\gamma,n_j} = \mu_i \ (= \mu_i^s)$. On the other hand the quantity $\left(\partial \mathcal{G}^s / \partial n_i^s \right)_{T,p,\gamma,n_j} = \zeta$ is not equal to the chemical potential, but differs from it by $\gamma \left(\partial \mathcal{A}/\partial n_i^s\right)_{T,p,\gamma,n_j} = \gamma \omega_i$ where ω_i is the partial molar surface area of component i, that is,

$$\mu_i = \mu_i^l = \mu_i^s = \zeta_i - \gamma \omega_i \tag{4.2.18}$$

The relationship (4.2.18) will be referred to in the Appendix (A5.2) to Chapter 5, where the treatment of adsorbed monolayers as two-dimensional solutions is discussed.

4.3   The Gibbs adsorption equation

Adsorption at an interface (fluid/fluid or solid/liquid) always gives rise to a change in interfacial tension, γ. The Gibbs equation is a thermodynamic expression that relates the amount of material adsorbed (or desorbed), that is, surface concentration Γ^s or surface excess concentration Γ^σ, to both the bulk activity of the adsorbing species (the *adsorbate*) and the interfacial tension. For systems with fluid/fluid interfaces, where interfacial tension is readily measurable experimentally, adsorption can be obtained from measured changes of tension with adsorbate activity in solution. For solid/liquid interfaces however, the interfacial tension (γ as defined in (4.2.14)) is not generally accessible, but if adsorption can be measured independently, over a range of adsorbate activity, the Gibbs equation allows *changes* in γ to be computed.

The so-called *Gibbs adsorption equation* is readily obtained from any of the differential expressions: (4.2.6) for dU^s, (4.2.8) for dA^s, (4.2.10) for dG^s, or (4.2.11) for dG^s. These expressions can be integrated for constant intensive variables (T, p, γ, and μ_i) and then differentiated, and the two expressions for dX^s ($X^s = U^s, A^s, G^s$, and G^s) compared. For example consider (4.2.11) for dG^s; integration for constant intensive variables yields (4.2.13), which on differentiation gives

$$dG^s = \sum_i n_i^s d\mu_i + \sum_i \mu_i dn_i^s \tag{4.3.1}$$

Comparison of (4.3.1) with (4.2.11) shows that

$$-\mathcal{A}d\gamma = S^s dT - V^s dp + \sum_i n_i^s d\mu_i \tag{4.3.2}$$

Written for unit area of surface, (4.3.2) becomes

$$-d\gamma = S_s dT - \tau dp + \sum_i \Gamma_i d\mu_i \tag{4.3.3}$$

where τ is the thickness of the surface and S_s is total or excess entropy of *unit area* of interface (depending on the surface model used). Like the preceding equations, (4.3.3) is valid for either model for the surface. For the surface phase model, τdp is negligibly small (because the surface is very thin) and Γ_i becomes Γ_i^s, the *surface concentration*, superscript s now denoting the choice of the surface phase model. If the Gibbs model is used, the surface is a plane, $\tau = 0$ and $\Gamma_i = \Gamma_i^\sigma$, the *surface excess concentration* of i. For constant temperature (4.3.3) becomes the Gibbs adsorption isotherm

$$-\mathrm{d}\gamma = \sum_i \Gamma_i \mathrm{d}\mu_i \qquad (4.3.4)$$

Some useful forms of the Gibbs (isotherm) equation (4.3.4) relevant to adsorption in surfactant systems are now considered (some in the Appendix A4).

4.3.1 Adsorption from a dilute solution of a single nonionic surfactant to the liquid/vapour interface

This is the simplest case and serves to illustrate the relationship between surface concentration and surface excess concentration of a surfactant. The solvent is denoted component 1 and the surfactant component 2. If the Gibbs model for the surface is adopted, and the dividing surface is placed so that the surface excess concentration of solvent, Γ_1^σ, is zero (4.3.4) becomes

$$-\mathrm{d}\gamma = \Gamma_2^{(1)}\mathrm{d}\mu_2 \qquad (4.3.5)$$

where $\Gamma_2^{(1)}$ is the surface excess concentration of surfactant, the superscript $^{(1)}$ denoting the choice of dividing surface where $\Gamma_1^\sigma = 0$. The surface excess $\Gamma_2^{(1)}$ is sometimes referred to as the *relative adsorption* of component 2. The chemical potential of surfactant in solution, μ_2, is given in terms of its activity, a_2, by

$$\mu_2 = \mu_2^o + RT \ln a_2$$

The standard chemical potential μ_2^o is independent of surfactant activity so $d\mu_2^o = 0$ and $\mathrm{d}\mu_2 = RT\,\mathrm{d}\ln a_2$ so that (4.3.5) becomes

$$\Gamma_2^{(1)} = -\frac{1}{RT}\left(\frac{\mathrm{d}\gamma}{\mathrm{d}\ln a_2}\right) \qquad (4.3.6)$$

For many dilute solutions a_2 can reasonably be approximated by the surfactant concentration (molarity, c, or mole fraction, x), and the surface excess concentration of surfactant can then be readily obtained from a knowledge of the variation of surface tension with the surfactant concentration in solution using (4.3.6).

If the interest is in the (total) surface concentration of surfactant, Γ_2^s, it is convenient to use the surface phase model. Equation (4.3.4) then becomes

$$-\mathrm{d}\gamma = \Gamma_1^s\mathrm{d}\mu_1 + \Gamma_2^s\mathrm{d}\mu_2 \qquad (4.3.7)$$

Noting the Gibbs–Duhem relationship (for constant T and p), $x_1 d\mu_1 + x_2 d\mu_2 = 0$, (where x_1 and x_2 are the mole fractions of solvent and surfactant in solution respectively), (4.3.7) becomes

$$-d\gamma = \left[\Gamma_2^s - \frac{x_2}{x_1}\Gamma_1^s\right] d\mu_2 \qquad (4.3.8)$$

and a comparison of (4.3.5) and (4.3.8) shows that

$$\Gamma_2^{(1)} = \left[\Gamma_2^s - \frac{x_2}{x_1}\Gamma_1^s\right] \qquad (4.3.9)$$

Although in general Γ_1^s is not zero (there will always be some solvent in the interface), for dilute surfactant solution $x_2 << x_1$ and so to a very good approximation, $\Gamma_2^s = \Gamma_2^{(1)}$.

4.3.2   Adsorption at an oil/water interface

For adsorption of a surfactant (component 3) at an oil/water interface (such as exists for example in an emulsion) there are two solvents in the system (components 1 and 2 forming phases α and β, respectively) and a dividing surface cannot be set such that the surface excesses of both are zero. If it is supposed that the two solvents are completely immiscible (as is effectively the case for alkanes and water) then it can be shown that

$$-d\gamma = \left(\Gamma_3^s - \frac{x_3^\beta}{x_2^\beta}\Gamma_2^s - \frac{x_3^\alpha}{x_1^\alpha}\Gamma_1^s\right) d\mu_3 \qquad (4.3.10)$$

For this system, $x_1^\alpha \approx x_2^\beta \approx 1$ and both x_3^α and x_3^β are much less than unity so to a good approximation (4.3.10) becomes $-d\gamma = \Gamma_3^s d\mu_3$.

Nonionic surfactants can distribute between oil and water phases, and indeed sometimes distribute in favour of the oil phase. At distribution equilibrium of surfactant, the chemical potential of surfactant is equal in each phase, that is, $\mu_3^\alpha = \mu_3^\beta$ and so as the surfactant concentration in the system changes, providing distribution equilibrium is maintained, $d\mu_3^\alpha = d\mu_3^\beta$ and $d\ln a_3^\alpha = d\ln a_3^\beta$. Thus in order to obtain the surface concentration of surfactant it is necessary to know how the surfactant activity (or concentration if in ideal dilute solution) changes in *one* of the phases at distribution equilibrium.

4.3.3   Adsorption of ionic surfactant from aqueous solution

Surfactants of the type A^+B^- are considered here; either the anion or cation can be the surface active ion as discussed in Chapter 1. By way of example the case where B^- is surface active, that is, where the surfactant is of the anionic type, is treated. It is necessary to take account in the Gibbs equation of both the species A^+ and B^- and for completeness also H^+, OH^-, and H_2O. If the surface phase model is used the Gibbs equation (4.3.4) can be written

$$-d\gamma = \Gamma^s_{A^+}d\mu_{A^+} + \Gamma^s_{B^-}d\mu_{B^-} + \Gamma^s_{H^+}d\mu_{H^+} + \Gamma^s_{OH^-}d\mu_{OH^-} + \Gamma^s_{H_2O}d\mu_{H_2O} \quad (4.3.11)$$

Except for the most dilute surfactant solutions the terms for H^+ and OH^- can be neglected. Then, the Gibbs–Duhem equation (Box 4.2) for this system,

$$x_{A^+}d\mu_{A^+} + x_{B^-}d\mu_{B^-} + x_{H_2O}d\mu_{H_2O} = 0 \quad (4.3.12)$$

can be combined with (4.3.11) to give

$$-d\gamma = \left(\Gamma^s_{A^+} - \frac{x_{A^+}}{x_{H_2O}}\Gamma^s_{H_2O}\right)d\mu_{A^+} + \left(\Gamma^s_{B^-} - \frac{x_{B^-}}{x_{H_2O}}\Gamma^s_{H_2O}\right)d\mu_{B^-} \quad (4.3.13)$$

where the x are mole fractions of the subscripted species. The mole fractions of A^+ and B^- are very small compared to x_{H_2O}, which is close to unity. Thus, although $\Gamma^s_{H_2O}$ is not zero the terms containing $\Gamma^s_{H_2O}$ are in the context negligible and so to a very good approximation

$$-d\gamma = \Gamma^s_{A^+}d\mu_{A^+} + \Gamma^s_{B^-}d\mu_{B^-} \quad (4.3.14)$$

Since the phases bounding the interface are homogeneous they are also electroneutral, and it follows that the interface itself must also be electroneutral, that is,

$$\Gamma^s_{A^+} = \Gamma^s_{B^-} = \Gamma_{AB} \quad (4.3.15)$$

This means that the counterions, A^+, neutralize the charge on the adsorbed surface active ions B^- within the surface (between planes AA and BB in Fig. 4.1). Changes in chemical potential are given by

$$d\mu_{A^+} = RT\,d\ln a_{A^+};\, d\,\mu_{B^-} = RT\,d\ln a_{B^-};\, d\,\mu_{A^+} + d\,\mu_{B^-} = 2RT\,d\ln a_{\pm} \quad (4.3.16)$$

where the a are ion activities in solution. The product $a_{A^+}a_{B^-} = a^2_{\pm}$, $a_{\pm}$ being the mean ion activity of the surfactant AB (product of surfactant concentration m_{AB} and mean ionic activity coefficient $f_{\pm}$). The Gibbs equation (4.3.14) can therefore be written

$$\Gamma^s_{AB} = -\frac{1}{2RT}\left(\frac{d\gamma}{d\ln a_{\pm}}\right) \quad (4.3.17)$$

which (remembering the near equivalence of surface concentration and surface excess concentration for surfactants) is similar to (4.3.6) for a nonionic surfactant, but now contains the factor 2. This means that the rate of reduction of surface tension with surfactant activity for a fully dissociated 1:1 electrolyte is twice that in a hypothetical nonionic surfactant system where $\Gamma^{(1)}_2 = \Gamma^s_{AB}$ and $a_{\pm} = a_2$.

Box 4.2   Gibbs–Duhem equation

Pierre Duhem (1861–1916), French physicist, mathematician, and philosopher of science.

For a bulk solution containing i components (cf. (4.2.11) for a surface)

$$dG = -SdT + Vdp + \sum_i \mu_i dn_i$$

and integration at constant intensive quantities* gives $G = \sum_i \mu_i n_i$ and so

$$dG = \sum_i n_i d\mu_i + \sum_i \mu_i dn_i.$$

Comparison of the two expressions for dG shows that, at constant T and p,

$$\sum_i n_i d\mu_i = 0,$$

which is a form of the Gibbs–Duhem equation. Dividing throughout by the total number of moles in the system gives the alternative expression $\sum_i x_i d\mu_i = 0$. The Gibbs–Duhem equation describes the relationship between changes in chemical potential for components in a system. For a two-component system consisting of components 1 and 2 for example the equation shows that $d\mu_2 = -(n_1/n_2)d\mu_1$. It can be seen that the Gibbs adsorption equation (e.g. (4.3.2)) is an analogue of the Gibbs–Duhem equation for a system which contains an interface.

*The process of integration at constant intensive quantities corresponds to building up the system by addition of small elements of material, all with the same intensive properties, that is, the added elements all have the same T, p, μ_i, and γ, which are also those of the final (i.e. integrated) system.

Often surfactant activities will not be available, but in dilute solution the simple Debye–Hückel limiting law gives reasonable estimates for mean ionic activity coefficients. In any event (4.3.17) may be written

$$-\frac{d\gamma}{d\ln m_{AB}} = 2RT\Gamma^s_{AB}\left(1 + \frac{d\ln f_\pm}{d\ln m_{AB}}\right) \tag{4.3.18}$$

The Debye–Hückel limiting law for a 1:1 electrolyte is

$$-\ln f_\pm = 2.303 A I^{0.5} \tag{4.3.19}$$

where I is the ionic strength (in this case equal to m_{AB}) and A is the Debye–Hückel constant for water.

4.3.4 Adsorption of ionic surfactant from aqueous solution of swamping inert electrolyte

In practical situations there is often surface inactive salt present with the ionic surfactant. Consider the case where the electrolyte A^+X^- is present with the anionic surfactant A^+B^-. In the context 'swamping' means that when the surfactant concentration is altered, the concentration of the counterion (A^+) remains essentially constant. Assuming the surfactant is unhydrolysed and the pH remains unchanged the Gibbs equation may be expressed (taking $d\mu_{H_2O} = 0$)

$$-d\gamma = \Gamma_{A^+}d\mu_{A^+} + \Gamma_{B^-}d\mu_{B^-} + \Gamma_{X^-}d\mu_{X^-} \tag{4.3.20}$$

Since A^+X^- is present at swamping concentration, when surfactant concentration is altered $d\mu_{A^+}$ and $d\mu_{X^-}$ are very small compared with $d\mu_{B^-}$ so that in effect

$$-d\gamma = \Gamma_{B^-}d\mu_{B^-} = RT\Gamma_{B^-}d\ln a_{B^-} \tag{4.3.21}$$

The system is virtually at constant ionic strength so that f_{B^-} is effectively constant, and since $m_{B^-} = m_{AB}$ (4.3.21) becomes

$$\Gamma_{AB} = -\frac{1}{RT}\left(\frac{d\gamma}{d\ln m_{AB}}\right) \tag{4.3.22}$$

The factor 2 no longer appears on the bottom of the right-hand side (cf. (4.3.17)), and m_{AB} replaces $a_{\pm}$.

Some other useful forms of the Gibbs equation covering, *inter alia*, intermediate salt concentrations are covered in Appendix A4.

4.4 Why does adsorption lower interfacial tension? The spreading pressure

It has been seen that adsorption lowers γ and desorption (negative adsorption) causes an increase in γ. It can be appreciated from the Gibbs adsorption equation that this is so. For the adsorption of, say, a nonionic surface active material the Gibbs equation is

$$-d\gamma = RT\Gamma d\ln m$$

and Γ is positive when $-(d\gamma/d\ln m)$ is positive, that is, when γ falls with increasing surfactant concentration m. This observation however does not immediately give much physical insight into the origins of the changes in tension.

Surfactants adsorb very strongly from dilute solution; as the surfactant chain length is increased the surfactant adsorbs more strongly and becomes less soluble in water, and in the limit becomes effectively insoluble. It can be imagined that the rate of exchange of surfactant molecules between surface and bulk water is very low compared to the rate of exchange of water molecules. So water molecules are taken up osmotically by the surface layer, generating a lateral pressure within the film. This pressure is termed the *spreading pressure* and is designated π. The surface tension of the surface is a contractile force; it is a negative two-dimensional pressure. Since the spreading pressure is expansile (positive), it causes a reduction in the tendency of the surface to contract, that is, the surface tension is lowered. The surface pressure is therefore defined as the lowering of γ caused by the presence of the surface film:

$$\pi = \gamma_0 - \gamma \tag{4.4.1}$$

where γ_0 is the interfacial tension in the absence of the surface film, that is, of the clean interface.

As will be seen in subsequent chapters, a surfactant can lower the surface tension of water, or the interfacial tension between nonpolar oil and water, by 40 mN m^{-1} or more. This pressure is exerted laterally by a film say 2 nm thick and, expressed as a three-dimensional pressure, it becomes equivalent to 40×10^{-3} (N)/2×10^{-9} (m^2) = 20×10^6 N m^{-2}, which is approximately 200 atm. This is a remarkably high pressure and it is this that is responsible for the movement of the debris on the surface of Mount Pond in Franklin's famous experiments, as described in Box 4.3.

Box 4.3   Benjamin Franklin 1706–1790, a most influential American

An early understanding of the behaviour of monolayers of surfactant-like molecules on the surface of water owes much to the observations of Benjamin Franklin (1706–1790) who, according to Isaacson (2004), was the most accomplished and influential American of his age. He was born in Boston, the tenth son of Josiah Franklin, a soap maker, and started his working life as a printer, apprenticed to his brother James. Relationships between Benjamin and James became strained and ultimately Benjamin ran away ending up in Philadelphia. Much later he was to became Governor of that city (1785–1788). At the age of 43 Franklin gave up printing and subsequently became the world-renowned inventor, scientist and diplomat the world remembers. When he died at the age of 84, 20,000 people attended his funeral.

Benjamin Franklin (1706–1790), American inventor, scientist, and statesman. Classic Stock/Alamy Stock Photo.

Box 4.3 *Continued*

Franklin had many friends in England, and when in London he famously placed a small amount of oil on the surface of Mount Pond on Clapham Common. He wrote of subsequent similar experiments (Franklin and Brownrigg, 1774) 'I fetched out a cruet of oil, and dropt a little of it on the water. I saw it spread itself with surprising swiftness upon the surface . . .'. The oil drop ultimately covered about half an acre (*ca* 0.2 hectare), making the water surface as 'smooth as glass'.

Figure 4.2 *(a) The pond on Clapham Common, London as it now is. (b) A similar photograph of the pond 10 minutes after deposition of a teaspoonful of olive oil on the surface close to the small island. The clearly visible extensive area of mirror-like surface is covered with a film of olive oil. Franklin's experiment was recreated, and these photographs taken, by the late Charles Giles of Strathclyde University in Glasgow.*

Franklin's experiment was recreated over 200 years later by the late Professor Charles Giles of Strathclyde University in Glasgow (Giles, 1969). He placed a teaspoonful of olive oil on the surface close to the island in Clapham Common Pond; about 10 minutes later he took the photograph shown on the right in Fig. 4.2; the area of very smooth water surface is clearly seen. Although Franklin did not make the calculation it is readily shown that the thickness of the oil film after it had completely spread was about 2 nm, roughly the length of a triglyceride molecule such as is found in olive oil. The spreading monolayer exerts a lateral two-dimensional pressure strong enough to have pushed aside small pieces of debris on the pond surface in Franklin's experiment. Without knowing it, Franklin had formed a monolayer at the air/water surface (Giles, 1969).

In the next chapter some experimental findings concerning the adsorption of surfactants at liquid/fluid interfaces are presented, and it is shown what can be simply deduced from these findings concerning the nature of adsorbed surfactant layers.

Appendix A4: Some further useful forms of the Gibbs adsorption equation

1. Adsorption of an anionic surfactant A^+B^- from aqueous solution containing an intermediate and constant concentration of inorganic electrolyte

Equation (4.3.20) for an anionic surfactant A^+B^- with added electrolyte A^+X^- is used as a starting point:

$$-d\gamma = \Gamma_{A^+}d\mu_{A^+} + \Gamma_{B^-}d\mu_{B^-} + \Gamma_{X^-}d\mu_{X^-} \qquad (4.3.20)$$

Since the adsorbed surfactant monolayer is anionic, the anion of the added electrolyte is desorbed and so it can be assumed that Γ_{X^-} is negligible. The salt concentration is constant as the surfactant concentration is varied so $dm_{X^-} = dm_{A^+}$ (although $m_{X^-} \neq m_{A^+}$). Also, from the requirement of electroneutrality in the surface, $\Gamma_{A^+} = \Gamma_{B^-} = \Gamma_{AB}$. Noting this, (4.3.20) becomes

$$-\frac{d\gamma}{RT} = \left(x + \frac{2d\ln f_{\pm}}{d\ln m_{AB}}\right)\Gamma_{AB}d\ln m_{AB} \qquad (A4.1)$$

where

$$x = \left(1 + \frac{m_{AB}}{m_{AB} + m_{AX}}\right)$$

For very large salt concentrations (relative to the surfactant concentration), $x \to 1$. Further, the ionic strength remains effectively constant with respect to surfactant concentration so that $d\ln f_{\pm}/d\ln m_{AB}$ is close to zero. For high salt concentration then, (A4.1) becomes equivalent to (4.3.22). At very low salt concentration, $x \to 2$ and (A4.1) is then equivalent to (4.3.18).

2. Adsorption of an anionic surfactant A^+B^- at constant concentration in aqueous solution containing varying amounts of inorganic electrolyte A^+X^-

Again, it is supposed that $\Gamma_{X^-} = 0$ and it is noted that $d\ln m_B = 0$ so that the Gibbs adsorption equation is

$$-\frac{d\gamma}{RT} = \Gamma_{A^+}d\ln m_{A^+} + \Gamma_{A^+}d\ln f_{A^+} + \Gamma_{B^-}d\ln f_{B^-} \qquad (A4.2)$$

Since $\Gamma_{B^-} = \Gamma_{A^+} = \Gamma_{AB}$ and $f_{A^+} = f_{B^-} = f_{\pm}$ (A4.2) may be written as

$$-\frac{d\gamma}{RT} = \Gamma_{AB}d\ln m_{A^+} + 2\Gamma_{AB}d\ln f_{\pm} \qquad (A4.3)$$

The way in which the interfacial tension varies with changes in salt concentration, that is, with m_{X^-}, is sought. From (A4.3)

$$-\frac{1}{RT}\frac{d\gamma}{d\ln m_{X^-}} = \Gamma_{AB}\frac{d\ln m_{A^+}}{d\ln m_{X^-}} + 2\Gamma_{AB}\frac{d\ln f_{\pm}}{d\ln m_{X^-}} \qquad (A4.4)$$

Since in general $d\ln m = d\,m/m$, and in this system $dm_{A^+} = dm_{X^-}$ (although $m_{A^+} \neq m_{X^-}$), (A4.4) becomes

$$-\frac{1}{RT}\frac{d\gamma}{d\ln m_{X^-}} = \Gamma_{AB}\frac{m_{X^-}}{m_{A^+}} + 2\Gamma_{AB}\frac{d\ln f_\pm}{d\ln m_{X^-}} \tag{A4.5}$$

The salt concentration m_{AX} is equal to m_{X^-} and m_{A^+} is equal to the sum of salt and surfactant concentrations and so (A4.5) can be expressed

$$-\frac{1}{RT}\frac{d\gamma}{d\ln m_{AX}} = \Gamma_{AB}\left[\left(\frac{m_{AX}}{m_{AX}+m_{AB}}\right) + \frac{2d\ln f_\pm}{d\ln m_{AX}}\right] \tag{A4.6}$$

Thus, from the variation of γ with the salt concentration at constant surfactant concentration the surface concentration of surfactant Γ_{AB} can be found as a function of the salt concentration.

3. System with a constant concentration of hydrolysable anionic surfactant A^+B^- (e.g. a sodium carboxylate) in which the pH is varied by addition of A^+OH^-. The ionic strength is maintained constant by addition of inert electrolyte A^+X^-.

For constant ionic strength $d\ln f_\pm = 0$ so that $d\ln a_\pm = d\ln m$, and $dm_{H^+} = -dm_{A^+}$. For constant surfactant concentration $d\ln m_{B^-} = 0$. Assuming, as before, that the surface concentrations of non-surface active anions (OH^- and X^- in this case) are zero, the Gibbs equation becomes

$$-\frac{d\gamma}{RT} = \Gamma_{H^+}d\ln m_{H^+} - \frac{m_{H^+}}{m_{A^+}}\Gamma_{A^+}d\ln m_{H^+} + \Gamma_{HB}d\ln m_{HB} \tag{A4.7}$$

where HB is the undissociated acid produced by the hydrolysis of the (say carboxylate) surfactant. The acid dissociation constant K_a is given by $[H^+][B^-]/[HB]$ so that $d\ln m_{H^+} + d\ln m_{B^-} = d\ln m_{HB}$. Further, it may reasonably be supposed that the ratio of the surface concentrations of H^+ and A^+ is equal to the ratio of the bulk concentrations of these species, that is, $\Gamma_{A^+}/\Gamma_{H^+} = m_{A^+}/m_{H^+}$, since the adsorption of both is a result of Coulombic rather than specific interactions. On this basis it follows therefore that (A4.7) may be written

$$d\gamma/RT = \Gamma_{HB}d\ln m_{H^+} \quad \text{or} \quad \frac{1}{2.303RT}\frac{d\gamma}{d(pH)} = \Gamma_{HB} \tag{A4.8}$$

Clearly if there is no surfactant hydrolysis in the surface $\Gamma_{HB} = 0$ and γ is independent of pH. Otherwise the value of Γ_{HB} may be obtained as a function of pH.

· ·

GENERAL READING

R. Defay, I. Progogine, A. Bellemans, and D.H. Everett, *Surface Tension and Adsorption*. Longmans, London, 1966.
E.A. Guggenheim, Thermodynamics, North Holland, Amsterdam, 1967.

....

REFERENCES

J.G. Crowther, *Famous American Men of Science*. Penguin Books, New York, 1944.
B. Franklin and W. Brownrigg, Of the stilling of waves by means of oil. *Phil. Trans. R. Soc. London*, 1774, **64**, 445–460.
C.H. Giles, Franklin's teaspoonful of oil—studies in the early history of surface chemistry, part 1. Chem. Ind., 1969, 1616–1624.
W. Isaacson, *Benjamin Franklin: An American Life*. Simon & Schuster, New York, 2004.

5

Adsorption of surfactants at liquid/vapour and liquid/liquid interfaces

5.1 Does the Gibbs adsorption equation give the right answer?

The Gibbs adsorption equation was introduced in Chapter 4 and it was shown how it links the amount of material adsorbed (or desorbed) in unit area of interface to the accompanying changes in the interfacial tension. Since the Gibbs adsorption equation is central to many aspects of surface chemistry, there have been attempts over the years to verify it by comparing values of surface concentration of a surfactant obtained from interfacial tensions, by use of the equation, with values obtained independently.

A convincing test has been carried out by Tajima et al. (1982) who measured the surface concentrations of a tritiated ionic surfactant, sodium tetradecyl-2,3-^{3}H,^{3}H-sulfate (^{3}HSTS), adsorbed from aqueous solution to the toluene/water interface, using a radiotracer method. Measurements were made over a range of surfactant concentration m_{AB} up to 1 mM, in the aqueous phase. The appropriate interfacial tensions were also measured, using the Wilhelmy plate method (see §A3.1). In the absence of added NaCl the relevant form of the Gibbs adsorption equation is

$$-\frac{d\gamma}{d\ln m_{AB}} = 2RT\Gamma^s_{AB}\left(1 + \frac{d\ln f_\pm}{d\ln m_{AB}}\right) \qquad (4.3.18)$$

The mean ionic activity coefficients, $f_\pm$, for the surfactant (a 1:1 electrolyte) were not available. For 1mM sodium chloride however (1 mM was the highest concentration of surfactant used), $d\ln f_\pm/d\ln m_{AB}$ is about -0.02 and the term in brackets in the Gibbs adsorption equation can reasonably be omitted. In the presence of 75 mM NaCl the appropriate form of the Gibbs adsorption equation is

Surfactants: In Solution, at Interfaces and in Colloidal Dispersions. Bob Aveyard. © Bob Aveyard 2019.
Published in 2019 by Oxford University Press. DOI: 10.1093/oso/9780198828600.001.0001

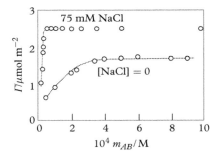

Figure 5.1 *Surface concentrations of ^3HSTS at the toluene/water interface at 30°C as a function of bulk aqueous phase surfactant concentration. Upper curve is for adsorption from 75 mM NaCl and the lower curve for zero [NaCl]. The points are obtained from radiotracer experiments and the dotted lines are calculated from the appropriate forms of the Gibbs adsorption equation. Redrawn from Tajima et al. (1982).*

$$\Gamma_{AB} = - \left(\frac{1}{RT} \right) \left(\frac{d\gamma}{d \ln m_{AB}} \right) \qquad (4.3.22)$$

The level of agreement between the surface concentrations of ^{3}HSTS obtained directly, and those calculated using the appropriate form of Gibbs adsorption equation is extremely good, as can be judged from Fig. 5.1. Subsequently, An et al. (1996) showed that surface concentrations obtained from surface tension and the Gibbs equation were in good agreement with those determined using neutron reflectivity (see §8.1).

5.2 Variation of interfacial tension with surfactant concentration

The way in which a surfactant lowers the interfacial tension of an aqueous solution in contact with air or water-insoluble nonpolar 'oil' (dodecane) is shown in Fig. 5.2a. The upper curve is for the air/water surface and the lower one for the oil/water interface. The surfactant is dodecyltrimethylammonium bromide but the curves are similar in shape for nonionic surfactants. At higher concentrations of surfactant the tension becomes more or less constant. Surface-active materials other than surfactants (e.g. long alkyl chain polar molecules such as alkanols and carboxylic acids) give curves of similar shape but the distinctive feature of surfactants is that they exhibit a critical micelle concentration (*cmc*), beyond which the tension changes only little and surfactant aggregates (micelles) form (see §1.1 and Chapter 9).

Plots of tensions in the form γ versus $\ln(m)$ are a little more revealing; the results for dodecyltrimethylammonium bromide are plotted in this way in Fig. 5.2b. At low concentrations the curves are relatively shallow with increasing slope (regions A) eventually becoming almost rectilinear (regions B). There is a distinct breakpoint at

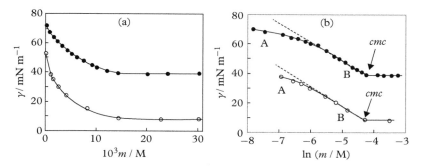

Figure 5.2 *(a) Plots of interfacial tension against aqueous phase concentration, m, of dodecyltrimethylammonium bromide for adsorption at (upper curve) the air/water interface and (lower curve) the dodecane/water interface at 25°C. (b) The same tensions plotted against ln m.*

the *cmc* and thereafter the tension changes only little. Surface concentrations can be obtained from the slopes of such curves below the *cmc* by use of the Gibbs adsorption equation.

The apparently linear regions are interesting. Neglecting the term $d \ln f_\pm / d \ln m_{AB}$ in (4.3.18), the linearity implies that the surface concentration Γ_{AB} is unchanging in this range of surfactant concentration. However, for the interfacial tension to fall adsorption must be increasing with surfactant concentration. Clearly then regions B are not strictly linear, and a small increase in Γ_{AB} gives a large reduction in tension. As will be seen later this indicates that the adsorbed surfactant monolayers are fairly incompressible in the 'linear' region of the γ–ln m curve. The approximate constancy of γ above the *cmc* shows that no further adsorption occurs and that an increase in concentration gives only more micellized surfactant. Thus, in this region, single surfactant entities in solution are in equilibrium with micellar surfactant and surfactant in the 'close-packed' adsorbed monolayer. A close-packed monolayer can in some ways be regarded as a planar surfactant aggregate.

Sometimes a surfactant can reach a solubility limit before a *cmc* can be attained. Above this limit surfactant crystals (say) are dispersed in the solution, and again the interfacial tension remains constant. Amphiphiles[1] other than surfactants can also reach a solubility limit of course, so again the tension will remain constant above the solubility limit. Obvious examples are long chain alkanols and alkanoic acids; they adsorb and form monolayers like surfactants, but do not form micelles.

---

[1] An 'amphiphile' is a species (surfactant or otherwise) with molecules that have a moiety which is hydrophilic (polar or ionic) and one which is hydrophobic (e.g. hydrocarbon or fluorocarbon chain). The term means 'loving' both nonpolar oil and water, and amphiphilicity is of course why the molecules are surface active.

5.3   What can be learned about adsorption from interfacial tensions?

Data such as that depicted in Fig. 5.2 can be used to obtain surface concentrations of surfactant (or other amphiphiles) as a function of the concentration in one or other of the contacting bulk phases. An *adsorption isotherm* is a plot of the surface concentration against the bulk phase concentration, and the shape for adsorption at a fluid/fluid interface has the general form illustrated in Fig. 5.3. At low concentrations the isotherm is linear (A), as indicated by the dashed line with slope equal to the initial slope. In the plateau region B the adsorbed monolayer is close-packed so no further adsorption is possible (see footnote 2, Chapter 1); the *cmc* of the surfactant occurs around the beginning of the plateau region.

Quite a lot can be learned from an adsorption isotherm. The initial slope is related to the 'strength' of adsorption of an isolated molecule. The overall shape reflects the behaviour of the monolayer as the surface concentration is varied, and the surface concentration in the plateau region gives an *effective* dimension of an adsorbed molecule in the plane of the film. Effective dimensions can include influences due to head group solvation and mutual electrical repulsion if present (see §5.3.2 and footnote 4, as well as §9.5.1).

5.3.1   The 'strength' of adsorption: standard free energies of adsorption

Adsorption is the process of distribution of a species between two phases, one of the phases being an interface. For two bulk liquids, constituting phases α and β, in contact and a solute X distributed at equilibrium between them the standard free energy of transfer of X from α to β, $\Delta_{trans}\mu^\circ$, is given by the well-known expression[2]

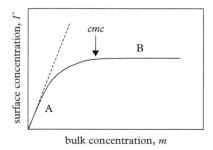

Figure 5.3 *General shape of an isotherm for the adsorption of a surfactant, or other amphiphile, at the air/water or oil/water interface.*

---

[2] Changes in free energy are often expressed as ΔG where G is a molar free energy. However μ and G are equivalent for a component in an ideal dilute solution.

$$\Delta_{trans}\mu^o = -RT\ln\frac{[X]_\beta}{[X]_\alpha} = -RT\ln K \qquad (5.3.1)$$

where $[X]_\alpha$ and $[X]_\beta$ are the concentrations (or activities) of X in phase α and β, respectively, at distribution equilibrium and K is the distribution ratio. An analogous expression exists for the standard free of adsorption, $\Delta_{ads}\mu^o$, but now the concentration of the adsorbed species must be expressed in different units for the bulk and surface.

An adsorbed surfactant layer can be considered either as a two-dimensional gas or as a two-dimensional solution (in which surfactant is mixed with solvent(s)), and this is considered later. For n mol of a three-dimensional ideal gas contained in a volume V at pressure p and absolute temperature T, the equation of state is: $pV = nRT$. The concentration of the gas is n/V which is proportional to p at constant temperature. A two-dimensional analogue of the ideal gas equation for a nonionic surfactant is (per mol)

$$\pi A = RT \qquad (5.3.2)$$

where π is the surface pressure (see §4.4) and A is the area available per mol in the adsorbed layer. This equation is only valid for systems which are very dilute in surfactant. The surface concentration Γ is equal to $1/A$ and so $\Gamma \propto \pi$ at constant T. Accordingly, the chemical potential, μ^s, of an adsorbed material in a very dilute film can be expressed as either

$$\mu^s = \mu_\pi^{o,s} + RT\ln\pi \ \text{ or } \ \mu^s = \mu_\Gamma^{o,s} + RT\ln\Gamma \qquad (5.3.3)$$

Here $\mu_\pi^{o,s}$ and $\mu_\Gamma^{o,s}$ are standard chemical potentials, equal to the value of μ^s when, respectively, $\pi = 1/\text{units}$ and $\Gamma = 1/\text{units}$. The units taken for surface pressure π are usually mN m^{-1} and for Γ, molecule nm^{-2} or mol m^{-2}. It is seen from (5.3.2) and (5.3.3) that

$$\mu_\pi^{o,s} = \mu_\Gamma^{o,s} - RT\ln RT \qquad (5.3.4)$$

At adsorption equilibrium, the chemical potentials of surfactant in bulk solution (μ^l) and in the surface (μ^s) are equal so that

$$\mu^l = \mu^{o,l} + RT\ln x = \mu^s = \mu_\pi^{o,s} + RT\ln\pi \qquad (5.3.5)$$

where $\mu^{o,l}$ is the standard chemical potential in bulk solution; here the solution concentration is expressed in mol fraction units, x. The standard free energy of adsorption is the difference between the standard chemical potentials of the surfactant in bulk and in surface so

$$\Delta_{ads}\mu_\pi^o = \mu_\pi^{o,s} - \mu^{o,l} = -RT\ln\left(\pi/x\right)_{x\to 0} \qquad (5.3.6)$$

The quantity $(\pi/x)_{x\to0}$ is the initial slope of the adsorption isotherm, expressed in the form of π versus x. A standard free energy of adsorption can also be defined as

$$\Delta_{ads}\mu_{\Gamma}^{o} = \mu_{\Gamma}^{o,s} - \mu^{0,l} = -RT\ln\left(\Gamma/x\right)_{x\to o} \tag{5.3.7}$$

and it is readily shown from what has been said that

$$\Delta_{ads}\mu_{\Gamma}^{o} = \Delta_{ads}\mu_{\pi}^{o} + RT\ln RT \tag{5.3.8}$$

The standard free energy of adsorption is the free energy change accompanying a hypothetical process in which 1 mol of material is transferred from its standard state in bulk solution to its standard state in the surface. The standard states for bulk and surface are unity in the chosen units. It is assumed that, whatever concentration scales are used, the bulk solution and adsorbed surface layer behave as they would at infinite dilution in all other respects than concentration. It should be stressed that the magnitude and sign of $\Delta_{ads}\mu^{o}$ are partly determined by the choice of standard states, that is, the concentration units employed. Thus the interpretation of a particular value of a standard free energy of adsorption can be quite subtle; relative values for, say, an homologous series of surfactants however can be simply interpreted, as will be seen.

Standard enthalpies and standard entropies of adsorption are related to the temperature dependence of the standard free energies of adsorption in the usual way, so that for example standard heats of adsorption, $\Delta_{ads}H^{o}$ are is given by

$$\Delta_{ads}H_{\pi}^{o} = \mathrm{d}\left(\Delta_{ads}\mu_{\pi}^{o}/T\right)/\mathrm{d}\left(1/T\right) \text{ and } \Delta_{ads}H_{\Gamma}^{o} = \mathrm{d}\left(\Delta_{ads}\mu_{\Gamma}^{o}/T\right)/\mathrm{d}\left(1/T\right) \tag{5.3.9}$$

The two standard enthalpies of adsorption can be shown to be related by

$$\Delta_{ads}H_{\pi}^{0} = \Delta_{ads}H_{\Gamma}^{o} + RT \tag{5.3.10}$$

Finally, in general, entropy changes are related to changes in free energy and enthalpy by $\Delta\mu = \Delta H - T\Delta S$. The two standard entropies of adsorption derived from the standard enthalpies and standard free energies are related by

$$\Delta_{ads}S_{\pi}^{o} = \Delta_{ads}S_{\Gamma}^{o} - R\left(\ln RT - 1\right) \tag{5.3.11}$$

In summary, it has been seen how some thermodynamic parameters of adsorption can be obtained for the adsorption of nonionic amphiphiles from very dilute solution to oil/water and air/water interfaces. Consideration will be given later to more concentrated films where lateral interactions within monolayers complicate matters. For example, hydrocarbon chains in adsorbed surfactant layers at the air/water interface are mutually attractive which leads to a reduction of the surface pressure in a way which varies with the surface concentration. Electrical interactions between adsorbed ionic surfactants are repulsive and give rise to an increase in surface pressure, which is again a function of surface concentration of surfactant. The finite size of adsorbed molecules must also be taken into account. All of these interactions have contributions to the form of realistic equations of state for adsorbed films; as mentioned, the ideal surface equation of state (5.3.2) is appropriate only for the lowest surface concentrations. For the moment, what can be learned about adsorbed films of amphiphiles from data for very dilute systems is considered.

Many surfactants are paraffin chain compounds with a hydrophilic group at one end, and so use of data for amphiphiles which are not surfactants, can be informative. There are good data available for normal alkanols, which are, in the context, excellent models for nonionic surfactants. Standard free energies of adsorption, $\Delta_{ads}\mu_\pi^o$,[3] are shown in Fig. 5.4 for adsorption of some short chain alkanols (ethanol to pentan-1-ol) from very dilute solution in water to the air/water interface (A).

The standard free energy decreases linearly with chain length, with slope -2.65 kJ $(mol\ CH_2)^{-1}$. Results are also shown for some longer chain length alkanols (butan-1-ol to heptan-1-ol) again adsorbed from very dilute aqueous solution but now to the dodecane/water interface (B). The standard free energies again decrease linearly with chain length, but with slope -3.20 kJ $(mol\ CH_2)^{-1}$. Line C represents data for long chain alkanols (decan-1-ol to octadecan-1-ol) adsorbed from dodecane to the dodecane/water interface and here the slope is effectively zero.

These results clearly show that, for adsorption from an aqueous phase, the alkyl chains are removed from water, since $\Delta_{ads}\mu_\pi^o$ is a linear function of chain length. The methylene

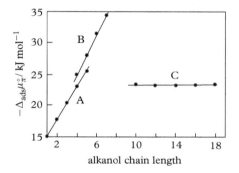

Figure 5.4 *Variation of standard free energies of adsorption of alkanols with chain length, at 25°C. A is for adsorption from water to the air/water interface, B for adsorption from water to the dodecane/water interface, and C for adsorption from dodecane to the dodecane/water interface.*

---

[3] The slopes of the plots of $\Delta_{ads}\mu_\Gamma^o$ against chain length are the same as those of plots of $\Delta_{ads}\mu_\pi^o$ against chain length but the lines are displaced by $RT\ \ln RT$, according to (5.3.8).

group decrement is different for adsorption to the air/and the oil/water interfaces, reflecting the difference in environment of the chains at the two types of interface. In contrast $\Delta_{ads}\mu_\pi^o$ for adsorption from dodecane to the oil/water interface is essentially independent of chain length indicating the chains remain in an oil environment in the adsorbed state. Adsorption from water is driven by transfer of chains from water to oil, the OH group remaining in an aqueous environment. The driving force for adsorption from oil is the transfer of the OH group from oil to water.

The standard free energies for adsorption to the air/water interface are consistent with the hydration free energies of alkanes and the standard free energies of adsorption of alkanes from vapour to the water/vapour interface, which were given in Chapter 2 (Table 2.1). As illustrated in Fig. 5.5, the hydration free energy of the CH$_2$ group ($+0.7$ kJ mol^{-1}) should be equal to the sum of the free energy of adsorption of the group from vapour to the water/vapour interface (-2.0 kJ mol^{-1}) and the free energy of *desorption* of the methylene group from the water/vapour interface into bulk water ($+2.65$ kJ mol^{-1}). Within the likely experimental errors on the various free energy changes this is seen to be the case.

Since the free energies of adsorption of the alkanes from vapour to the water/vapour interface are a linear function of chain length, it is likely that the alkane molecules assume a roughly parallel orientation at the water surface, and so it is reasonable to suppose that amphiphile chains are also flat along the air/water surface in dilute monolayers. It is not known from the adsorption free energies what the orientation of amphiphile chains is at the oil/water interface however. This is because the free energies of interaction of a CH$_2$ group with a water surface and with an alkane surface are very similar (Box 5.1). Thus, the standard chemical potential of a methylene group in an alkyl chain adsorbed at an alkane/water interface is similar in the cases where (i) the chain is immersed in alkane (e.g. when the chain is normal to the interface) and (ii) the chain lies flat along the interface and interacts partially with water. The near equivalence of the free energies of adsorption of a methylene group from vapour to the surface of water and of alkane suggests that, unlike the case for bulk water, alkyl chains do not induce water 'structuring' (see Chapter 2) in interfacial water. It is shown later how molecular orientations in adsorbed layers can sometimes be inferred from a consideration of surface equations of state, which, like free energies of adsorption are also obtained from interfacial tensions.

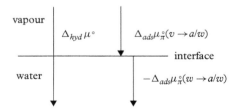

Figure 5.5 *Standard free energy changes for methylene group transfer.*

Box 5.1 The free energy of interaction of a CH_2 group with water and alkane surfaces

---

The methylene group in an alkyl chain interacts with interfacial water and interfacial alkane by dispersion forces. The surface tensions of liquid alkanes are a linear function of (1/chain length N_c); extrapolation to $1/N_c = 0$ gives a notional surface tension of liquid 'polymethylene' of 32.7 mJ m^{-2} at 298 K. Since, from molecular geometry, the cross-sectional area of a methylene group along an alkyl chain is about 0.059 nm^2, the free energy of 'adsorbing' a CH_2 group from bulk liquid hydrocarbon to the liquid/vapour interface is calculated to be 1.16 kJ mol^{-1}. The standard free energy of evaporation of a CH_2 in a liquid alkane is 3.04 kJ mol^{-1}, and it follows therefore that the standard free energy of taking a methylene group from vapour to the surface of a liquid alkane is about -1.9 kJ mol^{-1}. This is very similar to the standard free energy of adsorption from vapour to a water/vapour surface (-2.0 kJ mol^{-1}; see Table 2.1).

It is interesting that the enthalpy of adsorption of octanol from very dilute solution in dodecane to the dodecane/water interface, obtained from the temperature variation of $\Delta_{ads}\mu_\pi^o$ by the use of (5.3.9), is -31 kJ mol^{-1}. Enthalpies of transfer of alkanols butanol to hexanol from dilute solutions in alkanes to dilute solution in water, obtained by calorimetry, are about -30 kJ mol^{-1}, and it is believed that this enthalpy arises mainly from the transfer of the OH group. These observations are consistent with the driving force for adsorption of alkanols from oil to the oil/water interface arising from the transfer of the OH group from the nonpolar surrounding to water at the interface.

5.3.2 Description of the behaviour of adsorbed monolayers

Surface equations of state and adsorption isotherms

The thermodynamic parameters of adsorption of surfactants into infinitely dilute monolayers have been defined above. Behaviour in such infinitely dilute films is described by the ideal surface equation of state (5.3.2), which is analogous to the ideal gas equation for a bulk gas. A surface equation of state has the general form

$$\pi = kTf(\Gamma) \qquad (5.3.12)$$

and it relates the surface concentration, Γ, to the surface pressure, π exerted by the film. On the other hand an adsorption isotherm equation relates Γ (at adsorption equilibrium) to the bulk concentration c of the adsorbing species and has the general form

$$c = Kf'(\Gamma) \qquad (5.3.13)$$

The constant K, as will be seen, is related to a standard free energy of adsorption.

For a given model for the behaviour of an adsorbed monolayer (e.g. an ideal two-dimensional gas) there exists an equation of state and a corresponding adsorption isotherm equation. An equation of state can be converted into an isotherm equation by use of the Gibbs adsorption equation. Consider the simplest case of an ideal two-dimensional gas. The Gibbs adsorption equation for the adsorption of a nonionic surfactant from ideal dilute solution into a film with area $\mathcal{A}$ containing N molecules is (see §4.3.1)

$$-d\gamma = kT\Gamma d\ln c = kT\,(N/\mathcal{A})\,d\ln c \qquad (5.3.14)$$

where k is the Boltzmann constant, Γ the surface concentration in molecules per unit area, and c is the concentration of the surfactant in solution. The surface pressure π of the film is defined as $\gamma_0 - \gamma$ (see §4.4), where γ_0 is the interfacial tension of the clean interface, so that $-d\gamma = d\pi$ and hence the Gibbs adsorption equation may be written

$$d\pi = kT\,(N/\mathcal{A})\,d\ln c \qquad (5.3.15)$$

The ideal two-dimensional gas equation of state can be expressed

$$\pi\mathcal{A} = NkT \text{ or } \pi A = kT \qquad (A = \text{area per molecule}) \qquad (5.3.16)$$

from which, for a constant area of film,

$$d\pi = (kT/\mathcal{A})\,dN \qquad (5.3.17)$$

Comparison of (5.3.15) and (5.3.17) shows that the adsorption isotherm equation is

$$c = KN \qquad (5.3.18)$$

where K is a proportionality constant. So, for an ideal two-dimensional gas the surface concentration is a linear function of the bulk concentration of the adsorbing species. Since for an ideal film $N \propto \pi$ the isotherm equation may be written

$$c = K'\pi \qquad (5.3.19)$$

The standard free energy of adsorption $\Delta_{ads}\mu_\pi^o$ is equal to $-kT\ln\,(\pi/c)_{c\to o}$ (cf. (5.3.6)) and it is seen that the constant $K' = \exp\left(\Delta_{ads}\mu_\pi^o/kT\right)$. The isotherm equation may therefore be written

$$c = \exp\left(\Delta_{ads}\mu_\pi^o/kT\right)\pi \qquad (5.3.20)$$

An example of a linear isotherm is given in Fig. 5.6 which is a plot of the surface pressures of monolayers of dodecanol at the dodecane/water interface against dodecanol mole fraction in dodecane. Linearity is obtained up to surface pressures of around 1.6 mN m^{-1}, and thereafter the isotherm becomes curved (concave towards the abscissa).

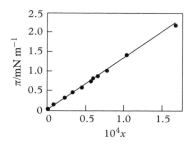

Figure 5.6 *Surface pressure π of monolayers of dodecanol at the dodecane/water interface as a function of mol fraction x of dodecanol in dodecane at 293 K.*

Effects of finite size of adsorbed molecules

As monolayers become more concentrated, the adsorbed molecules or ions begin to interact laterally within the film, which gives rise to deviations from ideal behaviour and the surface equations of state and adsorption isotherm equations must be modified. The finite size of adsorbed molecules results in a repulsive (excluded area) interaction as does the proximity of like charged head groups for example. On the other hand, for surfactant films at the air/water interface, the adsorbed chain groups can mutually attract through van der Waals forces. For a given area per molecule, repulsion between molecules or ions gives a positive contribution to the surface pressure whereas attractions lead to a negative contribution. Some of these observations are illustrated below by reference to experimental data.

In the absence of other interactions, the effect of finite size can often be accounted for rather simply. A molecular 'co-area' A_o (Box 5.2) is introduced to allow for the excluded area within the film which is unavailable to other molecules. Then, assuming the adsorbed film behaves as a mobile two-dimensional gas, the surface equation of state becomes (Aveyard and Haydon, 1973)

$$\pi \, (A - A_o) = kT \tag{5.3.21}$$

The equation is referred to as the Volmer equation. The value of A_o can be obtained from the (linear) plot of $1/\pi$ versus A; A_o is given as the intercept on the A axis at $1/\pi = 0$. In Fig. 5.7, π-A curves are shown for 1-octanol at the dodecane/water interface (filled symbols) and 1-dodecanol adsorbed at the octane/water interface (open symbols). Also included is the curve (full line) generated by the Volmer equation for $T = 298$ K and $A_o = 0.25$ nm^2. The Volmer curve fits the data well, and it is noteworthy that the π-A curves for octanol and dodecanol are very similar. This similarity, together with the value of A_o required to give the fit suggests that the adsorbed molecules at the oil/water interface are disposed roughly normal to the interface. The apparent lack of lateral attraction between adsorbed molecules arises since the mutual attraction between amphiphile chains and the attraction between the chains and oil molecules are very similar. Thus there is little change in net attractions with increasing surface coverage.

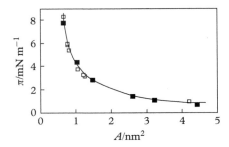

Figure 5.7 *π–A curves for octanol at the dodecane-water interface at 293 K (filled symbols) and dodecanol at the octane/water interface at 298 K (open symbols). The full line is from the Volmer equation for T = 298 K with $A_o = 0.25$ nm^2.*

Box 5.2 Significance of the molecular co-area A_o

That the area A_o is a rough measure of equilibrium 'molecular size' as can be seen as follows. Suppose that the alkyl chain molecules are cylinders disposed normal to the surface, so the cross-sectional area of an adsorbed molecule is that of a circular disc. It can be shown that the excluded area A_o is twice the cross-sectional area of a disc defined by the distance, d, of *closest* approach of centres. The *equilibrium* separation of centres, d_o, is however greater than d, and de Boer (1953) has suggested that $d_o = 1.37d$. Twice the area defined by d is therefore is equal to 1.06 x area defined by d_o. Thus, within a few %, A_o can be supposed to give the equilibrium cross-sectional area of a vertically oriented adsorbed cylindrical molecule.

As explained, it is possible to obtain an isotherm equation from a surface equation of state by use of the Gibbs adsorption equation. The isotherm corresponding to the Volmer equation (5.3.21) can be shown to be (Aveyard and Haydon, 1973)

$$c = K'' \left(\frac{A_o}{A - A_o} \right) \exp \left(\frac{A_o}{A - A_o} \right) \tag{5.3.22}$$

Like the Volmer equation, this isotherm is applicable to a mobile two-dimensional gas[4] and accounts for lateral repulsion between adsorbed molecules arising from the excluded area through the inclusion of the molecular co-area A_o. The constant K'' is related to a standard free energy of adsorption, $\Delta_{ads}\mu^o_{vol}$ by

$$\Delta_{ads}\mu^o_{vol} = RT \left(\ln K'' + 1 \right) \tag{5.3.23}$$

---

[4] In a *mobile* monolayer molecules are able to move freely along the surface. In contrast, a *localized* monolayer is one in which the surface molecules are present on *sites*. Molecules can move between sites but do so by desorbing and re-adsorbing on another site. See Appendix A5.1

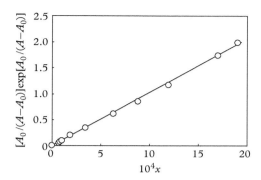

Figure 5.8 *Plot according to the isotherm equation (5.3.22) for the adsorption of decanol from dodecane to the dodecane/water interface at 293 K. The concentration in dodecane is expressed in mol fraction, x, and A_o is taken as 0.25 nm^2 molecule^{-1}.*

the standard states being unit concentration in solution and half coverage in the surface (i.e. $A_o/A = \theta = 0.5$). It can be shown that $\Delta_{ads}\mu_\pi^o$ (see (5.3.5)) is related to $\Delta_{ads}\mu_{vol}^o$ by

$$\Delta_{ads}\mu_\pi^o = \Delta_{ads}\mu_{vol}^o + RT\{\ln(A_o/kT) - 1\} \qquad (5.3.24)$$

Equation (5.3.22) is useful in obtaining a standard free energy of adsorption in appropriate systems where data for intermediate surface coverages are available. A plot according to (5.3.22) for the adsorption of decanol from dodecane to the oil/water interface at 293 K, with A_o set at 0.25 nm^2, is shown in Fig. 5.8. From the slope $\Delta_{ads}\mu_{vol}^o = -14.5$ kJ mol^{-1}, and conversion to $\Delta_{ads}\mu_\pi^o$ by use of (5.3.24) yields $\Delta_{ads}\mu_\pi^o = -23.7$ kJ mol^{-1}, consistent with the value shown in Fig. 5.4, which was obtained from the initial slope of a plot of surface pressure against solution concentration.

Lateral attraction between adsorbed molecules

If the argument concerning the apparent lack of chain interactions in monolayers at the *oil*/water interface is correct (see previous paragraphs), it should be possible to pick up chain-chain attractions in monolayers at the *air*/water surface. Such attractions are clearly indicated in Fig. 5.9 where π-A curves are compared for the cationic surfactant dodecyltrimethylammonium bromide adsorbed at the air/water (AW) and dodecane/water (OW) interfaces at 298 K. The dashed curve is given by the ideal two-dimensional gas equation. The curve for the air/water interface lies well below that for the oil/water surface indicating the attractive interactions between the alkyl chains in the monolayers at the air/water surface. Of course for this ionic surfactant there are repulsive interactions between the head groups (*vide infra*), but for a given value of A these can be assumed to be similar for monolayers at the oil/water and air/water interfaces.

Derivation of equations of state and adsorption isotherms that satisfactorily account for attractive interactions is not a simple matter. For example, it has sometimes been supposed that the difference in surface pressure for a given A in films of a given surfactant

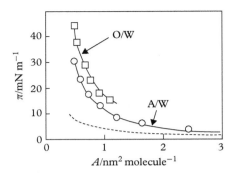

Figure 5.9 *π-A isotherms for adsorbed monolayers of dodecyltrimethylammonium bromide at the dodecane/water (OW) interface and at the air/water (AW) interface at 298 K.*

adsorbed at (i) an oil/water and (ii) an air/water interface is a measure of the attractive interactions between the chain groups at the air/water surface. However, as a film at the air/water surface becomes more concentrated it is possible that the chains progressively change from a roughly parallel disposition towards a more vertical orientation. If so, both A_o and attractive interactions change simultaneously with the surface concentration of surfactant (unlike the situation for monolayers at the oil/water interface). If the surfactant is ionic, the situation is even more complex as will be seen below because the electrical interactions are also changing with coverage of the surface.

The two-dimensional analogue of the van der Waals gas equation has sometimes been used to describe the behaviour of amphiphile monolayers in which intermolecular attractions are present. The surface equation of state may be expressed

$$\left(\pi + \left[\alpha/A^2\right]\right)(A - A_o) = kT \tag{5.3.25}$$

where α is a two-dimensional van der Waals interaction constant (positive for attraction). The corresponding isotherm equation is

$$c = K''' \left(\frac{A_o}{A - A_o}\right) \exp\left(\frac{A_o}{A - A_o} - \frac{2\alpha}{kTA}\right) \tag{5.3.26}$$

The two-dimensional van der Waals model is interesting since it gives a qualitative description of two-dimensional condensation in monolayers.[5] A plot of π against A calculated using (5.3.25) is shown in Fig. 5.10; the values of the constants used are given

---

[5] Two-dimensional condensation is most usually encountered with so-called 'insoluble' or 'spread' monolayers. These are monolayers formed by spreading a water-insoluble amphiphile (e.g. a long chain carboxylic acid or alkanol) from a suitable volatile solvent directly onto the surface of water between moving barriers on a Langmuir trough (see §6.2.2, and §17.4.1 for a description of the Langmuir trough).

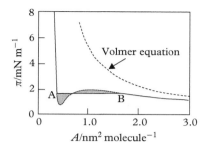

Figure 5.10 *Effects of attractive interactions at 298 K on π-A curve generated by two-dimensional van der Waals equation (5.3.23). Values of constants are: $A_o = 0.20 \ nm^2 \ molecule^{-1}$; $a = 3.1 \times 10^{-39}$ J $m^2 \ molecule^{-2}$.*

in the legend. Also shown in the figure is the curve generated by the Volmer equation of state (5.3.21) which, as discussed, only accounts for the excluded area due to finite molecular size. It is seen that the van der Waals equation produces the familiar loops; the horizontal line AB in the isotherm (corresponding to the two-dimensional saturated vapour pressure of the monolayer) is (schematically) drawn such that the two shaded areas are equal. The right hand limb represents the compression of a gaseous monolayer, and the steep limb at small areas the compression of condensed phase. At intermediate areas (between points A and B) two-dimensional gaseous and condensed phases coexist.

It was mentioned above, however, that for surfactants at the air/water surface lateral attractions vary in an unknown way with surface coverage (not simply with A^{-2}), as does also A_o. Thus, it is difficult to apply (5.3.25) and (5.3.26) meaningfully to experimental π-A data.

Electrical repulsion between charged surfactant head groups

For ionic surfactants, a term allowing for electrical repulsion between charged head groups in the adsorbed film must be incorporated into the surface equation of state. For monolayers of ionic surfactant adsorbed at an oil/water interface for example, account must be taken of the finite size of the molecules (through A_o) and a term included in the expression for surface pressure to allow for charging the film.

Although an ionic surfactant may be effectively completely ionized in dilute aqueous solution, it is likely that in an adsorbed monolayer (where the head groups are in much closer proximity than in the bulk solution) the degree of surfactant dissociation will be considerably smaller than unity. This means, with reference to Fig. 5.11, that some counterions (cations in the case illustrated) are 'bound' (adsorbed) to the anionic surfactant head groups. The layer of immobile counterions constitutes the Stern layer; beyond, there is a mixed, mobile layer of ions (in the *diffuse layer*) as discussed in §11.2.2. The definition of the degree of dissociation of surfactant in a monolayer is discussed in

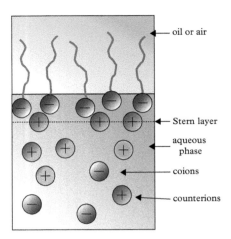

Figure 5.11 *Adsorbed layer of anionic surfactant at a water/fluid interface. Some cations (counterions) are 'associated' with the surfactant ions in the Stern layer, and non-surface active anions (coions) are desorbed.*

§9.6 (in connection with ionic micelles), but for the moment it is sufficient to note that only a fraction (possibly as low as a tenth) of the surfactant head groups carry a net charge.

Suppose that there are N surfactant entities in the film, of area $\mathcal{A}$, N_d of which are dissociated, and suppose also that the surfactant is a 1:1 electrolyte. The surface charge density $\sigma = N_d e^-/\mathcal{A} = \Gamma_d e^-$, where Γ_d is the surface concentration of surfactant ions and e^- the electronic charge. The electrical contribution to the surface pressure (i.e. the work of charging unit area), π_{el} is

$$\pi_{el} = \int_o^{N_d} \sigma\, \mathrm{d}\psi_o \qquad (5.3.27)$$

in which ψ_o is the mean electrical potential in the plane of the head groups. In order to obtain π_{el} a relationship between σ and ψ_o is needed. The Gouy–Chapman expression (see Box 5.3) is the simplest (see §11.2.2) and is

$$\sigma = (8N\,(\infty)\,\varepsilon_r\varepsilon_o kT)^{0.5} \sinh \frac{e^-\psi_o}{2kT} \qquad (5.3.28)$$

Here ε_o and ε_r are, respectively, the permittivity of free space and the relative permittivity (dielectric constant) of the solution, and $N(\infty)$ is the number density of ions (including any added inert 1:1 electrolyte) in bulk solution at an infinite distance from the surface.

Box 5.3 Some problems with the Gouy–Chapman theory and its application

---

Although in principle (5.3.30) and (5.3.31) might be expected to give an acceptable first approximation to the behaviour of adsorbed ionic surfactants, there are problems both with the theory itself and with its application. First, the Gouy-Chapman treatment makes several simplifying assumptions (see also §11.2.2. dealing with the electrical double layer). One assumption is that the surface charge is smeared out, the discreet nature and finite size of ions being ignored. Further, the possibility that the counterions (e.g. Na^+ for sodium octylsulfate) can penetrate further towards the oil phase than the plane of centres of the surfactant head group, is not considered. One of the main difficulties in applying the theory is that the adsorbed surfactant ions are only partially dissociated; the degree of dissociation (which varies with the surface concentration of surfactant) is often unknown. Thus the surface charge density is not available as a function of A. Further, the co-area A_o is not readily available either. For films obeying the Volmer equation (i.e. nonionic surfactants at the oil/water interface) A_o is obtained from π–A data as explained. It is clear from (5.3.30) however that the same cannot be done in the case of ionic films. It is noted that (5.3.30) and (5.3.31) do not allow for lateral attractions between surfactant molecules within the films.

From (5.3.24) and (5.3.25) it can be shown that

$$\pi_{el} = \frac{2kT}{e^-}(8N(\infty)\varepsilon_r\varepsilon_o kT)^{0.5}\left\{\left(\frac{\sigma^2}{8N(\infty)\varepsilon_r\varepsilon_o kT}+1\right)^{0.5}-1\right\} \tag{5.3.29}$$

Noting (5.3.21), (5.3.28), and (5.3.29), the surface equation of state becomes

$$\pi = \frac{kT}{(A-A_o)} + \frac{2kT}{e^-}(8N(\infty)\varepsilon_r\varepsilon_o kT)^{0.5}\left[\cosh\left(-\frac{e^-\psi_o}{2kT}\right)-1\right] \tag{5.3.30}$$

The isotherm equation which corresponds to (5.3.30) can be written

$$c = \left(\frac{A_o}{A-A_o}\right)\exp\left(\frac{A_o}{A-A_o}\right)\exp\left(\frac{\Delta_{ads}\mu_{vol}^o}{RT}-1\right)\exp\left(\frac{e^-\psi_o}{kT}\right) \tag{5.3.31}$$

More recent studies have made refinements of the earlier approaches described above. For example, in one such study (Kralchevsky et al, 1999) it has been shown how interfacial tensions obtained as a function of bulk concentrations of ionic surfactant and salt, can be treated so as to obtain the adsorption of both surfactant and counterions, and in addition, surface potentials and surface charge densities. It is clear from this study (and a number others) that counter-ion binding (i.e. Stern layer occupancy) can be very considerable.

Some other important models and approaches which have been widely used to account for the behaviour of monolayers of surfactants (but which have not been discussed above) are presented in the Appendix A5.1.

5.4 Effects of inorganic electrolytes on adsorption

Very often, surfactants are present in systems containing low or high concentrations of inorganic salt. The behaviour of ionic surfactants can be profoundly affected by the presence of very small amounts of salt, whilst large amounts of salt can have significant effects on both nonpolar and polar moieties of surfactants in aqueous surroundings.

5.4.1 Dilute electrolyte in systems with ionic surfactant

Inert electrolyte, such as NaCl, added to a system containing an ionic surfactant leads to screening of the charges on the head-groups. This results in an increase in surfactant adsorption (for a given bulk concentration of surfactant), as was clearly seen at the beginning of this chapter in Fig. 5.1. Added salt also gives obvious changes to the plots of interfacial tension against ln(surfactant concentration), as shown in Fig. 5.12a for decyltrimethylammonium bromide adsorbed at the decane/aqueous NaCl interface. For a given surfactant concentration c, presence of inert electrolyte lowers γ, and, as will be seen in Chapter 10, for some surfactants the oil/water interfacial tension at and above the critical micelle concentration (*cmc*) can be lowered to extremely low values. Salt effects on the π–A isotherms are however very much less marked (Fig 5.12b); for a given bulk surfactant concentration increase in [salt] increases π and decreases A in such a way that the π–A curves for different salt concentrations are effectively coincident. This is mentioned further in Appendix A5.1.

The Gouy–Chapman theory outlined above indicates a substantial effect of electrolyte on the surface potential (Fig. 5.13a), via $N(\infty)$ in (5.3.28). The corresponding effects on the electrical component of surface pressure (5.3.29) are illustrated in Fig. 5.13b. For a given surface concentration of (fully dissociated) surfactant both π_{elec} and ψ_o are reduced by addition of electrolyte.

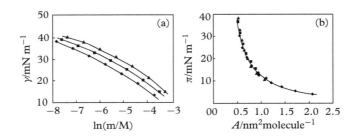

Figure 5.12 *(a) γ–ln c plots for decyltrimethylammonium bromide adsorbed from aqueous NaCl to the decane/aqueous solution interface at 293 K. The upper curve (▲) is for [NaCl] = 0.1 M, the middle curve (■) for 0.25 M, and the lower curve (◆) for 0.5 M. (b) The corresponding plots of π versus A. Adapted from data in Haydon and Taylor (1962).*

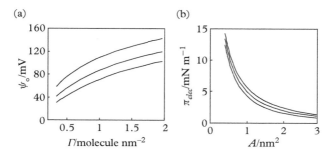

Figure 5.13 *(a) Variation of ψ_o with surface concentration for a fully dissociated 1:1 ionic surfactant. Concentrations of added 1:1 inert electrolyte are: upper curve, 0.1 M, middle curve, 0.25 M, and lower curve 0.5 M. (b) The variation of π_{elec} with area per molecule of 1:1 ionic surfactant. The curves are for (top) 0.1 M, (middle) 0.25 M, and (bottom) 0.5 M added inert 1:1 electrolyte.*

5.4.2 Concentrated electrolyte in systems with nonionic surfactant

Very low concentrations of electrolyte can drastically alter the adsorption of ionic surfactants, as seen, mainly as a result of the screening of the electrical repulsion between neighbouring surfactant head groups in the surface. Such small amounts of salt have very little influence on the adsorption of nonionic surfactants however. It is well-known that large concentrations of inorganic salts (say of the order of 1 M and higher) can markedly reduce the aqueous phase solubility of non-electrolytes, particularly nonpolar molecules and groups; this process is termed '*salting-out*' (Box 5.4). Since surfactant molecules contain a nonpolar group, high concentrations of salt are expected to influence the surfactant adsorption by a salting-out process.

Box 5.4   What is 'salting-out'?

In much of the early work on salt effects on nonelectrolytes in aqueous solution, interest centred on the way in which the solubility of sparingly soluble solutes was affected by addition of inorganic electrolytes. *Salting-out* occurs when the solubility of a solute is reduced by salt and *salting-in* when the solubility rises in the presence of salt. More generally, salting-out (salting-in) is said to occur when electrolyte causes an increase (decrease) in the standard chemical potential of the solute in solution. The solute need not be at, or even close to, its solubility limit.

Salt desorption from aqueous interfaces and its significance

Many inorganic electrolytes cause an increase, $\Delta\gamma$, in the tension of air/water and nonpolar oil/water interfaces; the increase is usually a linear function of electrolyte

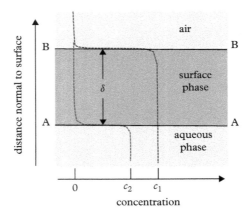

Figure 5.14 *Model for the surface of aqueous electrolyte with strongly hydrated cation. The surface phase lies between planes AA and BB and is ion-free (see text).*

concentration. Relative to the magnitudes of the decreases in tension caused by surfactants, the magnitude of the increases in tension caused by inorganic salts is very low; 1 mol dm^{-3} NaCl in water gives an increase in surface tension of only about 1.7 mN m^{-1} at 293 K. A similar increase in tension is given by NaCl for the alkane/water interface.

From the Gibbs adsorption equation, an increase in tension indicates that salt is desorbed, that is, excluded or repelled from the surface (Box 5.5). A crude but useful model for the interface in the kind of systems discussed here is depicted in Fig. 5.14. The surface phase is taken to be an ion-free water layer (between parallel planes AA and BB) of thickness δ; the salt and water concentrations, c_2 and c_1, respectively, fall abruptly to zero at planes AA and BB, respectively. Use of this model allows the value of δ to be calculated from interfacial tensions as a function of salt concentration. For LiCl, NaCl, and KCl, for example, desorbed from the air/water surface around room temperature, values of δ are calculated to be 0.32, 0.38, and 0.35 nm, respectively. These values are close to the diameter of a water molecule (about 0.32 nm) and to the hydration sheath thicknesses of the cations.

Box 5.5 **Surface potentials and the ionic double layer—disposition of ions normal to the surface**

From the Gibbs adsorption equation, an increase in tension indicates that salt as a whole is desorbed, that is, excluded or repelled from the surface. Changes in surface tensions however give no information on how anions and cations are disposed or segregated normal to the surface. For this, a knowledge of surface potentials, $\Delta \chi$ is needed. Positive (negative) values of $\Delta \chi$ indicate that the bulk of the solution is more positive (negative) than the surface. Positive experimental values of $\Delta \chi$ obtained for electrolyte solutions containing strongly hydrated

Box 5.5 *Continued*

cations and weakly hydrated anions have been interpreted as arising from the formation of an ionic double layer resulting from cations being excluded from the surface by their hydration sheaths; the anions are supposed to approach the surface more closely. There are of course electrolytes with strongly hydrated anions, and many electrolyte solutions have negative surface potentials. The surface potential however has contributions from dipole orientations as well as the presence of an ionic double layer at the surface (Aveyard and Haydon, 1973). The implicit assumption that the former contribution is unimportant however may not always be justified. A molecular dynamics simulation of aqueous solutions of KCl has shown that this salt does indeed give rise to a very high positive potential arising from the formation of an ionic double layer but that this is largely cancelled out by a negative contribution from induced dipoles, giving an overall surface potential of only +50 mV for 1M solution. Thus it is possible to have a strong segregation of ions normal to the surface but only record a small value of $\Delta\chi$ (Wick and Dang, 2006). The subject of salt desorption from aqueous solutions has been reviewed by dos Santos et al. (2010).

Salting-out of nonpolar moieties and molecules

To treat salting-out of nonpolar molecules or groups by salts with strongly solvated cations it is supposed that, as for a plane liquid surface, there is a salt free layer around the molecular entity, surface area Ω, dissolved in aqueous solution. The free energy change associated with salting-out of 1 mol solute, $\Delta_s\mu^o$, is then readily estimated as follows. The free energy of creating a cavity, surface area Ω, is $\gamma\Omega$ (for the salt solution) or $\gamma_o\Omega$ (for pure water). When the molecule is introduced into the cavity (from say the gas phase) it has interactions with surrounding molecules which, according to the ion-free layer model, are water molecules both in water and in salt solutions. Thus

$$\Delta_s\mu^o = N_A \left(\gamma - \gamma_o\right)\Omega = N_A\Delta\gamma\Omega \tag{5.4.1}$$

where N_A is the Avogadro number.

Usually salt effects have been discussed in terms of a salting constant, k_s. It is found for many (sparingly soluble) nonelectrolytes in aqueous salt solution, that the logarithm of the solubility (s) is linearly related to the salt molality m_2, that is,

$$\log\left(s^o/s\right) = k_s m_2 \tag{5.4.2}$$

where s^o is the solubility of nonelectrolyte in pure water. It is readily shown that, for $m_2 = 1$ mol kg^{-1},

$$\Delta_s\mu^o = 2.303RTk_s = N_A\Omega\Delta\gamma \left(m_2 = 1\,\text{mol}\,\text{kg}^{-1}\right) \tag{5.4.3}$$

Note that some molecules (or groups) are *salted-in* by electrolytes, that is, the standard chemical potential of the nonelectrolyte is more negative in salt solution than in water.

Since interest is centred on the effects of salt on the free energy of surfactant hydrocarbon chains, the simple approach described above is tested on the salting-out of normal alkanes, methane to butane, for which good solubility data exist (Morrison and Billet, 1952). To calculate $\Delta_s\mu^o$ using (5.4.3) values for Ω are required, and for this purpose the alkanes are treated as square prisms. In aligned liquid alkanes, the centres of the chains are separated laterally by 0.47 nm, and the C–C distance resolved along the chain is 0.126 nm. Thus for the alkanes of chain length N_c, $\Omega = 0.44 + 0.237N_c$ nm^2. Calculated values of $\Delta_s\mu^o$ for 1 mol kg^{-1} NaCl are shown in Fig. 5.15 as the line, and the points are experimental values determined from solubility measurements. For these systems the model described is seen to be quite successful and can reasonably be applied to effects of salts, with strongly hydrated cations, on surfactant chains.

Application to surfactant adsorption

When a surfactant molecule is adsorbed from aqueous solution to, say, an oil/water interface, the hydrophobic moiety leaves the aqueous surroundings and becomes immersed in oil; the head group also leaves bulk aqueous solution and takes up a position in interfacial water. The standard free energy of adsorption of the surfactant from both water, $\Delta_{ads}\mu^o_{water}$, and from salt solution, $\Delta_{ads}\mu^o_{salt}$, can readily be measured as previously described (§5.3.2), and the difference gives the effect of the salt on the adsorption. Thus, from what has been said,

$$\Delta_{ads}\mu^o_{salt} - \Delta_{ads}\mu^o_{water} = \Delta_s\mu^o_{surface} - \Delta_s\mu^o_{bulk} \tag{5.4.4}$$

where $\Delta_s\mu^o$ are the standard free energy changes accompanying salting-out (or salting-in) of the surfactant in the subscripted phases. The salt effect at the surface involves only the head group whereas the bulk salt effect involves both head and chain groups. The quantity $\Delta_s\mu^o_{surface}$ can be obtained from the effect of salt on the standard free energy of adsorption of the surfactant from the oil phase.

Using this approach it has been shown (Aveyard et al., 1998) for the sugar surfactant decyl β-glucoside that, assuming group additivity of salt effects, although the alkyl chain is salted-out from bulk aqueous solution by 1 mol dm^{-3} NaCl (free energy change

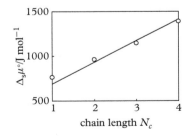

Figure 5.15 *Free energy changes, $\Delta_s\mu^o$, for the effects of 1 mol kg^{-1} NaCl on normal alkanes, chain length N_c. Full line is obtained by use of (5.4.3) with $\Delta\gamma = 1.64$ mN m^{-1} and Ω calculated as described in the text. Points are experimental values obtained from changes in solubility.*

$+2.2$ kJ mol^{-1}), the glucoside head group is salted-in, the free energy change for the latter being -1.2 kJ mol^{-1}. The salt effect on the (isolated) head group in the adsorbed state is not very different, with a free energy change of about -0.9 kJ mol^{-1}. It is possible that the observed salting-in effect on the head group is a result of cation complexation. It is worth noting that the ethyleneoxy head group ($-CH_2-CH_2-O-)_6$ in nonionic surfactants with *n*-alkyl chains of various chain length is salted-out by NaCl, presumably as a result of the presence of the methylene groups. The effects of salts on ionic surfactants in micelles is discussed in §9.6.

5.5 Adsorption from surfactant mixtures

Industrially and domestically, surfactants are usually present as mixtures, either because commercial surfactants are not pure and contain for example isomers and/or homologous chain components, or because mixtures of surfactants are more effective in their application (Rosen and Kunjappu, 2012). The synergistic behaviour is related in some way to the mutual interactions between the surfactants, both in solution (as micelles) and in adsorbed layers. Below, the co-adsorption at air/water (or oil/water) interfaces in systems with two surfactants present is considered. The surface composition with respect to the surfactants can be very different to that in bulk solution, both as monomers and in micelles. The measurement of surface composition with respect to the two surfactants is first discussed, and then it is shown how the surface behaviour can usefully be discussed, in a semi-empirical way, in terms of a surfactant interaction parameter. Micelle formation in mixed surfactant systems is treated in §9.7 and micellization in (surfactant + polymer) mixtures is discussed in §9.9.

Surface compositions can be determined from interfacial tensions and application of the Gibbs adsorption equation, as in systems with a single surfactant. Some of the best data however are those derived from neutron reflectivity measurements (see e.g. Penfold et al., 1995 and references in §8.1). A brief description of the technique is given in §8.1.

5.5.1 Adsorption from surfactant mixtures using surface tensions

As an example, adsorption of two anionic surfactants, A^+B^- and A^+D^- from swamping aqueous inorganic electrolyte A^+X^- (e.g. NaCl) at the air/solution interface is treated. The equations derived will also apply to mixtures of nonionic surfactants. There are various possibilities, of which two are considered here. The fraction of bulk phase concentrations, $m_{AB}/(m_{AB} + m_{AD})$, is denoted x and adsorption is considered in systems where (i) the total amount of surfactant (i.e. $m_{AB} + m_{AD} = m_t$) is varied at constant x and (ii) x is varied at constant m_t.

In swamping electrolyte the activities a_{A^+} and a_{X^-} remain effectively unchanged as does the activity of water (low surfactant concentrations), and so the Gibbs equation is

$$-d\gamma = RT\left(\Gamma_{B^-}d\ln m_{B^-} + \Gamma_{D^-}d\ln m_{D^-}\right) \tag{5.5.1}$$

For constant x, $d\ln m_t = d\ln m_{B^-} = d\ln m_{D^-}$ and so, from (5.5.1),

$$-\frac{1}{RT}\left(\frac{\partial\gamma}{\partial\ln m_t}\right)_x = \Gamma_{B^-} + \Gamma_{D^-} \tag{5.5.2}$$

Then, say, Γ_{B^-} alone can be obtained by measuring tensions as m_{B^-} is varied at constant m_{D^-} (through x at the appropriate m_t) and applying (5.5.1) with $d\ln m_{D^-} = 0$.

In the second case, x is varied at constant m_t and an expression for $(\partial\gamma/\partial x)_{m_t}$ is required. For constant m_t, $dm_{B^-} = -dm_{D^-}$ and $dm_{B^-} = m_t dx$, and so

$$(\partial\gamma/\partial x)_{m_t} = (\partial\gamma/\partial m_{B^-})_{m_t}m_t \tag{5.5.3}$$

Noting (5.5.1) it is easy to show that

$$-\left(\frac{\partial\gamma}{\partial m_{B^-}}\right)_{m_t} = RT\left(\frac{\Gamma_{B^-}}{m_{B^-}} - \frac{\Gamma_{D^-}}{m_{D^-}}\right) \tag{5.5.4}$$

So, from (5.5.3) and (5.5.4) the required expression is obtained:

$$-\left(\frac{\partial\gamma}{\partial x}\right)_{m_t} = RT\left(\frac{\Gamma_{B^-}}{x} - \frac{\Gamma_{D^-}}{1-x}\right) \tag{5.5.5}$$

Then, Γ_{B^-} alone can be determined from tensions in appropriate systems in which m_{D^-} is held constant as m_{B^-} is varied. The mole fraction of, say, B^- in the surface, $x_{B^-}^s$, is defined as $\Gamma_{B^-}/(\Gamma_{B^-} + \Gamma_{D^-})$.

It may be that the specific interest is in surface compositions in systems at the *cmc* (say to compare them with *micellar* composition at the *cmc*). The tension at the *cmc* is denoted γ_c, (Fig. 5.16); the tension varies with the mol fraction x of surfactants in bulk solution at constant m_t, and with m_t at constant x. In the present context $m_t = cmc$, so the variation of γ_c with x can be expressed

$$\frac{d\gamma_c}{dx} = \left(\frac{\partial\gamma}{\partial\ln m_t}\right)_x\frac{d\ln cmc}{dx} + \left(\frac{\partial\gamma}{\partial x}\right)_{m_t} \tag{5.5.6}$$

Combination of (5.5.6) with (5.5.2) and (5.5.5) gives the expression

$$\frac{d\gamma_c}{dx} = -RT\left(\Gamma_{B^-} + \Gamma_{D^-}\right)\left[\frac{x^s}{x} - \frac{1-x^s}{1-x} + \frac{d\ln cmc}{dx}\right] \tag{5.5.7}$$

From a knowledge of the variation of γ_c with x (see Fig. 5.16b for an experimental example), values of $(\Gamma_{B^-} + \Gamma_{D^-})$ at the *cmc* (obtained using (5.5.2)) and the way in which the *cmc* varies with x, values of surfactant mol fractions in the surface, x^s can be

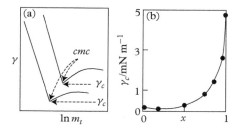

Figure 5.16 *(a) Schematic representation of γ versus ln m_t curves for two different values of x. The minimum tensions, γ_c, occur at the cmc. (b) Values of γ_c as a function of aqueous phase mol fraction (see text) of SDS in mixtures with AOT. Surfactants were adsorbed from 0.1 mol dm^{-3} aqueous NaCl to the interface with heptane at 298 K. Redrawn from Aveyard et al. (1988).*

computed. It is shown below how such information can be used to describe the adsorbed surfactant mixtures in terms of the regular solution theory.

5.5.2 Adsorption from surfactant mixtures using neutron reflectivity

The technique of neutron reflection for the study of adsorption is described in §8.1. Measurement of adsorption in mixed surfactant films is exemplified here by the work of Penfold et al. (1995) who studied the co-adsorption, to the air/null reflecting water (NRW) interface, of the nonionic surfactants $C_{12}E_3$ (component 1) and $C_{12}E_8$ (component 2), where C_{12} denotes a dodecyl chain and E the ethyleneoxy group. Reflectivities $R(Q)$ were measured as a function of momentum transfer $Q = 4\pi\sin\theta/\lambda$, where θ is the (fixed) angle of incidence of the neutron beam with respect to the surface (1.5°) and λ is neutron wavelength. Adsorbed surfactant was taken to be a homogeneous monolayer, and the scattering length density ρ and layer thickness d were determined. The area per molecule, A, in an adsorbed layer with only one surfactant is given by

$$A = 1/\Gamma = b/(\rho d) \tag{5.5.8}$$

in which b is the scattering length density of an adsorbed molecule. For a mixed layer containing two surfactants, 1 and 2,

$$\rho = b_1/(A_1 d) + (b_2/A_2 d) \tag{5.5.9}$$

Equimolar mixtures of the two surfactants were studied in the concentration range 10^{-5}–10^{-2} mol dm^{-3} in NRW. For each concentration reflectivities were determined for two combinations of surfactant labelling: d-$C_{12}E_3/d$-$C_{12}E_8$ and d-$C_{12}E_3/h$-$C_{12}E_8$. In this way, by the use of (5.5.9), Γ_1 and Γ_2 were obtained separately at each concentration (Fig. 5.17) and so the mole fractions in the surface could be calculated.

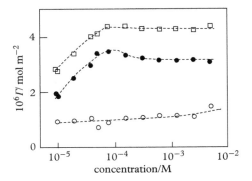

Figure 5.17 *Surface excesses of $C_{12}E_8$ (open circles), $C_{12}E_3$ (filled circles), and their equimolar mixtures (squares), as a function of surfactant concentration in null reflecting water. Redrawn from Penfold et al. (1995).*

5.5.3   Treatment of films with two surfactants using a regular solution approach

As described above, surface compositions of mixed surfactant layers with respect to the two surfactants can be obtained over a concentration range. Just as with bulk mixtures, there is an obvious interest in exploring how surface mixtures of surfactants can be described thermodynamically. One widely used approach has been to treat binary surfactant mixtures (mainly in micelles but also in adsorbed layers) as if they were regular solutions (Box 5.6). In what follows it is supposed, rather crudely, that the monolayer consists only of surfactant(s), that is, that water is absent from the film. The non-idealities resulting from the interactions between the surfactants are characterized by an interaction parameter β which is negative for attractive interactions, such as might exist between anionic and cationic surfactants for example. (N.B. in Appendix, A5.2, a monolayer containing a single surfactant is treated as a two dimensional solution of adsorbed surfactant and solvent.)

Box 5.6   What is a regular solution?

In a regular solution of say molecules of 1 (mole fraction x_1) and 2 (mole fraction $x_2 = 1 - x_1$), although there is a finite heat of mixing the entropy of mixing is supposed to be ideal, that is, mixing is taken to be completely random. The solution can be approximated by a quasi-crystalline lattice and the model is likely to be most useful when (i) molecules of 1 and 2 are roughly the same size and shape so that they can move between sites without changing the lattice structure, and (ii) the interactions between unlike molecules are not too strong relative to kT (Denbigh, 1961). An interaction parameter w is defined as $\varepsilon_{12} - \varepsilon_{11}/2 - \varepsilon_{22}/2$, where ε are potential energies of interaction between the subscripted pairs. The parameter β used in

Box 5.6 *Continued*

the treatment above is defined as $\beta = zN_A w/RT$ where z is the average number of neighbours a molecule has, and N_A is the Avogadro number. The excess molar free energy of mixing in a regular solution is

$$\Delta_{mix}G^E = RT\,(x_1 \ln f_1 + x_2 \ln f_2) = RT\beta x_1 x_2 = \Delta_{mix}H^E$$

where the f are activity coefficients given by

$$\ln f_1 = \beta x_2^2; \quad \ln f_2 = \beta x_1^2$$

and $\Delta_{mix}H^E$ is the excess enthalpy of mixing.

The starting point is the expressions for the chemical potential of any component i in a surface (μ_i^s) (see Appendix, (A5.7)) and in bulk solution (μ_i^l). In the surface,

$$\mu_i^s = \mu_i^{o,s} + RT \ln x_i^s f_i^s - \gamma \omega_i \tag{5.5.10}$$

in which ω_i is the partial molar surface area of i, f_i^s is its mole fraction activity coefficient in the surface and x_i^s is the surface mol fraction defined by

$$x_i^s = \Gamma_i/(\Gamma_1 + \Gamma_2)$$

For i in a bulk solution (assumed for simplicity of presentation to be ideal), using molar concentrations c,

$$\mu_i^l = \mu_i^{o,l} + RT \ln c_i = \mu_i^{o,l} + RT \ln (\alpha_i c_{12}) \tag{5.5.11}$$

where c_{12} is the sum of the molar concentrations, c_1 and c_2 of surfactants 1 and 2, respectively, in bulk mixed solution and α_i is defined as c_i/c_{12}. At adsorption equilibrium, $\mu_i^l = \mu_i^s$ so from (5.5.10) and (5.5.11) one can write, for example, for surfactant 1 in the mixed system:

$$\mu_1^{o,s} - \mu_1^{o,l} = RT \ln (\alpha_1 c_{12}) - RT \ln x_1^s f_1^s + \gamma_{12}\omega_1 \tag{5.5.12}$$

In (5.5.12), γ_{12} is the interfacial tension in the systems with both surfactants present. For a system with only one surfactant (say 1) present, (5.5.12) indicates that

$$\mu_1^{o,s} - \mu_1^{o,l} = RT \ln c_1^o + \gamma_1^o \omega_1^o \tag{5.5.13}$$

in which the superscript ^{o} denotes a system with only one surfactant present. From (5.5.12) and (5.5.11),

$$\gamma_{12}\omega_1 - \gamma_1^o\omega_1^o = RT \left(\ln c_1^o - \ln \alpha_1 c_{12} + \ln x_1^s f_1^s\right) \tag{5.5.14}$$

Consider two systems, one with only surfactant 1 present and the other with both surfactants present but with the same interfacial tension, so that $\gamma_{12} = \gamma_1^o$. Also assume that the partial molar surface area of surfactant 1 in the mixed surfactant systems is the same as that for pure 1, that is, $\omega_1 = \omega_1^o$. It then follows from (5.5.14) that

$$x_1^s = \alpha_1 c_{12}/c_1^o f_1^s; \quad x_2^s = \alpha_2 c_{12}/c_2^o f_2^s \tag{5.5.15}$$

Since $x_2^s = 1 - x_1^s$,

$$\frac{1}{c_{12}} = \frac{\alpha_1}{c_1^o f_1^s} + \frac{\alpha_2}{c_2^o f_2^s} \tag{5.5.16}$$

For regular solutions the surface activity coefficients are given by

$$\ln f_1^s = \beta\left(x_2^s\right)^2; \quad \ln f_2^s = \beta\left(x_1^s\right)^2 \tag{5.5.17}$$

in which β is the parameter which reflects the interactions between the two surfactants in the surface. As mentioned, β is negative for attractive interactions between the two surfactants in the adsorbed layer.

Using the above approach it is possible to obtain a value for β from an analysis of interfacial tensions obtained for systems with the single surfactants and with mixtures of the two surfactants. This can be exemplified by reference to Fig. 5.18. Tension versus total concentration curves, each obtained for a fixed bulk phase mole fraction x^l, are obtained. In the figure, curves are shown for SDS and AOT alone, and for mixtures with a bulk mole fraction of SDS of 0.756. Adsorption is from 0.1 mol dm^{-3} aqueous NaCl solution to the heptane/solution interface at 298 K. Surface mol fractions of the surfactants can be calculated from the tensions, as described at the beginning of the section, for given bulk mole fractions. Also, for a given interfacial tension, c_1^o, c_2^o (corresponding to points A and B in Fig. 5.18) are known. Then, by use of (5.5.16) and (5.5.17) it is possible to calculate a value of c_{12} corresponding to the same interfacial tension in the mixed surfactant system (point C in the figure) by varying β until the calculated and experimental concentrations agree. This process is repeated for other fixed tensions and the tension–ln c for the mixed surfactant system is constructed. In Fig. 5.18 the dashed line (3) has been calculated in this way using $\beta = -0.97$, which gives good agreement with the experimental tension curve over the whole range of concentrations shown.

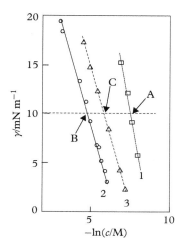

Figure 5.18 *Interfacial tensions at the heptane/0.1 mol dm^{-3} aqueous NaCl interface as a function of total surfactant concentration in systems with SDS (curve 1), AOT (curve 2), and mixtures of SDS and AOT, with $\alpha_{SDS} = 0.756$ (see text). The dashed line represents calculated tensions using regular solution theory with $\beta = -0.97$. Adapted from data in Aveyard et al. (1988).*

5.5.4 Treatment of adsorbed films using a two-component van der Waals model

This section is concluded by pointing briefly to an interesting application of the van der Waals model to adsorbed surfactant mixtures, proposed by Kralchevsky and co-workers (2003, 2004). The use of the two-dimensional analogue of the van der Waals gas equation to describe the behaviour of single surfactant monolayers in which mutual intermolecular attractions are present was discussed in §5.3.2. The surface equation of state, which is appropriate for a non-localized (i.e. mobile) monolayer, was expressed

$$\left(\pi + \left[\alpha/A^2\right]\right)(A - A_o) = kT \tag{5.3.25}$$

which, noting that $\Gamma = 1/A$ and multiplying both sides by Γ, can be equivalently written

$$\left(\pi + \alpha\Gamma^2\right)(1 - \Gamma A_o) = \Gamma kT \tag{5.3.25a}$$

in which α is a parameter reflecting the attraction between the adsorbed surfactant molecules/ions (positive for attraction)[6] and A_o is the surfactant molecular co-area. It

---

[6] The usual sign convention for interaction parameters is that attractive energies are negative (as with the β parameter in the regular solution theory). However, in the van der Waals treatment the opposite convention is

will be recalled that the contribution to the surface pressure from repulsive interactions between charged head groups was calculated separately (see (5.3.26)).

The equation obtained by Kralchevsky, Danov and co-workers for adsorbed surfactant *mixtures* (of components 1 and 2) has the same form as (5.3.25a) and is, using their nomenclature (see footnote 6),

$$\left(\pi_a + \beta \Gamma^2\right)(1 - \alpha\Gamma) = \Gamma kT \tag{5.5.18}$$

The total surface pressure of the monolayer is π_s; it is the sum of π_a occurring in the van der Waals equation and π_d arising from the diffuse electrical double layer. The adsorption Γ is the sum of the adsorptions of the two surfactants, that is, $\Gamma_1 + \Gamma_2$. The parameter β is related to the parameters β_{11} and β_{22} for the two pure surfactants by

$$\beta \equiv \beta_{11}\left(x_1^s\right)^2 + 2\beta_{12}x_1^s x_2^s + \beta_{22}\left(x_2^s\right)^2$$

Similarly α, a compound excluded area is defined by

$$\alpha \equiv \alpha_{11}\left(x_1^s\right)^2 + 2\alpha_{12}x_1^s x_2^s + \alpha_{22}\left(x_2^s\right)^2$$

The surface mole fractions x_i^s are defined as earlier, that is,

$$x_i^s = \Gamma_i/(\Gamma_1 + \Gamma_2)$$

The electrical contribution to surface pressure, π_{elec} is obtained using

$$\pi_{elec} = \frac{2kT}{e^-}(8N(\infty)\,\varepsilon_r\varepsilon_0 kT)^{0.5}\left[\cosh\left(-\frac{e^-\psi_o}{2kT}\right) - 1\right] \tag{5.5.19}$$

which is the same as (5.3.29).[7]

Fitting experimental data using the van der Waals approach is not trivial; there is a set of nine parameters whose values are to be specified and some of the constants involved are floated to give best fits of surface tension isotherms. It is possible however to obtain excellent fits of tension-concentration curves for surfactant mixtures with constants that are physically reasonable. For example, some fits (full lines) are shown in Fig. 5.19 of

---

used, and positive α implies attractive interactions. In the context a further potential confusion arises since in the van der Waals equation α is used as the interaction parameter whereas in the regular solution theory, β is the interaction parameter. In this section on the van der Waals equation of state for mixtures the nomenclature used by Kralchevsky and co-workers (2003, 2004) is employed, that is, β is used for the interaction parameter, and this is positive for attraction between surfactant molecules. The equivalent area to A_o for mixtures is here designated α.

[7] Kralchevsky and co-workers did not use the rationalized systems of electrical units employed here and so their equation for π_{elec} has a different form to (5.5.19). Also, they used ionic strength I rather than the number density of ions $N(\infty)$, but the two quantities are of course readily interconvertible.

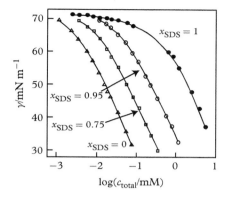

Figure 5.19 *Surface tensions of solutions of SDS and betaine in 0.01 mol dm^{-3} aqueous NaCl as a function of total surfactant concentration. The curves are for different bulk mol fractions of SDS (as indicated). Some curves for other values of x_{SDS} have been omitted for clarity. Redrawn from Danov et al. (2004).*

experimental data for mixtures of sodium dodecylsulfate (SDS, component 1) and a zwitterionic (betaine) surfactant (component 2) adsorbed from 0.01 mol dm^{-3} aqueous NaCl at the air/solution interface. The various curves are for different mole fractions of SDS in the bulk solutions, defined as concentration of SDS divided by total surfactant concentration. The values of, for example, excluded molecular areas involved in the fits (for all the curves) are $\alpha_{11} = 29.76$ Å^2, $\alpha_{22} = 27.78$ Å^2, and $\alpha_{12} = 28.77$ Å^2, all of which are physically realistic.

5.6 Mixed surfactant–hydrocarbon layers at liquid/vapour interfaces

It can reasonably be assumed that alkane molecules penetrate into the chain region of surfactant monolayers at alkane/water interfaces, although it is not clear to what extent. For a given surfactant however, it is believed that the shorter chain alkanes penetrate more strongly (in terms of volume) than do longer chain alkanes. This will have major consequences on preferred curvature of close-packed surfactant films at oil/water interfaces, and this can lead to very low interfacial tensions and, if low enough, to the formation of microemulsions. This is discussed in Chapter 10.

Although not necessarily directly relevant to behaviour at the oil/water interface, the uptake of hydrocarbons by surfactant monolayers adsorbed at the air/water interface can be determined experimentally, indicating trends with respect to changes in alkane and surfactant chain lengths. The simplest approach experimentally is to determine the effects on surface tension of the uptake of hydrocarbon by a surfactant monolayer, since adsorption causes a decrease in tension which can readily be measured. Tensiometry will not of course give any detailed information about structure of the mixed film. For this it is necessary to turn to neutron reflection measurements, which can give, for example, information about the relative disposition of hydrocarbon and surfactant molecules normal to the surface.

Alkanes of varying chain length have been introduced into monolayers of a homologous series of alkyltrimethylammonium bromides (C_nTAB, where n is the alkyl chain length) adsorbed at the air/water surface, and the lowering of surface tension determined (Aveyard et al., 1990). In such experiments, the surface concentration of surfactant can be adjusted by varying the bulk aqueous phase concentration. Alkane can be introduced into the films via small lenses of either the pure alkane (mole fraction x_a of alkane = 1) or as solutions in the non-adsorbing high molecular weight hydrocarbon squalane (mole fraction activity a_a of alkane varies). Activity coefficients of alkanes over a range of mole fractions in squalane are known, and so the uptake of alkane into the surfactant films can also be determined over a range of alkane activities.

Pure alkane lowers the surface tension of surfactant solutions over a wide concentration range, as can be seen from Fig. 5.20, where results for the effects of pure dodecane on the tensions of solutions of $C_{12}TAB$ are shown as a function of surfactant concentration m_s. The upper points were obtained in the absence of dodecane and the lower ones are for the surfaces saturated with the alkane. The surface concentrations Γ_s of surfactant prior to alkane addition to the surface are obtained using the approximate form of the Gibbs equation:[8]

$$-\frac{d\gamma}{d\ln m_s} = 2RT\Gamma_s \tag{5.6.1}$$

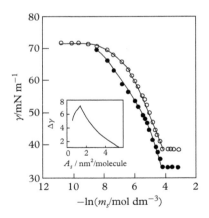

Figure 5.20 *Variation of surface tension of aqueous $C_{12}TAB$ solutions at 298 K with concentration m_s in the absence of dodecane lenses (open points) and with dodecane present (filled points). The inset shows the variation of $\Delta\gamma$ with area A_s; see text. Redrawn from Aveyard et al. (1990).*

---

[8] In writing (5.6.1) the term $(1 + \partial \ln f_{\pm}/\partial \ln m)$ which occurs in (4.3.18) has been omitted for simplicity. This gives rise to a maximum error in Γ_s up to the *cmc* of about 5%.

The inset in Fig. 5.20 shows the lowering of tension, $\Delta\gamma$, caused by alkane adsorption as a function of the area per surfactant molecule ($A_s = 1/\Gamma_s$). Since $\Delta\gamma$ is approximately proportional to the adsorption of alkane Γ_a, (as argued below), it is clear that as the surfactant film is diluted from the *cmc* alkane uptake initially increases ($\Delta\gamma$ rises), until the area per surfactant is about twice that at the *cmc*. As the surfactant monolayer is expanded further however, it becomes progressively too dilute to incorporate alkane, which adsorbs only very little on a pure water surface.

The surface concentrations of alkane, Γ_a, can also be obtained by use of the Gibbs equation, which for a system with surfactant, alkane and water (w) is

$$-\mathrm{d}\gamma = \Gamma_a\,\mathrm{d}\mu_a + \Gamma_s\,\mathrm{d}\mu_s + \Gamma_w\,\mathrm{d}\mu_w \qquad (5.6.2)$$

At and below the *cmc*, where alkane dissolution in the bulk aqueous phase is absent, addition of alkane does not change μ_s and μ_w and so

$$\Gamma_a = -(\partial\gamma/\partial\mu_a)_{\mu_w\mu_s} = (a_a/RT)\,(\mathrm{d}\Delta\gamma/\mathrm{d}a_a) \qquad (5.6.3)$$

For a fixed surfactant concentration, $\Delta\gamma$ is found to be an approximately linear function of the mole fraction x_a of dodecane in squalane which does not differ much from a_a. So, from (5.5.3), for a pure alkane lens on surfactant solution $\Gamma_a \approx \Delta\gamma/RT$.

As mentioned above, it is believed that shorter chain alkanes penetrate into surfactant monolayers more strongly than do longer chain alkanes, and that this has an important influence on the preferred curvature of close-packed surfactant monolayers at alkane/water interfaces. It is certainly true that for a given surfactant chain length, N_s, Γ_a increases as alkane chain length N_a *decreases*; also for a given alkane, adsorption is greater the *longer* the surfactant chain. This can be appreciated from the plot of Γ_a against the difference in surfactant and alkane chain lengths, $(N_s - N_a)$ in Fig. 5.21a. However, for a given alkane to swell out the monolayer chain region more than another alkane, the *volume* adsorbed must be greater. As N_a increases, although Γ_a decreases, the alkane molecular volume increases also, so it is not immediately obvious that shorter chain lengths swell the surfactant chains more than do longer alkanes. It can be shown that they do as follows, assuming the monolayer chain region is liquid-like. The volume of a CH_2 group in liquid hydrocarbon is 0.027 nm^3 and that of the CH_3 group is double this. Therefore the volume of hydrocarbon, v, adsorbed per unit area of surface is

$$v\left(\mathrm{nm}^3\mathrm{nm}^{-2}\right) = 0.027\Gamma_a\,(N_a + 2) = 6.534 \times 10^{-3}\Delta\gamma\,(N_a + 2) \qquad (5.6.4)$$

In (5.6.4) Γ_a is in molecule nm^{-2} and $\Delta\gamma$ in mN m^{-1}. Knowing v and Γ_s the volume fraction ϕ_a of alkane in the hydrocarbon part of the mixed monolayer can be calculated. The way in which ϕ_a varies with N_a in monolayers of C_{10}TAB above the *cmc* is shown in Fig. 5.21b, which indeed illustrates that, the shorter the alkane the greater the volume adsorbed.

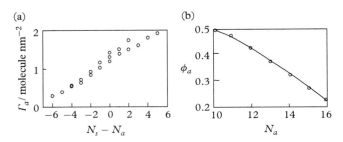

Figure 5.21 *(a) Surface concentrations of alkane against the difference $(N_s - N_a)$ in surfactant and alkane chain lengths. $N_a = 10$–16; $N_s = 10$, 12, and 16. Surfactant concentrations are respectively 81, 20, and 1.6 mmol dm^{-3}, all above the cmc. (b) Alkane volume fraction in chain region of C_{10}TAB monolayer on 81 mmol dm^{-3} surfactant solution as a function of N_a. Adapted from data in Aveyard et al. (1990).*

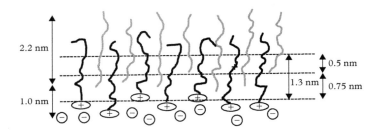

Figure 5.22 *Representation of the structure, derived from neutron reflection measurements, of a mixed monolayer of C_{14}TAB and dodecane on 2.6 mmol dm^{-3} aqueous surfactant solution. Redrawn from Lu et al. (1992).*

It is not known from investigations of the type already described where the alkanes reside within the mixed monolayer. Such information is however available from the study of neutron reflection from the mixed surfaces. The picture emerging for dodecane adsorbed into a monolayer of C_{14}TAB is shown in Fig. 5.22 from the work of Lu et al. (1992). The concentration of surfactant in the solution was 2.6 mmol dm^{-3}, which is below the *cmc* (3.7 mmol dm^{-3} in the absence of dodecane). Broadly, the findings are that the surfactant chains are a little more extended when the dodecane is present and the dodecane forms a layer which is significantly thicker (2.2 nm) than its fully extended chain length, which is 1.7 nm. There is a separation between the centres of the surfactant chain and dodecane distributions of about 0.5 nm at this surfactant concentration. The two chain components are separated from the water in the sub-phase by 0.75 nm (surfactant chains) and 1.3 nm (dodecane molecules). It is seen that the surfactant molecules are staggered vertically, presumably lowering the electrical head group repulsion in this way. It also appears that the surface of the interface between the monolayer and air phase is somewhat like that of pure dodecane.

The effects of hydrocarbon chain penetration into surfactant monolayers will be revisited in the consideration of ultralow interfacial tensions and the formation of microemulsions in Chapter 10.

Appendix A5: Further commonly used isotherms and surface equations of state

A5.1 Isotherms for localized monolayers

The adsorption isotherms and surface equations of state (also called surface tension isotherms) introduced in §5.3.2 were all appropriate for *mobile* monolayers, that is monolayers in which the surfactant molecules are free to translate laterally within the surface (Fig. A5.1a). The assumption of mobility is appropriate for monolayers at liquid/fluid interfaces considered in this chapter. For systems involving adsorption of surfactants onto solid surfaces it may be that, in some cases, the monolayer is *localized*, that is, one in which the surfactant molecules cannot freely translate along the surface because they are strongly held on adsorption sites (Fig. A5.1b). Molecules can however change sites by desorbing and re-adsorbing on different sites. The model on which the well-known *Langmuir adsorption isotherm* is based is one in which the adsorbed molecules are localized, and are assumed not to mutually interact laterally. It is essentially the localized counterpart of the Volmer model for mobile, non-interacting monolayers. The Langmuir equation was originally derived for chemisorption of gases onto solids, where it is known the chemisorbed layers are localized. Using a similar format to that used earlier in the chapter, the Langmuir adsorption isotherm can be written (cf. (5.3.19))

$$c = \left(\frac{A_o}{A - A_o} \right) \exp \left(\frac{\Delta_{ads} \mu^o_{Langmuir}}{RT} \right) \tag{A5.1}$$

The standard states to which the standard free energy of adsorption $\Delta_{ads}\mu^o_{Langmuir}$ relates are, for the bulk unit concentration c of surfactant, and for the surface half coverage, that is, $\theta = A_o/A = 0.5$. The surface equation of state which corresponds to (A5.1) is

$$\pi = \frac{kT}{A_o} \ln \left(\frac{A}{A - A_o} \right) \tag{A5.2}$$

The *Frumkin model* allows for lateral interaction between adsorbed molecules in the localized monolayer, and the resulting adsorption isotherm (A5.3) and surface equation of state (A5.4) are

$$c = K \left(\frac{A_o}{A - A_o} \right) \exp \left(\frac{xVA_o}{kTA} \right) \tag{A5.3}$$

$$\pi = \frac{kT}{A_o} \ln \left(\frac{A}{A - A_o} \right) + \frac{xVA_o}{2A^2} \tag{A5.4}$$

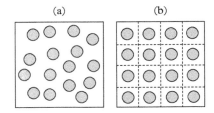

Figure A5.1 *(a) Snapshot of a mobile film. The adsorbed molecules (circles) are randomly arranged over the surface. (b) A localized film, in which the adsorbed molecules occupy regularly arranged sites and are not present between sites. In the Frumkin model (localized adsorption with lateral attractions between molecules) it is assumed, in a statistical derivation of the isotherm, that the chance of adsorption on a site next to an occupied site is the same as on a site with no nearest neighbour occupied sites.*

In these equations x is the number of nearest neighbour sites per site and V the interaction energy of a pair of molecules on nearest neighbour sites; V is negative for attraction between molecules (unlike α in (5.3.22) and (5.3.23) which is positive for attractive interactions). The constant K contains a standard free energy of adsorption. Although the Langmuir and Frumkin models are based on localization of adsorbed molecules it is shown below that isotherms with the same form can be obtained without the assumption of localization.

A5.2 Treatment of adsorbed surfactant layers as two-dimensional solutions

Up to now specific models for adsorbed monolayers have been considered. In principle, if adsorption data are well-represented by an equation of state or adsorption isotherm, something is known about the nature of the monolayer. For example, attempts to understand the nature of electrical interactions in monolayers of ionic surfactants have been described in §5.3.2. However another broad, thermodynamic approach to the representation of experimental findings has been used by various workers, in which it is recognized a surfactant monolayer is a two-dimensional *solution* that contains both surfactant (ionic or nonionic) and solvent molecules.[9] In this case, the surface pressure π is regarded as a two-dimensional osmotic pressure rather than as a two-dimensional gas pressure. As an illustration of the two-dimensional solution approach the adsorption of a single nonionic surfactant (component 2) from water (component 1) is considered.

For a component i in a bulk liquid mixture, the chemical potential of i, $\mu_i^l (= \mu_i)$, is related to (say) the mole fraction of i, x_i^l in the mixture by

$$\mu_i^l = \mu_i^{o,l} + RT \ln x_i^l f_i^l \tag{A5.5}$$

where f_i^l is the mole fraction activity coefficient of i (unity for $x_i^l = 1$). The chemical potential of i in the surface however, depends not only on its composition in the surface but also on the surface tension. This fact is expressed by the relationship, presented in §4.2,

---

[9] Mixtures of *two surfactants* were treated in §5.5.3 using the regular solution approach, but there no account was taken of the presence of solvent(s) in the adsorbed layers.

$$\mu_i = \mu_i^l = \mu_i^s = \zeta_i - \gamma\omega_i \tag{4.2.18}$$

where ω_i is the partial molar surface area of i. The quantity ζ_i contains the dependence of μ_i^s on the surface composition (e.g. expressed as mole fraction x_i^s) by

$$\zeta_i = \mu_i^{o,s} + RT\ln x_i^s f_i^s \tag{A5.6}$$

in which f_i^s is the activity coefficient of i in the surface. From (A5.6) and (4.2.18) it is apparent that

$$\mu_i^s = \mu_i^{o,s} + RT\ln x_i^s f_i^s - \gamma\omega_i \tag{5.5.10}$$

At adsorption equilibrium, from (A5.5), (4.2.18), and (5.5.10)

$$\mu_i^{o,s} + RT\ln x_i^s f_i^s - \gamma\omega_i = \mu_i^{o,l} + RT\ln x_i^l f_i^l \tag{A5.7}$$

For pure solvent (component 1) $x_1^l = x_1^s = f_1^l = f_1^s = 1$ and so from (A5.7)

$$\mu_1^{o,s} - \mu_1^{o,l} = \gamma_o\omega_1; \quad \gamma_o = \frac{\mu_1^{o,s} - \mu_1^{o,l}}{\omega_1} \tag{A5.8}$$

where γ_o is the surface tension of pure solvent and ω_1 the molar surface area of the solvent. (Note in passing that the right hand form of (A5.8) gives the definition of the surface tension of a pure liquid as the free energy of forming unit area of interface.)

Combination of (A5.7) and (A5.8), written for the solvent, gives

$$\gamma_o - \gamma = \pi = -\frac{RT}{\omega_1}\ln\left(\frac{x_1^s f_1^s}{x_1^l f_1^l}\right) \tag{A5.9}$$

which is essentially the required surface equation of state. Also,

$$x_1^s = \Gamma_1/(\Gamma_1 + \Gamma_2) \tag{A5.10}$$

At this stage it is necessary to make some assumptions or adopt some convention to relate Γ_1 and Γ_2. For example some workers have assumed that the surface phase is a monolayer and used the approximate expression

$$\Gamma_1^s\omega_1 + \Gamma_2^s\omega_2 = 1 \tag{A5.11}$$

where Γ_i^s are surface concentrations. An alternative approach (see Fainerman and Lucassen-Reynders, 2002) has been to consider the surface phase to be a Gibbs dividing surface and to adopt the convention that

$$\Gamma_1^\sigma + \Gamma_2^\sigma = \Gamma_2^\infty = 1/\omega_2 \tag{A5.12}$$

where Γ_i^σ are surface excesses and Γ_2^∞ is the limiting adsorption of the surfactant at high surfactant concentration. In this convention the partial molar surface area, ω ($= \omega_2$), of solvent and solute are the same. Combination of (A5.9), (A5.10), and (A5.12), for an ideal dilute surfactant solution $(x_1^l f_1^l \approx 1)$, yields

$$\pi = \frac{RT}{\omega} \ln \left(\frac{A}{A - \omega} \right) - RT \ln f_1^s \qquad (A5.13)$$

For a perfect surface solution $f_1^\sigma = 1$, and noting that $\omega \equiv A_0$, (A5.13) reduces to (A5.2); no assumption of localized adsorption has been made in the derivation of (A5.13).

It is simple to show, using the monolayer model and assuming that the partial molar surface areas for solvent (ω_1) and surfactant (ω_2) are different, that

$$\pi = \frac{RT}{\omega_1} \ln \left(\frac{A - \omega_2 + \omega_1}{A - \omega_2} \right) - RT \ln f_1^s \qquad (A5.14)$$

For $\omega_1 = \omega_2$ and $f_1^s = 1$, (A5.14) becomes equivalent to (A5.2).

The above derivations are for systems with a nonionic surfactant. Using the Gibbs approach, Lucassen-Reynders (see Fainerman and Lucassen-Reynders, 2002), considered the situation for ionic surfactants both with and without added inorganic electrolytes. For a 1:1 ionic surfactant for example the right hand side of (A5.13) has to be multiplied by 2, as might be expected. It also turns out that the two-dimensional solutions formed by ionic surfactants at oil/water interfaces often appear to be perfect (i.e. $f_1^s = 1$) with and without added inorganic electrolyte. Use of the Gibbs dividing surface for the surface phase avoids employing any specific model for the surface (e.g. an electrical model as in §5.3.2) since the dividing surface allows for all deviations from bulk behaviour and is overall electrically neutral. Since this is the case, all ions must adsorb as electroneutral combinations and adsorption is the sum of the surface excesses in the monolayer and the underlying electrical double layer (see Chapter 11).

Some experimental results for adsorbed layers of sodium octylsulfate (SOS) and of decyltrimethylammonium bromide (DTAB) at the decane/water interface, each in systems with

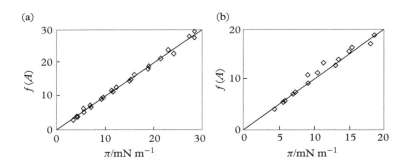

Figure A5.2 *Plots of f(A) vs π for (a) sodium octylsulfate (SOS) and (b) decyltrimethylammonium bromide (DTAB) adsorbed at the decane/water interface at 293 K. Plots include data obtained at three concentrations of NaCl (between 0.05 M and 0.25 M) for each surfactant. The ordinate f(A) = 2(RT/ω)ln(A/[A − ω]) and ω has been taken as 42 and 58 Å²/molecule for SOS and DTAB, respectively. Full straight lines each have slope of unity. Adapted from data in Haydon and Taylor (1962).*

NaCl ranging in concentration from 0.05 to 0.25 M, are depicted in Fig.A5.2. The plots are made according to (A5.13) with $f_1^s = 1$ and the factor 2 included; the right hand side of the resulting equation is designated $f(A)$ and is the ordinate in the graphs in Fig. A5.2. As seen, experimental points for all salt concentrations fall close to straight lines of slope unity up to high surface pressures, as required by (A5.13); values used for ω are, for SOS and DTAB, respectively, 0.42 and 0.58 nm^2 molecule^{-1}. The addition of salt has no effect on the ideality of the surface layer or on the surface pressure for a given value of surfactant adsorption (i.e. on the π–A curve). It appears that the effect of added salt on the adsorption from solution (as illustrated in Fig. 5.1) is a result of salting-out of the surfactant from *bulk* solution.

REFERENCES

S.W. An, J.R. Lu, R.K. Thomas, and J. Penfold, Apparent anomalies in surface excesses determined from neutron reflection and the Gibbs equation in anionic surfactants with particular reference to perfluorooctanoates at the air/water interface. *Langmuir*, 1996, **12**, 2446–2453.

R. Aveyard and D.A. Haydon, *Introduction to the Principles of Surface Chemistry*. Cambridge University Press, Cambridge, 1973.

R. Aveyard, B.P. Binks, J. Mead, and J.H. Clint, Interfacial tensions and microemulsion formation in heptane–aqueous NaCl systems containing aerosol OT and sodium dodecyl sulphate. *J. Chem. Soc., Faraday Trans. 1*, 1988, **84**, 675–686.

R. Aveyard, P. Cooper, and P.D I. Fletcher, Solubilisation of hydrocarbons in surfactant monolayers. *J. Chem. Soc., Faraday Trans.*, 1990, **86**, 3623–3629.

R. Aveyard, B.P. Binks, J. Chen, J. Esquena, P.D.I. Fletcher, R. Buscall, and S. Davies, Surface and colloid chemistry of systems containing pure sugar surfactant. *Langmuir*, 1998, **14**, 4699–4709.

K.D. Danov, S.D. Kralchevska, P.A. Kralchevsky, K.P. Ananthapadmanabhan, and A. Lips, Mixed solutions of anionic and zwitterionic surfactant (betaine): surface-tension isotherms, adsorption, and relaxation kinetics. *Langmuir*, 2004, **20**, 5445–5453.

J.H. de Boer, *The Dynamical Character of Adsorption*. Oxford University Press, Oxford, 1953, p.173.

K. Denbigh, *The Principles of Chemical Equilibrium*. Cambridge University Press, Cambridge, 1961, Chapter 14.

V.B. Fainerman and E.H. Lucassen-Reynders, Adsorption of single and mixed ionic surfactants at fluid interfaces. *Adv. Colloid Interface Sci.*, 2002, **96**, 295–323.

D.A. Haydon and F.H. Taylor, Adsorption of sodium octyl and decyl sulphates and octyl and decyl trimethylammonium bromides at the decane–water interface. *Trans. Faraday Soc.*, 1962, **58**, 1233–1250.

P.A. Kralchevsky, K.D. Danov, G. Broze, and A Mehreteab, Thermodynamics of ionic surfactant adsorption with account for the counterion binding: effect of salts of various valency. *Langmuir*, 1999, **15**, 2351–2365.

P.A. Kralchevsky, K.D. Danov, V.L. Kolev, G. Broze, and A. Mehreteab, Effect of nonionic admixtures on the adsorption of ionic surfactants at fluid interfaces. 1. Sodium dodecyl sulfate and dodecanol. *Langmuir*, 2003, **19**, 5004–5018.

J.R. Lu, R.K. Thomas, R. Aveyard, B.P. Binks, P. Cooper, P.D.I. Fletcher, A. Sokolowsky, and J. Penfold, Structure and composition of dodecane layers spread on aqueous solutions of tetradecyltrimethylammonium bromide: neutron reflection and surface tension measurements. *J. Phys Chem.*, 1992, **96**, 10971–10978.

T.J. Morrison and F. Billett, The salting-out of non-electrolytes. Part II. The effect of variation in non-electrolyte. *J. Chem. Soc.*, 1952, 3819–3822.

J. Penfold, E. Staples, L. Thompson, and I. Tucker, The composition of non-ionic surfactant mixtures at the air/water interface as determined by neutron reflectivity. *Colloids Surf. A*, 1995, **102**, 127–132.

M.J. Rosen and J.T. Kunjappu, *Surfactants and Interfacial Phenomena*, 4th edition. Wiley, New York, 2012.

A.P. dos Santos, A. Diehl, and Y. Levin, Surface tensions, surface potentials, and the Hofmeister series of electrolyte solutions. *Langmuir* 2010, **26**, 10778–10783.

K. Tajima, H. Murata, and T. Tsutsui, Direct measurement of adsorbed amount of sodium tetradecyl-2,3-^{3}H,^{3}H-sulfate at the oil–water interface by radiotracer method. *J. Colloid Interface Sci.*, 1982, **85**, 534–539.

C.D. Wick and L.X. Dang, Simulated surface potentials at the vapor–water interface for the KCl aqueous electrolyte solution. *J. Chem. Phys.*, 2006, **125**, 024706.

6

Dynamic aspects of liquid interfaces

6.1 Dynamic interfacial tension

In the discussion so far, systems at adsorption equilibrium have been considered. In very many contexts however (e.g. in industry and biology) systems are not at equilibrium. For example in high speed liquid coating processes new interface is created rapidly, as it is when liquid is sprayed from a nozzle in the formation of an aerosol. In systems which contain surfactants, a finite time elapses after interface formation before the surfactant can adsorb and reach adsorption equilibrium, during which the interfacial tension falls. When droplets impinge upon a surface (as for example in spray painting) and subsequently coalesce, the surface is rapidly compressed and in this case surfactant must desorb in order for the surface to attain equilibrium, and during this time the interfacial tension will rise. It is clear therefore that in many instances knowledge of how interfacial tensions change with time is sought, that is, *dynamic* tensions are required. As will be seen, knowledge of the variation of dynamic interfacial tension with time provides the means to (i) understand dynamic aspects of the adsorption process and (ii) probe the interfacial rheology of surfactant solutions. Timescales of interest range from a few milliseconds (for attainment of the equilibrium interfacial tension of a pure liquid/vapour interface) through seconds (for systems with conventional small surfactant molecules) up to hours or even days (for systems with large molar mass polymers or highly surface active trace impurities including polyvalent ions).

A useful way to illustrate dynamic effects is to consider a surfactant solution in a container with a moveable barrier in the interface, as depicted in Fig. 6.1. If the barrier is moved rapidly to the left as shown, the surfactant monolayer on the left is compressed and desorption of surfactant follows. At the same time the surface on the right is expanded and adsorption from bulk solution subsequently occurs. The (dynamic) tension of the interface on the left, γ_1, is lower than the equilibrium tension γ_e, and that on the right (γ_2) is higher than γ_e. The changes of γ_1 and γ_2 with time after the movement of the barrier are shown schematically in Fig. 6.2.

Surfactants: In Solution, at Interfaces and in Colloidal Dispersions. Bob Aveyard. © Bob Aveyard 2019.
Published in 2019 by Oxford University Press. DOI: 10.1093/oso/9780198828600.001.0001

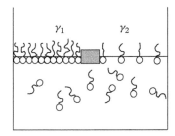

Figure 6.1 *Compression and expansion of (say) liquid/vapour interfaces containing soluble surfactant.*

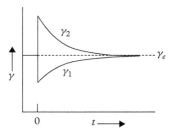

Figure 6.2 *Schematic representation of changes in interfacial tensions γ with time t. Rapid movement of the central barrier in Fig. 6.1 occurs at $t = 0$.*

6.2 Dynamic interfacial tensions impart viscoelastic properties to interfaces: interfacial rheology

Liquid interfaces with monolayers of surfactants present exhibit viscoelastic behaviour. The nature of this behaviour depends somewhat on the system, for example, whether or not the surfactant is soluble in the bulk phases or whether adsorption equilibrium exists. Broadly, an interface can be *sheared* (changing its shape) or *dilated* (changing its area). It is possible then to define and measure interfacial shear and dilational *moduli* and their related viscosities.[1] It is important to realize that, when a surfactant monolayer is perturbed in some way, the solvent in adjacent layers is coupled with the surfactant and the 'interface' consists of a thin layer of surfactant together with neighbouring solvent(s), and this is referred to later.

Suppose a force is applied to deform an interface in some way. If, when the applied force is removed, the interface reverts to its original state it is said to exhibit a purely *elastic* response. On the other hand, some of the energy gain of the interface may be dissipated, in which case the interface exhibits interfacial *viscosity* in addition to elasticity, that is, the response is *viscoelastic*. As will be seen, insoluble monolayers may give a purely elastic

---

[1] A *modulus* is a coefficient associated with a particular physical property of a material, such as elasticity (see also footnote 3, Chapter 15).

(dilational) response, but in general the surfaces of surfactant solutions show viscoelastic behaviour.

6.2.1   Interfacial shear properties

Interfacial shear rheology describes the resistance that a fluid interface offers to a shear force within the interface. Imagine a square area of interface, with sides of length l as illustrated in Fig. 6.3a. Suppose for the moment that the interface is an elastic two-dimensional solid and that a force F is applied along one of the edges in the x-direction (the opposite edge remaining anchored) causing a deformation defined by the angle α. The interfacial (elastic, in this case) *shear modulus*, G, is defined as the ratio of *shear stress* (τ) to *shear strain* ($\boldsymbol{\gamma}$),[2] that is,

$$G = \tau/\boldsymbol{\gamma} = (F/l) / \tan\alpha = (F/l) / (\Delta l/l) = F/\Delta l \qquad (6.2.1)$$

For a solid obeying Hooke's law G remains constant, irrespective of the stress or strain applied since stress and strain are linearly related.

For a liquid interface G is not constant. If a constant velocity (stress) is applied to one edge of the interface, a velocity gradient (denoted by the arrows in Fig. 6.3b) is set up within the interface in a direction normal to that edge. The strain increases with time, but if the change in velocity v with distance normal to the edge (y-direction) is linear (i.e. Newtonian behaviour, as shown in Fig. 6.3b) the *rate of change* of the strain with respect to time is constant. This rate is called the *shear rate* or *strain rate* and is denoted $\dot{\boldsymbol{\gamma}}$. It is equal to the rate of change of velocity v with y. The ratio of shear stress to shear strain rate is *the shear viscosity, η_s* and so

$$\eta_s = \tau/\dot{\boldsymbol{\gamma}} = (F/l) / (\mathrm{d}v/\mathrm{d}y) \qquad (6.2.2)$$

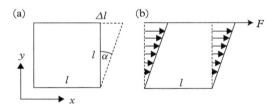

Figure 6.3 *(a) Shear deformation of a square area of interface, side l. The shear strain is tan α. (b) A force F acts on a line length l causing a velocity gradient normal to the line within the interface.*

---

[2] The lower case Greek symbol gamma is commonly used to denote strain in rheology; unfortunately γ is also widely used (as here) to denote interfacial tension. To distinguish the two quantities, a bold non-italic symbol, $\boldsymbol{\gamma}$, is used for strain, rather than the italic, non-bold γ used for interfacial tension.

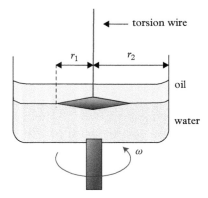

Figure 6.4 *Biconical bob shear rheometer.*

A common way to investigate interfacial shear viscosity of soluble surfactant layers is to use a biconical bob suspended by a torsion wire, as illustrated in Fig. 6.4 for an oil/water interface. The bicone, radius r_1, is suspended centrally in a circular dish, radius r_2, with the edge of the cone located at the oil/water (or air/water) interface. The dish, resting centrally on a rotating arm, is rotated at a constant angular velocity, ω, and this causes a deflection (angle θ) of the bob. A blank experiment is also performed with the same solvents but without the surfactant layer present; this gives a deflection of θ', resulting from the effects on the bob of the solvents alone. The interfacial shear viscosity η_s can then be calculated, for Newtonian flow within the interface (i.e. when shear stress is related linearly to shear strain rate), using

$$\eta_s = \frac{K\left(\theta - \theta'\right)}{4\pi\omega}\left(\frac{r_2^2 - r_1^2}{r_1^2 r_2^2}\right) \tag{6.2.3}$$

in which K is the torsional constant of the wire suspending the bob and π is the mathematical constant.[3] The technique is sufficiently sensitive for the determination of low interfacial viscosities and is also used for the measurement of high viscosities (say of films of polymeric surfactant). In the case of non-Newtonian films, the shear strain rate is not constant across the surface and so the viscosity obtained by the use of (6.2.3) is an apparent viscosity only, and relates to some average strain rate between the edge of the bob and the dish wall. The method can also be used in an oscillatory mode to yield both storage and loss moduli of the interface. This will not be covered here, but it will be shown below how oscillatory measurements in dilational rheology can yield information about relaxation processes in surfactant films.

---

[3] The symbol π is also used in this book for surface pressure; it will be obvious in a given context to which quantity π refers.

6.2.2   Interfacial dilational properties

If the area $\mathcal{A}$ of the interface of a surfactant solution is suddenly increased or decreased there will be a change in the interfacial tension, which will then revert over time to the equilibrium value, as discussed earlier. An interfacial *dilational modulus*, ε^*, is defined as the change of interfacial tension (stress) with relative strain $(d\varepsilon/\varepsilon)$:

$$e^* = \mathcal{A}d\gamma/d\mathcal{A} = d\gamma/d\ln\mathcal{A} = -\mathcal{A}d\pi/d\mathcal{A} \tag{6.2.4}$$

where π is the surface pressure (lowering of interfacial tension). The units of ε^* are seen to be those of interfacial tension, that is, force length^{-1}. The corresponding interfacial *dilational viscosity*, η_d, is the ratio of the increase in tension, $\Delta\gamma$ (i.e. the stress), caused by a finite change in area, to the *rate of change* (with respect to time, t) of the relative surface strain, that is,

$$\eta_d = \Delta\gamma/(d\ln\mathcal{A}/dt) \tag{6.2.5}$$

If the interfacial area of a bulk surfactant solution is altered at *equilibrium* (i.e. sufficiently slowly) the interfacial tension remains unchanged, and ε^* is zero. If however the area of an *insoluble* surfactant layer is altered (assuming no subsequent collapse or molecular rearrangements), extension of the interface gives an increase in interfacial tension and a decrease in the surface pressure and so ε^* (which in this case is simply the *elastic* modulus) is now positive. For such a system, following a method of Maxwell, the elastic modulus can be obtained in a very simple geometric way, as pointed out by Ross and Morrison (1988). With reference to the schematic surface pressure–surface area $(\pi–A)$ isotherm in Fig. 6.5 (where A is the area per molecule) a tangent is drawn at any point P and extended to the π-axis, meeting it at Y; a horizontal line PX is also drawn as shown. The quantity $d\pi/dA$ in (6.2.4) is equivalent to XY/XP and A is equivalent to XP so that ε^* at point P is simply represented by the length XY.

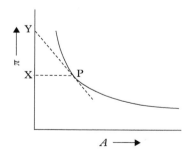

Figure 6.5 *Schematic π–A isotherm for an insoluble monolayer; A is area per molecule.*

It is clear that for an insoluble monolayer ε^* is determined completely by the equation of state, since $d\pi/dA$ is given by the equation of state. For example, if the π–A data for an insoluble monolayer are described by the two-dimensional van der Waals equation (5.3.22) then

$$\varepsilon^* = kT \left\{ \frac{A}{(A - A_o)^2} - \frac{2\alpha}{kTA^2} \right\} = kT\Gamma \left\{ \frac{\Gamma_\infty}{(\Gamma_\infty - \Gamma)^2} - \frac{2\alpha\Gamma}{kT} \right\} \tag{6.2.6}$$

The quantity $\Gamma_\infty = 1/A_o$ is the value of Γ in the close-packed monolayer.

If the surface area of a solution of *soluble* surfactant is rapidly expanded (say by extruding from a jet or by atomization), the interfacial tension will initially increase and then relax back to its equilibrium value over a period of time (see Fig. 6.2). This produces an interfacial dilational elasticity since the increase in interfacial tension tends to oppose the expansion. This tendency however is reduced as surfactant adsorbs to the newly formed interface from the bulk solution, giving rise to interfacial dilational viscosity. When the area of an interface is varied in an oscillatory fashion, the dilational viscoelastic modulus is a function of the frequency of oscillation, ν ($= \omega/2\pi$, where ω is the angular frequency in rad s^{-1}).

One technique for the study of dilational behaviour of soluble surfactant monolayers employs the Langmuir trough (Box 6.1). The trough consists of a dish, with water-repellent edges, which is filled so as to give a convex meniscus at the edges. Two hydrophobic barriers rest in the surface, one (or possibly both) of which is moved in an oscillatory fashion (with frequency ν), and the interfacial tension is measured between the barriers (e.g. by a Wilhelmy plate—§A3.1) as shown in Fig. 6.6. When the interface is compressed from an initial area A (by ΔA) the tension falls relative to the initial tension γ (by $\Delta\gamma$), and on expansion of the interface the tension rises. As a result of the finite time required for surfactant desorption and adsorption, there is a lag between area change and tension change, and in order to determine the dilational interfacial elasticity (the in-phase component) and viscosity (out-of-phase component), the phase difference, φ, between ΔA and $\Delta\gamma$ must be known.

Box 6.1 Insoluble monolayers and the Langmuir trough

Surfactants are usually soluble in water and, below the *cmc*, an increase in bulk concentration gives an increase in interfacial concentration and a reduction in the value of A. If however the chain length of a surfactant is sufficiently large the surfactant becomes effectively water-insoluble so it cannot be adsorbed from an aqueous solution. It can nonetheless be spread on the surface of water from a suitable solvent and the π–A characteristics studied on a Langmuir trough (see Fig. 6.6 and §17.4.1). Many amphiphilic materials, such as long alkyl chain alkanols and alkanoic acids, which are not surfactants, form insoluble monolayers.

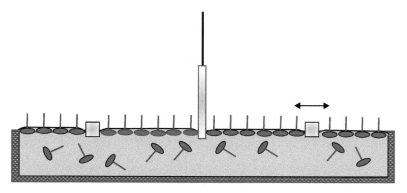

Figure 6.6 *Langmuir trough for making oscillatory dilational measurements on a soluble surfactant film. One barrier oscillates and the surface (interfacial) tension is measured using a Wilhelmy plate.*

Suppose the (small) periodic area and tension changes are sinusoidal.[4] The quantities $\Delta\mathcal{A}$ and $\Delta\gamma$ can be plotted on the same graph as a function of angle and the phase difference (φ) observed as illustrated in Fig. 6.7a. For such a case a plot of $\Delta\gamma$ versus $\Delta\mathcal{A}$ is an ellipse, as seen in Fig. 6.7b irrespective of the relative amplitudes of $\Delta\mathcal{A}$ and $\Delta\gamma$. With reference to the ellipse, the phase shift φ is simply determined from the lengths of the lines AB and CD, which are obtained from the points where the diagonals of the box containing the ellipse intersect the ellipse:

$$\varphi = 2\tan^{-1}\left(\frac{AB}{CD}\right) \tag{6.2.7}$$

The complex dilational viscoelastic modulus, ε^*, has real and imaginary parts and is expressed

$$\varepsilon^* = \varepsilon' + i\varepsilon'' = |\varepsilon|\cos\varphi + i|\varepsilon|\sin\varphi \tag{6.2.8}$$

where the absolute value of ε^*, $|\varepsilon|$, is equal to $\mathcal{A}\Delta\gamma_{max}/\Delta\mathcal{A}_{max}$, the subscript $_{max}$ denoting the maximum value of the change. The storage (elastic) modulus, ε', is the real part of ε^*, that is, $|\varepsilon|\cos\varphi$. The loss modulus, $\varepsilon'' = |\varepsilon|\sin\varphi$ is the imaginary part of ε^* and is equal to $\omega\eta_d$ where η_d is the interfacial dilational viscosity. As mentioned, ω is the angular

---

[4] It is supposed that, for small amplitude perturbations the interfacial dilational elasticity and viscosity are independent of the strain and stress involved in their measurement. Such a regime of stress and strain is said to be linear. Further, it is assumed that the surface concentration at any instant is uniform over the surface and that the surfactant molecules are at local equilibrium. This means that the time scale for any conformational change (e.g. of a polymeric surfactant) in the interface is much smaller than that for diffusion of molecules to and from the bulk solution.

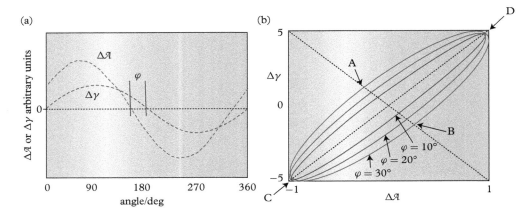

Figure 6.7 *(a) Sinusoidal changes in surface area and interfacial tension as a function of angle (see text); the phase shift φ illustrated is 30°. (b) Changes in tension plotted against changes in interfacial area. Ellipses are given and the phase shift is obtained from AB/CD as discussed in the text.*

frequency of the harmonic oscillation of the surface, and both storage and loss moduli are frequency dependent, as well as depending on the surface concentration of surfactant.

To give an idea of the kind of data that can be obtained using oscillatory experiments, work of Lucassen and Giles (1975) is referred to here. They studied the dynamic interfacial properties, at the air/solution interface, of pure single ethoxylated nonionic surfactants of the type C_nE_m (where C_n denotes C_nH_{2n+1} and E_m represents $(OCH_2CH_2)_m$). Some results are illustrated in Fig. 6.8 for $C_{12}E_6$. For this simple low molar mass nonionic surfactant the viscoelastic modulus $|\varepsilon|$ below the critical micelle concentration (*cmc*) is of the order of 10 mN m^{-1} and it increases with increasing frequency. The results below the *cmc* could be accounted for entirely on the basis of diffusional exchange of surfactant between bulk and surface. Diffusional exchange and dynamic surface tension are discussed below.

An alternative to varying frequency experimentally is to extract the frequency dependence of ε' and ε'' from a single relaxation curve of interfacial tension versus time t (such as that in Fig. 6.2) following a rapid expansion of the surface. Changes in tension can be considered either in the frequency domain or the time domain. Here ε' and ε'' are obtained as a function of frequency v ($= \omega/2\pi$) starting from the single trace of tension $\gamma(t)$ versus time t.

The Fourier (Box 6.2) transform is employed to change the continuous time signal into the required frequency domain, and it turns out that, assuming the time interval for producing the change in area is much smaller than the time of subsequent relaxation,

$$\varepsilon' = \frac{\omega \mathcal{A}}{\Delta \mathcal{A}} \int_0^\infty \Delta \gamma(t) \sin \omega t \, dt$$

$$\varepsilon'' = \frac{\omega \mathcal{A}}{\Delta \mathcal{A}} \int_0^\infty \Delta \gamma(t) \cos \omega t \, dt$$

(6.2.9)

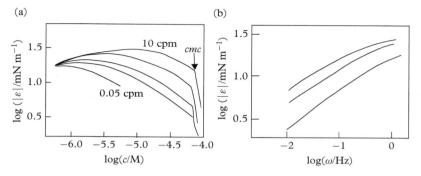

Figure 6.8 *(a) Viscoelastic modulus at 298 K as a function of c, measured at various frequencies ranging from 10 cycles per minute (cpm) down to 0.05 cpm, for $C_{12}E_6$ adsorbed at the air/solution interface. The concentrations studied span the cmc. (b) The modulus for the same system plotted against frequency for 3 different concentrations: 10^{-5} M (top curve); 2×10^{-5} M; 5×10^{-5} M (bottom curve). Curves for some frequencies in (a) omitted for clarity. Redrawn from Lucassen and Giles (1975).*

Box 6.2 Jean Baptiste Joseph Fourier, 1768–1830

Jean Baptiste Joseph Fourier, the French mathematician, originally trained for the priesthood but in the event did not take up holy orders. He studied under Lagrange and Laplace, and was involved in politics during the French Revolution, joining a local revolutionary committee. He ended up in prison, but fortunately he was later freed. Subsequently he became a scientific advisor in Napoleon's army in Egypt and was a friend of Napoleon. The Fourier transform is used very widely in the physical sciences and engineering. It converts a signal expressed as intensity vs time (e.g. γ vs t) into a spectrum of amplitude versus frequency (e.g. ε^* versus ω). The first two published articles of William Thomson (later to become Lord Kelvin—see §3.4.2), which appeared when he was 16 and 17 years old, constituted a defence of Fourier's work, which was being attacked by British scientists at the time. Thomson was the first to promote the idea that Fourier's mathematics, although applied by Fourier to the flow of heat, could be applied in the study of flow of other forms of energy, for example, electrical flow in a wire.

This method for determination of ε' and ε'' as a function of frequency was introduced by Loglio et al. (1979), who carried out experiments on aqueous solutions of dimethyldo-decylphosphine oxide ($DC_{12}PO$) below the *cmc*. In general, for systems with a single low molar mass surfactant, plots of ε' and ε'' against (log of) frequency have the appearance shown schematically in Fig. 6.9a. The loss modulus ε'' passes through a maximum while the elastic modulus only rises, reaching a plateau value at high frequency. The plot of ε' against frequency has an inflection point where it crosses the ε'' curve at its maximum; the frequency at this point is the characteristic frequency (ω_c) for the relaxation process; the characteristic relaxation time, $\tau_c = 1/\omega_c$.

(a)   (b)

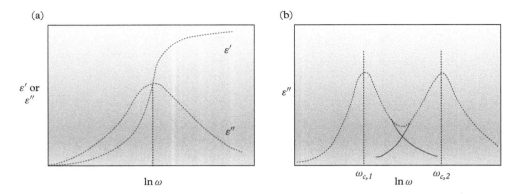

Figure 6.9 *(a) Variation of ε' and ε'' with frequency for a pure low molar mass surfactant film. (b) The loss modulus as a function of frequency for a system with two relaxation processes.*

Various relaxation processes are possible after expansion of an interface. Diffusion of surfactant will occur from the bulk phase(s), and this is likely to be the dominant process in the case of nonionic systems with a single low molar mass surfactant. For higher molar mass materials such as proteins, and polymers generally, molecular rearrangements can take place in the monolayer and in some cases collapse of the monolayer may occur. In impure systems there may be many relaxation processes occurring in overlapping regions of frequency, in which case the ε'' curve becomes broad and cannot be resolved readily into the constituent curves. In Fig. 6.9b a simple case is illustrated where there are only 2 relaxation processes with widely separated characteristic frequencies.

To indicate the magnitude of some of the quantities referred to, some results for dimethyldodecylphosphine oxide obtained by Loglio et al. (1979) are shown in Table 6.1. The characteristic frequencies, ω_c, for the two sub-micellar solution concentrations, c, represented are of the order of 10^{-3} Hz. The quantity ω_c is related to the diffusion coefficient, D, of the surfactant. The quantity ε_o is the limiting (plateau) value of the elastic modulus ε'' at high frequency. In this regime, there is insufficient time for surfactant to adsorb or desorb during the compression and dilation cycles and so the monolayer behaves as if it were insoluble. With reference to (6.2.4) it follows that (since for an insoluble film $A \propto \mathcal{A}$, and $\Gamma = 1/A$)

$$\varepsilon_o = d\gamma/d \ln A = -d\gamma/d \ln \Gamma \qquad (6.2.10)$$

The elastic modulus ε_o is sometimes referred to as the Gibbs elasticity, although it is not equivalent to the quantity defined by Gibbs (see §12.4).

Marangoni effects. When part of the surface of a surfactant solution is expanded rapidly, and the surface concentration of surfactant in that area is reduced, there is in addition to adsorption from the bulk, a flow of surfactant *along* the surface tending to equalize the surface concentrations. Liquid just below the surface is dragged along with surfactant and this movement of liquid is termed *Marangoni flow* (Box 6.3) As will be seen, it is of great importance in conferring stability on thin liquid soap films such as those making up foams (Chapter 14).

Table 6.1 *Some surface rheological quantities for the air/solution interface of dimethyldodecylphosphine oxide below the cmc at 293 K (results of Loglio et al., 1979).*

c/M	$\varepsilon''_{max}/mN\ m^{-1}$	ω_c/Hz	$\eta_d/mN\ m^{-1}\ s^*$	$\varepsilon_0/mN\ m^{-1}$
2×10^{-6}	~6	1.5×10^{-3}	4×10^3	15.7
4×10^{-6}	~9	6×10^{-3}	1.5×10^3	27.8

*At $\omega = \omega_c$

Box 6.3 Carlo Marangoni, 1840–1925

Carlo Marangoni, an Italian physicist, studied at the ancient university of Pavia in Italy, graduating in 1865 with his dissertation entitled 'On the spreading of liquid droplets', which was concerned with phenomena in liquids resulting from the action of surface tension. He moved to Florence in 1868, and in 1869 joined the Liceo Classico Dante (a high school institute) and worked there as a teacher and researcher for over 40 years. In Italy at that time a great deal of high quality scientific research was conducted in the laboratories of such institutes, which were comparable to those in universities. Phenomena arising from interfacial tension gradients are collectively referred to as Marangoni effects; Marangoni published a series of papers in this area between 1871 and 1878.

Figure 6.10 *Teardrops formed on the side of a sherry glass.*

Marangoni flow is responsible for the familiar phenomenon of formation of 'teardrops' on the sides of a glass containing, say, sherry (Fig. 6.10), as described by James Thomson, the elder brother of William Thomson (Lord Kelvin). Ethanol is surface active and is more volatile than water. The drink wets and rises up the sides of the glass, and ethanol preferentially evaporates from the liquid film covering the glass. This causes local increases in surface tension and ethanol diffuses along the surface to equalize the surface concentration. This movement causes a migration of liquid beneath the surface (Marangoni flow) and results in the formation of the 'tear' drops, which run down the glass.

6.3   The variation of tension with time indicates what factors influence the adsorption process: kinetics of adsorption

Obviously, adsorption of surfactant from bulk solution to an interface takes a finite time. The question addressed here is what processes can influence the rate of adsorption from bulk solution. Surfactants can have a low or a high molar mass, and be ionic or nonionic. Further, the solution concentration can be below or above the *cmc*. Low molar mass nonionic surfactants at concentrations below the *cmc* present the simplest case and it has already been mentioned that (some) results can be accounted for entirely on the basis of diffusional exchange of surfactant between bulk and surface. If the surfactant is present in micellar form also, then micellar diffusion and the kinetics of exchange of molecules between micelles and solution must be accounted for. With ionic surfactants that are (at least partly) dissociated at the surface, one might expect that the charge presents a barrier to further adsorption. With polymer surfactants there may also be relaxation processes occurring in the region of the interface after adsorption has occurred that can affect the interfacial tension.

6.3.1   Adsorption below the *cmc*

A central equation in the treatment of diffusion controlled adsorption to a plane interface of constant area was derived by Ward and Tordai in 1946. The relevant model is illustrated in Fig. 6.11. Surfactant diffuses from bulk solution (where its concentration is c_e) to the subsurface (concentration c_s), which is taken to be a plane. Adsorption then occurs rapidly to the surface so that the rate determining step is surfactant diffusion from bulk to subsurface. Initially the surface is empty but as the surface concentration rises, the back diffusion of surfactant to the bulk must be accounted for, that is,

$$d\Gamma(t)/dt = j_{ads} - j_{des} \tag{6.3.1}$$

where the j are fluxes of adsorption and desorption, as subscripted. On this basis, Ward and Tordai (1946) arrived at the equation for the change of the surface concentration, $\Gamma(t)$, with time t:

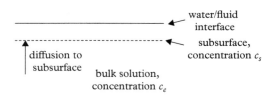

Figure 6.11 *Monomer surfactant molecules diffuse to the subsurface. Once in the subsurface adsorption can either be very rapid, in which case the overall adsorption is diffusion-controlled. Alternatively, there may be a barrier to adsorption to the surface from the subsurface.*

$$\Gamma(t) = \Gamma(0) + 2c_e\sqrt{\frac{Dt}{\pi}} - 2\sqrt{\frac{D}{\pi}}\int_0^{\sqrt{t}} c_s\mathrm{d}\left(\sqrt{t-\xi}\right) \qquad (6.3.2)$$

in which D is the diffusion coefficient of the surfactant in bulk solution and ξ is a dummy variable. The term containing the integral on the right hand side of (6.3.2) accounts for surfactant desorption. It was supposed that the adsorption at $t = 0$, $\Gamma(0)$, is zero but this assumption is not necessary (Horozov and Arnaudov, 2000). The regimes normally discussed are those for short time and long time, for which (6.3.2) has simple, useful asymptotic forms which do not contain the integral (see e.g. Fainerman et al., 1994).

As t approaches zero, desorption can be neglected and so (6.3.2) becomes

$$\Gamma(t) \cong \Gamma(0) + 2c_e\sqrt{\frac{Dt}{\pi}} \qquad (6.3.3)$$

A plot of $\Gamma(t)$, for a given c_e, against $\sqrt{t}$ gives $\Gamma(0)$ as the intercept on the ordinate and D can be obtained from the slope. This equation can be converted so as to express the change in tension with time if $\Gamma(t)$ can be related to $\gamma(t)$; for this an appropriate surface equation of state is required. If $\Gamma(0) = 0$, and at sufficiently short times it may be that the adsorbed film behaves ideally, in which case the surface pressure $(\gamma_o - \gamma)$ is given by (see §5.3.2)

$$\gamma_o - \gamma = z\Gamma RT \text{ per mol} \qquad (6.3.4)$$

where γ_o is the tension of the interface in the absence of surfactant, and, in the present context γ is equal to $\gamma(t)$. For a nonionic surfactant z in (6.3.4) is 1, and for a *fully ionized* 1,1 electrolyte ionic surfactant $z = 2$ (but see §9.6). Previously surfaces at equilibrium were considered when introducing surface equations of state; here, although dynamic tensions are being considered, it is assumed that *local* equilibrium exists between surface and subsurface so that (6.3.4) is applicable. Combination of (6.3.3) and (6.3.4) yields, for a nonionic surfactant,

$$\gamma(t) = \gamma_o - 2RTc_e\sqrt{\frac{Dt}{\pi}} \qquad (6.3.5)$$

The (linear) plot of $\gamma(t)$ against $\sqrt{t}$ has an intercept on the ordinate of γ_o and D is obtained from the slope. The validity of the above treatment depends on the validity of using the ideal two-dimensional gas equation of state (6.3.4) to describe the monolayer behaviour. Simple nonionic surface active materials such as straight chain alkanols only exhibit a linear relationship between c_e and surface pressure up to surface pressures of around 2 mN m^{-1} at the oil/water interface (Fig. 5.6). The linear region is likely to be even smaller for films at the air/water surface, as a result of lateral intermolecular attractions between the adsorbed alkyl chains. More realistic equations of state can be used, which for mobile monolayers at liquid interfaces could be the Volmer and van der

Waals equations (see §5.3.2) or the equation of state corresponding to the Langmuir isotherm (based on a two-dimensional solution model) (A5.2).

For long times ($t \rightarrow \infty$) it is reasonable to suppose that c_s is approximately constant so that it may be taken from inside the integral in (6.3.2) making the integration simple. Noting the Gibbs equation for the equilibrium adsorption Γ_e,

$$\Gamma_e = -\frac{1}{RT}\frac{d\gamma_e}{d\ln c_e} \tag{4.3.6}$$

Horozov and Arnaudov (2000) obtained, from (4.3.6) and (6.3.2), the long-time asymptote

$$\gamma(t) = \gamma_e + \left(1 - \frac{\Gamma(0)}{\Gamma_e}\right)\frac{RT\Gamma_e^2}{2c_e}\sqrt{\frac{\pi}{Dt}} \tag{6.3.6}$$

The linear plot of $\gamma(t)$ against $1/\sqrt{t}$ yields γ_e from the intercept for the particular value of c_e, and D is obtained from the slope, assuming Γ_e and $\Gamma(0)$ are available. The value for $\Gamma(0)$ can be obtained by the use of (6.3.3) with data obtained in the short time regime. Finite values of $\Gamma(0)$ lead to a reduction in the slope of the plot according to (6.3.6); the greater $\Gamma(0)$ the lower the slope. If $\Gamma(0)$ is not zero but is disregarded, the calculated diffusion coefficient will be overestimated.

By way of illustration of the treatment of dynamic tension data, some dynamic surface tensions of aqueous solutions ($c_e = 5 \times 10^{-4}$ mol dm^{-3}) of the di-chain nonionic glucamide surfactant (C_6H_{13})$_2$C[CH$_2$NHCO(CHOH)$_4$CH$_2$OH]$_2$ (di-(C$_6$-Glu)), obtained by Eastoe et al. (1998), are shown in Fig. 6.12. Tensions are presented as a function of time in Fig. 6.12a for a range of temperatures between 10 and 50°C, and in Fig. 6.12b the same data are plotted according to (6.3.6) (Eastoe et al. implicitly assumed $\Gamma(0) = 0$). The *cmc* of this surfactant, 1.3×10^{-3} mol dm^{-3}, is effectively constant over the temperature range studied.

The straight lines shown in the plots of tension versus $t^{-1/2}$ are least squares fits of the data for $t^{-1/2} < 2$. The intercepts on the ordinate give the equilibrium tensions of the solutions. If the adsorption is purely diffusion controlled then (in the present case with $\Gamma(0) = 0$) the diffusion coefficient of the surfactant molecule can be obtained from the gradient (*grad*) using

$$D = D_{eff} = \left(\frac{RT\Gamma_e^2\pi^{0.5}}{grad\ 2c_e}\right)^2 \tag{6.3.7}$$

If the adsorption is not purely diffusion controlled then the diffusion coefficient obtained from (6.3.7) is an effective diffusion coefficient, D_{eff}, which is discussed below. The true diffusion coefficient, D, can of course be obtained independently, and for this surfactant D at 25°C, obtained using NMR, is 2.7×10^{-10} m^2 s^{-1}. From this, approximate values of D at other temperatures can be estimated using the Stokes-Einstein equation

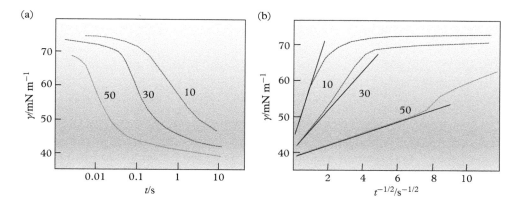

Figure 6.12 *(a) Dynamic surface tensions of aqueous di-(C$_6$-glu) at various temperatures (°C) as indicated. (b) Same data plotted according to (6.3.6); full straight lines are drawn through the points for long times. Individual data points omitted for clarity. Redrawn from Eastoe et al. (1998).*

$$D \approx \frac{kT}{6\pi \eta r} \tag{6.3.8}$$

in which η and r are, respectively, the solvent viscosity and molecular radius of the diffusing surfactant molecule (assumed to be spherical).

Inspection of Fig. 6.12b shows that the plot for 50°C is linear over a much wider time range than for the other plots at lower temperatures. The diffusion coefficient, D_{eff}, obtained from (6.3.7) is 2.3×10^{-10} m^2 s^{-1} at 50°C, as compared to 2.9×10^{-10} m^2 s^{-1} obtained using the Stokes–Einstein equation and the true D at 25°C. Essentially, adsorption at 50°C appears to be diffusion controlled. For the lower temperatures however, the discrepancies between D and D_{eff} become progressively larger as the temperature falls, indicating the existence of an activation barrier to adsorption. Values of D and D_{eff} are given in Table 6.2. and, assuming an Arrhenius type of relationship exists between them, one may write

$$D_{eff} = D \exp\left(-E_a/RT\right) \tag{6.3.9}$$

and an activation barrier E_a can be calculated. Application of (6.3.9) to the data in Table 6.2 gives $E_a = 78$ kJ mol^{-1}.

For this mixed 'diffusion–activation' adsorption, the energy barrier can arise in various ways, depending on the surface active material being adsorbed. It was suggested in the case of the glucamide sugar surfactant that the barrier could arise since the surfactant must be favourably oriented for adsorption to occur, although it is probable that such conclusions cannot be securely arrived at from dynamic tensions alone. For adsorbed

Table 6.2 *Estimated values of the true diffusion coefficient,*
D, and D_{eff} obtained from dynamic surface tensions of
5×10^{-4} mol dm^{-3} aqueous di-(C$_6$-Glu) (see text).

$T/^\circ$C	10	20	30	40	50
$10^{10}D$/m^2 s^{-1}	2.56	2.65	2.74	2.83	2.93
$10^{10}D_{eff}$/m^2 s^{-1}	0.032	0.86	0.20	0.66	2.32

ionic surfactants (in the absence of added electrolyte) an electrostatic energy barrier to adsorption could exist.

6.3.2   Adsorption above the *cmc*

So far adsorption in systems without micelles present has been considered, but the question naturally arises as to what effect micelles might be expected to have on the kinetics of decay of dynamic interfacial tension, and under what circumstances. The subject is rather involved and only a few basic ideas are mentioned below.

Although it has often been supposed that micelles are not themselves surface active, they may well be involved in making monomers available for adsorption and thus influence adsorption kinetics (see §9.10). In the subsurface region, monomer adsorption reduces the surfactant concentration below the *cmc*. Because the local equilibrium between monomers and micelles is perturbed, some kind of micellar dissociation occurs, and this can lead to an increase in monomer adsorption. Treatments of dynamic tensions in micellar solutions have differed in the way in which micellar dissociation is supposed to occur. In some studies micelles are taken to be monodisperse, that is, all micelles have the same aggregation number. This means that if a monomer leaves a micelle the micelle completely breaks down and the number of micelles per unit volume near the interface changes. An alternative is to employ the more realistic model of Anianson (§9.10) in which micelle growth involves a stepwise equilibrium leading to micellar polydispersity (§9.4). Micelle breakdown now has two characteristic relaxation times, a short time τ_1 corresponding to the lifetime of a molecule in a micelle, and a long time τ_2 corresponding to the overall breakdown of a micelle (which is presumably similar to the relaxation time in the simpler model). The two relaxation times can differ by several orders of magnitude. Assuming the Anianson model, micelles can enter the diffusion zone in the subsurface, release a small number of monomers (in time of the order of τ_1) and then leave.

For the model of monodisperse micelles (with only one, long, relaxation time), if the micellar lifetime is longer than the time needed for the dynamic tension to reach its equilibrium value, the micelles cannot act as a source of monomers for adsorption and so the interfacial tension will decay more slowly than it otherwise would. Using a model which incorporates micelle polydispersity, the existence of the short relaxation τ_1 means that monomer surfactant can be supplied to the subsurface region much more rapidly.

· ·

REFERENCES

J. Eastoe, J.S. Dalton, and P.G.A. Rogueda, Evidence for activation–diffusion controlled dynamic surface tension with a nonionic surfactant. *Langmuir*, 1998, **14**, 979–981.

V.B. Fainerman, A.V. Malievski, and R. Miller, The analysis of dynamic surface tension of sodium alkyl sulphate solutions, based on asymptotic equations of adsorption kinetic theory. *Colloids Surf. A*, 1994, **87**, 61–75.

T. Horozov and L Arnaudov, Adsorption kinetics of some polyethylene glycol octylphenyl ethers studied by the fast formed drop technique. *J. Colloid Interface Sci.*, 2000, **222**, 146–155.

G. Loglio, U. Tesei, and R. Cini, Spectral data of surface viscoelastic modulus acquired via digital Fourier transformation. *J. Colloid Interface Sci.*, 1979, **71**, 316–320.

J. Lucassen and D. Giles, Dynamic surface properties of nonionic surfactant solutions. *J. Chem. Soc., Faraday Trans. I*, 1975, **171**, 217–232.

S. Ross and I.D. Morrison, *Colloidal Systems and Interfaces*. John Wiley & Sons, New York, 1988, p.167.

A. Ward and L. Tordai, Time-dependence of boundary tensions of solutions I. The role of diffusion in time-effects. *J. Chem. Phys.*, 1946, **14**, 453–461.

7

Adsorption of surfactants at solid/liquid interfaces

In principle there is much in common between adsorption at liquid/fluid interfaces and at solid/liquid interfaces. However, in the case of solids it is not possible to measure interfacial tensions directly. Consequently, the most common method for the determination of surfactant adsorption (onto powders) involves the measurement of changes in solution concentration of surfactant accompanying adsorption. Further, the nature of solid/liquid interfaces is much more varied than that of liquid/fluid interfaces and can involve porosity, chemical heterogeneity and state of electrical charge. The charge can be positive, negative or zero often depending on (aqueous) solution conditions, for example, pH. The adsorption of surfactants, which can themselves be nonionic, anionic, cationic, or amphoteric, is very dependent on such factors, which influence the interactions between the solid surface and surfactant molecules/ions.

Much of the interest in adsorption of surfactants at the solid/liquid interface arises since surfactants modify the wettability of solid surfaces (Chapter 16) and they also confer stability on dispersions of solid particles in liquids (Chapters 11 and 13). Some direct physical methods for probing the structure of adsorbed surfactant layers at solid/liquid interfaces are discussed in Chapter 8, and results obtained using force microscopy (AFM and STM) are presented later in this chapter (§7.5.3).

7.1 How can adsorption of surfactant from solution onto solid powders be measured simply?

Probably the most common method for the determination of adsorption at *liquid* interfaces is, as seen in §4.3, to analyse the variation of interfacial tension with the concentration of surfactant in solution. This is not an option however for measurement of adsorption onto solid surfaces since the surface tension of solids is not readily measurable. A very common method for the determination of adsorption from solution onto powdered solids is to measure the fall in solution concentration of surfactant when the solution is brought into contact, and equilibrated with a solid powder. The method is often referred to as the *solution depletion* method.

Surfactants: In Solution, at Interfaces and in Colloidal Dispersions. Bob Aveyard. © Bob Aveyard 2019.
Published in 2019 by Oxford University Press. DOI: 10.1093/oso/9780198828600.001.0001

It can be shown by mass balance that for a mass m of solid at adsorption equilibrium with a binary solution of components 1 and 2 (mole fractions x_1 and x_2 at adsorption equilibrium)

$$\frac{n_o \Delta x_2}{m} = n_2^s x_1 - n_1^s x_2 \tag{7.1.1}$$

where n_o is the total number of moles of 1 and 2 in the system, Δx_2 is the change in mole fraction of 2 in solution caused by adsorption, and n_1^s and n_2^s are the numbers of moles of 1 and 2, respectively, at the surface of unit mass of solid at equilibrium. If the equation is divided throughout by Σ, the specific surface area of the solid,[1] then it is seen that

$$\frac{n_o \Delta x_2}{\Sigma m} = \Gamma_2^s x_1 - \Gamma_1^s x_2 \tag{7.1.2}$$

in which the Γ^s are surface concentrations of the subscripted components (see §4.1). For dilute solutions of surfactant (2) in solvent (1), $x_1 \sim 1$ and $x_2 \sim 0$ and so to a very good approximation the surface concentration of surfactant is given by

$$\Gamma_2^s = \frac{n_o \Delta x_2}{\Sigma m} \tag{7.1.3}$$

The use of (7.1.3) does not imply that Γ_1^s is zero; the surface of the solid is always completely covered with a mixture of solvent and solute, and when Γ_2^s is low therefore, Γ_1^s is correspondingly high. Adsorption of surfactant causes desorption of solvent and adsorption is an exchange process.

For the solution depletion method to be appropriate, Σ must be known and be sufficiently large, otherwise Δx_2 is too small to be measured precisely. The method is good for investigation of adsorption onto porous solids, providing the pore entrances have diameters large enough for surfactant molecules to enter and adsorb. Changes in surfactant concentration accompanying adsorption (i.e. Δx_2) have been measured in a variety of ways. Techniques used include differential refractometry and IR and UV absorbance (if a suitable absorbing group is present in the surfactant molecule). It is also possible to obtain surfactant concentration by measuring the surface tension of the equilibrated solution. HPLC, using both refractive index and UV detection can be used to monitor simultaneous adsorption from surfactant mixtures. Other, more sophisticated, methods for probing adsorption and adsorbed layer properties are discussed in Chapter 8.

Although interfacial tensions of solid/fluid interfaces are not conveniently measurable, *changes* in tension caused by adsorption (i.e. surface pressures, π, of adsorbed surfactant

---

[1] The specific surface area, Σ, is the surface area of unit mass of the solid. It is often determined by the adsorption of nitrogen onto the solid using the so-called BET (Brunauer, Emmett, and Teller) method. The point in the adsorption isotherm that corresponds to the completion of a monolayer of adsorbed gas is identified and, from the known cross-sectional area of the N_2 molecule, the surface area of the solid can be computed.

layers) can be obtained readily from the measurements described above. The Gibbs adsorption equation for, say, the adsorption of a nonionic surfactant, concentration c_2, from ideal dilute solution can be expressed

$$-d\gamma = d\pi = RT\Gamma_2^s d\ln c_2 \qquad (7.1.4)$$

Then, combination of (7.1.3) and (7.1.4) yields the expression for π:

$$\pi = RT \int_o^{c_2} \frac{n_o \Delta x_2}{m\Sigma} d\ln c_2 \qquad (7.1.5)$$

With values of surface pressure and surface concentration available, it is possible to test the applicability of surface equations of state to experimental adsorption data.

7.2   What characteristics of solid/liquid interfaces can influence surfactant adsorption?

In many reported studies of surfactant adsorption, the nature of the solid/liquid interface has remained undefined. Nonetheless, the surface structure and nature of chemical groupings present can have a profound effect on the adsorption process. An initial classification of solid surfaces can be made in terms of the wettability by water. Hydrophobic surfaces are nonpolar in nature and include surfaces of some carbonaceous materials and some polymers. Hydrophilic surfaces on the other hand are polar, rendering them wetted by water. They contain polar groups or ionogenic sites and include for example silicates, inorganic oxides and hydroxides as well as natural fibres and proteinaceous materials. The surface properties can depend on the mode of preparation, as well as subsequent treatment and the history of the surface, and can vary over the surface. Heating can remove certain groups, as for example with carbon where oxygen complexes are progressively removed. Heat treatment can also remove OH groups from the surface of silica, so that the original hydrophilic surface becomes hydrophobic.

Surface charge on inorganic materials and minerals in water has elicited much interest. The charge can arise, for example, from the preferential dissolution of ions or from hydrolysis of surface species. Following such hydrolysis, pH-dependent dissociation of surface hydroxyl groups can occur. Dissociation at surfaces of oxides (such as alumina, titania, and silica) in water can be usefully represented by the simple scheme

$$\mathrm{MOH_2^+} \underset{\mathrm{H^+}}{\rightleftharpoons} \mathrm{MOH} \underset{\mathrm{H^+}}{\rightleftharpoons} \mathrm{MO^-} \qquad (7.2.1)$$

From this, it is seen that at low pH the surface becomes positively charged and at high pH it is negative. There exists an intermediate pH for which the surface is neutral; this pH is termed the *point of zero charge* (*p.z.c.*). Since $\mathrm{H^+}$ and $\mathrm{OH^-}$ determine the surface charge, they are called *potential determining ions*.

The surface charge on simple inorganic solids, such as silver halides, has been widely studied. For example, for AgI the charge arises from preferential dissolution of lattice ions; the potential determining ions in this case are Ag^+ and I^-. For high concentrations of Ag^+ ions in aqueous solution in contact with solid AgI the surface becomes positive, and conversely for high $[I^-]$ the surface is negative. At some value of pAg^+ or pI^- (which are related through the solubility product for AgI in water) the surface becomes electrically neutral.

Clay minerals (Box 7.1) constitute a very important class of solids, not least in the past because of their relevance in surfactant adsorption that occurs during surfactant flooding of oil wells, used to enhance the production of crude oil. Those clay minerals which consist of tetrahedral sheets of SiO_4 together with octahedral sheets of AlO_6 linked by shared oxygens often carry a negative surface charge as a result of, say, the substitution of Al^{3+} for Si^{4+}. This *face* charge does not depend on solution conditions, unlike the charge of the *edges* of clay particles which do carry a pH-dependent charge. The *p.z.c.* of clay particles is determined by the sum of the edge and face charges and in general both edges and faces carry a charge (of opposite sign) at the *p.z.c.*

Box 7.1   Clay minerals—the Chinese connection

The building units of clay minerals are (i) tetrahedral sheets composed of Si or Al in tetrahedral coordination with four oxygens and (ii) octahedral layers of cations (Mg, Fe(II), Fe(III), or Al) in octahedral coordination with six oxygens. When there is one tetrahedral and one octahedral sheet the clay is referred to as a 1:1 clay. A 2:1 clay has two tetrahedral sheets and one central octahedral layer. If the layers are charged (say by the isomorphous substitution of Al^{3+} for Si^{4+}) the net charge is compensated by interlayer cations (e.g. Na^+). Kaolinite is a 1:1 clay mineral and has no layer charge. Examples of 2:1 clays, which do carry a layer charge, are montmorillonite (from the smectite group of clays), vermiculite and muscovite (mica group). The layer surface charge density for muscovite, for example, is 0.343 C m^{-2}. Kaolinite, mined as kaolin (or china clay) in a number of areas across the world, is widely used in medicine (often to settle the stomach), ceramics, and in the manufacture of glossy paper, toothpaste, and cosmetics. It is also used as an insect repellent sprayed onto vegetables. Its name is derived from Gaoling (meaning 'High Hill') in Jingde Town in China—hence china clay.

In the present context, the *p.z.c.* is obviously an important property since the adsorption of ionic surfactants is influenced by surface charge on the solid. (The *p.z.c.* is not to be confused with the *isoelectric point (i.e.p.)*, which is introduced in §7.3.1). Values of the *p.z.c.* values of some commonly encountered solids are given in Table 7.1 (taken from Lyklema, 1987). Ways of measuring surface charge are discussed by Hunter (1993).

The possibility that solids can exhibit porosity has been referred to. Porosity is often required since it results in high specific surface areas (Σ) and hence a high capacity for adsorption. Pores are often classified according to the pore radius r, the pores for this purpose being considered to be idealized cylinders. On this basis macropores have radii

Table 7.1 *Points of zero charge of some minerals at room temperature*

solid	p.z.c. (pH)	solid	p.z.c. (pH)
SiO_2 (precipitated)	2 to 3	α-Fe_2O_3 (haematite)	8.5–9.5
SiO_2 (quartz)	3.7	α-$FeO.OH$ (goethite)	8.4–9.4
TiO_2 (anatase)	6.2	Al_2O_3 (corundum)	9.0
TiO_2 (rutile)	5.7 to 5.8	edges of clay plates	6–7

of around 50 nm or above, for mesopores r is of the order of a few nm and micropores have radii smaller than about 2 nm. For a given volume of pores, micropores obviously exhibit much larger surface areas than mesopores or macropores.

Carbons of various kinds are widely used in industry. So-called 'active' carbons, for example, are used as adsorbents and have specific surface areas Σ of say 500 to 1000 m^2 g^{-1}. For such samples the pore volumes of micropores and macropores are comparable, with only about 1 m^2 g^{-1} being contributed by macropores, the remainder coming mainly from micropores. However, Σ is usually measured by the adsorption of small gas molecules (e.g. N_2), and much of this area may not be accessible to larger surfactant molecules, particularly polymer surfactants. A relatively nonporous form of carbon is Graphon; it has a fairly (chemically) homogeneous surface and Σ is less than about 100 m^2 g^{-1}. Surface areas and porosity of many minerals are small compared to say those of carbons. For example, values of Σ of a range of silicate minerals are reported by Brantley and Mellott (2000), and generally fall below 1 m^2 g^{-1}.

As mentioned, Σ is conveniently measured using the BET gas adsorption method. Micro- and mesoporosity can also be determined using gas adsorption since capillary condensation (§3.4.2) occurs in these pores, from which information about pore size, pore size distribution and pore volume can be extracted. Macropores are often investigated using mercury porosimetry (see Box 3.3, page 36) in which the pressure required to force mercury into the pores of an outgassed (non-wetting) solid gives the pore radius. The volume of mercury entering the pores gives the pore volume. Gregg and Sing (1982) give a thorough description of methods used to determine the surface area and porosity of solids (see §3.4.2).

7.3 What interactions are responsible for adsorption?

From what has been said about the nature of solid surfaces and the chemical structure of surfactant molecules, which can be low molar mass or polymeric materials, it is clear that a range of interactions between surfactants and solid surfaces is possible. Surfactant-solid interactions can be 'electrical' (Coulombic) or 'specific' in nature. Specific interactions have sometimes been referred to as 'chemical' although this does not necessarily imply that chemical bonds exist between surfactant and solid surface. Mutual lateral interactions between adsorbed surfactant molecules can also exist and influence the adsorption process, particularly for concentrated films.

7.3.1 Electrical interactions

Possible origins of charge on the surface of a solid in contact with water were discussed in §7.2. The charge in unit area of interface (i.e. the surface charge density) is denoted σ_o, and the electrical potential at the solid surface is ψ_o (Fig. 7.1). The charge on the solid is neutralized over a short distance normal to the surface by a mixture of neighbouring charged species in the (aqueous) solution (H^+, OH^-, and ions from added electrolyte, including ionic surfactant). The array of positive and negative charges normal to the interface is termed the *electrical double layer* (see §11.2.2 for a fuller discussion).

Suppose, for example, a solid with a positively charged surface is in contact with water. Some anions, possibly surfactant ions, may be specifically adsorbed, and such ions are likely to be unsolvated, or at least only partly solvated (see Fig. 7.1). The plane through the centres of specifically adsorbed ions is the *inner Helmholtz plane* (i.H.p.), where the surface charge density and surface potential are, respectively, σ_s and ψ_s. Non-specifically adsorbed ions in the *outer Helmholtz plane* (o.H.p.), also called the *Stern plane*, retain their hydration sheath and the electrical potential in this plane (through the ion centres) is designated ψ_d. The *diffuse* part of the double layer, consisting of a mixture of anions and cations, extends from the Stern plane, strictly out to infinity, and the net charge, σ_d (expressed per unit area of surface) in this layer completes the neutralization of the surface charge. The distribution of anions and cations in the diffuse layer is determined by the competing influences of Coulombic forces, tending to give an uneven distribution, and thermal effects which tend to give a random distribution.

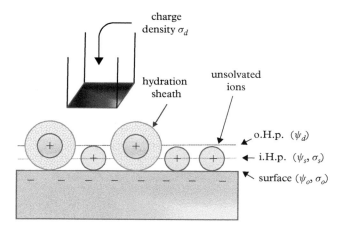

Figure 7.1 *Representation of a solid/liquid interface showing surface charge densities, electrical potentials and positions of adsorbed ions. The outer Helmholtz plane is also called the Stern plane. σ_d is the charge density in the diffuse part of the double layer.*

Since the net charge over the region of the double layer is zero (i.e. the ions on the solution side of the charged solid surface neutralize the surface charge on the solid itself) it follows that

$$\sigma_o + \sigma_s + \sigma_d = 0 \tag{7.3.1}$$

A system is at its *i.e.p.* when $\sigma_o + \sigma_s = 0$, that is, when specifically adsorbed ions in the i.H.p. exactly neutralize the surface charge on the solid. It will be recalled that the *p.z.c.* corresponds to $\sigma_o = 0$. The electrical potential ψ_d at the o.H.p. is often taken to be equal, or close to the *zeta potential*, ζ, which is the potential at the plane of shear in electrophoretic measurements (Hunter, 1981); at the *i.e.p.* ψ_d (ζ) $= 0$. The *i.e.p.* and *p.z.c.* coincide if $\sigma_o = 0$, that is, for systems with no specific adsorption.

Obviously, for a positively charged solid surface, Coulombic forces give rise to an attractive (negative) component to the free energy of adsorption of an anionic surfactant; a negatively charged surface tends to attract a cationic surfactant.

7.3.2   Specific interactions

Here the term specific is taken to mean all interactions involved in surfactant adsorption other than those arising directly from Coulombic forces. Surfactant head groups can interact with a solid surface through polar forces, an extreme form being the formation of hydrogen bonds. Chemical bond formation between surfactant and surface is also sometimes possible, for example, in the adsorption of oleate anions onto fluorite (see Hanna and Somasundaran, 1977).

Surfactant chains can play a very important role in the adsorption of surfactants from aqueous solution, the effects depending on the surface concentration of surfactant. For hydrophobic solids, hydrocarbon chains in dilute films can interact with the solid surface through hydrophobic bonding, as illustrated in Fig. 7.2a; in the case shown the head group–solid interaction is relatively weak. As the film becomes more concentrated, the chains of adsorbed molecules will begin to interact mutually (again through hydrophobic bonding). This occurs below the critical micelle concentration (*cmc*) of the surfactant in the bulk solution since the chains are brought into close proximity as a result of adsorption (Fig. 7.2b). At relatively high *surface* concentrations, aggregates can form at the solid/liquid interface, as discussed §7.5.

(a)   (b)

Figure 7.2 *Adsorbed surfactant molecules at the interface between a hydrophobic solid and water. (a) is for a dilute adsorbed film, and (b) represents the situation as the surface concentration increases, giving rise to chain–chain interactions.*

7.4   Adsorption isotherms and adsorption free energies

Adsorption free energies encapsulate information about the strength of adsorption, as discussed in §5.3. Adsorbed surfactant films at liquid/fluid interfaces are non-localized (mobile), whereas films at the solid/liquid interface are often considered to be localized (see Appendix A5). For very dilute films however, the isotherms for non (laterally) interacting mobile films and localized films reduce to the form for the adsorption of an ideal two-dimensional gas. Thus for very dilute nonionic surfactant films at the solid/liquid interface the appropriate isotherm is (see §5.3.2)

$$c = KN \qquad (5.3.15)$$

where c is the bulk surfactant concentration at adsorption equilibrium, N is the number of adsorbed surfactant molecules in the film, and K is a constant containing the standard free energy of adsorption of the surfactant. The surface concentration of surfactant, Γ, is proportional to N (in a film of fixed area) and so

$$c = K'\Gamma \qquad (7.4.1)$$

A standard free energy of adsorption of surfactant, $\Delta_{ads}\mu_\Gamma^o$ is related to K' by

$$\Delta_{ads}\mu_\Gamma^o = RT \ln K' = -RT \ln \left(\frac{\Gamma}{c}\right)_{c \to 0} \qquad (7.4.2)$$

The standard states for $\Delta_{ads}\mu_\Gamma^o$ are unit c and Γ in whatever units that are being used.

Equation (7.4.2) can usefully be employed to compute adsorption free energies if sufficient, good quality data are available to determine the initial rise in an isotherm (of Γ versus c). Interpretation of the absolute value of an adsorption free energy, which depends on the chosen standard states, is not altogether straightforward. However, values of standard free energies of adsorption are often used to *compare* the strength of adsorption in two or more systems. For this purpose it is of course essential that the free energies refer to the same standard states for all the systems; in this case the *differences* do not contain contributions from choice of standard states and reflect only the different interactions of isolated molecules with the surface. It is worth bearing in mind also that if the solid surface is heterogeneous, the standard free energies of adsorption obtained using (7.4.2) refer to adsorption on the sites which give rise to the most negative free energies (i.e. the 'highest energy' sites).

For more concentrated localized films of nonionic surfactant, in the absence of lateral interactions between surfactant molecules, the Langmuir isotherm is applicable (see the Appendix to Chapter 5):

$$c = \left(\frac{\theta}{1 - \theta}\right) \exp\left(\frac{\Delta_{ads}\mu_{Langmuir}^o}{RT}\right) \qquad (A5.1)$$

in which θ is the surface coverage (i.e. the fraction of surface sites occupied by adsorbed molecules). In real systems, lateral interactions are likely to be present, and if the surfactant is ionic and the surface charged, then electrical interactions must also be accounted for. The latter interactions include those between surfactant head group and surface and laterally between surfactant head groups. Surface heterogeneity if present complicates matters further. A simple modification of (A5.1), which accounts for specific interactions between surfactant and solid surface (through a standard free energy contribution $\Delta_{ads}\mu_{spec}^{o}$) and electrical interactions normal to the surface (but not laterally between surfactant head groups), is

$$\ln\left(\frac{c\,(1-\theta)}{\theta}\right) = \frac{1}{RT}\left(\Delta_{ads}\mu_{spec}^{o} + N_{A}ze\psi_{s}\right) \tag{7.4.3}$$

Here z is the valence of the ionic head group and ψ_{s} the electrical potential at the i.H.p. (see Fig. 7.1). Equation (7.4.3) is often referred to as the *Stern–Langmuir equation* (Box 7.2), and since lateral interactions are unaccounted for it can only be expected to describe data for relatively low surface coverages.

In the study of adsorption, experimental determination of the adsorption isotherm is usually the first (and necessary) step. Attempts to fit the experimentally determined adsorption data to a supposedly appropriate theoretical isotherm however can often be unfruitful. This is in part because the solid surfaces are frequently ill-defined and heterogeneous in nature. The use of, *inter alia*, adsorption calorimetry has been very useful in understanding the processes occurring during adsorption, and atomic force microscopy (AFM) has been particularly effective in probing the disposition of surfactant molecules in adsorbed layers. This is illustrated in the next section.

Box 7.2 Irving Langmuir: Nobel Laureate and scientific polymath

Irving Langmuir (1881–1957) was an American physicist and chemist, born in New York. After studying metallurgical engineering at the Columbia School of Mines (because it 'was strong in chemistry ... had more physics than the chemical course, and more mathematics than the course in physics—and I wanted all three.') and gaining a Ph.D. in Gottingen, Germany (under Nernst on the Nernst Glower, a kind of electric lamp) he taught chemistry at Stevens Institute of Technology in Hoboken, New Jersey. Thereafter, his scientific career was spent with General Electric in Schenectady, New York, where he ultimately became Associate Director.

His research covered an astonishing range of subjects and he worked on, amongst other things, chemical reactions, atomic structure, plasmas, heat transfer, thermionic phenomena, meteorology and atmospheric science. A major part of Langmuir's efforts however was directed towards molecular behaviour, both chemical (catalysis) and physical, at solid and liquid interfaces. He studied insoluble monomolecular films on water using the 'Langmuir trough', and (with Kathleen Blodgett) so-called Langmuir-Blodgett multilayers deposited

Box 7.2 *Continued*

Irving Langmuir, 1881–1957. Granger Historical Picture Archive/Alamy Stock Photo.

on solids. He was awarded the 1932 Nobel Prize in chemistry 'for his discoveries and investigations in surface chemistry'.[†] The foundations of the work leading to the award were reported in a paper published in 1917 on the physical chemistry of oil films on water (Langmuir, 1917). Langmuir hypothesized that amphiphilic materials (such as alkanols and alkanoic acids) are present on the surface of water as monolayers with their hydrophilic head groups dipping into the water and the hydrophobic chains in air (see §5.3). The thickness of a monomolecular film can be calculated from its area and the amount of film material deposited on the surface (see §4.4 where the experimental observations of Benjamin Franklin are described).

Reference is made above to the *Stern–Langmuir isotherm* ((7.4.3). Otto Stern (1888–1969) was a German physicist who was awarded the 1943 Nobel Prize in Physics 'for his contribution to the development of the molecular ray method and his discovery of the magnetic moment of the proton'.

[†] There is a remarkable film of Langmuir explaining his ideas on monolayers on water and describing his work with Kathleen Blodgett on multilayers deposited on solids.: *Irving Langmuir GE film on surface chemistry 1939*. The film was made in celebration of his Nobel Prize, and can be seen for example on YouTube.

7.5   Some experimental results

There is a bewildering array of reported shapes of experimentally observed adsorption isotherms, reflecting the complexities of many of the systems investigated (see e.g. Parfitt and Rochester, 1983). At the risk of oversimplification, studies are selected here which give relatively unambiguous insights into the interactions and adsorbed layer structures involved in adsorption onto some well-defined solid surfaces.

7.5.1 Hydrophobic surfaces

Investigations of adsorption from aqueous solution onto the relatively well-defined hydrophobic surfaces of graphite have given some good insights. Isotherms for the adsorption of the anionic surfactant sodium dodecylsulfate (SDS) onto two different samples of hydrophobic graphitized carbon samples (Graphon and Sterling MTG) are coincident, as seen in Fig. 7.3. There is an inflexion point in the isotherm which starts to rise a second time when the area per SDS molecule at the surface, A, is about 0.72 nm^2; along the final plateau, which commences at the *cmc* of SDS in water (about 8 mM), $A = 0.43$ nm^2. This latter value is very similar to that for SDS in a close-packed vertically oriented monolayer at the air/solution interface. Thus, very broadly it can be supposed that in the low concentration regime of the isotherm the surfactant molecules lie flat along the graphite surface and that in the final plateau region the adsorbed molecules are more nearly vertical. The situation for cationic surfactants, such as dodecyltrimethylammonium bromide, adsorbed on hydrophobic surfaces is very similar to that for anionic surfactants. But, from the shape of isotherms alone it is not possible to say much more.

The adsorption of sodium decylsulfate (SDeS) onto the graphite/water interface has been studied by Kiraly et al. (2001) using microcalorimetry. The shape of the isotherm at 298 K is very similar to that shown below for SDS as expected, with the adsorption at the plateau being a little lower than for SDS (i.e. about 3.5 compared to 3.8 μmol m^{-2} for SDS). The surfactant has a high affinity for the surface, that is, the initial slope of the isotherm is very high, and in this region it is the strong attraction between the alkyl chains of the surfactant with the graphite surface which drive the adsorption. There is a near perfect match between the chains in the all trans configuration and the carbon hexagons of the graphite basal planes. The differential molar enthalpy of adsorption in the early stages of the isotherm is found to depend strongly on the surfactant chain length, but to be essentially independent of surface coverage. This implies that the adsorbed layers are ideal, and since there is no reason to suppose a mixed film of surfactant and water is ideal, it appears likely that the adsorbed monolayers are separated into two equilibrium

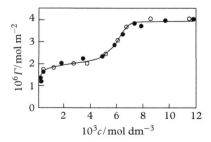

Figure 7.3 *Isotherm for the adsorption of SDS from aqueous solution onto Graphon (filled symbols) and Sterling MTG (open symbols) at 298 K. Redrawn from Parfitt and Picton (1968).*

phases, one mainly patches of aligned surfactant ions and the other water. The relative ratio of the two phases changes with increased adsorption.

At higher surfactant concentrations, approaching the *cmc* in bulk solution, the calorimetric data were consistent with the formation of hemicylindrical aggregates at the surface, facilitated by the underlying (largely) hydrophobic adsorbed monolayer, as illustrated in Fig. 7.4. For the monolayer to act as a template for this aggregation, the surfactant in the monolayers must be arranged in a head-to-head chain-to-chain configuration. Below, in §7.5.3, more direct evidence for the formation of surfactant aggregates of various geometries at solid surfaces is discussed.

There are very many types of nonionic surfactant and so it is difficult to make generalizations about the shapes of adsorption isotherms. Nonetheless, the shape of some isotherms for the adsorption of nonionic surfactants onto hydrophobic surfaces is rather similar to that for ionic surfactants described above, and it can be supposed the adsorption sequence is similar to that described for the ionic surfactants. An example, reported by Corkill et al. (1967) is shown in Fig. 7.5 for octylsulfinylethanol adsorbed on Graphon.

The effects of temperature on the adsorption of a nonionic surfactant can be very marked if the bulk solution is fairly close to a lower consolute temperature (LCT),

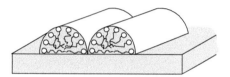

Figure 7.4 *Schematic representation of parallel hemicylindrical surfactant aggregates at the surface of a hydrophobic solid in contact with aqueous surfactant solution.*

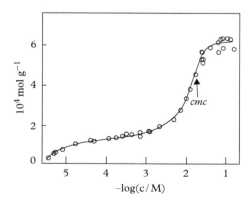

Figure 7.5 *Isotherm for the adsorption of octylsulfinylethanol from aqueous solution onto Graphon at 298 K. Redrawn from Corkill et al. (1967).*

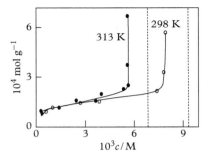

Figure 7.6 *Adsorption isotherms for the adsorption of C_8E_3 from aqueous solution onto Graphon at 298 K and 313 K. Redrawn from Corkill et al. (1966).*

as illustrated in Fig. 7.6.[2] The figure shows isotherms for the adsorption of *n*-octyl trioxyethyleneglycol monoether (C_8E_3) at two temperatures. The isotherms (for 298 K and 313 K) both show a very sharp rise in 'adsorption' at a temperature fairly close to (but lower than) the LCT of the surfactant + water mixtures. For example, consider the adsorption data for 313 K. The left of the two dashed vertical lines is drawn at a bulk surfactant concentration for which the LCT is 313 K. It is believed that a nucleation of a surfactant-rich phase is induced by the surface at a temperature at which the solution is normally homogeneous. Since clouding in bulk solution is associated with surfactant–water interactions, it appears that such interactions are also involved in adsorption.

7.5.2 Hydrophilic surfaces

As discussed in §7.2, hydrophilic surfaces in water often carry a charge which is dependent on the concentration of potential determining ions in solution, and so charge–charge (Coulombic) interactions can play an important role in the adsorption of ionic surfactants onto such surfaces. For say silica in aqueous solution (*p.z.c.* at pH = 2–4 depending on method of production—see Table 7.1) at pH above around 2–4 the surface is negatively charged and so tends to attract cations and repel anions. On the other hand, the surface of alumina (*p.z.c.* around pH 9) is positively charged at pH values below 9 and so in this regime the surface attracts anions and repels cations. As examples of adsorption of ionic surfactants, the adsorption of a cationic surfactant on hydrophilic silica and an anionic surfactant on alumina, where charge–charge interactions can be expected to favour adsorption, are considered.

Much attention has been given over the years to the adsorption of cationic surfactants from aqueous solution onto the silica/water interface. An isotherm for the adsorption of benzyldimethyldodecylammonium bromide ($BDDA^+B^-$) at the silica/water interface is

---

[2] Occurrence of phase separation or 'clouding' in aqueous solutions of nonionic surfactants is discussed in §9.8.3. The temperature of phase separation, the so-called *cloud point*, depends on the concentration of the aqueous surfactant solution.

shown in Fig. 7.7, together with the adsorption isotherm for benzyltrimethylammonium bromide (BTMA$^+$B$^-$), which is effectively the head group of the surfactant.

By studying this pair of materials, Trompette et al. (1994) were able to distinguish the role of the surfactant hydrocarbon chain in the adsorption process. The surfactant adsorption is very much greater than that of BTMA$^+$B$^-$ over the whole concentration range, and it is clear from this that the removal of the surfactant chains from water in bulk together with chain–chain interactions at the surface play a major role in determining the extent of adsorption of the surfactant.

If the surface charge on the solid particles is altered by adsorption of ionic surfactant, this will be reflected in the electrophoretic mobility of the particles dispersed in the surfactant solution. Trompette et al. (1994) report such mobilities, for the systems represented in Fig. 7.7, and these are reproduced in Fig. 7.8. For the surfactant cation BDDA$^+$ there is a range at very low concentration (up to 0.08 mmol kg^{-1} at point A) in which the mobility remains constant, but in which the adsorption rises sharply (Fig. 7.7). This indicates that in this regime the adsorption is an ion exchange process between BDDA$^+$ ions and other, pre-adsorbed cations (shown to be mainly sodium ions) present at the silica/water interface. From 0.08 mmol kg^{-1} up to just below the *cmc* the mobility of the particles rises almost linearly, and adsorption rises continuously also; at 0.41 mmol kg^{-1} (point B) the electrophoretic mobility is zero. The cause of the increase in adsorption beyond the ion exchange region is mainly the removal of hydrophobic chains from the aqueous environment and subsequent mutual cohesion in the adsorbed phase. Above the *cmc* the adsorption and mobility remain unchanged with concentration. Of course, Coulombic forces oppose adsorption when the mobility is positive.

The electrophoretic mobility of the silica particles in the presence of the 'head group' BTMAB$^+$ alone remains unchanged over much of the concentration range, supporting the conclusion that the rise in surfactant adsorption above the ion exchange region is due to interactions involving the surfactant chains. The small rise in mobility in the presence of BTMAB$^+$, above concentrations of about 2.5 mmol kg^{-1}, is probably due to the increase in ionic strength.

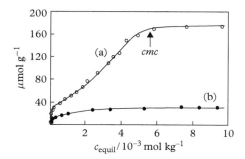

Figure 7.7 *Isotherms for the adsorption of (a) BDDA$^+$B$^-$ and (b) BTMA$^+$B$^-$ from water at 298 K and natural pH onto nonporous precipitated silica with specific surface area 40 m^2 g^{-1}. Redrawn from Trompette et al. (1994).*

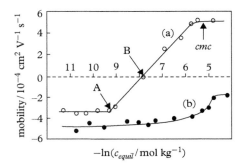

Figure 7.8 *Electrophoretic mobility of silica particles as a function of the equilibrium solution concentration of (a) BDDA⁺B⁻ and (b) BTMA⁺B⁻ under the same conditions as represented in Fig. 7.7. Redrawn from Trompette et al. (1994).*

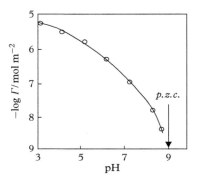

Figure 7.9 *Adsorption of sodium dodecylsulfonate from aqueous NaCl onto alumina. Surfactant concentration is 10^{-4} mol dm^{-3} and ionic strength is 2×10^{-3} mol dm^{-3}. The p.z.c. is at pH 9.1. Redrawn from Fuerstenau (1971).*

Wakamatsu and Fuerstenau (1968) proposed a similar scheme for the adsorption of an homologous series of sodium alkylsulfonates onto alumina at pH 7.2, where the surface charge is positive (*p.z.c.* at pH = 9.1). For a given surfactant concentration in the region where adsorption is largely driven by chain cohesion at the surface, adsorption is greater the longer the alkyl chain. The effects of surface charge on adsorption can be clearly seen by varying the pH at constant ionic strength and surfactant concentration. Adsorption of sodium dodecylsulfonate from aqueous NaCl (concentration adjusted to give total ionic strength of 2×10^{-3} mol dm^{-3}) onto alumina over a range of pH up to the *p.z.c.* is shown in Fig. 7.9. At pH 7.2 the adsorption is about 1×10^{-7} mol m^{-2}, which is in the region where chain–chain cohesion is thought to be a major cause of adsorption, and where the surface charge is still positive. Nonetheless, adsorption is strongly reduced as the surface charge falls (pH rises) and, conversely, as the charge density rises so does the adsorption. Thus the interplay between effects of surface charge and of chain–chain cohesion on adsorption is clearly seen.

As pointed out by Clunie and Ingram (1983), there are over 250 types of nonionic surfactant and often samples are heterogeneous mixtures. Thus, the example given for the adsorption of nonionic surfactants onto charged solids cannot be taken as representative of nonionic surfactants as a whole. Surfactants with structure $R_n(OCH_2CH_2)_mOH$, (designated C_nE_m), where R_n denotes a normal alkyl chain with n carbon atoms, have been widely used in academic studies since pure samples are available with known values of n and m, which can be varied. Isotherms for adsorption of C_8E_6, $C_{10}E_6$, and $C_{12}E_6$ onto silver iodide particles, taken from Mathai and Ottewill (1966), are shown in Fig. 7.10. The aqueous solutions contain 10^{-4} mol dm^{-3} KI; I$^-$ ions are potential determining in this system and the AgI particles carry a negative charge. For all the surfactants there is a plateau in the isotherms at the lower concentrations represented, followed by a sharp rise around the *cmc* for each of the surfactants.

The adsorption values at the low concentration plateaux correspond to about those expected for a complete monolayer of vertically oriented surfactant molecules (area/molecule A ca. 0.60 nm^2); the polarizable head groups are presumably in contact with the solid surface. The second plateaux (A ca. 0.2 nm^2) obviously correspond to very much more than monolayer coverage, and since the AgI dispersions were well-stabilized at these higher surfactant concentrations it appears that the polar E_6 groups in the adsorbed layers are in contact with the aqueous phase, rendering the particles wetted by water.

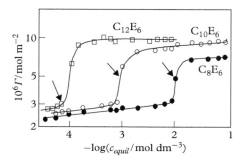

Figure 7.10 *Isotherms for the adsorption of C_nE_6 onto silver iodide particles from aqueous 10^{-4} mol dm^{-3} KI at 298 K. The arrows denote the cmcs. Redrawn from Mathai and Ottewill (1966).*

7.5.3   Surface-induced aggregation of surfactants at solid/liquid interfaces

As discussed, the study of adsorption isotherms alone cannot give detailed information about the disposition and structure of surfactant entities at solid surfaces. Nonetheless, aggregation in incomplete adsorbed surfactant layers had been proposed by Gaudin and Fuerstenau (1955) for both anionic and cationic surfactants adsorbed on quartz. The understanding of surfactant behaviour at solid/liquid interfaces has been considerably enhanced in more recent years, when it has become possible to probe adsorbed layers on solids directly, using physical techniques such as neutron reflectivity, ellipsometry, fluorescent probe methods, and particularly AFM and scanning tunnelling microscopy (STM) (see Chapter 8). In this section, the discussion is restricted to some findings furnished by AFM and STM.

The formation of surfactant aggregates in bulk aqueous solution, having a wide range of structures, has been extensively studied (Chapters 9 and 10). Examples of structures formed include isolated spherical and rod-like micelles, and phases formed with these aggregates as building blocks, such as hexagonal and cubic phases. Much of the process of primary aggregation in solution is determined by the curvature requirements of cohering surfactant layers. It can be expected therefore that such curvature requirements may well play a role in the formation of surfactant aggregates at solid/water interfaces, but that this self-organization will be perturbed to a greater or lesser extent by contact with the surface, which can be hydrophobic or hydrophilic.

Aggregate formation on hydrophobic graphite surfaces, which can readily be cleaved to yield clean molecularly smooth surfaces, is considered first. It was mentioned in §7.5.1 that for an adsorbed surfactant monolayer to act as a template on graphite for the formation of hemimicelles (as illustrated in Fig. 7.4), the arrangement of surfactant molecules would have to be head to head and tail to tail. Such an arrangement has been directly observed using STM by Yin et al. (2005) for adsorbed monolayers of sodium *n*-hexadecylsulfonate (SHS) on highly oriented pyrolytic graphite (HOPG). An adsorbed layer was prepared by placing a drop of aqueous surfactant solution on the HOPG surface, and allowing time for the first layer to form, after which the drop was blown away; the HOPG sample was then allowed to dry in a nitrogen atmosphere. An image at atomic resolution for a monolayer of SHS on HOPG is shown in Fig. 7.11. The aligned sulfonate head groups appear as rows of large white spots, and the alkyl chains are disposed flat on the graphite surface at about 90° to a line through adjacent head groups. The small bright spots along the length of the alkyl chains are attributed to H atoms in the methylene groups. The repeat unit of the monolayer, at right angles to the direction of the rows, is 4.6 nm which is twice the extended molecular length of the surfactant. To maintain electroneutrality in the region of the opposing sulfonate head groups in the monolayer, counterions (Na^+) must be present and it may also be that the polarity of water molecules plays a role in reducing head group repulsion.

Adsorbed layer structures of ionic, nonionic and zwitterionic surfactants on graphite have been probed using AFM. The bulk concentration of surfactant has usually been

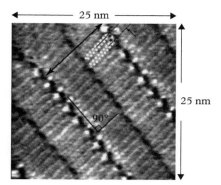

Figure 7.11 *The STM image (25 nm square) shows a monolayer of SHS lying flat on the surface of HOPG. The large bright spots are the sulfonate head groups, and the alkyl chains are oriented at about 90° to the lines through the head groups. Reprinted with permission from X.L. Yin, L.J. Wan, Z.Y. Yang, and J.Y. Yu, Appl. Surf. Sci., 2005, 240, 13.*

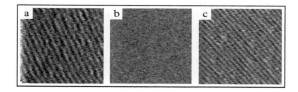

Figure 7.12 *AFM images of adsorbed layers of $C_{12}DMAO$ on freshly cleaved graphite surfaces in contact with aqueous solution (c > bulk cmc) at (a) pH = 8.0, (b) pH = 4, and (c) pH = 1.3. The surfactant becomes progressively ionized as the pH falls. The images are 100 nm × 100 nm. Reprinted with permission from H. Kawasaki, M. Shinoda, M. Miyahara, and H. Maeda, Colloid Polym. Sci., 2005, 283, 359.*

well above the *cmc*. For (normal) alkyl chain lengths of about 12 and up, long parallel stripes taken to be hemicylinders are often observed at the surface (Warr, 2000). An example (taken from Kawasaki et al., 2005) is given in Fig. 7.12 for adsorbed layers of dodecyldimethylamine oxide ($C_{12}DMAO$). For solution pH = 8.0 it is expected that the adsorbed surfactant is unionized, and, as seen in Fig. 7.12a, hemicylindrical micelles are formed. The spacing between stripes arises from the diameter D of the micelles (about 4 nm, which is twice the extended molecular length) together with the space X between the surfaces of the micelles (about 5 nm in the present case), as illustrated in Fig. 7.13. For pH = 4.0, the adsorbed layer is partially ionized, that is, a mixture of $C_{12}H_{25}(CH_3)_2N{\rightarrow}O$ and $C_{12}H_{25}(CH_3)_2N^+OH$, and, as shown in Fig. 7.12b, is laterally homogeneous. Further reduction in pH, however, leads to greater ionization, and at pH = 1.3 aligned hemimicelles are again observed (Fig. 7.12c).

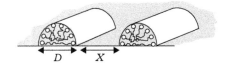

Figure 7.13 *Aligned hemicylindrical micelles, diameter D, with space X between their surfaces.*

As discussed below, and in Chapters 9 and 10, the shapes of surfactant aggregates in bulk systems are strongly influenced by the curvature requirements of close-packed surfactant monolayers. In simple terms, if the effective cross-sectional area of a surfactant head group is greater than that of the chain moiety, positive curvature of the layer results, that is, that for which the head groups are on the 'outside', as shown for the hemicylindrical micelles depicted in Fig. 7.13. Presumably, the intrinsic molecular size and hydration of the polar amine oxide head group render its cross-sectional area greater than that of the dodecyl chain group. In mixed monolayers of the nonionic and ionic forms of the surfactant, obtained at intermediate pH, it has been proposed that hydrogen bonding occurs between the ionic and nonionic head groups (i.e. $-N^+-OH\ldots O-N-$). Since in this regime a homogeneous (planar) monolayer is observed, it appears that the mean head group size is less than that of the unionized head group alone. The 'attraction' (i.e. H-bonding) between head groups reduces the effective mean head group size. At low pH the head groups become more fully ionized, which gives rise to repulsion between head groups, increasing their effective cross-sectional area, and the more positive monolayer curvature results again in the formation of hemicylindrical micelles. However, it is stressed that, for the adsorption of alkyl chain surfactants onto graphite from water, the very strong attractive interactions between the chains and graphite surface are very important. It is these strong attractive interactions that result in the formation of hemicylinders rather than complete cylinders.

Adsorption on water-wetted surfaces is now considered. Mica is a model hydrophilic surface and strong chain-surface interactions no longer perturb the structures of surface aggregates. Indeed, surfactant aggregate behaviour at mica/water surfaces can be quite similar to that observed in bulk solution although, as will be seen, interactions between charged surfactant head groups and the charged mica surface can modify the aggregate curvature and structure.

It is useful here briefly to anticipate some ideas concerning the *surfactant packing factor*, denoted by P, to be discussed in more detail in Chapters 9 and 10. P is defined by

$$P = v/a_h\, l_c = a_c/a_h \tag{7.5.1}$$

in which v is the volume of the alkyl chain and l_c its length; v/l_c is the chain cross-sectional area a_c. The effective cross-sectional area of the head group moiety, a_h, is determined not only by its molecular size but also by its solvation and by the attraction and electrical repulsion between neighbouring head groups. The shape and mean curvature of surfactant aggregates in bulk solution is determined by the way in which neighbouring surfactant molecules can pack most efficiently, and as mentioned above, the group with the larger cross-sectional area will be on the 'outside' of the curved surfactant layer surface. It transpires that for $P < 1/3$ spherical aggregates result, for P between 1/3 and 1/2 (less curved) rod-like micelles are formed and for $P > 1/2$ bilayers are given. In systems containing both oil and water, it will be seen (in Chapter 10) that inverse structures are possible (e.g. water-in-oil microemulsion droplets), and here, $P > 1$.

Patrick et al. (1999) have investigated the adsorption of various quaternary ammonium cationic surfactants onto freshly cleaved mica, which carries a negative charge (1 electronic charge per 0.5 nm^2) in contact with water at normal pH. The alkyl chain surfactants (chain length n) had the general structure $C_nNR_3^+X^-$ in which R is a Me, Et, Pr, or Bu moiety, and X^- is Cl^- or Br^-. In bulk aqueous solution P for these surfactants is less than 1/3 and spherical micelles are formed above the *cmc*. Adsorption of the cationic surfactants at the negatively charged mica/water interface gives a reduction in mutual head group repulsion which results in a reduction in a_h and an increase in P so that, for C_{12} and $C_{14}NMe_3Br$ for example, only rod-shaped micelles are formed at the surface. Increasing the head group size by increasing the size of R gives an increase in a_h and a reduction in P, and for example $C_{12}Et_3NBr$ and $C_{14}Et_3NBr$ form spherical micelles at the mica/water interface. Increasing the surfactant chain length gives an increase in a_c, and an increase in P. Accordingly $C_{12}NMe_3Cl$ for example gives spherical micelles at the mica/water interface whereas $C_{14}NMe_3Cl$ gives cylindrical micelles. A schematic representation of a spherical and a cylindrical adsorbed micelle is given in Fig. 7.14.

Surfactant molecular geometry, and hence micellar structure and curvature, can be varied in interesting ways by the use of Gemini (dimeric) surfactants (see §1.3.6). Manne et al. (1997) have investigated the aggregation of some straight chain di-cationic quaternary ammonium surfactants at mica/water and graphite/water interfaces using AFM. The general structure of the surfactants used is

$$(C_nH_{2n+1})\left[N^+(CH_3)_2\right](CH_2)_s\left[N^+(CH_3)_2\right](C_mH_{2m+1})$$

The spacer group is $(CH_2)_s$ and the overall structure is abbreviated to $C_{n\text{-}s\text{-}m}$ in which n and m are the chain lengths of the two alkyl groups. For symmetric surfactants ($n = m$) the head group size increases with s; repulsion between the two centres of positive charge causes the spacer group to be extended. Symmetric Gemini surfactants with small spacer groups ($s = 2$ or 3) give cylindrical micelles in solution. Larger spacer groups give rise to spherical micelles in solution. Highly asymmetric Gemini surfactants, especially

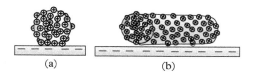

(a)   (b)

Figure 7.14 *Schematic representation of (a) a spherical cationic micelle and (b) a cylindrical (rod) cationic micelle at a negatively charged solid/aqueous solution interface.*

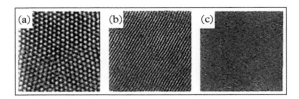

Figure 7.15 *AFM images (150 nm × 150 nm) of Gemini surfactants adsorbed at the mica–solution interface. Solutions were all above the bulk cmc. (a) Hexagonally packed spherical micelles of C_{18-3-1}; solution concentration = 3.0 mM. (b) Parallel cylinders of $C_{12-4-12}$; solution concentration = 2.2 mM. (c) Featureless bilayer of $C_{12-2-12}$; solution concentration = 1.0 mM. Reprinted with permission from S. Manne, T.E. Schäffer, Q. Huo, P.K. Hansma, D.E. Morse, G.D. Stucky, and I. A. Aksay, Langmuir, 1997, 13, 6382. Copyright 1997 American Chemical Society.*

Table 7.2 *Aggregate shapes of some Gemini surfactants in bulk solution and at interfaces*

surfactant	aqueous solution	mica/solution	graphite/solution
C_{18-3-1}	spheres	hexagonally packed spheres	parallel half cylinders
$C_{12-4-12}$	spheres and spheroids	parallel cylinders	parallel half cylinders
$C_{12-2-12}$	cylinders	bilayer	parallel half cylinders

when *m* is unity or very small, behave somewhat like conventional ('monomer') cationic surfactants with large repulsive head groups and consequently form spherical micelles in solution.

Some AFM images of surface micelles of Gemini surfactants at mica/solution interfaces, obtained by Manne et al. (1997), are reproduced in Fig. 7.15. The aqueous solutions are all above the *cmc* of the surfactants. The left-hand image shows hexagonally packed spherical micelles of the asymmetric surfactant C_{18-3-1}. The nearest neighbour distance is 8.8 nm, which is more than 50% greater than the micelle diameter. The centre image shows parallel cylindrical micelles (aligned with the mica symmetry axes) of the symmetric surfactant $C_{12-4-12}$. The spacing in this case (4.3 nm) is about twice the length of the dodecyl chains. The right hand image for $C_{12-2-12}$ is featureless and the adsorbed layer is taken to be a planar bilayer. The micelle structures for some surfactants in bulk aqueous solution and at the mica/water and graphite/water interfaces are given in Table 7.2.

In summary, adsorption (above the *cmc*) of cationic surfactants at the mica/water interface gives micelles whose shape is largely determined by surfactant curvature requirements, but the effective area of the surfactant head group is reduced by its interaction with the surface. This means that micelles at the surface have a lower net curvature than those formed in bulk solution. In the case of adsorption onto graphite, the chain-surface interaction has a drastic effect on the surface micelle structure. Thus, for example, hemicylindrical micelles rather than full cylinders are formed.

7.6   Adsorption of polymeric surfactants

Many polymeric materials adsorb at surfaces for a variety of reasons (Holmberg et al., 2003). When a polymer molecule adsorbs it displaces solvent molecules in contact with the surface, which each gain three degrees of translational freedom giving rise to a positive change in entropy. The polymer molecules on the other hand lose a little more than three degrees of freedom, and since one polymer molecule displaces many solvent molecules the net change in entropy accompanying adsorption is positive, giving a negative contribution to the free energy of adsorption. The larger the polymer molecule the larger will be this entropic effect. If a polymer is only slightly soluble in the solvent a solid surface can provide nucleation sites for 'precipitation' of the polymer, providing a further driving force for adsorption. Finally, as for low molar mass surfactants, there can be specific interactions between surfactant moieties and the solid surface.

It is a little arbitrary what one chooses to call a polymeric surfactant. Some authors include only those polymers which, in addition to adsorbing at the solid/liquid interface, also adsorb at the liquid/vapour and liquid/liquid interfaces and aggregate in aqueous solution (see §1.1 and e.g. Porter, 1994)). Perhaps the most useful polymer surfactants satisfying these criteria are block and graft copolymers (§1.3.4). Other authors however also include homopolymers, for example, which do not adsorb strongly at liquid surfaces but can adsorb quite strongly at a solid/liquid interface. Although not amphiphilic in the normal sense of the word, a homopolymer can contain moieties which interact specifically with a solid interface. Here, only those polymers which contain both hydrophilic and hydrophobic groups and which adsorb strongly at liquid (as well as solid/liquid) surfaces and tend to form micelles in aqueous solution are considered.

Molecules of block and graft copolymers at the solid/liquid interface are shown schematically in Fig. 7.16. In all cases block B is adsorbed at the interface and A groups extend into the solvent. The graft copolymer has what is often referred to as a 'comb structure', for obvious reasons. The block and graft copolymer surfactants are widely used to stabilize dispersions, which is in large measure why they are of such scientific interest and practical importance.

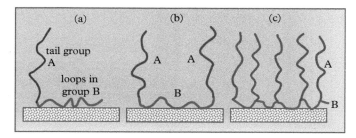

Figure 7.16 *Polymeric surfactants adsorbed at a solid/liquid interface. (a) AB block copolymer. (b) ABA block copolymer. (c) Graft copolymer with chains A grafted onto backbone B.*

Graft copolymers can have hydrophilic backbones with hydrophobic side chains, or vice versa. Examples of those with a hydrophilic backbone include lipopolysaccharides, (which occur naturally as major components of the outer membrane of Gram-negative bacteria) and hydrophobized starch (Fig. 7.17 in which R is an alkyl group). The silicone polyether surfactant illustrated in Fig. 1.9 is an example of a graft copolymer with a hydrophobic backbone and hydrophilic side chains. Another example is where the backbone is a poly(methyl methacrylate) chain with hydrophilic polyethylene oxide (PEO) side chains.

Block copolymers with PEO blocks and polypropylene oxide (PPO) blocks have been widely available commercially for many years, and more recently polymer surfactants in which the PPO block is replaced with a poly(1,2-butylene oxide) (PBO) block have also become available. The PEO blocks are hydrophilic and the PPO and PBO blocks are hydrophobic. The size of the various blocks can be adjusted as can the overall molar mass, making the surfactants rather versatile. With reference to Fig. 7.16, blocks A can be the hydrophilic or hydrophobic groups and B groups are then hydrophobic or hydrophilic, respectively.

The experimental adsorption isotherms for polymeric surfactants are generally 'Langmuirian' in shape; there is a more or less sharp rise in adsorption at low concentrations in solution, and then the adsorption tends to a plateau. Examples are

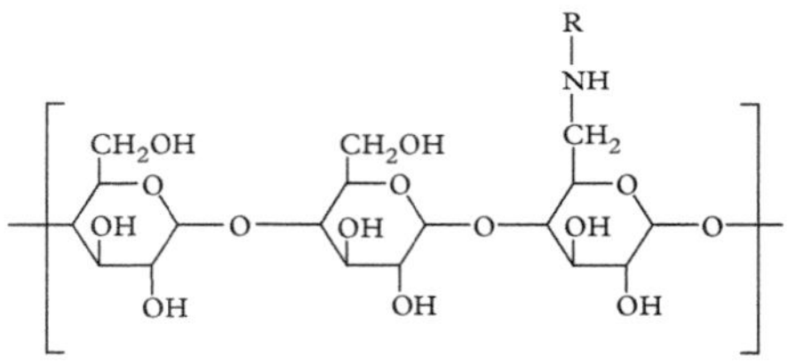

Figure 7.17 *Unit of a hydrophobized starch molecule; R is an alkyl group.*

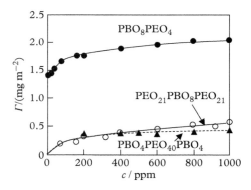

Figure 7.18 *Adsorption isotherms for PEP/PBO di- and triblock copolymers (as indicated) on hydrophobic methylated silica surfaces. Redrawn from Schillén et al. (1997).*

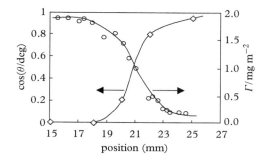

Figure 7.19 *Left ordinate: cosine of contact angle of water on hydrophobized silica as a function of the position along the surface; surface is hydrophobic on the left ($\theta = 90°$) and hydrophilic on the right. Right ordinate: adsorption of $PEO_{37}PPO_{56}PEO_{37}$ as a function of position on the solid surface. Redrawn from van de Steeg and Gölander, (1991).*

given in Fig. 7.18 where isotherms are shown for the adsorption of PEO/PBO di- and triblock copolymers onto hydrophobized silica, obtained by Schillén et al. (1997) using ellipsometry. Equilibrium solution concentrations c are given in parts per million by weight (ppm). The surfactants are anchored to the surface by the PBO moieties and the hydrophilic PEO groups extend into the aqueous solution.

The strength of adsorption of block copolymers from water depends markedly on just how hydrophobic or hydrophilic a surface is as illustrated by the work of van de Steeg and Gölander (1991), who studied the adsorption of PEO/PPO block copolymers ('Pluronics') onto solid surfaces with graded hydrophobicity (silicon wafers treated using a diffusion technique with dimethyldichlorosilane). The surface wettability was determined using receding contact angles (θ) with water (see §16.3.2); cos θ is plotted against position along the surface in Fig. 7.19. The plate is hydrophobic on the left and hydrophilic on the right. The adsorption of the triblock copolymer $PEO_{37}PPO_{56}PEO_{37}$ from water is also shown as a function of position on the plate in Fig. 7.19. As is clearly seen, the adsorption is strongest on the hydrophobic parts of the plate and falls off rapidly as the plate becomes more hydrophilic. It is obvious that the PEO moieties therefore do not adsorb significantly onto the hydrophilic surface.

As mentioned, adsorbed polymeric surfactants are widely used to confer stability on particulate dispersions, both in polar and nonpolar media. The way in which surfactants, including polymeric surfactants, influence the forces between dispersed particles is discussed in Chapters 11 and 13.

REFERENCES

S.L. Brantley and N.P. Mellott, Surface area and porosity of primary silicate minerals. *Am. Mineral.*, 2000, 85, 1767–1783.

J.S. Clunie and B.T. Ingram, Adsorption of nonionic surfactants, in *Adsorption from Solution at Solid/Liquid Interface* (ed. D.G. Parfitt and C.H. Rochester), Chapter 3. Academic Press, London, 1983.

J.M. Corkill, J.F. Goodman, and J.R. Tate, Adsorption of non-ionic surface-active agents at the Graphon/solution interface. *Trans. Faraday Soc.*, 1966, **62**, 979–986.

J.M. Corkill, J.F. Goodman, and J.R. Tate, Adsorption of alkylsulphinylalkanols on Graphon. *Trans. Faraday Soc.*, 1967, **63**, 2264–2269.

D.W. Fuerstenau, in *The Chemistry of Biosurfaces*, Vol. 1 (ed. M.L. Hair), Chapter 4. Marcel Dekker, New York, 1971.

M. Gaudin and D.W. Fuerstenau, Streaming potential studies. Quartz flotation with cationic collectors. *Trans. AIME*, 1955, **1**, 958–962.

S.J. Gregg and K.S.W. Sing, *Adsorption, Surface Area, and Porosity*. Academic Press, New York, 1982.

W.S. Hanna and P. Somasundaran, in *Improved Oil Recovery by Surfactant and Polymer Flooding* (ed. D.O. Shah and R.S. Schechter), p.253. Academic Press, New York, 1977.

R.J. Hunter, *Zeta Potential in Colloid Science*. Academic Press, London, 1981.

R.J. Hunter, *Introduction to Modern Colloid Science*. Oxford Science Publications, Oxford, 1993, Chapter 8.

K. Holmberg, B. Jőnsson, B. Kronberg, and B. Lindman, *Surfactant and Polymers in Aqueous Solution*, 2nd edition. John Wiley & Sons Ltd, 2003.

H. Kawasaki, M. Shinoda, M. Miyahara, and H. Maeda, Reversible pH-induced transformation of micellar aggregates between hemicylinders and laterally homogeneous layers at graphite–solution interfaces. *Colloid Polym. Sci.*, 2005, **283**, 359–366.

Z. Kiraly, G. Findenegg, E. Klumpp, H. Schlimper, and I. Dékány, Adsorption calorimetric study of the organization of sodium n-decyl sulfate at the graphite/solution interface. *Langmuir* 2001, **17**, 2420–2425.

I. Langmuir, The constitution and fundamental properties of solids and liquids. II. Liquids. *J. Am. Chem. Soc.*, 1917, **39**, 1848–1906.

J. Lyklema, in *Solid/Liquid Dispersions* (ed. T.F. Tadros), Chapter 3. Academic Press, London, 1987.

S. Manne, T.E. Schäffer, Q. Huo, P.K. Hansma, D.E. Morse, G.D. Stucky, and I.A. Aksay, Gemini surfactants at solid–liquid interfaces: control of interfacial aggregate geometry. *Langmuir*, 1997, **13**, 6382–6387.

K.G. Mathai and R.H. Ottewill, Stability of hydrophobic sols in the presence of non-ionic surface-active agents. Part 1. Electrokinetic and adsorption measurements on silver iodide sols and suspensions. *Trans. Faraday Soc.*, 1966, **62**, 750–758.

G.D. Parfitt and N.H. Picton, Stability of dispersions of graphitized carbon blacks in aqueous solutions of sodium dodecyl sulphate. *Trans. Faraday Soc.*, 1968, **64**, 1955–1964.

G.D. Parfitt and C.H. Rochester (eds.), *Adsorption from Solution at the Solid/Liquid Interface*. Academic Press, London, 1983.

H.N. Patrick, G.G. Warr, S. Manne, and I. A. Aksay, Surface micellization patterns of quaternary ammonium surfactants on mica. *Langmuir*, 1999, **15**, 1685–1692.

M.R. Porter, *Handbook of Surfactants*, 2nd edition. Blackie, London, 1994.

K. Schillén, P.M. Claesson, M. Malmsten, P. Linse, and C. Booth, Properties of poly (ethylene oxide)–poly (butylene oxide) diblock copolymers at the interface between hydrophobic surfaces and water. *J. Phys. Chem. B*, 1997, **101**, 4238–4252.

J.L. Trompette, J. Zajac, E. Keh, and S. Partyka, Scanning of the cationic surfactant adsorption on a hydrophilic silica surface at low surface coverages. *Langmuir*, 1994, **10**, 812–818.

L.M.A. van de Steeg and C.-G. Gölander, Adsorption of ethylene oxide/propylene oxide copolymers on hydrophobicity gradient surfaces. *Colloids Surf.*, 1991, 55, 105–119.

T. Wakamatsu and D.W. Fuerstenau, The effect of hydrocarbon chain length on the adsorption of sulfonates at the solid/water interface. *Adv. Chem. Ser.*, 1968, 79, 161–172.

G.G. Warr, Surfactant adsorbed layer structure at solid/solution interfaces: impact and implications of AFM imaging studies. *Curr. Opin. Colloid Interface Sci.*, 2000, 5, 88–94.

X.L. Yin, L.J. Wan, Z.Y. Yang, and J.Y. Yu, Self-organization of surfactant molecules on solid surface: an STM study of sodium alkyl sulfonates. *Appl. Surf. Sci.*, 2005, **240**, 13–18.

8

Direct characterization of adsorbed surfactant layers

Much of the information presented so far about the nature of adsorbed surfactant layers has been obtained indirectly. For example, in the case of adsorption at liquid/vapour and liquid/liquid interfaces, aspects of the structure of surfactant monolayers were deduced from the way in which the adsorption lowers the interfacial tension as a function of the surfactant concentration in a bulk phase. From this, surface pressures, π, and surface concentrations, Γ, can be obtained which allows surface equations of state (or adsorption isotherm equations) based on assumed models of monolayer behaviour to be tested. Such an indirect approach, mixed with some intuition and a little common sense, can reveal a fair amount but is inevitably limited, and many details of structure are unobtainable in this way. In the last few decades however a number of experimental techniques have become available which allow more direct probing of adsorbed surfactant layers at liquid/fluid and solid/liquid interfaces, some of these techniques being relatively cheap and laboratory based. Below, brief overviews of some of the more useful modern techniques are presented.

There are of course other important methods of characterization of surfactant layers than are presented here (e.g. surface light scattering, X-ray reflection, nonlinear optical methods, etc.) and the interested reader can find out more in a valuable book edited by Binks (1999).

8.1 Neutron reflection

Neutron reflection is obviously not one of the laboratory-based methods, but it is probably the most powerful technique that has been employed for probing the details of adsorbed surfactant film structures. It has been widely used to study adsorbed layers at air/liquid and solid/liquid interfaces. The neutron beam has to pass through one or other of the phases to reach the interface, air in the case of the liquid/air interface. Crystalline solids are sufficiently transparent to neutrons, which can therefore approach the

Surfactants: In Solution, at Interfaces and in Colloidal Dispersions. Bob Aveyard. © Bob Aveyard 2019.
Published in 2019 by Oxford University Press. DOI: 10.1093/oso/9780198828600.001.0001

solid/liquid interface through the solid phase. Liquids, however, are not very transparent and so it is rather difficult to study films at, for example, the hydrocarbon/water interface. The oil film needs to be very thin and of constant thickness during the experiment, and unless a null reflecting oil is used, reflection from the oil/air interface is picked up in addition to reflection from the oil/water interface.

A great deal of pioneering work on the use of neutron reflection in the study of adsorbed surfactant layers has been done over the years by R.K. Thomas and co-workers, who have written various comprehensive reviews on the subject (e.g. Penfold and Thomas, 1990, Lu and Thomas, 1998, Thomas, 1999, 2004). The method is particularly important because it can reveal a great deal about the layer structure including the composition of the layer, the surface concentration of surfactant (or surfactants in mixed systems) and importantly, the separation of the different groupings within a molecule normal to the plane of the surface. As Thomas points out (e.g. Thomas, 1999), although models are often used in the interpretation of experimental results, the calculated quantities depend very little on the chosen model.

In a neutron reflection experiment, a collimated (i.e. parallel) beam of neutrons is specularly reflected (Box 8.1) from (between 1 and 50 cm^2 of) a flat surface at glancing angles θ of less than about 5°. The reflected beam intensity is a function of both θ and the neutron wavelength λ as well as of incident beam intensity. A reflectivity curve is obtained which is a plot of reflectivity $R(Q)$ against the so-called *momentum transfer Q* (modulus of the wave vector—see (A10.4) defined as

$$Q = 4\pi \sin\theta/\lambda \qquad (8.1.1)$$

Reflectivity, $R(Q)$, is the intensity of the reflected beam divided by that of the incident beam. The wavelength λ for neutrons is typically about 0.1–1 nm, which is about 10^{-3} of the wavelength of visible light. Consequently neutrons can be used to probe films with thicknesses of 10^{-3} of the thicknesses of films (e.g. soap films) that can usefully be studied using visible light. The neutron reflection experiment essentially senses the changes in neutron refractive index through and normal to the film, and the neutron refractive index can be related to composition through the film. The same is not true for light where the relationship between composition and refractive index is not well-enough understood.

Box 8.1 Specular reflection

In *specular* reflection (which occurs from mirror-like surfaces—*speculum* is Latin for mirror) the angle of incidence of the beam is equal to the angle of reflection. In contrast, *diffuse* reflection is given from rough (matt) surfaces and occurs over a broad band of angles. In the reflection experiment the angle usually referred to is that of the beam with respect to the plane of the surface rather than the surface normal. This glancing angle is low for grazing incidence.

The neutron refractive index n is given by

$$n^2 = 1 - \frac{\lambda^2 \rho}{\pi} \qquad (8.1.2)$$

in which ρ, the *scattering length density* of the material in the film, is defined as

$$\rho = \sum_i b_i n_i \qquad (8.1.3)$$

In (8.1.3) b_i is the scattering length of the nucleus i (an empirically determined, known quantity which depends on the amplitude of the neutron scattering given by the nucleus) and n_i is the number density of nuclear species i. The chemical formula and bulk density of a material together with the scattering lengths allow the calculation of the refractive index of a material for neutrons of a given wavelength. Values of b for O, H, and D are, respectively, 5.8×10^{-6} nm, -3.74×10^{-6} nm, and 6.67×10^{-6} nm and since n_i for water molecules is 33 nm^{-3} values of ρ for H_2O and D_2O are, respectively, -0.56×10^{-4} nm^{-2} and $+6.3 \times 10^{-4}$ nm^{-2}. For neutrons with $\lambda = 1$ nm, say, it follows that the refractive index of H_2O is 1.000009 and of D_2O is 0.99990. Since these refractive indices are so close to unity it is usual to work with scattering length densities rather than neutron refractive indices.

The fact that the values of ρ for H_2O and D_2O have opposite signs can be used to great advantage in neutron reflection. A mixture of D_2O and H_2O with a mole fraction of $D_2O = 0.088$ has a zero scattering length density and a neutron refractive index of unity, the same as that of air (for which n_i in (8.1.3) is effectively zero). This means that neutrons are not reflected from the interface between air and this water mixture, termed *null reflecting water* (NRW). If therefore surfactant is adsorbed at the NRW/air interface (or any other interface between bulk phases for which the refractive indices have been matched) neutron reflection will occur only from the surfactant layer. In addition, specific parts of a surfactant molecule can be H/D labelled giving the possibility of measuring the thickness of individual fragments of a monolayer.

Since, in general, neutron refractive indices of different materials differ by only very little, the reflected intensity from an interface is low except at grazing incidence. Further, many neutrons pass into the sub-phase and are scattered, causing a background signal to the reflected beam. An additional factor to be considered in the interpretation of reflection data for fluid interfaces is surface roughness (Box 8.2), which is obviously relevant when considering the thickness of a surface layer.

Box 8.2 It's rough at the top

Although free liquid/fluid interfaces appear smooth to the naked eye, as a result of thermal motion of molecules there are spontaneously excited *capillary waves* in the interface which give rise to surface roughness (see e.g. Aarts et al., 2005). The interface tends towards its equilibrium flat state as a result of the restoring forces arising from interfacial tension γ and gravity. For a given projected plane surface area, the area of a rough interface exceeds that

Box 8.2 *Continued*

of a flat one and so therefore does its surface free energy. Gravitational forces arise from the vertical displacement of material in the waves. The surface waves are damped as a result of the viscosity of the adjacent bulk phases. The interfacial tension of an interface can be obtained from the spectrum of the (laser) light scattered by the rough interface (Pouchelon et al., 1980). The surface roughness is proportional to $(kT/\gamma)^{0.5}$ and the amplitudes of the capillary waves are of the order of a few Å. It is argued that capillary waves play an important role in the rupture of thin liquid films such as those present in foams and emulsions (see e.g. §12.4).

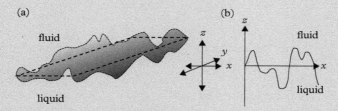

Representation of a thermally rough fluid interface. (a) A rectangular area of interface. (b) Schematic interface profile for a vertical cut in the x direction. After Aarts et al. (2005).

To give an idea of how information is obtained, the simple case of reflection from a planar uniform surfactant film (1), thickness d, at the smooth interface between phases 0 and 2, as illustrated in Fig. 8.1, is considered. A relationship between $R(Q)$ and Q is needed which involves a required quantity, say film thickness d or surface concentration Γ. An approximate relationship is (Thomas, 1999)

$$R(Q) = \frac{16\pi^2}{Q^2}|\rho(Q)|^2 \tag{8.1.4}$$

in which $\rho(Q)$ is the one-dimensional Fourier transform of $\rho(z)$, the mean scattering length density profile normal (direction z) to the surface, that is,

$$\rho(Q) = \int_{-\infty}^{\infty} \rho(z) \exp{(-iQz)} \, dz \tag{8.1.5}$$

For the case where the scattering length densities ρ_o and ρ_2 of the two bulk phases are matched, $(\rho_o = \rho_2 = \rho)$, it can be shown that

$$R(Q) = \frac{16\pi^2}{Q^4}\left[4(\rho_1 - \rho)^2\sin^2\left(\frac{Qd}{2}\right)\right] \tag{8.1.6}$$

where ρ_1 is the scattering length density of the uniform film. The equation describes interference fringes (amplitude $\rho_1 - \rho$) on top of a rapidly decaying reflectivity resulting from the dependence on $1/Q^4$. The film thickness d and ρ_1 can be obtained by fitting experimental reflectivity curves. For a smooth interface in the absence of an adsorbed film the reflectivity is given by

$$R(Q) = \frac{16\pi^2}{Q^4}(\Delta\rho)^2 \tag{8.1.7}$$

in which $\Delta\rho$ is the difference in scattering length densities of the two contacting bulk phases.

Two simulated reflectivity curves, generated using (8.1.6), are shown in Fig. 8.2 for surfactant layers at null reflecting interfaces ($\rho_0 = \rho_2$). Also shown is the reflectivity curve (a) for the H_2O/air surface in the absence of an adsorbed layer, obtained using (8.1.7). Curve (b) is typical of that given by an adsorbed monolayer of low molar mass surfactant at the NRW/air interface; values of d and ρ_1 used (2 nm and 4.7×10^{-4} nm^{-2}, respectively) are those reported for monolayers of fully deuterated tetradecyltrimethylammonium bromide adsorbed from an aqueous solution above the *cmc* (Simister et al., 1992). For thicker films (such as surfactant bilayers), interference fringes appear in the simulated reflectivities, as seen in curve (c). The latter was generated using d and ρ_1 appropriate for the adsorption of a bilayer of the fully deuterated alcohol ethoxylate $C_{12}E_6$ (see Chapter 1 for nomenclature) at the null reflecting interface between water and crystalline quartz, that is, $d = 4$ nm and $\rho_0 = \rho_2 = 4.2 \times 10^{-4}$ nm^{-2}. In practice, because of background scattering the accessible range of $R(Q)$ goes down to only about 10^{-6} or 10^{-7}.

In conclusion, it is indicated how the film thickness and the surface concentration Γ in an adsorbed film can be obtained from reflectivities. As discussed, both d and ρ_1 for a surfactant film can be obtained (dependently) by fitting the neutron reflectivity curve. The scattering length density of the (uniform) film, ρ_1, is related to the scattering length b_1 and number density n_1 of the whole molecule by $\rho_1 = b_1 n_1$, and the surface concentration Γ_1 of surfactant is given by

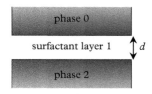

Figure 8.1 *A smooth planar surfactant layer 1, thickness d, between phases 0 and 2.*

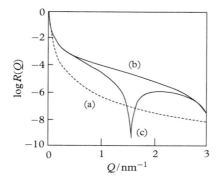

Figure 8.2 *Simulated neutron reflectivity curves. (a) Reflectivity at the air/H_2O surface in the absence of adsorbed layer. (b) Reflectivity from a homogeneous layer, thickness 2 nm and $\rho_1 = 4.7 \times 10^{-4}$ nm^{-2}, at the NRW/air surface. (c) Reflectivity for a layer, thickness 4 nm and $\rho_1 = 6.5 \times 10^{-4}$ nm^{-2}, at a null reflecting interface with $\rho_0 = \rho_2 = 4.2 \times 10^{-4}$ nm^{-2}.*

$$\Gamma_1 = dn_1 \tag{8.1.8}$$

If the reflectivity curve exhibits fringe(s) (Fig. 8.2, curve (c)) then of course the film thickness is known independently of Γ_1 by the position of, say, the first minimum.[1] This situation is not likely to be the case for a monolayer of low molar mass surfactant. However it is possible to obtain the surface concentration independently of d by a simple extrapolation procedure. As Q tends to zero, $\sin(Qd/2)$ in (8.1.6) tends to $Qd/2$ and so for a monolayer at the NRW/air interface ($\rho = 0$), (8.1.6) can be rewritten, in the limit of zero Q,

$$\lim_{Q \to 0} R(Q)Q^2 = 16\pi^2 d^2 \rho_1^2 = 16\pi^2 d^2 b_1^2 n_1^2 = 16\pi^2 b_1^2 \Gamma^2 \tag{8.1.9}$$

For fully deuterated tetradecyltrimethylammonium bromide (dC$_{14}$dTAB) above its *cmc*, for example, values of ρ_1 and d obtained by fitting experimental reflectivities to (8.1.6) were found to be 4.7×10^{-4} nm^{-2} and 2.0 nm, respectively (Simister et al., 1992). In a plot of $\log(R(Q)Q^2)$ versus Q (Fig. 8.3) generated using (8.1.6) and these values of ρ_1 and d, the intercept at zero Q corresponds to 1.4×10^{-4} nm^{-2}. From (8.1.9) this intercept is seen to be equal to $16\pi^2 b_1^2 \Gamma^2$. Since b_1 is known independently of the reflectivity experiment (and is equal to 4.05×10^{-4} nm), Γ can be calculated, and is found to be 2.27 molecule nm^{-2}, corresponding to an area per molecule $A = 0.44$ nm^2.

Much more information than described above can be gleaned from more sophisticated experiments (e.g. involving the use of selectively deuterated surfactants) and more

---

[1] In general minima appear when $\sin\theta = n\lambda/2d$, where n is an integer, so that from the position of the first minimum (n = 1), $d = 2\pi/Q$. Thus with reference to Fig. 8.2 curve (c) the minimum reflectivity occurs for $Q \sim 1.5708$ so that $d = 4$ nm, which was an input for d in generating the reflectivity curve.

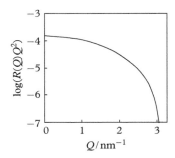

Figure 8.3 *Plot of log(R(Q)Q²) versus Q. The ordinate has been calculated using (8.1.6) with $\rho_1 = 4.7 \times 10^{-4}$ nm^{-2}, $\rho = 0$, and $d = 2.0$ nm. The intercept corresponds to $R(Q)Q^2 = 1.4 \times 10^{-4}$ nm^{-2}.*

detailed analysis (e.g. by treating the surfactant film as being made up of several layers of different composition). The interested reader should consult the reviews by Thomas and co-workers to get a better feel for the remarkable scope of the neutron reflection technique in probing the detailed nature of adsorbed surfactant layers.

8.2 Scanning probe microscopy

Some examples of the remarkable images that can be obtained using scanning probe microscopy (SPM) were presented in §7.5.3. SPM techniques include, *inter alia*, scanning tunnelling microscopy (STM) and atomic force microscopy (AFM), and commercial instruments are available for both. To set these techniques into some context, it is noted that the unaided eye can see objects with dimensions down to about 100 μm, the typical thickness of a human hair. Optical microscopes can image objects down to sizes of the order of 1 μm (e.g. bacteria and blood cells). The maximum resolutions of the transmission electron microscope (TEM) and scanning electron microscope (SEM) are, respectively, atomic dimensions and ca. 1 nm. Samples in both the latter cases need to be present in a vacuum. AFM in contrast can operate with imaging environments of air, liquid or vacuum, and the maximum resolution is down to atomic sizes. Instruments are small and, relative to say SEM or TEM, are inexpensive.

8.2.1 Scanning tunnelling microscopy

The scanning tunnelling microscope was invented by Binnig and Rohrer (Box 8.3) in 1981 (Binnig and Rohrer, 1982), and is capable of giving atomic scale images of the surfaces of electrically conducting or semi-conducting materials. The principle of the method involves applying a small bias voltage (a few mV up to 3 V) between an atomically sharp tip and the surface under investigation. When the tip and surface approach to within about 1 nm and less a tunnelling current of the order of pA to nA flows. Tunnelling is a quantum mechanical effect, and the current between the tip and the surface exhibits an exponential decay with separation. Very small changes in separation give rise to very large changes in tunnelling current resulting in atomic resolution when the tip is moved across the surface.

Box 8.3 Founders of scanning probe microscopies

Gerd Binnig (a German) and Heinrich Rohrer (Swiss) worked at the research laboratory of IBM in Zurich, Switzerland, and they shared the 1986 Nobel Prize in Physics for the development of STM (with Ernst Ruska who designed the first electron microscope). Earlier, the first successful SPM, the *topografiner*, was developed by Russell Young and colleagues (between 1965 and 1971) at the National Bureau of Standards (NBS) in America (Young, 1971).

An atomically sharp metal wire tip (of e.g. tungsten) is raster scanned across the sample surface (or the surface moved relative to the probe) using a piezoelectric scanning device (Fig. 8.4). In the *constant current* mode, feedback electronics using the tunnelling current maintains a constant distance between the tip and the surface so the tip follows the surface profile. STM is very useful for investigating structures of thin surfactant layers (with thickness less than the tunnelling distance) at solid/liquid interfaces (Rabe, 1999). Organic molecules can be oxidized or reduced at bias voltages of 1 V or more so voltages less than this must be employed, when the tunnelling distances are between 1 and 2 nm. Surfaces with flat adsorbed molecules are therefore well suited for study. For thicker layers, AFM can be employed.

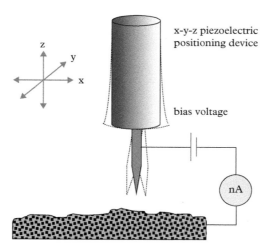

Figure 8.4 *Schematic representation of the operation of the STM. The atomically sharp tip can be moved in the x, y, and z directions across the surface, and the tunnelling current measured.*

8.2.2 Atomic force microscopy

AFM, which evolved from STM, appeared in 1986 (Binnig et al., 1986), and it has been used very widely in surfactant science and elsewhere. Since it can be applied to insulating substrates as well as surfaces of conducting and semi-conducting materials, it is extremely versatile and a number of commercial instruments are available. It can operate in ambient conditions or under liquid, and gives direct topographic information about interfaces, often with atomic resolution. Some images of surfactants adsorbed at solid/liquid interfaces obtained using AFM have been presented in Figs. 7.12 and 7.15. In addition to providing high resolution images of surfaces, AFM is also capable of the direct measurement of surface forces; such measurements are discussed in §11.5.

In an AFM experiment the sample rests on a piezo-electric positioning device that can be controlled in the x, y and z directions, with sub-nanometre resolution, as shown in Fig. 8.5. The sample is raster (line by line) scanned under a micro cantilever-tip assembly, usually made from Si or Si_3N_4. The apical tip radius is typically around 10 nm. The vertical (and possibly torsional) position of the tip is determined by reflecting laser light from the back of the cantilever assembly into a position sensitive photodiode detector. For the investigation of hard surfaces (e.g. of inorganic materials), which are not readily damaged by the passage of the tip, the instrument can be operated in the so-called *contact* or *constant force* mode. The surface is pushed against the tip using the piezoelectric positioning device (in the *z*-direction) until the required repulsive force

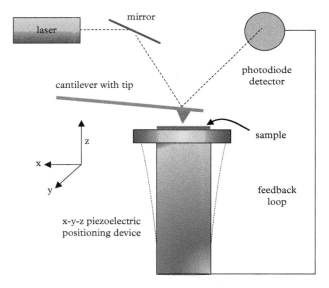

Figure 8.5 *Schematic representation of an AFM setup. The piezoelectric drive raster scans the sample over the tip on a cantilever. The cantilever/tip deflection is monitored by a laser beam reflected from the back of the cantilever onto a photodiode detector. The electronic feedback loop responds to the photodiode signal (see text).*

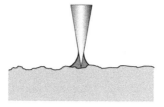

Figure 8.6 *Capillary-condensed water bridge between tip and surface.*

between tip and sample surface is achieved. This force might be of the order of 10^{-9} N, and tip and surface are in 'contact'. Any change in the force is accompanied by a change in deflection of the cantilever. If, during raster scanning of the surface by the tip, the measured deflection changes, the feedback loop applies a voltage to the piezo-electric positioning device, which raises or lowers the sample to restore the original deflection (force). The feedback voltage applied is a measure of the height of the surface under the tip.

Surface damage can be reduced by reducing the contact force but there are practical limits to how low a repulsive force can be controlled. Further, when operating in ambient conditions water vapour is present which can condense between tip and solid surface forming a liquid bridge, as illustrated in Fig. 8.6. The presence of this bridge gives rise to an *attractive* capillary force between tip and surface. The magnitude of the force depends on the mean curvature of the meniscus surface, which in turn depends on the tip geometry. The force however can be of the order of 10^{-7} N. To avoid this problem the experiment can be carried out under liquid if this is appropriate.

As mentioned, surfactant layers are 'soft' and are likely to be damaged by the tip when AFM is operated in the contact mode. Such problems can be overcome by using the *tapping* mode. The simple principle is to place the tip in contact with the surface (giving high resolution) and then lifting it to avoid dragging it across the surface during scanning. In air, the tip oscillation frequency is typically between 50 and 500 kHz and the amplitude of the vibration is greater than 20 nm. When the tip reaches, say, a high point in the surface during a scan the amplitude of the tip vibration is reduced. A feedback loop is used to maintain a constant amplitude during scanning by moving the piezo-electric table in the z-direction, thus reproducing the topographical features of the surface.

8.3   Ellipsometry

Ellipsometry is an optical technique for probing the properties of surfaces and thin films by observing the changes in polarization of light resulting from reflection from the interface. It is used to calculate both thickness and refractive index of surfactant layers, from which surface composition (with respect to surfactant and solvent) can be estimated.

Light has both electric and magnetic fields, but in the present context only the electric field is relevant, and here a plane wave of monochromatic light is considered. The electric field vector can be considered in terms of two orthogonal components designated x and y at right angles to the direction z of propagation of the light wave. For a simple harmonic wave the x and y components have the same frequency but can differ both in phase and amplitude. The polarization state of the light is described in terms of the projection of the tip of the electric vector on a plane at right angles to the z-direction. For linear polarized light, the x and y components are in phase, the direction of the electric vector is constant, and the projection on the plane is a straight line (Fig. 8.7a, dotted line). The direction of the line (i.e. of polarization) is determined by the relative amplitudes of the two components (not shown). If the x and y components have the same amplitude but differ in phase by 90°, the time evolution of the electric vector is a circular spiral, its projection is a circle, and the polarization is termed circular polarization (Fig. 8.7b). Polarization other than linear or circular (i.e. when phase and amplitude are arbitrary) is elliptical; the electric vector traces out an ellipse, as illustrated in Fig. 8.7c.

It is usual to speak of s- and p-directions rather than the x and y directions so far discussed. A plane of incidence of light falling on a planar sample is defined as the plane determined by the direction of the light beam and the surface normal, as illustrated in Fig. 8.8a.

Then the p-direction is parallel to the plane of incidence and the s-direction is parallel to the surface. Correspondingly the electric field E can be resolved into its p and s-components.

As mentioned, reflection from a surface changes the ellipticity of monochromatic light. The relative *amplitudes* of the p and s-components change on reflection as does the *phase difference* between the two components. The ellipsometric experiment described below yields the complex amplitude reflection ratio, ρ, defined as (see e.g. Jenkins, 1999)

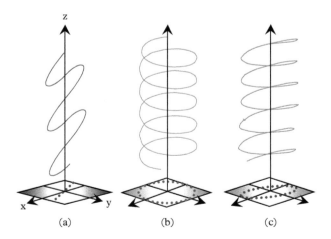

Figure 8.7 *Electric field vectors in polarized monochromatic light: (a) linearly polarized; (b) circularly polarized; (c) elliptically polarized.*

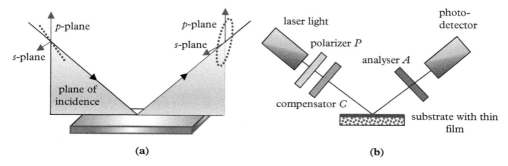

Figure 8.8 *(a) The incidence plane is defined by the direction of the light beam and the normal to the surface. The direction p is parallel to the incidence plane and the s-plane is parallel to the plane sample surface. Directions p and s replace x and y used in Fig. 8.7. In the example shown the incident wave is linearly polarized and the specularly reflected wave is elliptically polarized (heavy dashed straight line and ellipse respectively). (b) A schematic possible setup for ellipsometry. The elements of the ellipsometer are a laser light source, a polarizer, a quarter-wave plate (compensator), a planar sample surface, an analyser, and a photo-detector.*

$$\rho = r_p / r_s \tag{8.3.1}$$

The quantities r_s and r_p are the reflection amplitude ratios for s- and p-polarized light, respectively. They are complex quantities and can be written

$$r_s = | r_s | \exp(i\delta_{rs}); r_p = | r_p | \exp(i\delta_{rp}) \tag{8.3.2}$$

where δ_{rs} and δ_{rp} are phase shifts upon reflection of the s and p components, respectively. From (8.3.1) and (8.3.2),

$$\rho = (|r_p|/|r_s|) \exp i (\delta_{rp} - \delta_{rs}) = (|r_p|/|r_s|) \exp(i\Delta) \tag{8.3.3}$$

The ellipsometric parameter $\Delta = (\delta_{rp} - \delta_{rs})$ is the change in the phase difference between p and s-components caused by reflection. A further ellipsometric parameter ψ is defined as

$$| r_p | / | r_s | = \tan \psi \tag{8.3.4}$$

and so

$$\rho = \tan \psi \exp(i\Delta) \tag{8.3.5}$$

There are various possible configurations for an ellipsometric experiment, and an example is given in Fig. 8.8b. Unpolarized monochromatic (laser) light is passed through a polarizer (P) to give linearly polarized light and then through a compensator C (a quarter-wave plate—a birefringent crystal such as calcite) giving elliptically polarized light. The ellipticity is adjusted using P and C in such a way that light reflected from the sample surface is linearly polarized. The reflected light is passed through an analyser A (also a linear polarizer) and onto a photo-detector. The angular setting of A causes extinction of the light (minimum signal from the detector) when its axis is at 90° to the direction of polarization of the reflected light.

A common method of conducting an ellipsometric experiment is to fix the compensator position at some suitable angle (the plane of rotation being perpendicular to the plane of incidence) and rotate only the polarizer and analyser. P is rotated to give minimum signal at the detector then A is rotated, keeping P at its minimum signal position, to give minimum signal. The procedure is repeated iteratively to give results of the required accuracy. From the orientations of P, C, and A which give the minimum (null) signal it is possible to calculate the ellipsometric parameters ψ and Δ defined above.

In order to obtain the required film properties (thickness, refractive index) it is necessary to carry out some optical modelling and hence calculate theoretical values of ψ and Δ which closely match those obtained by experiment. The simplest possibility is where the film is modelled as a single homogeneous layer between two bulk phases (e.g. a liquid and a solid substrate) (Fig. 8.9a). Some values of ψ and Δ calculated using such a model are shown in Fig. 8.9b where it has been assumed that $n_1 = 1.418$ (similar to dodecane) and $n_3 = 1.750$ (appropriate for alumina). The angle of incidence of the laser light ($\lambda = 632.8$ nm) is taken as 70°. A given curve corresponds to a fixed value of n_2 (as indicated) and is generated by varying the film thickness d_2 (between 0 and about 6 nm in the example shown). Then, essentially, a curve is generated by selection of a suitable value of n_2 such that the experimentally obtained ψ and Δ fall on that curve. Thus n_2 and d_2 are obtained for the adsorbed film. For a commercial instrument this process

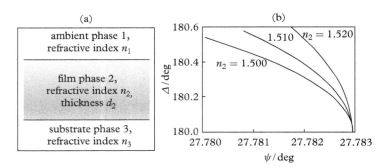

Figure 8.9 *Fig. 8.9 (a) Representation of a single layer model for a thin film, 2, thickness d_2 between two bulk media 1 and 3. (b) Relationship between ψ and Δ for this model. Each curve is generated by assuming values for n_1 (1.418 for dodecane), n_2 (for surfactant as indicated in the diagram), and n_3 (1.750 corresponding to alumina) and varying the film thickness.*

is carried out by computer using included software. If the refractive index of the pure surfactant is known independently, and the way in which the refractive index of solutions of surfactant in liquid 1 is known (or assumed) the composition of the adsorbed layer can be estimated from the value of n_2 obtained from ellipsometry.

The example given above is the simplest case, and models that involve more layers can be employed. It should be noted that if any of the phases involved is not transparent the refractive index, $\tilde{n}$, is a complex quantity expressed as

$$\tilde{n} = n - i\mathcal{K} \tag{8.3.6}$$

where i is the square root of -1 and $\mathcal{K}$ is the extinction coefficient which accounts for loss due to absorption of light. The method becomes more versatile when variable angle of incidence of the light is employed and measurements are made over a range of wavelengths (spectroscopic ellipsometry).

8.4 Quartz crystal microbalance

The basic principle of the quartz crystal microbalance (QCM) relies on the observation that the resonant frequency of oscillation of a piezoelectric (often) quartz crystal (see Box 8.4) depends on the mass of an adsorbed or deposited film present on its surface. Usually, as illustrated in Fig. 8.10, a QCM consists of a circular wafer of AT cut quartz with thin (say 200 nm) gold electrodes with diameter of the order of a cm, one on each face. The active area (over which adsorption is measured) in a QCM experiment is that where the electrodes overlap. The crystal is caused to oscillate (in the MHz region), in the shear-thickness mode (as illustrated in Fig. 8.12a), by an alternating voltage signal applied across the electrodes at a characteristic resonant frequency, ν_o. For adsorption from a vapour phase, where the adsorbed film is likely to be rigid and hence oscillate in phase with the crystal, the change in frequency, $\Delta\nu$, caused by the mass of the adsorbed film on a perfectly smooth surface, is given by the Sauerbrey (1959) equation

$$\Delta\nu = -\left[2n\nu_o^2\left[\mu_q\rho_q\right]^{-0.5}\right]\Delta m_u \tag{8.4.1}$$

Here Δm_u is the mass change on the active part of the crystal per unit area, and μ_q is the shear modulus and ρ_q the density of quartz, 2.947×10^{11} g cm^{-1} and 2.648 g cm^{-3}, respectively. The quantity n is the number (1 or 2) of sides of the crystal on which adsorption occurs. If the electrode surface has a degree of roughness then the actual surface area will exceed the geometrical area of the electrode(s).

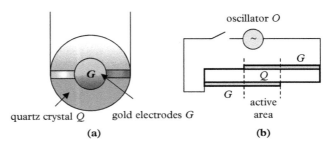

Figure 8.10 *Quartz crystal microbalance (QCM). (a) Gold electrodes G on upper (dark shading) and lower sides of the quartz crystal Q. The circular parts of the electrodes overlap on the two sides. (b) Edge view of electrodes G on quartz crystal Q. The electrodes are connected to an oscillator O.*

Box 8.4 The piezoelectric effect

The piezoelectric effect was discovered by Pierre Curie and his elder brother Jacques who carried out a series of experiments in Paris between 1878 and 1882. The name derives from the Greek *piezein*, to press; electric polarity is induced in crystals with no centre of symmetry by compression and dilation. Conversely when a voltage is applied to the crystal it becomes deformed due to lattice strains, and the strains are reversed when the polarity is reversed.

'AT cut' is a commercial designation and refers to the angle at which the quartz plate is cut relative to the optic axis of the crystal. AT cut quartz wafers have good temperature stability around room temperature and vibrate in the thickness-shear mode (see Fig. 8.12a). The fundamental frequency of a shear-thickness resonator is inversely related to its thickness.

To give an idea of the sensitivity of the method, changes in frequency, $\Delta \nu$, with time are shown in Fig. 8.11 for the adsorption of decane vapour onto an octadecanethiol layer coating a gold electrode on a quartz crystal (Aveyard et al., 1999); at adsorption equilibrium the frequency has fallen by about 53 Hz. The surface concentration of decane at equilibrium is about five molecules nm^{-2} which means that an adsorbed mass of ca. 2.5 ng cm^{-2} reduces the frequency by 1 Hz.

In the example just quoted, the gold electrode surfaces were rendered hydrophobic by treating with octadecanethiol. In general however, the sensor surface can be covered with any material that can be applied as a thin film that adheres to the underlying surface. A range of pre-coated surfaces can be obtained commercially, including a number of metals, metal oxides and spin coated polymers.

For a crystal oscillating in a liquid there will be some coupling between the crystal surfaces and liquid, for which the Sauerbrey equation does not account. The change in frequency $\Delta \nu'$ of a crystal when it is transferred from a vacuum (or air) into a liquid depends on the viscosity, η_l, and density, ρ_l, of the liquid according to (Kanazawa and Gordon, 1985)

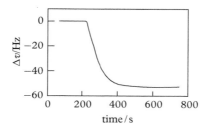

Figure 8.11 *Change in resonance frequency of quartz crystal following adsorption of decane vapour on octadecanethiol layer coating the gold electrode. Decane relative pressure = 0.93. Redrawn from Aveyard et al. (1999).*

$$\Delta v' = -n v_o^{3/2} (\eta_l \rho_l)^{1/2} (\pi \mu_q \rho_q)^{-1/2} \qquad (8.4.2)$$

This can make the use of the QCM in its simple form a little uncertain for the study of adsorption of surfactants from solution. Nonetheless, for Newtonian liquids with low viscosity and for sufficiently thin adsorbed films, a linear relationship between adsorbed mass and frequency change may still exist. In these cases however, the thickness obtained by use of the Sauerbrey equation is higher than the thickness obtained by other techniques. It should be noted that the QCM technique detects solvent bound to the adsorbed molecules.

For macromolecules, for example, proteins, adsorbed layers are not rigid (as required for the correct application of (8.4.1)), and they may well contain 'bound' solvent molecules. In addition adsorbed macromolecules may slip on an oscillating crystal surface (Höök et al., 2001). Clearly such 'soft' films are not fully coupled with the oscillating surface of a quartz crystal.

The viscoelasticity can however be probed by measuring the dissipation (damping) of the crystal shear movement (Fig. 8.12a) when the signal to the crystal is removed (the technique now being referred to as QCM-D). The dissipation shift, ΔD, is the change in the absolute dissipation caused by adsorption of the film and is defined by

$$\Delta D = \frac{E_{dis}}{2\pi E_{stored}} \qquad (8.4.3)$$

where E_{dis} and E_{stored} are, respectively, the dissipated and the stored energy per oscillation. With regard to the dissipation, adsorption introduces 2 new interfaces, one being the film/substrate interface and the other the interface between the film and solution. The crystal/solvent interface is removed. It can be argued that slippage at the new interfaces would be expected to give a decrease in the dissipation (ΔD negative). If an *increase* in dissipation is observed (see results for protein adsorption in Fig. 8.13) this can be attributed to deformation of the film during oscillation. By studying this damping and employing frequencies at several overtones it is possible, using an appropriate

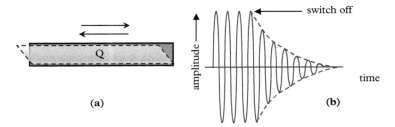

(a)   **(b)**

Figure 8.12 *(a) Thickness-shear motions of oscillating quartz crystal. (b) Decay of the amplitude of crystal oscillation (damping) after removal of signal from oscillator.*

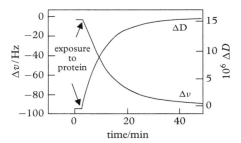

Figure 8.13 *Δv and ΔD obtained by QCM-D for the adsorption of the mussel adhesive protein Mefp-1 from an acetate buffer solution onto a hydrophobized gold electrode surface. Redrawn from Höök et al. (2001).*

model of film viscoelasticity, to obtain an estimate of film thickness as well as rheological properties of the films (Höök et al., 2001).

8.5   Brewster angle microscopy (BAM)

This imaging technique is based on the observation that the reflectivity coefficient of p-polarized light (electric field vector parallel to the plane of incidence—see §8.3) vanishes at a certain angle of incidence to a surface, termed the Brewster angle (θ_B) (after the Scottish scientist and inventor Sir David Brewster (1781–1868)). The angle (which is measured relative to the surface normal) is related to the refractive indices of the contacting phases, say air (n_1) and water (n_2) by $\theta_B = \tan^{-1}(n_2/n_1)$. Suppose p-polarized light impinges at the Brewster angle on the surface of water with a phase-separated adsorbed monolayer present. The monolayer consists of regions of dilute two-dimensional gas in contact with condensed regions of surfactant. The dilute regions have little effect on the light and since there is no reflection at the Brewster angle, these regions appear as dark background. The much more concentrated condensed regions (domain size say 20 μm and up with refractive index different from those of the surrounding

Figure 8.14 *BAM image of a condensed region in an adsorbed film of N-dodecyl-γ-hydroxybutyric acid amide at the air/water interface about 55 min after adsorption at 15°C. Reprinted with permission from D. Vollhardt and V. Melzer, J. Phys. Chem. B, 1997, 101, 3370. Copyright 1997 American Chemical Society.*

phases) do however reflect light to some extent (ca. 10^{-6} intensity of incident light), and this can be used to form an image of the condensed phase regions. A BAM image of a condensed region of a soluble film of N-dodecyl-γ-hydroxybutyric acid amide at the air/water interface is shown in Fig. 8.14.

REFERENCES

D.G.A.L. Aarts, M. Schmidt, H.N.W Lekkerkerker, and K.R. Mecke, Microscopy on thermal capillary waves in demixed colloid–polymer systems. *Adv. Solid State Phys.* 2005, **45**, 15–27.

R. Aveyard, B.D. Beake, and J.H. Clint, Alkane adsorption from vapour onto hydrophobic solid/vapour and hydrophobic solid/water interfaces. *Phys. Chem. Chem. Phys.*, 1999, **1**, 2513–2520.

B.P. Binks (Ed.), *Modern Characterisation Methods of Surfactant Systems*. Surfactant Science Series, Volume 83. Marcel Dekker, New York, 1999.

G. Binnig and H. Rohrer, Scanning tunneling microscopy. *Helv. Phys. Acta*, 1982, **55**, 726–735.

G. Binnig, C.F. Quate, and C. Gerber, Atomisk kraftmikroskop. *Phys. Rev. Lett.*, 1986, **56**, 930–933.

F. Höök, B. Kasemo, T. Nylander, C. Fant, K. Sott, and H Elwing, Variations in coupled water, viscoelastic properties, and film thickness of a Mefp-1 protein film during adsorption and cross-linking: a quartz crystal microbalance with dissipation monitoring, ellipsometry, and surface plasmon resonance study. *Anal. Chem.*, 2001, **73**, 5796–5804.

T.E. Jenkins, Multiple-angle-of-incidence ellipsometry. *J. Phys. D: Appl. Phys.* 1999, **32**, R45.

K.K. Kanazawa and J.G. Gordon, The oscillation frequency of a quartz resonator in contact with liquid. *Anal. Chim. Acta*, 1985, **175**, 99–105.

J.R. Lu and R.K. Thomas, Neutron reflection from wet interfaces. *J. Chem. Soc., Faraday Trans.*, 1998, **94**, 995–1018.

J. Penfold and R.K. Thomas, The application of the specular reflection of neutrons to the study of surfaces and interfaces. *J. Phys.: Condens. Matter*, 1990, **2**, 1369.

A. Pouchelon, J. Meunier, D. Langevin, and A.M. Cazabat, Light scattering from oil–water interfaces: measurements of low interfacial tensions. *J. Phys. Lett.*, 1980, **41**, 239–242.

J.P. Rabe, in *Modern Characterisation Methods of Surfactant Systems* (ed. B.P. Binks), Chapter 2. Surfactant Science Series, Volume 83. Marcel Dekker, New York, 1999.

G.Z. Sauerbrey, The use of quartz oscillators for weighing thin layers and for microweighing. *Z. Phys.*, 1959, **155**, 206.

E.A. Simister, R.K. Thomas, J. Penfold, R. Aveyard, B.P. Binks, P. Cooper, P.D.I. Fletcher, J.R. Lu, and A. Sokolowski, Comparison of neutron reflection and surface tension measurements of the surface excess of tetradecyltrimethylammonium bromide layers at the air/water interface. *J. Phys. Chem.*, 1992, **96**, 1383–1388.

R.K. Thomas, in *Modern Characterisation Methods of Surfactant Systems* (ed. B.P. Binks), Chapter 11. Surfactant Science Series, Volume 83. Marcel Dekker, New York, 1999.

R.K. Thomas, Neutron reflection from liquid interfaces. *Ann. Rev. Phys. Chem.*, 2004, **55**, 391–426.

D. Vollhardt and V. Melzer, Phase transition in adsorption layers at the air–water interface: bridging to Langmuir monolayers. *J. Phys. Chem. B*, 1997, **101**, 3370–3375.

R.D. Young, Surface microtopography. *Phys. Today*, 1971, **24**, 42.

Section III

Aggregation of surfactants in solution

Above a critical concentration in aqueous solution, surfactants aggregate to give *micelles* (Chapter 9). If a similar volume of a nonpolar oil phase exists in equilibrium with the aqueous phase, surfactant can distribute between the phases, and aggregation can take place in either phase, depending on conditions such as temperature, salinity of the aqueous phase and chemical structure of the oil (Chapter 10). The size and shape of surfactant aggregates, which can be microemulsion droplets in systems with both oil and water present, depend on surfactant molecular structure and solvation, which govern the packing of surfactant molecules within aggregates. At high surfactant concentrations, interactions between aggregates result in the formation of various kinds of surfactant-rich mesophases. Some dynamic aspects of micelle and microemulsion droplet behaviour are explained in Chapters 9 and 10, respectively.

9

Aggregation of surfactants in aqueous systems

9.1 Micelle formation

It has already been seen that in very dilute aqueous surfactant solutions, adsorption to the air/solution interface increases with increasing surfactant concentration, m. The adsorption is reflected in a reduction in the surface tension, as shown schematically in Fig. 9.1a. In the (almost) linear region A in a plot of surface tension γ against ln m, the adsorbed film is almost close-packed and is fairly incompressible (§5.2). At point X there is a sharp break and at higher concentrations the tension remains essentially constant. Since the tension no longer falls there is no further adsorption, and so presumably no further increase in single surfactant molecule ('monomer') concentration. When further surfactant is added to the solution it is in some other state that does not adsorb, and the monomer activity remains constant.[1]

It is now well established that in aqueous solutions of pure surfactants, aggregates called micelles (Box 9.1) begin to form abruptly at a well-defined concentration, referred to as the *critical micelle concentration* (*cmc*). Beyond (and close to) the *cmc* the monomer concentration remains essentially constant and surfactant added to the solution forms only further micelles. Given this, it is to be expected that colligative properties (i.e. those depending on the *number* of solute entities present rather their *type*) will change abruptly at the *cmc*. This is illustrated in Fig 9.1b, with a schematic plot of the osmotic coefficient, g, against $m^{0.5}$. The osmotic coefficient is the ratio of an observed colligative property (e.g. freezing point lowering) to the theoretical (ideal) value. As seen, there is a precipitate fall in g above the *cmc* because the number of 'particles' in the system becomes considerably less than the number of 'monomer' molecules added.

---

[1] Although in the present context it is reasonable to suppose the monomer concentration is fairly constant close to, and above the *cmc*, over a wider range of concentrations above the *cmc*, the monomer concentration can vary significantly (see §9.10.1).

Surfactants: In Solution, at Interfaces and in Colloidal Dispersions. Bob Aveyard. © Bob Aveyard 2019.
Published in 2019 by Oxford University Press. DOI: 10.1093/oso/9780198828600.001.0001

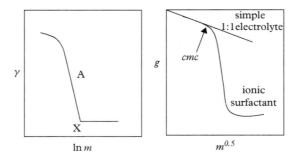

Figure 9.1 *(a) Surface tension versus ln m showing the breakpoint X at which m = cmc. (b) Schematic plot of osmotic coefficients for an ionic surfactant against $m^{0.5}$. The cmc corresponds to the point where the curve for the surfactant departs from that for a simple 1:1 electrolyte.*

Box 9.1   Nonsense McBain!

J.W. McBain and co-workers performed a number of early seminal studies of the behaviour of aqueous surfactant solutions (Vincent, 2014). In a paper published in 1914, McBain and Martin noted that aqueous solutions of sodium hexadecanoate exhibited unexpected behaviour, which in the view of the authors pointed to the possibility of surfactant aggregation (McBain and Martin, 1914). When such a proposal was made by McBain earlier, in a lecture at the 1913 Faraday meeting in London, the chairman is said (possibly apocryphally) to have muttered 'Nonsense McBain'. Of course, with the benefit of hindsight, it was not nonsense. The term 'micelle', deriving from the Latin *mica* meaning a crumb, that is, a small aggregate in the present context, appears to have been first used in the scientific literature by the botanist Nägeli in 1858 to refer to aggregates of starch and cellulose in water.

There has been an enormous effort over the years to understand various aspects of the behaviour of micellar systems, not only because of the considerable industrial importance of surfactants but also because micelle formation and behaviour are often relevant to the interests of scientists in disciplines other than physical and colloid chemistry. There are various questions that quite naturally arise in any consideration of the aggregation process. An obvious one is why do surfactants aggregate at all? Although, as seen, it is favourable to remove hydrophobic tails from water to some part in a micelle where water is absent, the surfactant head groups must come into close mutual proximity on the micelle surface. This may well be an energetically unfavourable process because of, say, the hydration of the head groups and/or because they are charged. There will also be an entropy cost of bringing a number of kinetically free monomers together to make a single micellar unit. Another important question is: when ionic surfactants, which are strong electrolytes in dilute aqueous solution, aggregate into micelles do the counterions (e.g. Na^+ or Br^-) remain free or is the micelle only partially ionized?

There are also other issues that arise. Why for example do micelles begin to form over a very narrow concentration range, that is, why is micellization apparently a cooperative phenomenon? What shape and size are micelles, and why? McBain in his early studies envisaged micelles as having a lamellar structure (Latin *lamina* = thin plate), as illustrated in Fig. 9.2a; later Hartley proposed a spherical structure as shown in cross-section in Fig. 9.2b (see e.g. Menger, 1979). If the lamellar structure were to be correct, when would it stop growing, that is, what would be the aggregation number? The interior of a micelle must not contain free space or presumably water. It is usually said the interior is akin to liquid hydrocarbon. Given that the hydrocarbon interior is liquid, and that the surfactant head groups must be somehow accommodated on the surface, how do these geometrical requirements dictate the micelle shape and size? It is also necessary to be able to model the aggregation process and hence obtain thermodynamic parameters associated with micelle formation. In what follows, answers to some of these questions are presented.

Micelles form in dilute systems, but as surfactant concentration continues to increase other larger surfactant structures appear, and this is discussed later; in the present context micelles are the simplest of the aggregates which form.

A schematic temperature-composition phase diagram for a fairly dilute ionic surfactant system is shown in Fig. 9.3. On the low temperature side of the solubility curve hydrated crystalline solid exists in the aqueous phase above the solubility limit. At higher temperatures, micelles form above the *cmc*, and the temperature variation of the *cmc* is indicated in the diagram. The temperature at which the *cmc* curve and the solubility curve meet is termed the *Krafft temperature*, T_K; at this temperature the surfactant solubility begins to rise rapidly. It should be noted that the surface tension curve in Fig. 9.1a would have the same form both above and below T_K. In the former case point X represents the *cmc* and in the latter case it reflects the solubility limit, and crystals form at higher concentrations rather than micelles.

Aqueous micellar solutions of nonionic surfactants become cloudy on heating, at the so-called *cloud point* (*CPt*). As will be seen (§9.11), phase separation, which causes the clouding, occurs into a surfactant-rich and a surfactant-lean phase at a lower consolute boundary.

(a)   (b)

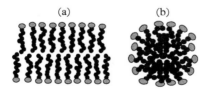

Figure 9.2 *Possible micelle structures. (a) Lamellar structure originally proposed by McBain. (b) Cross-section of a spherical micelle such as proposed by Hartley.*

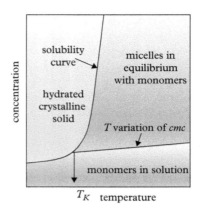

Figure 9.3 *Schematic phase diagram for relatively dilute aqueous ionic surfactant solutions. T_K is the Krafft point (see text).*

9.2   Simple models for micelle formation and some thermodynamic quantities

9.2.1   Mass action model

Micellization is a rather complex process but it can be simply and usefully treated using the so-called mass action model. Monomers associate to give aggregates and this can be represented by

$$D + D \leftrightarrow D_2 + D \leftrightarrow D_3 \cdots\cdots D_n + D \leftrightarrow \cdots\cdots \tag{9.2.1}$$

where D denotes a surfactant monomer and D_i is an associated entity containing i monomers. Each of the association steps has an equilibrium constant K_i associated with it, which is related to the standard free energy of formation of that associated species. It turns out that the size distribution of micelles is quite narrow (see §9.4) and so for present purposes the equilibrium involving only one type of aggregate (closed-association model) can be considered:

$$nD \rightleftharpoons D_n \tag{9.2.2}$$

Suppose a fraction f of a nonionic surfactant in solution, overall concentration m, is in micellar form then the equilibrium constant for the association represented in (9.2.2) is

$$K_n = [D_n] / [D]^n = (mf/n) / [m(1-f)]^n \tag{9.2.3}$$

This expression can be used to show why micelles might abruptly form as the concentration is increased, provided that the aggregation number, n, is sufficiently large. Values of

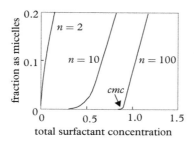

Figure 9.4 *Fraction of surfactant present as micelles, with aggregation numbers shown, as a function of total surfactant concentration, calculated using (9.2.3) with K_n arbitrarily set at unity.*

the fraction f as a function of total surfactant concentration, calculated using (9.2.3), are shown in Fig. 9.4 for several aggregation numbers, n. It has been assumed that $K_n = 1$, although the conclusions arrived at are independent of the choice of K. For $n = 2$ (i.e. formation of dimers), it is evident that there is no sharply defined concentration at which dimers suddenly start to form. For $n = 10$, micelles start to appear over a fairly broad range of surfactant concentrations. A sharp *cmc* is observed, however, for $n = 100$. Note that, although the model predicts a sharp *cmc* for high enough n, it does not explain *why* micelles form. That question is addressed later, both in terms of a molecular thermo-dynamic approach (§9.4) and the effective molecular shape of surfactant molecules/ions (§9.5).

A standard free energy of micelle formation is readily obtained from a knowledge of the *cmc*. In general, a standard free energy change $\Delta\mu^o$ is related to the equilibrium constant K for a process by the van 't Hoff isotherm

$$\Delta\mu^o = -kT \ln K \qquad (9.2.4)$$

Noting (9.2.3) and (9.2.4), the standard free energy of micelle formation $\Delta_{mic}\mu^o$ (per *micelle*)[2] for a nonionic surfactant is therefore

$$\Delta_{mic}\mu^o = -kT \ln [D_n] + nkT \ln [D] \qquad (9.2.5)$$

where $[D_n]$ and $[D]$ are the concentrations of micelles and monomers respectively. Above but close to the *cmc*, $[D] \sim cmc$, and for large n the term in $[D_n]$ in (9.2.5) is negligible so that to a reasonable approximation

$$\Delta_{mic}\mu^o = nkT \ln cmc \qquad (9.2.6)$$

---

[2] The free energy change expressed per *molecule* is $\Delta_{mic}\mu^o/n = kT \ln cmc$; the quantity $\Delta_{mic}G^o = RT \ln cmc$, values of which are quoted in Table 9.1, is equal to $N_A \Delta_{mic}\mu^o/n$ and is the molar standard free energy of micelle formation.

For an ionic surfactant account must be taken of the counterions in the equilibrium between monomers and micelles; only a fraction, α, of surfactant in micellar form is ionized, that is, some of the counterions are 'bound' to the micelles. For, say, a 1,1-anionic surfactant Na^+D^- where m counterions are bound to a micelle, the equilibrium may be expressed

$$nD^- + mNa^+ \leftrightarrow D_n^{(n-m)-} \tag{9.2.7}$$

From this, taking mean ionic activity coefficients to be unity, the standard free energy of micelle formation per 'molecule' of surfactant is

$$\frac{\Delta_{mic}\mu^o}{n} = -\frac{kT}{n}\ln [D_n] + kT\ln [D^-] + \frac{m}{n}kT\ln [Na^+] \tag{9.2.8}$$

For a fully dissociated surfactant at the *cmc*, $[D^-] = [Na^+] = cmc$ and, neglecting the first term on the right hand side of (9.2.8) (for large n)

$$\frac{\Delta_{mic}\mu^o}{n} = \left(1 + \frac{m}{n}\right) kT\ln cmc \tag{9.2.9}$$

The degree of dissociation α is $(n - m)/n = (1 - m/n)$ and so (9.2.9) can be written

$$\Delta_{mic}\mu^o = (2 - \alpha)nkT\ln cmc \tag{9.2.10}$$

If the micelles are present in swamping inert electrolyte it turns out that the expression for the standard free energy becomes equivalent to (9.2.6) for a nonionic surfactant. Electrolyte screens the repulsion between charged surfactant head groups at the micelle surface and if present in sufficient amount, causes the surfactant to behave as if it were nonionic. An analogous situation pertains for the adsorption of an ionic surfactant to a liquid surface in the presence of inert electrolyte. The Gibbs equation in the absence of salt has a factor of 1/2 (see (4.3.17)) which disappears when swamping electrolyte is present (see (4.3.22)).

It is possible in principle to obtain standard entropies and enthalpies of micelle formation of nonionic surfactants, $\Delta_{mic}S^o$ and $\Delta_{mic}H^o$, from a knowledge of the temperature variation of $\Delta_{mic}\mu^o$ in the usual way, that is,

$$\Delta_{mic}H^o = -nkT^2\frac{\partial \ln cmc}{\partial T} \text{ and } \Delta_{mic}S^o = \frac{\Delta_{mic}H^o - \Delta_{mic}\mu^o}{T} \tag{9.2.11}$$

For ionic surfactants, the degree of dissociation may change significantly with temperature. It is often preferable therefore to obtain enthalpies (hence standard entropies) of micellization calorimetrically.

9.2.2 Separate phase approach

An alternative to the mass action model of micelle formation is the so-called separate phase approach. As mentioned, micellization is a cooperative phenomenon and is in some ways analogous to the separation of a liquid phase (as droplets) from a solution of a sparingly soluble liquid which is at (or above) its solubility limit. The *cmc* is analogous to the saturation solubility. There is of course an obvious difference between micelle formation and bulk condensation. In micelle formation the 'condensed phase' cannot grow indefinitely whereas liquid droplets can. There is a definite (average) aggregation number for surfactant micelles governed, as will be seen, by packing constraints of surfactant molecules in an aggregate.

The (standard) chemical potential of a nonionic surfactant molecule in the micellar phase can be written μ^o_{mic} and the chemical potential of a monomer in ideal dilute solution, μ_{mon} is $\mu^o_{mon} + kT \ln(m/m_s)$ where m is the surfactant concentration in the solution phase, and m_s a standard concentration (taken here as usual to be unity in the same units as m). At and above the *cmc*, $m = cmc$ and at equilibrium between solution and micellar phases therefore

$$\mu^o_{mic} = \mu^o_{mon} + kT \ln cmc; \quad \frac{\Delta_{mic}\mu^o}{n} = \mu^o_{mic} - \mu^o_{mon} = kT \ln cmc \qquad (9.2.12)$$

which is equivalent to (9.2.6).

9.3 How does the *cmc* depend upon surfactant molecular structure?

Understandably surfactant molecular structure has one of the greatest effects on *cmc*. Standard free energies of micellization (derived from *cmc*s by the use of (9.2.6)) vary greatly both with chain length and head group structure of a surfactant molecule. Some *cmc* values are listed in Table 9.1 together with the derived standard molar free energy of micelle formation, $\Delta_{mic}G^o$ (see footnote 3) for nonionic surfactants. The absolute values of the standard free energies depend on the concentration units used (molar concentrations in the present case), but since the same units are used in all cases, *relative* values for different surfactants can be discussed meaningfully.

From data for the homologous glucosides it is seen that the contribution of a chain methylene group to the standard free energy of micellization is -3.0 kJ mol^{-1}. This is very similar to the CH$_2$ group contribution to the standard free energy of adsorption of alkyl chain surfactants from water to the oil/water interface (-3.2 kJ mol^{-1}), consistent with the notion that the micelle interior is liquid-like hydrocarbon. For the C$_{12}$E$_m$ series, the *cmc* tends to rise a little as m increases indicating that the monomers are

Table 9.1 *Values of the cmc at room temperature and standard free energies of micelle formation per mol monomers,* $\Delta_{mic}G^{\circ} = N_A\Delta_{mic}\mu^{\circ}/n = RT\ln cmc$ *(see (9.2.6)).* N_A *is the Avogadro number.*

Surfactant (see §1.3.1)	cmc/mol dm^{-3}	$-\Delta_{mic}G^{\circ}$/kJ mol^{-1}
Nonionic		
$C_{12}E_5$	6.4×10^{-5}	23.92
$C_{12}E_6$	8.7×10^{-5}	23.16
$C_{12}E_7$	8.2×10^{-5}	23.31
$C_{12}E_8$	1.1×10^{-4}	22.58
n-C_8-β-D-glucoside	2.5×10^{-2}	9.14
n-C_{10}-β-D-glucoside	2.2×10^{-3}	15.16
n-C_{12}-β-D-glucoside	1.9×10^{-4}	21.23
Cationic		
$C_{12}H_{25}N(CH_3)_3{}^+Br^-$ (DoTAB)	1.6×10^{-2}	
Anionic		
$C_{12}H_{25}SO_4{}^-Na^+$ (SDS)	8.2×10^{-3}	
$C_{12}H_{25}OC_2H_4SO_4{}^-Na^+$	3.9×10^{-3}	
$C_{12}H_{25}(OC_2H_4)_2SO_4{}^-Na^+$	2.9×10^{-3}	
Zwitterionic		
$C_{12}H_{25}N^+(CH_3)_2CH_2COO^-$	1.8×10^{-3}	

more favourably solvated in the aqueous phase and/or that it becomes more difficult to accommodate the head groups at the micelle surface. The *cmc* values for $C_{12}E_8$ and dodecyl-β-D-glucoside are fairly similar suggesting that the hydrophilicity of the glucoside and E_8 head groups are comparable. When head groups carry a charge, it becomes more difficult to bring them together in close proximity on the micelle surface. For this reason the *cmc*s of the cationic surfactant DoTAB and anionic surfactant SDS are much higher than those for the nonionic surfactants with the same (dodecyl) chain.

The data presented in Fig. 9.5 for the thermodynamic parameters of micelle formation of an homologous series of nonionic surfactants C_nE_6 allow estimates to be obtained for the methylene group contributions to the enthalpy and the entropy of micelle formation. The enthalpies were obtained calorimetrically. The free energy contribution is -3.1 kJ mol^{-1} and is, as expected, very similar to that obtained for the glucosides mentioned above. The enthalpy and entropy contributions are respectively -1.5 kJ mol^{-1} and $+5.2$ J mol^{-1} K^{-1}. These latter values are reasonably consistent with those calculated from the corresponding thermodynamic parameters for the hydration and the vaporization of the

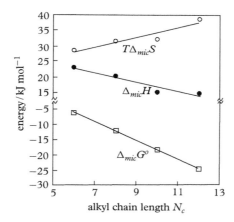

Figure 9.5 *Thermodynamic parameters of micelle formation of C_nE_6 surfactants in water at 298 K. Adapted from data in Andersson and Olufsson (1988).*

Table 9.2 *Methylene group contributions to the thermodynamic parameters of micelle formation at room temperature.*

ΔG°/kJ mol^{-1}	ΔH°/kJ mol^{-1}	ΔS°/J mol^{-1} K^{-1}
vaporization		
2.96	5.0	6.8
hydration		
0.7	−2.9	−11.7
micellization calculated		
−3.7	−2.1	4.9
micellization experimental		
−3.1	−1.5	5.2

The calculated quantities ($\Delta_{mic}X^o$) for micelle formation are obtained using $-\Delta_{mic}X^o = \Delta_{hyd}X^o + \Delta_{vap}X^o$. The vaporization parameters are calculated from data for alkanes pentane to nonane, and quantities for hydration are from data for ethane, propane, and butane.

methylene group, as seen from Table 9.2. It is interesting that the enthalpies of micelle formation for all the C_nE_6 surfactants represented in Fig. 9.5 are positive, but the CH_2 contribution to this is a negative one. The process of bringing the head groups from bulk solution to the micelle surface involves a large positive enthalpy change which militates against micellization. It is apparent that a major driving force for micelle formation is the positive entropy change associated with the transfer of the hydrocarbon chains from solution to the micelle interior.

9.4 How a simple molecular thermodynamic treatment of micellization predicts a favoured distribution of micelle sizes

There have been a number of molecular thermodynamic approaches to micelle formation that build on the pioneering contributions of Tanford (1980) and of Israelachvili et al. (1976). As seen, the driving force for micellization comes from the transfer of hydrophobic chains from aqueous solution into the nonpolar liquid-like core of the micelle. However, there are other effects which tend to oppose the aggregation. The micelle surface exposes some hydrocarbon to water; also the process of taking the head groups, which are well-spaced in solution, and placing them in close proximity at the micelle surface is energetically unfavourable. It is the interplay between these various factors that gives rise to a favoured distribution of micelle sizes at equilibrium.

Conceptually a micelle can be formed by bringing n surfactant chains, detached from their head groups, to form a micelle core and then subsequently attaching the head groups back onto the chains at the surface of the micelle. At the *cmc*, the surfactant hydrocarbon chains in aqueous solution are notionally supersaturated since the *cmc* is greater than the saturation solubility, c_s, of the parent hydrocarbon in water.[3] By analogy with the condensation of supersaturated vapour to form a liquid droplet, the free energy change accompanying the formation of the hydrocarbon core of a micelle containing n chains is given approximately by

$$\Delta\mu \approx -nkT\ln\left(\frac{cmc}{c_s}\right) + n\Delta_{surf}\mu^o \tag{9.4.1}$$

where $\Delta_{surf}\mu^o$ is the (standard) free energy change *per molecule* associated with the formation of hydrocarbon/water interface at the micelle surface. Note that $\Delta\mu$ is a change in free energy, and contains a contribution from a change in concentration, from the *cmc* to c_s. The *standard* chemical potential, μ^o, of a molecule reflects its interactions with the local surroundings and $\Delta_{surf}\mu^o$ is a change in free energy resulting from a change in environment of a molecule.

Comparison of (9.4.1) with (3.4.11) (in §3.4.2) shows that the supersaturated vapour pressure p in (3.4.11) is replaced by the *cmc*, and saturated vapour pressure p^o by c_s. The relationship between $\Delta\mu$ and n in (9.4.1) shows a maximum at a low, critical value of n, and thereafter the free energy falls continuously with increasing n (see Figure 3.10). However, when the surfactant heads are reattached to the chains at the micelle surface a term $\Delta_{head}\mu^o$ must be added to (9.4.1) to account for head group repulsion. This leads to a minimum in $\Delta\mu$ ($= \Delta_{mic}\mu$) at a higher value of n and so micelle growth is limited and there is a favoured distribution of micelle sizes at equilibrium. For micelles containing n

---

[3] As an approximation, the chain solubility is taken to be equal to that of the parent alkane. In other contexts, for example, when assigning a molecular volume, the surfactant chain is taken to be the alkyl group.

molecules in equilibrium with aqueous surfactant solution at the *cmc*, $\Delta_{mic}\mu$ is zero and, noting (9.2.6) and (9.4.1) it follows that for a nonionic surfactant

$$nkT \ln cmc = \Delta_{mic}\mu^o \approx nkT \ln c_s + n\Delta_{surf}\mu^o + n\Delta_{head}\mu^o \tag{9.4.2}$$

where $\Delta_{mic}\mu^o$ is expressed per micelle (see footnote 3).

To exemplify, and extend this approach it is helpful to focus on a particular system containing the nonionic surfactant dodecylpentaoxyethylene glycol monoether ($C_{12}E_5$), which at low concentrations in aqueous solution at room temperature has been shown to give spherical micelles with an aggregation number ($n = n_{agg}$) of 54 (Velinova et al., 2011); the experimentally determined *cmc* is 6.2×10^{-5} mol dm^{-3}. The quantity c_s is taken to be the saturation solubility of dodecane in water (1.2×10^{-8} mol dm^{-3}), which has been experimentally determined by Tolls et al. (2002). The term $kT \ln c_s$ appearing in (9.4.2) is the standard free energy change accompanying the removal of a surfactant chain from water and placing it in *bulk* liquid alkane. Hydrocarbon chains confined within a micelle core however are more restricted than in bulk liquid and calculation of the free energy of transfer of a chain from solution to a micelle core should allow for this. The change, $\Delta_{pack}\mu^o$, associated with chain deformation in the micelle core, radius R, is given as (Nagarajan, 2002)

$$\Delta_{pack}\mu^o = \frac{3kT\pi^2 R^2}{80 n_{seg} l_{seg}^2} = 0.3701 kT \left(\frac{R^2}{l_{seg}^2 n_{seg}} \right) \tag{9.4.3}$$

where n_{seg} is the number of segments within a tail group and l_{seg} is a characteristic segment length, taken here to be 0.47nm, the diameter of an extended alkyl chain. The quantities n_{seg} and l_{seg} are related through $n_{seg} l_{seg}^3 = v$ where v is the molecular volume of the dodecyl tail group (0.35 nm^3). The core radius R is given by

$$R = \left(\frac{3nv}{4\pi} \right)^{1/3} \tag{9.4.4}$$

The free energy associated with the hydrocarbon/water interface at the micelle surface, $\Delta_{surf}\mu^o$ (per molecule), is the product of the area of hydrocarbon interface exposed and its interfacial tension, γ_{ow}, which is 0.0528 N m^{-1} for the plane dodecane/water interface. Thus

$$\Delta_{surf}\mu^o = (a - a_{cs}) \gamma_{ow} \tag{9.4.5}$$

in which a is the area available to a head group at the micelle surface ($= 4\pi R^2/n$) and a_{cs} is the cross-sectional area of a head group parallel to the micelle surface. The latter

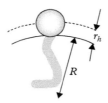

Figure 9.6 *A surfactant molecule in a micelle with a core radius R. The surface pressure is supposed to act in the plane (dashed line) through the centres of the head groups, a vertical distance r_h from the core surface.*

is equal to half of the molecular co-area, A_o, referred to below (see also §5.3.2); for $C_{12}E_5$ $A_o = 0.53$ nm^2.[4]

There have been various methods proposed for the evaluation of $\Delta_{head}\mu^o$. Here a simple approach is proposed in which the energy of head group repulsion is set equal to $\pi_v a_h$, where π_v is the surface pressure exerted by the heads in the plane through their centres (dashed line in Fig. 9.6) and a_h is the area available per head in the same plane. Supposing the head group is spherical, with radius r_h, then

$$a_h = \frac{4\pi (R + r_h)^2}{n}; r_h = (A_o/2\pi)^{1/2} = (a_{cs}/\pi)^{1/2} \tag{9.4.6}$$

For the E_5 head group, $r_h = 0.29$ nm.

By analogy with the treatment of mobile nonionic surfactant monolayers at a plane oil/water interface (§5.3.2), the surface pressure is supposed to be given by the Volmer two-dimensional equation of state, which in the present context is written (per molecule) as

$$\pi_v (a_h - A_o) = kT \tag{9.4.7}$$

From the above, the following expressions are obtained for $\Delta_{mic}\mu$ and $\Delta_{mic}\mu^o$:

$$\frac{\Delta_{mic}\mu}{kT} = -n\ln cmc + 0.3701n \left(\frac{R^2}{l_{seg}^2 n_{seg}} \right) + n\ln c_s + \frac{n\gamma_{ow}}{kT} \left(a - \frac{A_o}{2} \right) + \frac{na_h}{(a_h - A_o)} \tag{9.4.8}$$

$$\Delta_{mic}\mu^o/nkT = 0.3701 \left(\frac{R^2}{l_{seg}^2 n_{seg}} \right) + \ln c_s + \frac{\gamma_{ow}}{kT} \left(a - \frac{A_o}{2} \right) + \frac{a_h}{(a_h - A_o)} \tag{9.4.9}$$

[4] The co-area is taken to be the reciprocal of the limiting adsorption, Γ^∞, at a fluid/fluid interface reported by Stoyanov et al. (2004).

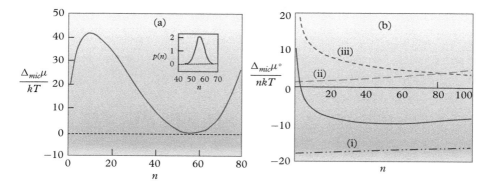

Figure 9.7 *(a) Free energy of formation of $C_{12}E_5$ micelles as a function of number of molecules per micelle. The minimum occurs at $n = n_{agg} = 57$, when the free energy is effectively zero. Inset is the micelle size distribution curve, p(n) vs n (see text). (b) Standard free energy of micelle formation, expressed per molecule, as a function of n (full line). Dashed curves (i), (ii), and (iii) represent the contributions from, respectively, chain transfer, head group repulsion and core surface formation. Values of the various parameters used in the calculations are given in the text.*

The free energy of micelle formation for $C_{12}E_5$, obtained using (9.4.8) and the values of the various parameters given above, is plotted against n in Fig. 9.7a. The inset shows the micellar distribution $p(n)$ plotted against n, where $p(n)$ is given by

$$p(n) = \exp\left(-\frac{\Delta_{mic}\mu}{kT}\right) \tag{9.4.10}$$

There is a minimum value of $\Delta_{mic}\mu/kT$ (close to zero), and a maximum value of $p(n)$, for $n = n_{agg} = 57$, similar to the value of 54 given by Velinova et al. (2011). The predicted value of the *cmc* can be obtained from the minimum value of $\Delta_{mic}\mu^o/nkT$ ($= \ln cmc$) as a function of n, shown in Fig. 9.7b; the *cmc* is given as 6.1×10^{-5} mol dm^{-3} (cf. the experimental value of 6.2×10^{-5} mol dm^{-3}). Also shown in the figure are the contributions to $\Delta_{mic}\mu^o/nkT$ arising from chain transfer, head group repulsion at the micelle surface and formation of the hydrocarbon/water interface of the micelle core. The overall micelle diameter ($=(R + 2r_h)$) obtained using the present treatment is 4.5 nm.

Although the simple treatment described predicts values of n_{agg} and *cmc* for $C_{12}E_5$ which are in good accord with experimental and other data, this level of agreement may be considered somewhat serendipitous. In using the Volmer equation of state for head group interactions, the micelle curvature has been neglected. There are also uncertainties in the values of the various input parameters (e.g. A_o, r_h, a_{cs}). Nonetheless, the molecular thermodynamic approach outlined provides a reasonable insight into the energetic factors involved in micelle formation. Having considered the *size* of (spherical) micelles, attention is now focused on the *shape* of micelles, which is discussed in terms of the effective molecular geometry of surfactant molecules.

9.5   The shape of aggregates and the packing factor *P* of a surfactant molecule

It has been shown above that there is an optimum size for a spherical surfactant micelle. Other shapes of micelles can be energetically favoured, and shapes and sizes can depend on prevailing conditions such as salt concentration and temperature. In the publications by Israelachvili et al. (1976), Tanford (1980), and Mitchell and Ninham (1981) referred to earlier, it was shown how surfactant molecular geometry determines the size and shape of micelles and other surfactant aggregates. A simple summary of this approach follows.

9.5.1   Effective molecular shape and the packing factor *P*

The basic requirement of surfactant packing in a micelle is that the surfactant chains are able to form the liquid-like micelle core, with constant density across the diameter, and that the head groups have enough space to be accommodated at the surface of the core. So, the volume v of a chain, as well as its 'critical' length l_c (80–90% of its fully extended length[5]) are important, as is the 'optimal' area for a head group, a_o (equivalent to a in the previous section). In the context (zeroth order theory) optimal means the area in a bilayer is independent of the size and shape of the micelle. The formation of a spherical micelle with core radius R is taken as an example; the tail length l_c must obviously be $\leq R$. The volume V_{core} of the micelle core is related to the micelle aggregation number n_{agg} by

$$n_{agg} = V_{core}/v = \left[(4/3)\,\pi R^3\right]/v \tag{9.5.1}$$

At the same time n_{agg} is related to the core surface area A_{core} by

$$n_{agg} = A_{core}/a_o = \left[4\pi R^2\right]/a_o \tag{9.5.2}$$

It follows therefore, equating (9.5.1) and (9.5.2) that

$$v/a_o R = 1/3 \tag{9.5.3}$$

Since $l_c \leq R$, for a spherical micelle to form packing constrains require that

$$v/a_o l_c \leq 1/3 \tag{9.5.4}$$

---

[5] Chains are not fully extended in the liquid state; full extension would imply a crystalline structure.

The quantity $v/a_0 l_c$ is often termed the *packing factor*, *P*. By extension of this treatment, ranges of values of *P* for the formation of other aggregate shapes are obtained and shown in Table 9.3.

The cross-sectional area of a surfactant head group at the surface of an aggregate, a_0, is determined not only by its molecular size (such as could be obtained from molecular models), but also by mutual interactions between neighbouring moieties at the micelle surface. These interactions are governed by, for example, electrical repulsion between charged heads, which in turn depends on the concentration of inert electrolyte in the system. Head group solvation also plays a role in determining the effective head area. In systems with oil present, surfactant chains are 'solvated' by penetration of oil molecules into the surfactant chain region; this is discussed further in the section on microemulsion formation. The important point to note here is that *P* is not simply a property of a surfactant molecule, but rather that of a surfactant molecule coupled with the system in which it exists.

Table 9.3 *Aggregate shapes in relation to surfactant molecular structure and values of P. The surfactant head groups are represented by the light grey circles.*

Values of *P*	Notional surfactant molecular shape	Surfactant type
$P < 1/3$	a_0 l_c	Single chain surfactant with large head group. e.g. SDS. Gives spherical micelles.
$1/3 < P < 1/2$	$v = $ volume of truncated cone	Single chain surfactant with smaller head group. e.g. nonionic or ionic in presence of electrolyte. Gives cylindrical/rod shaped micelles.
$1/2 < P < 1$		Double chain surfactants with large head group and flexible chains. Gives flexible bilayers and liposomes
$P \sim 1$	$v = $ volume of cylinder	Double chain surfactants with small head group and/or rigid chains. Gives planar bilayers.
$P > 1$	a_0 l_c	Double chain surfactants with small head group and bulky chains. Gives inverted structures.

Although only spherical micelles have been discussed so far, reference is made to various other geometries of surfactant aggregates in Table 9.3. Some of these geometries (e.g. cylindrical micelles and bilayers) often exist in assemblies of aggregates in more concentrated surfactant solutions, and such systems are discussed later. It is worth mentioning here, however, two particular structures which have elicited much interest: liposomes and wormlike micelles. It turns out that liposomes and wormlike micelles can often be interconverted.

9.5.2 Liposomes

Liposomes are closed, uni- or multilamellar spherical structures; liposomes consisting of a single bilayer (as illustrated schematically in Fig. 9.8) are referred to as vesicles. The word liposome is derived from *lipid*, that is, fat, and *soma*, Greek meaning body. They are often fluid-filled, containing the same solution as that in which they are dispersed. They were first described by A.D. Bangham and co-workers in the early 1960s (see e.g. Bangham et al., 1965) and were proposed as models for biological cell membranes. Natural liposomes contain phospholipid and cholesterol, both of which have very low solubility in water; they can be prepared by sonication of the constituent materials in water. Liposomes can also be prepared from mixtures of pure surfactants and other low molar mass components, for example, hexadecylpyridinium chloride + sodium salicylate (see e.g. Rehage and Hoffmann, 1991) or mixed anionic and cationic surfactants (Kaler et al., 1989).

A major reason for the enduring interest in liposomes has been their use in drug delivery systems. There has long been an interest in formulating drugs in such a way as to target them to specific tissues within the body. Paul Erlich coined the phrase 'magic bullet' in this connection at the beginning of the twentieth century (see Gabizon, 2001). Drugs can be incorporated in liposomes, often within the aqueous phase interior, in which case the lipid bilayer is a barrier to the outward flow of the contents. Although liposomes themselves are non-toxic, when injected into the bloodstream they

Figure 9.8 *Schematic cross-section of a uni-lamellar liposome (vesicle). The aqueous solution phase fills the central cavity; the shaded area is occupied by the lipid tails.*

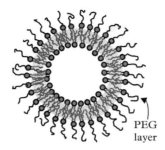

Figure 9.9 *Cross-section of a stealth liposome showing the outer layer of PEG groups.*

are instantly recognized as foreign particles by the reticuloendothelial system (RES).[6] The liposomes rapidly bind to plasma proteins and become unstable, releasing drugs into the bloodstream. Those liposomes which survive are sequestered by the macrophages in the liver and spleen. What is required is that the liposomes remain stable for long periods of time in the bloodstream and ultimately collect by some means at the site of a target tumour and release their contents there. Polyethylene glycol (PEG) chains covalently attached to the outer lipid layer of liposomes (Fig. 9.9) inhibit protein adsorption and hence interfere with recognition by the RES. Such coated entities have been called 'stealth' liposomes, and the highly hydrated PEG groups sterically inhibit interactions (see §13.3) with various blood components. In effect, the RES does not recognize the coated liposomes which become 'invisible'. Polymers other than PEG can also protect liposomes in this way.

The interior surface of blood vessels is lined with a thin layer of cells (the endothelium). In normal blood vessels the endothelium is continuous and liposomes do not pass through, that is, extravasate. In tumours, however, there are gaps in the endothelium through which liposomes, if they are in the correct size range (diameters less than 600 nm), can pass. In liposomes with an effectively solid bilayer at body temperature, drug delivery at a tumour site can be substantially enhanced by heating the site to above the solid-liquid transition temperature of the liposome after delivery.

9.5.3 Wormlike micelles

It has been seen (Table 9.3) that for a value of the packing factor P between 1/3 and 1/2, formation of cylindrical micelles is to be expected at concentrations above the *cmc*. The ends of the cylinder are presumably spherically curved as shown in Fig. 9.10. The values of P that favour spherical curvature are less than those for cylindrical curvature and the free energy per surfactant molecule of forming the caps exceeds that for formation of

---

[6] The RES is a collection of cells within the body exhibiting the property of phagocytosis, the ingestion and destruction of foreign materials such as microorganisms. Macrophages (from Greek meaning 'big eaters'!) are phagocytes.

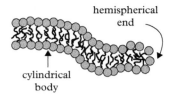

Figure 9.10 *Cross-section of part of a worm-like micelle showing one of the hemispherical ends.*

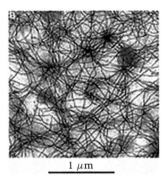

Figure 9.11 *Cryo-TEM image of wormlike micelles formed from 1% by weight diblock copolymer (see text) in water. Reprinted with permission from Y.-Y. Won, H.T. Davis, and F.S. Bates, Science, 1999, 283, 960.*

the cylindrical body. There is evidence that the ends can be swollen, with a diameter greater than that of the cylinder (see e.g. Ezrahi et al., 2006). In a simple way it can be supposed that the length of a micelle is related to the difference in free energy of forming the spherically curved caps and the cylindrical body (Menge et al., 1999). That having been said, the length of cylindrical micelles can be remarkably large, up to 1 μm or even greater. A cryo-transmission electron microscopic (TEM) image of such giant micelles formed from 1 wt % of a low molar mass di-block copolymer of polyethylene oxide and polybutadiene in water is shown in Fig. 9.11 (taken from Won et al., 1999).

These large cylindrical aggregates are termed *wormlike* micelles, and can be formed from a wide range of surfactant systems (Acharya and Kunieda, 2006), among them mixtures of ionic surfactant with a cosurfactant or added salt, mixed nonionic surfactants, and mixed anionic–cationic surfactants in water. Reverse wormlike micelles in organic liquids have also been observed (Tung et al., 2006).

Wormlike micelles are long and flexible and in some respects like polymer chains in solution. Important differences however are that wormlike micelles can break up into smaller units and then recombine, and of course surfactant molecules can exchange with the solution phase. Wormlike micelles have sometimes been referred to as 'living' polymers. As a result of the viscoelastic properties of wormlike micellar solutions resulting from entanglement of the micelles (as discussed in §15.4.5), the solutions find

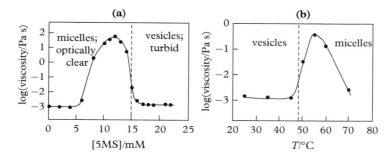

Figure 9.12 *Zero shear viscosities in aqueous solutions of CTAB and 5MS. (a) Variation of viscosity with concentration of 5MS in systems with 12.5 mM CTAB. (b) Variation of viscosity with temperature in systems with 12.5 mM CTAB and 20 mM 5MS. Redrawn from Davies et al. (2006).*

applications in a number of areas (Davies et al., 2006) and the rheology of these solutions has received considerable attention in recent years.

Wormlike micelles can be formed from dispersions of unilamellar vesicles in water. The inter-conversion can be induced for example by changing the system composition or the temperature (and hence the surfactant packing factor). Davies et al. (2006) investigated aqueous systems containing the cationic surfactant hexadecyltrimethylammonium bromide (CTAB) and 5-methylsalicylic acid (5MS). Some zero shear viscosities, η, obtained from steady state shear experiments are illustrated in Fig. 9.12. In systems with 12.5 mM CTAB and 5MS at less than 15 mmol dm^{-3}, micelles are present. For low concentrations of 5MS the viscosity is low (similar to that of water) but starts to rise rapidly when [5MS] reaches about 5 mmol dm^{-3}. The viscosity peaks for an equimolar ratio of CTAB to 5MS; the peak viscosity is 5 orders of magnitude greater than that of the CTAB solution alone. The viscous solutions exhibit viscoelasticity (see §15.4.5), and for [5MS] $\leq$ 15 mmol dm^{-3} the systems are optically clear and contain micelles. For [5MS] $\geq$ 15 mmol dm^{-3} the solutions contain vesicles rather than micelles, which scatter light rendering the solutions turbid. It is possible to take systems with vesicles and effect a transition to micelles by increasing the temperature as shown in Fig. 9.12b. The system here has [CTAB] = 12.5 mmol dm^{-3} and [5MS] = 20 mmol dm^{-3}. Again the micellar (i.e. high temperature) systems are optically clear and the vesicle dispersions are turbid.

9.6   Effects of inert electrolytes on the *cmc* of ionic surfactants

Suppose an inert electrolyte (e.g. NaCl) is added to a micellar solution of an ionic surfactant with a common counterion, such as SDS. It is expected that the electrical repulsion between the charged head groups on the micelle surface will be screened, and that the *cmc* will therefore fall. This can be appreciated from the equilibrium expressed in (9.2.7), from which the equilibrium constant for micelle formation, K_{mic}, is

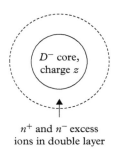

n^+ and n^- excess
ions in double layer

Figure 9.13 *Core of an anionic micelle, containing r surfactant entities z of which are charged. This charge is neutralized in the region surrounding the micelle core so that overall the micelle is neutral.*

$$K_{mic} = \frac{\left[D_n^{(n-m)-}\right]}{[D^-]^n[Na^+]^m} \tag{9.6.1}$$

Addition of counterions (i.e. Na^+) will tend to cause a decrease in $[D^-]$ ($= cmc$) and an increase in the micelle concentration $\left[D_n^{(n-m)-}\right]$. At the same time, the micellar charge $((n-m)^-)$ and aggregation number (n) can also change. With regard to the latter, the packing factor P (see Table 9.3) will increase as a result of a reduction in the effective head group area at the micelle surface, and so the mean curvature of the micelles will fall and the aggregates will grow and become more elongated and therefore n will increase. An increase in $[Na^+]$ may well cause an increase in the number of counterions, m, 'bound' to the micelle surface, and if so the degree of dissociation of micellized surfactant will fall.

How to define a degree of dissociation of a micelle is not altogether obvious. One definition that has commonly been used (or implied) is arrived at as follows. Suppose a micelle is made up of r surfactant entities, z of which carry a (say negative) charge so that the degree of dissociation of surfactant in the micelle, α, is equal to z/r. The region surrounding the micelle core in which the charge z is neutralized consists of both coions (anions in this case) and counterions (cations) such that the micelle overall is electrically neutral (Fig. 9.13). The *excess* number of counterions in this region is denoted n^+. This is the excess over the number of cations in similar volume of bulk solution far from the micelle. Similarly the excess number of coions is designated n^-; this excess is negative because anions are repelled by the negatively charged micelle core. Since the micelle is overall electrically neutral $n^+ - n^- = z$.

It is now supposed that the concentration profiles of n^+ and n^- normal to the micelle surface are mirror images, in which case $n^+ = -n^-$. On this basis,

$$\frac{n^+}{r} - \frac{n^-}{r} = \frac{z}{r} = \alpha; \tag{9.6.2}$$

That is, the degree of dissociation is minus twice the (negative) surface excess of coions per surfactant entity in the micelle. The obvious drawback to this definition is that the ion distributions normal to the micelle surface are not mirror images except in the limit of zero surface potential (see §11.2.2 and Fig. 11.6.). Using the above definition of α, Hall (1983) showed that the variation of *cmc* with counterion concentration m_{Na^+} (arising from both added electrolyte and surfactant) is given by

$$-\frac{d \ln cmc}{d \ln m_{Na^+}} = (1 - \alpha) + (2 - \alpha)\, \partial \ln f_{\pm}/d \ln m_{Na^+} \qquad (9.6.3)$$

where $f_{\pm}$ is the mean ionic activity coefficient in the solution. For SDS in the presence of sodium chloride at room temperature $d\ln cmc/d\ln m_{Na^+}$ is roughly constant at about -0.59 in the region of [NaCl] from 0.01 to 0.1 mol dm^{-3}. The values of α calculated using (9.6.3) and experimentally obtained values of $f_{\pm}$ for NaCl are shown in Fig. 9.14. In this case α is about 0.3, that is, SDS micelles are only about 30% dissociated.

A definition of α ($=\alpha_c$) which has been used in studies of micelle dissociation using, for example, conductivity measurements is (again taking the example of SDS with NaCl present)

$$\alpha_c = \frac{[\text{free Na}^+ \text{ from micellised SDS}]}{[\text{micellised SDS}]} = \frac{m_{Na^+ (free)} - cmc - m_{NaCl}}{m_{SDS} - cmc} \qquad (9.6.4)$$

It is supposed that the counterions (Na$^+$) are either bound to the head groups of the surfactant in the micelle or are free. In (9.6.4) $m_{Na^+ (free)}$ is the total concentration of free sodium ions, arising from both surfactant and NaCl. The free surfactant monomer (anion) concentration is equal to the *cmc* and m_{NaCl} and m_{SDS} are the total concentrations of added NaCl and SDS in the system. The relationship between α and α_c is complicated

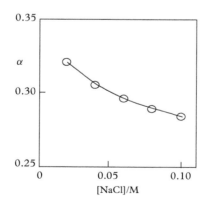

Figure 9.14 *Degree of dissociation of SDS micelles in the presence of NaCl at room temperature;* α *was calculated using (9.6.3).*

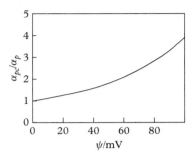

Figure 9.15 *Ratio of the degrees of dissociation for plane ionic surfactant monolayers given by (9.6.5).*

because the micelle surface is highly curved, but it can be shown that the analogous degrees of dissociation for *plane* monolayers (α_p and α_{cp} respectively) are approximately related by (Aveyard et al., 1999)

$$\frac{\alpha_{cp}}{\alpha_p} = -\left(\frac{\sinh x}{\exp(-x) - 1}\right) \qquad (9.6.5)$$

in which $x = (e^- \psi / 2kT)$. Here e^- is the electronic charge and ψ the electrical potential at the outer Helmholtz plane (o.H.p., see Fig 7.1); k and T have their usual significance. A plot of (α_{cp}/α_p) against ψ (for values of ψ in a range typical for ionic surfactant monolayers) is shown in Fig. 9.15. As expected, (α_{pc}/α_p) approaches unity as $\psi \to 0$, but for higher values of ψ the degrees of dissociation can differ very markedly. It can be expected that α defined by (9.6.3) and (9.6.4) for micelles can also differ considerably. In terms of understanding the physical significance of micelle dissociation, it is probable that α_c is a more useful quantity than α, although α is the easier to obtain experimentally.

9.7 What happens to the *cmc* when two surfactants are mixed?

Adsorption from mixed surfactant solutions to liquid interfaces has been treated in §5.5. It was seen how the approach of regular solutions for bulk mixtures can be usefully (if a little crudely) applied to mixed adsorbed planar monolayers. Micelles can loosely be regarded as *curved* monolayers and similar interactions between different surfactants exist in micelles as in planar monolayers. For example interactions between oppositely charged ionic surfactants are attractive (synergistic) and between like charged surfactants repulsive (antagonistic). As a result of attractive interactions it is possible to have the *cmc* in a binary mixed surfactant system below the *cmc* values of the two pure components.

In order to relate the *cmc* values of pure surfactants to the *cmc* in mixed surfactant systems (Clint, 1974; Holland and Rubingh, 1983), the following nomenclature will be used, in which subscript i refers to species i in a multicomponent system:

μ_i^o = standard chemical potential of monomeric i in solution;

μ_i^{mic} = chemical potential of i in mixed micelles;

$\mu_i^{mic,o}$ = chemical potential of i in pure micelles of i;

m_i^{mon} = concentration of monomers of component i in mixed solution;

x_i = mole fraction i in mixed micelle;

cmc_i = cmc of pure i;

cmc_{mix} = cmc in mixed system;

α_i = mole fraction of i in the total mixed surfactant (on a surfactant only basis).

For a monomer i in ideal solution the chemical potential, μ_i, is expressed

$$\mu_i = \mu_i^o + kT \ln m_i^{mon} \tag{9.7.1}$$

The chemical potential of i in a mixed micelle, considered as a separate (pseudo) phase is

$$\mu_i^{mic} = \mu_i^{mic,o} + kT \ln f_i x_i \tag{9.7.2}$$

where f_i is the activity coefficient of i in the micelle. For *pure* micelles of i, for which $f_i x_i = 1$, it is seen from (9.7.2) that $\mu_i^{mic} = \mu_i^{mic,o}$ which in turn is equal to the chemical potential of i in solution, i.e. μ_i. Since in a pure system at and above the cmc $m_i^{mon} = cmc_i$ from (9.7.1) and (9.7.2)

$$\mu_i^{mic,o} = \mu_i^o + kT \ln cmc_i \tag{9.7.3}$$

At equilibrium in a *mixed* micellar solution $\mu_i^{mic} = \mu_i$ and combination of (9.7.1), (9.7.2) and (9.7.3) shows that

$$m_i^{mon} = x_i f_i cmc_i \tag{9.7.4}$$

Also, *at the cmc* in a mixed system

$$m_i^{mon} = \alpha_i cmc_{mix} \tag{9.7.5}$$

where α_i is the mole fraction of i in the total mixed surfactant (on a surfactant only basis) i.e. the concentration of i in solution divided by the total concentration of all surfactants present. Since the sum of all mole fractions of i in mixed micelles is unity, i.e. $\sum_i x_i = 1$, from (9.7.4) and (9.7.5) it follows that

$$\frac{1}{cmc_{mix}} = \sum_i \frac{\alpha_i}{f_i cmc_i} \qquad (9.7.6)$$

For a binary surfactant system, denoting $\alpha_1 = 1 - \alpha_2 = \alpha$, (9.7.6) can be expressed

$$\frac{1}{cmc_{mix}} = \frac{a}{f_1 cmc_1} + \frac{(1-\alpha)}{f_2 cmc_2} \qquad (9.7.7)$$

or

$$cmc_{mix} = \frac{f_1 cmc_1 f_2 cmc_2}{cmc_1 f_1 (1-\alpha) + cmc_2 f_2 \alpha} \qquad (9.7.8)$$

As seen in §5.5 (5.5.17), for regular solutions the activity coefficients can be written

$$\ln f_1 = \beta(1-x)^2 \qquad \ln f_2 = \beta(x)^2 \qquad (9.7.9)$$

where x is the mole fraction of surfactant 1 in the binary micelle and $(1-x)$ that of surfactant 2. For ideal mixed micelles $f_1 = f_2 = 1$. It should be noted that the interaction parameters β for a given system are not equal for mixtures in a monolayer (see (5.5.15)) and in a micelle.

For a binary surfactant system a value of β (which is the same for all compositions) can be obtained from a known *cmc* of a mixed system for a given α, together with the *cmc* in each of the two pure surfactant solutions, as follows. At the cmc, from (9.7.4), (9.7.5) and (9.7.9)

$$\beta = \frac{\ln [\alpha cmc_{mix}/x cmc_1]}{(1-x)^2} = \frac{\ln [(1-\alpha) cmc_{mix}/(1-x) cmc_2]}{x^2} \qquad (9.7.10)$$

The value of x is obtained iteratively from (9.7.10) and β calculated. The mixed *cmc* values can then be calculated using (9.7.7) or (9.7.8). The mole fractions of the two components in the mixed micelles (at the *cmc*) over a range of α can be computed in the following way. From (9.7.4), (9.7.5) and (9.7.8) it can be shown that

$$x = \frac{\alpha f_2 cmc_2}{(1-\alpha) f_1 cmc_1 + \alpha f_2 cmc_2} \qquad (9.7.11)$$

A range of x values is assumed and the right hand side of (9.7.11) calculated over this range. The required x is that for which the right hand side is equal to the assumed x. This procedure can be repeated for a range of α values.

Some results are shown in Fig. 9.16 for mixtures of sodium decylsulfate and the nonionic surfactant C_8E_4 in 0.05 M NaBr at 23°C. In Fig. 9.16a experimental *cmc* values (filled symbols), obtained by Holland and Rubingh (1983) are shown together with the

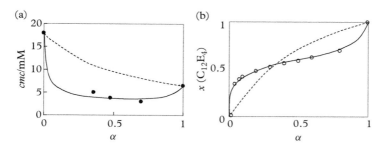

Figure 9.16 *(a)* cmc *values in binary mixtures of $C_{10}OSO_3Na$ and C_8E_4 (in 0.05 M NaBr at 23°C) as a function of the overall mole fraction of C_8E_4. The* cmcs *of aqueous pure C_8E_4 and $C_{10}OSO_3Na$ were taken as 6.5 and 18 mmol dm^{-3}, respectively. The full line is generated using $\beta = -4.1$ (see text) and the points are experimental. The dashed line is for ideal mixing in the micelle ($\beta = 0$). (b) Calculated micellar mole fractions (at the* cmc*) of C_8E_4 plotted against the overall mole fraction. The dashed line is calculated for ideal mixing. Experimental points in (a) are taken from Holland and Rubingh (1983).*

calculated *cmc* curve (full line) obtained by use of (9.7.7) with $\beta = -4.1$; the dashed line is for an ideal system ($\beta = 0$) with the same values of cmc_1 and cmc_2. In Fig. 9.16b, calculated micellar mole fractions of C_8E_4 are plotted against the overall mole fractions α. The nonideality leads to a sigmoid curve, not observed for an ideal system (dashed curve).

Rosen and Kunjappu (2012) list values of β for a range of binary surfactant systems, both for mixed adsorbed monolayers and mixed micelles. Largest negative values (corresponding to attractive interactions) occur as expected for mixtures of anionic and cationic surfactants. Mixtures of nonionic and ionic surfactants give smaller negative values and some positive values are known for mixtures of surfactants carrying like charges. Some values, taken from Rosen and Kunjappu (2012), are listed in Table 9.4 both for mixed monolayers at the air/water interface and for mixed micelles. It is again noted that, although the regular solution approach to treating surfactant mixtures can be very useful in a qualitative way, it is only semi-empirical at best.

Above, micellar compositions have been considered at the *cmc*. As the overall concentration m of the mixed system increases however, x must approach α. It can be shown from a mass balance that above the *cmc* (at a total concentration m) in a mixed binary system x is given by (Holland and Rubingh, 1983)

$$x = \frac{\alpha m}{cmc_1 f_1 - cmc_2 f_2 + (1 - \alpha)/(1 - x)} \tag{9.7.12}$$

and x can be calculated, for given α and m, in a manner similar to that described above. The way in which x (for C_8E_4) and $(1 - x)$ (for $C_{10}SO_4Na$) vary with overall surfactant concentration for $\alpha = 0.5$ is shown in Fig. 9.17a. Since, for given α, micelle composition changes with overall surfactant concentration the monomer concentrations

Table 9.4 *Values of β for some binary surfactant systems, both for mixed adsorbed monolayers and mixed micelles. Values taken from Rosen and Kunjappu (2012).*

Surfactant mixture	β (monolayer)	β (micelle)
Anionic–cationic		
$C_{12}SO_4^-Na^+ + C_{12}N^+Me_3Br^-$	−27.8	−25.5
Anionic–zwitterionic		
$C_{14}SO_4^-Na^+ + C_{12}N^+H_2(CH_2)_2COO^-$	−15.5	−15.5
Anionic–nonionic		
$C_{12}SO_4^-Na^+ + C_{12}E_8$	−2.7	−4.1
Anionic–anionic		
$C_7F_{15}COO^-Na^+ + C_{10}SO_4^-Na^+$ (in 0.1M NaCl)	+0.8	+0.3

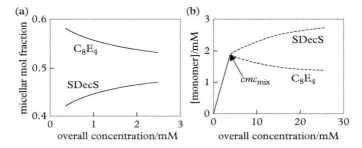

Figure 9.17 *(a) Variation of micellar mole fraction of C_8E_4 (x) and $C_{10}SO_4Na$, SDecS (1 − x) with overall surfactant concentration in equimolar mixtures (α = 0.5). (b) Variation of the monomer concentrations with overall surfactant concentration in the same system. Below the cmc the concentrations of the two surfactants are equal since α = 0.5.*

m_i^{mon} in solution, given by (9.7.4), also change with m; this is shown in Fig. 9.17b. The surfactant C_8E_4 has a lower *cmc* than SDecS and above the mixed *cmc* the monomer concentration of the nonionic surfactant falls with m whereas that of the ionic surfactant rises.

In practical use, as mentioned, surfactants are present as mixtures either by design or because commercial surfactants are impure. In various applications the adsorption of surfactants at various kinds of interface is central to their effectiveness. Adsorption, as seen elsewhere, is related to the monomer chemical potential and hence activity in solution. It is obvious therefore that information, such as that contained in Fig. 9.17, can be very valuable in designing formulations for specific purposes.

9.8 Can materials be dissolved in micelles? Solubilization and what happens to the micelles

Micelles in water have liquid-like interiors composed in many cases of hydrocarbon moieties of the surfactant molecules. It can be expected therefore that these hydrocarbon cores can dissolve nonpolar molecules. It has been seen above that micelles of one surfactant can take up molecules of another surfactant, and so it can be expected that micelles can also incorporate (*solubilize*) other amphiphilic materials which are not themselves surfactants, such as alkanols, alkanoic acids and long chain esters. Solubilization has been widely studied for many decades, in part for its practical importance (see e.g. Christian and Scamehorn, 1995). It occurs not only in micelles but also in other surfactant aggregates, for example, lamellar liquid crystalline phases, and surfactant monolayers at liquid interfaces (as described in §5.5). The importance of vesicles as drug delivery systems has already been touched upon in §9.5.2. As will be seen, surfactants dissolved in hydrocarbons can solubilize water in the centre of inverted surfactant aggregates.

9.8.1 Some empirical observations

Unsurprisingly, the extent of solubilization at saturation depends on the structure of the surfactant and the solubilizate; some examples are given in Fig. 9.18. The uptake of alkanes by micelles of sodium oleate falls as the alkane chain length increases (Fig. 9.18a); this parallels the uptake of alkanes by surfactant monolayers illustrated in Fig. 5.21b. Increase in the chain length of surfactant, however, has the opposite effect on solubilization, as can be judged from the solubilization of ethylbenzene by a homologous series of potassium alkanoates. Again, this is paralleled by uptake of a given solute by homologous surfactant monolayers. The surfactant head group can also play an important role. For example, the solubilization of hydrocortisone in a series of commercial nonionic surfactants $C_{16}E_n$ depends strongly on the (mean) value of n for n between 17 and 63. Solubilization falls with increasing n, as seen in Fig. 9.18c.

9.8.2 Effect of solubilization on the *cmc*

The *cmc* of a surfactant is lowered by solubilization. Hall (1983) has shown, for 1:1 ionic surfactants such as SDS or CTAB in supporting inert 1:1 electrolyte at constant T and p and in the presence of a sparingly soluble nonionic solubilizate (component 3), that

$$-d\ln cmc = \frac{N_3}{N_1}d\ln m_3 + \{(1-\alpha) + (2-\alpha)\,\partial\ln f_\pm/\partial\ln m_2\}\,d\ln m_2 \qquad (9.8.1)$$

in which the surfactant ion is species 1, and 2 denotes the total counterion. The degree of dissociation α of the surfactant is that defined in (9.6.2), and $f_\pm$ is the mean ionic activity coefficient in the solution; N_3 and N_1 are, respectively, the mean numbers of solubilizate molecules and surfactant ions in a micelle. In the absence of added electrolyte $d\ln m_2 = d\ln cmc$ and from (9.8.1)

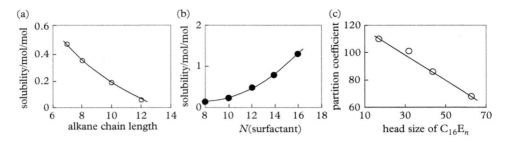

Figure 9.18 *Solubilization in micellar solutions. (a) Solubilization of normal alkanes in 0.1 M sodium oleate at 25° C. Results from McBain and Richards (1946). (b) Solubilization of ethylbenzene in a homologous series of potassium alkanoates at 25° C. Results from Klevens (1950). (c) Solubilization of hydrocortisone in n-alkyl polyoxyethylene surfactants $C_{16}E_n$ (n = 17, 32, 44, 63). The partition coefficient is the quotient of the difference between the total solubility of solubilizate in 1 molar surfactant solution and that in water alone divided by that in water alone. Redrawn from Barry and El Eini (1976).*

$$\frac{d \ln cmc}{d \ln m_3} = \frac{-N_3/N_1}{(2 - \alpha)\,[1 + (\partial \ln f_{\pm}/\partial \ln cmc)]} \qquad (9.8.2)$$

In the case of fixed concentration of swamping electrolyte $d\ln m_2 = 0$ and (9.8.1) becomes

$$\left(\frac{\partial \ln cmc}{\partial \ln m_3}\right)_{m_2} = -N_3/N_1 \qquad (9.8.3)$$

Equation (9.8.3) shows that the composition of mixed micelles, in equilibrium with surfactant monomer at the *cmc*, can be obtained from a knowledge of the variation of the *cmc* with added solubilizate. Also, for constant solubilizate concentration, m_3, it can be shown from (9.8.1) that the variation of the *cmc* with salt concentration is given by

$$-\left(\frac{\partial \ln cmc}{\partial \ln m_2}\right)_{m_3} = (1 - \alpha) + (2 - \alpha)\,\partial \ln f_{\pm}/\partial \ln m_2 \qquad (9.8.4)$$

Thus, by the use of (9.8.4) the effects of solubilized (nonionic) additive on micellar dissociation can be calculated from the variation of the *cmc* with salt concentration, in the presence of a constant concentration of additive. Note that (9.8.4) has the same form as (9.6.3). The degree of micellar dissociation increases with increased solubilization of polar material such as alkanols. Presumably, the head groups of the solubilized molecules lie between the charged surfactant heads at the micelle surface and screen the electrical repulsion, which reduces counterion binding.

As an example of the extent of solubilization of a sparingly water-soluble additive, the way in which the *cmc* of the di-chain anionic surfactant AOT in aqueous NaCl

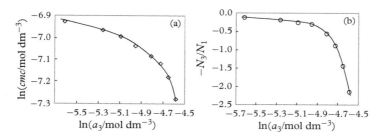

Figure 9.19 *(a) Variation, with dodecanol activity a_3 in heptane, of the cmc of AOT in 0.0171 mol dm^{-3} aqueous NaCl in equilibrium with heptane phase at 303 K. (b) Molecules of dodecane per AOT in micelles in the aqueous phase at 303 K as a function of dodecanol activity in heptane.*

$(0.0171 \text{ mol dm}^{-3})$ varies with the activity of added dodecanol is shown in Fig. 9.19a. The alkanol, which is only very sparingly soluble in the aqueous phase, is dissolved in a contacting phase of heptane which acts as a reservoir for dodecanol in the system. The variation of dodecanol activity in the aqueous phase is proportional to that in heptane at distribution equilibrium. The micellar compositions, calculated as described, are shown in Fig. 9.19b; at the higher dodecanol activities, there are about 2 molecules of dodecanol per micellized surfactant molecule. Of course in this system some heptane can dissolve in the aqueous phase micelles (see §9.8.3), but its activity in the system remains effectively constant since the alkanol activities are very low.

9.8.3 Solubilization of large amounts of oil (water) in water (oil) using small amounts of surfactant

It is a common goal to solubilize as much water (or oil) in a nonpolar oil (or water) by using as little surfactant as possible. If sufficient oil (water) is solubilized a *microemulsion* results, and this is discussed in greater detail in Chapter 10. Microemulsions, say water drops in an oil medium, can be used to dissolve water soluble materials in a nonpolar phase, which has obvious practical uses.

First, the ability of an aqueous solution of a nonionic surfactant, say of the type C_nE_m, to take up a nonpolar oil such as an alkane is considered. Aqueous solutions of C_nE_m (and many other nonionic surfactants) exhibit lower consolute boundaries. On heating, the clear solution suddenly becomes cloudy at a well-defined temperature (the *CPt*) which is dependent on the surfactant concentration. As will be seen, cloud points can be very sensitive to the presence of additives, including surfactants of various kinds, other amphiphilic materials, and nonpolar oil. With reference to the phase diagram shown in Fig. 9.20a, consider a clear aqueous solution (in the shaded area) containing nonionic surfactant and nonpolar oil, represented by point A. When the solution is heated a phase separation occurs at the *CPt*, on the *cloud point curve*. The coexisting phases above the *CPt* are a surfactant-lean micellar phase and a surfactant-rich phase containing clusters of interacting surfactant aggregates, that is, the phase separation is driven by

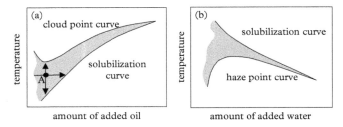

Figure 9.20 *Schematic representations of solubilization behaviour in (a) water-rich region and (b) oil-rich region. Shaded areas represent clear single phase regions. The systems contain a fixed overall % surfactant, for example, 5 wt % in a typical practical system.*

attractive interactions between micelles. It is believed that hydration of the $-E_m$ head groups stabilizes micelles against micellar aggregation; increasing the temperature will lead to progressive dehydration. At the same time, dehydration is expected to reduce the effective size of the head groups. This will cause an increase in the surfactant packing factor P (§9.5) which results in the formation of larger micelles with reduced curvature (rod or disc shape rather than spherical). The area of contact between elongated micelles is larger than that between spherical micelles and hence the attraction between micelles is increased, ultimately causing phase separation. In summary then, variation in T induces interrelated changes in micellar shape, size, and intermicellar interactions, which results in the existence of the CPt.

As seen from Fig. 9.20a, solubilization of nonpolar oil in the surfactant solution causes changes in the CPt. In the case of alkanes, the shape of the CPt curve depends on the alkane chain length. From what has been said, changes in CPt are expected to be related to changes in micellar shape accompanying the solubilization process. As an example, the CPt curve for the addition of octane to aqueous solutions of $C_{12}E_5$ is shown in Fig. 9.21; the CPt first falls with the addition of octane and then begins to rise. A simple indicator of micellar shape in dilute systems is the relative viscosity η_R of the solution (viscosity of solution/viscosity of solvent). As micelles become more asymmetrical the relative viscosity rises and an inverse relationship is therefore expected between changes in CPt and η_R. This is as observed in Fig. 9.21.

A plausible explanation for these findings runs as follows. For small additions of octane to the aqueous surfactant solutions, the alkane is solubilized in the chain region of the micelles, thus giving an increase in the effective cross-sectional area of the chains. This results in an increase in the packing factor P and an elongation of the micelles, which in turn causes a fall in CPt and an increase in η_R. When the surfactant chain regions within the micelles become saturated with alkane, further additions begin to form a liquid hydrocarbon core and the micelles become progressively more spherical, ultimately becoming microemulsion droplets. This process is accompanied by a fall in η_R and a (slow) rise in CPt. The propensity of alkanes to mix with surfactant chains within micelles depends very much on the alkane chain length. Higher alkanes mix less well and liquid cores are formed for smaller additions. Thus, hexadecane initially causes only an increase in CPt and a concomitant decrease in η_R.

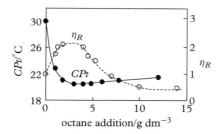

Figure 9.21 *Variation of relative viscosity (at 20° C) and cloud point with addition of octane to 2 wt % $C_{12}E_5$ in water.*

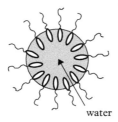

Figure 9.22 *Inverted micelle with water core.*

Returning now to a consideration of the solution at point A in Fig. 9.20a, if more alkane is added or if the temperature is lowered, another phase boundary is reached, termed the *solubilization curve*. At this boundary, the micelles become saturated and the excess added solubilizate forms a separate phase consisting mainly of oil, with dissolved monomeric surfactant. If sufficient oil has been incorporated, the surfactant aggregates are spherical microemulsion droplets and the phase separation occurs because the surfactant layers coating the droplets have attained their *preferred curvature*, a concept discussed further in Chapter 10.

Suppose now that the nonionic surfactant is dissolved in oil, and that water is added to the solution; this situation is illustrated in Fig. 9.20b. As for the aqueous systems, there is a single phase micellar region bounded by two phase boundaries. The micelles in this region are *inverted*, that is water forms the interior, in which the surfactant head groups are immersed, as illustrated in Fig. 9. 22. In this case however, the higher temperature boundary is the solubilization curve and the lower boundary, corresponding to the cloud point curve in aqueous systems, is termed the *haze point curve* and arises from intermicellar attractions.

There is an interest in extending the single phase regions of solubilization by shifting the two phase boundaries further apart. For example the cloud point curve can be shifted to higher temperatures by introducing a small amount of charge (repulsion) into the micelles by incorporating a small proportion of ionic surfactant. This is illustrated in Fig. 9.23 (for systems without added oil). Cloud points are given for 24 mM $C_{12}E_5$ aqueous solution with 1.5 mol % sodium dodecylsulfate (SDS) added. The SDS elevates

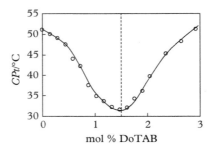

Figure 9.23 *Cloud points of solutions of 24 mM C12E5 containing 1.5 mol % SDS, as a function of the amount (mol %) of added DoTAB.*

the *CPt* of the nonionic surfactant solution from 31.5°C to 51.2°C. When, to this mixed solution of nonionic and anionic surfactant, aqueous dodecyltrimethylammonium bromide (DoTAB) is added, the *CPt* is progressively lowered. When the system contains equimolar amounts of anionic and cationic surfactant the *CPt* attains a minimum value of 31.5°C, equal to the *CPt* of the $C_{12}E_5$ solution in the absence of ionic surfactant. After this equivalence point the *CPt* again rises as the micelles are charged up by the DoTAB. The action of the anionic and cationic surfactants is entirely equivalent.

In order to shift the solubilization curves, the preferred curvature of the surfactant monolayers stabilizing the micelles can be modified. This can be achieved by incorporating other surfactants or amphiphiles into the micelles. Of course, a given additive can in principle modify both phase boundaries simultaneously, thus complicating matters somewhat.

9.9   The interactions of surfactants with polymers

Surfactants are used, for example, as wetting agents, emulsion stabilizers, foaming agents, and solubilizing agents for water-insoluble additives. On the other hand polymers can be used as thickening agents and for rheology control in general. In many formulations and processes surfactants and polymers are therefore present together, for example, in oil recovery systems, foodstuffs, cosmetics and detergents, paints, and pesticides, and they can interact synergistically, antagonistically or in some cases not at all. It is obviously important therefore to understand how (or if), the two components mutually interact. The discussion below is restricted to aqueous systems.

Surfactant that is bound to a polymer in some way becomes unavailable for adsorption at, say, the air/water interface. Therefore it can be expected that binding will be reflected in plots of the surface tension of a polymer–surfactant solution against the total concentration of surfactant, for a constant polymer concentration. Such a plot is shown in Fig 9.24 for solutions of the anionic surfactant SDS containing the nonionic polymer polyvinylpyrrolidone (PVP). The curve for solutions of SDS in the absence of PVP (open symbols) has a single breakpoint (A) at the *cmc* of pure SDS (~8 mmol dm^{-3}) as

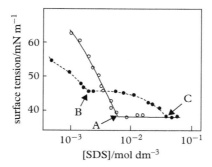

Figure 9.24 *Surface tension of aqueous solutions of SDS (open symbols) and of solutions of SDS in aqueous PVP (10 g dm⁻³) (filled symbols) as a function of the SDS concentration. Redrawn from Goddard (1986), using original data of Lange (1971).*

discussed in §9.1. In the presence of the (fixed amount of) polymer however a second breakpoint (B) occurs at a lower surfactant concentration, and higher surface tension. This is interpreted as being a second critical aggregation concentration (*cac*) at which surfactant micelles begin to form in association with the polymer. As the surfactant concentration is increased beyond B the surface tension initially remains constant as further micelles form on the polymer. Ultimately however the tension begins to fall as the concentration of free surfactant monomers again rises, until point C is reached. Here polymer-free micelles begin to appear and the tension again becomes constant. The process of aggregation of surfactant on polymer molecules can be followed by the direct measurement of the concentration of monomeric surfactant entities; Gilanyi and Wolfram (1981) used potentiometric measurements and ion selective electrodes. In the case of SDS and PVP it is reported that the aggregates that form along the polymer chain are smaller than ordinary SDS micelles, containing around 12 surfactant ions. This number increases with counterion concentration and in the presence of 0.1 mol dm⁻³ NaNO$_3$ the aggregation number is between 40 and 50. It is interesting that the standard free energy of micelle formation (§9.2) of surfactant associated with the polymer chain is very close to that for pure micelles. This suggests that the major part of the driving force for aggregation on the polymer is the cooperative interactions between surfactant entities (as in the case of pure micelles) although obviously the interactions between micelles and polymer are important.

The solubilization of materials by free surfactant micelles was considered in §9.8; the question arises as to whether micelles associated with polymer molecules can also be involved in solubilization. Data for the solubilization of the water-insoluble dye Orange OT in SDS micelles, both in the presence and absence of PVP, are shown in Fig. 9.25. For a given overall SDS concentration, the solubilization in the presence of the polymer greatly exceeds that in its absence, over a wide range of SDS concentrations. The polymer induces micelle formation at much lower surfactant concentrations than would otherwise be the case and the polymer-associated micelles clearly solubilize the dye just as free micelles do.

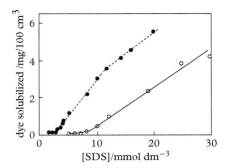

Figure 9.25 *Solubilization of Orange OT in aqueous solutions of SDS in the absence (open symbols) and presence (filled symbols) of 3 g dm^{-3} PVP, as a function of surfactant concentration. Redrawn from Goddard (1986), using original data of Lange (1971).*

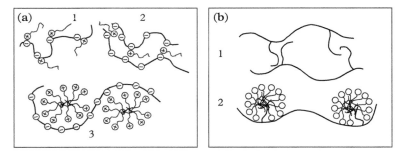

Figure 9.26 *Representations of polymer–surfactant aggregation structures. (a) (1) Charged polymer with ionic surfactant of opposite charge at low surfactant concentration; (2) surfactant ions bridging polymer molecules; (3) surfactant micelles associated with polymer molecule. (b) (1) Two weakly associated hydrophobically modified polymer molecules in the absence of surfactant; (2) surfactant micelles formed around side chains of hydrophobically modified polymer.*

Various morphologies can be imagined for the entities formed by association of polymer and surfactant, depending on the chemical nature of the two species, and some examples are given in Fig. 9.26. In the case of a nonionic polymer and ionic surfactant, such as SDS with PVP, the interactions between surfactant and polymer presumably involve the charge on the surfactant head groups with the polar moieties of the polymer chain giving a structure somewhat like that for an ionic polymer with ionic surfactant of opposite charge shown in Fig. 9.26a (3). Also shown in Fig. 9.26a (1) are associated species between ionic surfactant and ionic polymer at surfactant concentrations below the *cac*. At intermediate concentrations the surfactant ions can bridge polymer molecules (Fig. 26a (2)).

The illustration in Fig. 9.26b relates to a water-soluble nonionic polymer with hydrophobic side chains, such as hydrophobically modified ethylhydroxyethylcellulose

(HM-EHEC). Such polymers can mutually associate in the absence of surfactant via the hydrophobic side chains, as shown in Fig. 9.26b (1). When surfactant is progressively added, single surfactant entities can bind via their hydrophobic groups to the hydrophobic side groups on the polymer. At higher surfactant concentrations micelles which incorporate the side chains on adjacent polymer molecules can form, thus creating bridges between polymer molecules. As the surfactant concentration increases further however, micelles form on individual polymer side chains as illustrated in Fig. 9.26b (2) and bridging is absent.

Such a sequence of events is reflected in viscosity changes in the solutions (Holmberg et al., 2003), with systems containing bridged polymers (see also Box 9.2) exhibiting relatively high viscosity. An example is shown in Fig. 9.27 for systems with SDS and the sodium salt of poly(acrylic acid) with an average molecular weight of 500,000 (PAA-500) and with the hydrophobically modified material (PAA-500-1-C18) which contains

---

Box 9.2 Polymers can self-assemble as well

For the most part, the material in this chapter is concerned with low molar mass surfactants and their self-assembly, that is, their aggregation into micelles and larger structures. However, it is well known that block copolymers can also self-assemble and in so doing lead to an astonishing array of supermicellar structures (Lunn et al., 2015), hence linking traditional surfactant science with the broader fields of nanoscience (see §17.1) and soft condensed matter. The defining property of soft condensed matter is the ease with which it is distorted and flows as a result of relatively small external stresses. Systems include colloidal dispersions, various biomaterials, polymers and their composites, lyotropic and thermotropic liquid crystalline phases, etc.

The simplest kind of block copolymer is the AB diblock type consisting of two different polymer segments. If one block is soluble in (say) water and the other is not, then the (amphiphilic) copolymer will self-assemble above a critical aggregation concentration in the solution, with the hydrophobic group forming the micelle core and the hydrophilic group the outside 'corona', in much the same way as an amphiphilic low molar mass surfactant. The polymer blocks can be responsive to various stimuli such as pH, temperature, solvent and ionic strength. Groups containing acidic or basic entities for example respond to changes in pH; poly(methacrylic acid) accepts protons at low pH and releases them at neutral and high pH. As well as amphiphilic block copolymers it is also possible to have copolymers in which both blocks are hydrophilic but which respond differently to say pH. A range of aggregate structures including spherical, flower-shaped, wormlike, and vesicle (hollow) micelles have been formed by amphiphilic and double hydrophilic block copolymers by employing changes in pH (Kocak et al., 2017).

The aggregates formed from block copolymers can be used as basic units in the formation of complex hierarchical structures (Lunn et al., 2015). For self-assembly of basic aggregates into larger structures, the aggregates need to interact mutually and therefore exhibit some form of anisotropy of shape or surface functionality. Such micelles have been termed 'patchy'.

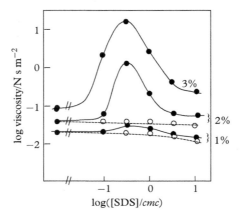

Figure 9.27 *Viscosity of solutions containing PAA-500 (dashed lines) and hydrophobically modified polymer PAA-500-1-C18 (full lines; see text) as a function of the concentration of added SDS. The points on the left hand side of the graph below the break (∥) are for systems in the absence of SDS. The wt % of polymer in the systems is indicated on the graph. Redrawn from Iliopoulos et al. (1991).*

1 mol % octadecyl chains randomly spread over the PAA chain. The viscosity of systems with 1 and 2 wt % PAA-500 changes (falls) only little with surfactant concentration up to 10× the *cmc* of pure SDS. The polymer (an anionic *polyelectrolyte*) and surfactant both carry a negative charge and are not expected to associate mutually. In contrast, systems with hydrophobically modified polymer exhibit large changes in viscosity as SDS is progressively added. There is a maximum, for all (fixed) polymer concentrations at about 0.3*cmc* of pure SDS, the maximum being higher the greater the polymer concentration. The SDS concentration for the onset of the rise in viscosity is well below the position of the maximum and is dependent on polymer concentration. It appears therefore that, in the case of the hydrophobically modified polymer, even though polymer and surfactant carry the same sign charge, repulsion between polymer and surfactant is outweighed by the favourable mutual hydrophobic interactions between surfactant chains and between surfactant chains and the polymer side chains.

9.10   Micelles are dynamic structures that continually break and reform

Many of the applications of surfactants, for example, in wetting and in emulsion and foam formation (where large areas of surfactant-covered interfaces are rapidly formed),

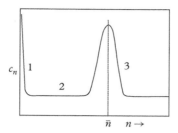

Figure 9.28 *Schematic representation of size distribution of aggregates, concentration c_n, in a micellar solution. Peak 1 results from monomers and possibly some small associated species. Peak 3 gives the distribution of micellar aggregates, and has been drawn as a Gaussian (normal) distribution. In region 2, there are very low concentrations of aggregates of intermediate sizes.*

depend on surfactant adsorption at interfaces, and the rate at which adsorption can occur is therefore an important consideration. In Chapter 6, the dynamics of adsorption and the way in which in micellar solutions surfactant might become available for adsorption were discussed. Here, the dynamics of micelle formation and breakdown are considered in a little more detail. If micelles in a solution were all the same size (i.e. monodisperse), if one molecule were to leave a micelle, the micelle would completely disintegrate into monomers. More realistically however, it can be supposed that micelles have a degree of polydispersity, that is, a range of sizes, and indeed micelle formation has been represented as a stepwise process in (9.2.1), in which aggregates of intermediate sizes exist. It is also clear from the way in which the free energy of micelle formation varies with aggregation number (Fig. 9.7a, inset) that a size distribution exists. A distribution of aggregate sizes is shown schematically in Fig. 9.28, in which the ordinate c_n is the concentration of species with aggregation number n. Suppose that a micellar solution is rapidly diluted at constant temperature, pressure, and overall surfactant content, such that the final solution is still above the *cmc*. Initially monomers would quickly attain a quasi-equilibrium with the micelles since the diffusion of monomers to the micelles is relatively rapid. However to reach equilibrium proper, since the final solution is more dilute, micelles must disintegrate, via aggregates with sizes which are present at only low concentrations (region 2 in Fig. 9.28). This means that the disintegration process is slow relative to the exchange of single surfactant molecules with micelles.

Using rapid kinetic techniques, for example, pressure (Box 9.3) and temperature jump and stopped-flow measurements, two relaxation times can be identified; τ_1 reflects the rate at which monomers exchange between micelles and solution and τ_2 is a measure of the rate of formation and disintegration of micelles. From what has been said it can be appreciated that $\tau_1 < \tau_2$.

Box 9.3 The pressure's on

A pressurized surfactant solution has a higher *cmc* than that at atmospheric pressure. If for example the surfactant is ionic the conductivity rises as the *cmc* rises (because more fully dissociated monomers are present). The relaxation of a solution (involving the *formation* of micelles) can therefore be followed by the change in conductivity with time following the release of the pressure. In the case of nonionic surfactants relaxation can be observed spectrophotometrically following dilution of micelles containing solubilized dye (Patist et al., 2002). In this case micellar *breakdown* is followed, and dye is released into water. Differences in absorption spectra of solubilized and free dye are exploited.

9.10.1   The fast relaxation time

Aniansson and co-workers (see e.g. Aniansson et al., 1976) were the first to derive expressions for the two micelle relaxation times. For the fast time τ_1 the expression is, for a nonionic surfactant,

$$\frac{1}{\tau_1} = \frac{k^-}{\sigma^2} + \frac{k^-}{n}\left(\frac{c - m_1}{m_1}\right) \tag{9.10.1}$$

The constant k^- is the rate constant for the dissociation of a monomer from a micelle in the stepwise equilibrium:

$$D_1 + D_{n-1} \underset{k_n^-}{\overset{k_n^+}{\rightleftharpoons}} D_n \quad n = 2, 3, 4, \ldots . \tag{9.10.2}$$

it being assumed that all k_n^- are equal, and equal to k^-. The total solution concentration is c and m_1 is the surfactant monomer concentration. The distribution (Box 9.4) of micelle sizes has a standard deviation σ. In the case of ionic surfactants the expression for the fast relaxation time becomes

$$\frac{1}{\tau_1} = \frac{k^-}{\sigma^2} + \frac{k^-}{n}\left[1 + \frac{(1-\alpha)^2}{(1+\alpha a)}\right]a \tag{9.10.3}$$

In (9.10.3) α is the degree of micellar dissociation (as defined in (9.6.4)) and $a = (c - m_1)/m_1$.

Box 9.4 Gaussian or normal distribution

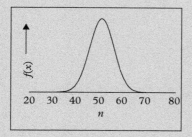 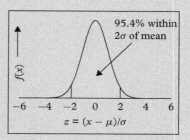

Left: Gaussian distribution of micelle size ($\mu = 50, \sigma = 5$). Right: Corresponding standard normalized distribution (i.e. with $\mu = 0$ and $\sigma = 1$).

The micelle peak (aggregate concentration *versus* aggregation number) in Fig. 9.28 has been drawn (schematically) with a Gaussian distribution (also called a *normal distribution*), named after Carl Friedrich Gauss (1777–1855), the great German mathematician and astronomer. The normal distribution is given by

$$f(x) = \frac{1}{\sigma\sqrt{2\pi}} \exp\left\{-\frac{1}{2}\frac{(x-\mu)^2}{\sigma^2}\right\}$$

where (in the present case) x is the aggregation number n, μ is the mean value of x ($\bar{n}$ here) and σ is the *standard deviation*, i.e. a measure of the spread of x (n) values. The *variance*, σ^2, is the average of the squared differences from the mean. In the left hand figure a Gaussian distribution of micelle concentration (represented by $f(x)$) against aggregation number has been calculated using a mean aggregation number of 50 and standard deviation of 5. A standardized form is shown on the right where the abscissa is the number of standard deviations z. For a Gaussian distribution 95.4% of data lies within two standard deviations of the mean. Normal distributions are often used in the natural sciences (and elsewhere) for random variables with unknown distributions.

The quantity m_1 above the *cmc* has often been taken to be equal to the *cmc*, but Gharibi et al. (1991), using ion selective electrodes, measured m_1 up to concentrations $5\times$ *cmc* for a range of homologous alkyltrimethylammonium bromides (C_nTAB) and sodium alkylsulfates (SC_nS). Some results are shown in Fig. 9.29 for m_1 in solutions of C_{12}TAB and SC_{10}S, where it is seen that m_1 falls substantially above the *cmc*.[7] Plots

---

[7] Elsewhere in this book it has been supposed that the monomer concentration remains essentially constant above the *cmc*. In the contexts, where the concern has been with micellar solutions close to the *cmc*, this is justified. However for such high concentrations discussed above (up to 5 times the *cmc*), m_1 is clearly dependent on overall concentration.

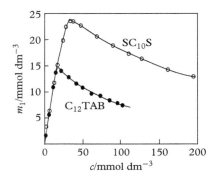

Figure 9.29 *Variation of monomer concentration m_1 with overall surfactant concentration, c, for aqueous solutions of sodium decylsulfate and dodecyltrimethylammonium bromide. Redrawn from Gharibi et al. (1991).*

according to (9.10.1) and (9.10.3) are shown in Fig. 9.30, where $1/\tau_1$ is plotted against the bracketed concentration terms in the equations. Plots are linear and values of k^-/n and k^-/σ^2 can be obtained from slopes and intercepts respectively. Since aggregation numbers are available independently, it is possible to obtain k^- and σ from the relaxation times obtained over a range of surfactant concentration. Note that n/k^- is the residence time of a monomer in a micelle. Some values of the various parameters obtained by Gharibi et al. (1991) are shown in Table 9.5. The aggregation numbers and degrees of micellar dissociation are given beneath the surfactant designation. Method (a) analysis uses (9.10.1), setting $m_1 = cmc$; method (b) also uses (9.10.1) but with experimentally determined values of m_1 (such as those shown in Fig. 9.29). Method (c) involves the use of (9.10.3) which allows for micellar dissociation. For the most part, values of k^- obtained using methods (b) and (c) are very similar showing that micellar dissociation appears not to influence k^- very much. Setting $m_1 = cmc$ (method (a)) on the other hand gives substantially higher values of k^-. Increasing surfactant chain length causes a decrease in k^-, that is, reduces the rate of the monomer exchange process.

In summary then, account must be taken of the change in monomer concentration above the *cmc* to analyse fast relaxation data correctly, although allowing for dissociation of ionic micelles appears to be relatively unimportant. Fast relaxation is dependent on surfactant molecular structure, the monomer exchange being slower the longer the surfactant chain.

9.10.2   The slow relaxation time

The slow relaxation times, τ_2, for micelles are several orders of magnitude greater than τ_1 and as mentioned relate to micelle breakdown rather than monomer exchange with micelles. Micelles can be supposed to form or disintegrate in two ways, depending on conditions. One way, at low concentrations, is described by the stepwise process envisaged in (9.10.2) in which, for micellar breakdown, a monomer leaves an aggregate,

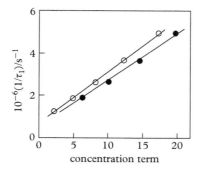

Figure 9.30 *Plots according to (9.10.1) (open circles) and (9.10.3) with $\alpha = 0.2$ (filled circles) for $C_{12}TAB$ solutions in 10^{-4} mol dm^{-3} NaBr. Redrawn from Gharibi et al. (1991).*

Table 9.5 *Values of kinetic parameters associated with monomer–micelle exchange for some anionic and cationic surfactants. Results taken from Gharibi et al. (1991).**

Surfactant	Method	$(k^-/n)/s^{-1}$	$(k^-/\sigma^2)/s^{-1}$	k^-/s^{-1}	σ ($\pm 50\%$)
$C_{12}TAB$	a	7.3×10^5	3.5×10^5	3.1×10^7	10
($n = 49$)	b	2.1×10^5	8.3×10^5	1.0×10^7	3
($\alpha = 0.25$)	c	1.9×10^5	6.3×10^5	9.5×10^6	4
$C_{14}TAB$	a	4.8×10^4	1.8×10^5	3.2×10^6	4
($n = 66$)	b	1.0×10^4	3.6×10^5	6.7×10^5	1
($\alpha = 0.12$)	c	1.0×10^4	3.3×10^5	6.6×10^5	1
$SC_{10}S$	a	2.1×10^6	1.2×10^4	8.5×10^7	83
($n = 41$)	b	6.2×10^5	1.0×10^5	2.6×10^7	5
($\alpha = 0.4$)	c	6.1×10^5	7.6×10^5	2.5×10^7	6
$SC_{14}S$	a	1.4×10^4	1.7×10^3	8.8×10^5	23
($n = 64$)	b	8.0×10^3	2.2×10^3	5.1×10^5	15
($\alpha = 0.26$)	c	6.7×10^3	1.3×10^3	4.3×10^5	18

*Aggregation numbers n and degrees of micellar dissociation α are given below the surfactant designation. Methods of analysis: (a) eqn. (9.10.1) with $m_1 = cmc$; (b) eqn. (9.10.1) with experimentally determined values of m_1; (c) eqn. (9.10.3) with experimentally determined values of m_1.

D_n to give D_{n-1} and so on. Another possibility at higher concentrations is a process of fusion or fission of sub-micellar aggregates, which can be represented by

$$D_j + D_k \leftrightarrow D_n \quad j + k = n \tag{9.10.4}$$

For the low concentration region, stepwise breakdown occurs and Aniansson and co-workers obtained the expression for the slow relaxation time, which is designated here τ_{21}:

$$\frac{1}{\tau_{21}} = \frac{n^2}{m_1 R} \left(\frac{n}{n + a\sigma^2} \right) \tag{9.10.5}$$

The quantities n and a have the same significance as before, and R is given by

$$R = \sum_{n_1+1}^{n_2} \left(\frac{1}{k_n^- [D_n]} \right) \tag{9.10.6}$$

Aniansson (1985) showed that the average micellar lifetime, T_m, is related to τ_{21} by

$$T_m = \frac{\tau_{21} n^2 a}{n + \sigma^2 a} \tag{9.10.7}$$

The significance of the quantity R in (9.10.6) can be understood as follows. The region 2 in the size distribution illustrated in Fig. 9.28 acts as a 'resistance' to the stepwise breakdown of micelles, since the aggregates in this region are present only in very low concentration. The quantity in round brackets in (9.10.6) is an inverse rate and the summation is taken from size n_1+1 at the beginning of region 2 to n_2 at the end of region 2.

The slow relaxation times of ionic micelles (e.g. SDS micelles) often show a maximum with respect to the surfactant concentration. This occurs at 200 mmol dm^{-3} for SDS, which is roughly $25 \times cmc$ (Fig. 9.31). Maxima are not predicted by the stepwise breakdown discussed above. They can however be explained by taking into account a mechanism of fission for micellar breakdown at the higher surfactant concentrations (Kahlweit, 1982). At low concentrations ionic micelles are mutually repulsive, but as the surfactant concentration increases, the counterions (Na$^+$ in the case of SDS) are able to screen this repulsion so that fusion/fission can occur. The relaxation time associated with this mechanism is designated τ_{22}. Kahlweit proposes that the formation (breakdown) reaction path of micelles can be likened to two resistors in parallel so that the formation of micelles is analogous to the discharge of a capacitor through the two parallel resistors, as illustrated in Fig. 9.32. The resistance R_1 is the same as R in (9.10.5) and (9.10.6).

At lower surfactant concentrations, τ_{21} rises with overall surfactant concentration according to (9.10.5) and micelle breakdown due to fission is very low (R_2 very high). As the surfactant (counterion) concentration rises further however R_2 falls and when

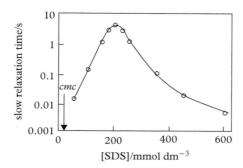

Figure 9.31 *Slow relaxation times of SDS micelles at 298 K as a function of surfactant concentration. The maximum occurs at 200 mmol dm^{-3}, and the cmc is at 8.3 mmol dm^{-3}. Redrawn from Patist et al. (2002).*

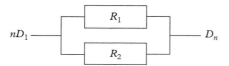

Figure 9.32 *Representation of two slow micelle breakdown reaction paths, likened to the resistors R_1 and R_2 in parallel.*

$R_1 = R_2$, the reciprocal of the slow relaxation time is minimum and the relaxation time is maximum (at 200 mmol dm^{-3} for SDS as shown in Fig. 9.31).

Data have been given above for the micellar breakdown of ionic surfactants. Values of τ_2 for nonionic surfactants are often substantially higher than those for ionic surfactants. Indeed commercial samples can have τ_2 equal to 100 s or more. Such high relaxation times can be associated with impurities in the samples; it is known that for example the incorporation of long chain alkanols into the micelles can result in substantial increases in τ_2 (Patist et al., 2002). For the pure nonionic surfactants $C_{12}E_5$ and $C_{12}E_8$ at concentrations of the order of 10 times the *cmc*, τ_2 values are, respectively, 10 and 4 seconds. Presumably it is the absence of micellar charge that renders nonionic micelles more 'stable' (larger τ_2) than ionic micelles.

9.11 Concentrated systems: phase behaviour and lyotropic liquid crystals

As the surfactant concentration in a micellar solution is progressively increased, the concentration of micelles is increased (which can be accompanied by changes in micelle size and shape) and so the separation between micelles decreases. Consequently intermicellar interactions become increasingly significant, which results in the formation of a variety of *lyotropic* liquid crystalline phases (*mesophases*).[8] It can be imagined that the various phases arise by stripping water from micellar solutions. Liquids have only short-range order which is manifest over only 1–3 molecular lengths, whereas solid crystals have much longer-range order over hundreds of molecular diameters. Lyotropic liquid crystals are intermediate in that they can be liquid-like in some dimensions but have long-range order in others. For example, lamellar phases consisting of parallel sheets of surfactant molecules (see Fig. 9.34d) may have only short range order within a (liquid-like) sheet, but have long range crystal-like order in the direction normal to the sheets.

Before discussing surfactant mesophases further, it is useful to describe how phase behaviour is represented in phase diagrams, having temperature as the ordinate and composition as the abscissa. The Gibbs phase rule applies to the systems at equilibrium, and it is worth noting that micellar solutions are single phase systems, any pure surfactant (ionic or nonionic) is a single component, and if a surfactant undergoes hydrolysis, the hydrolysis products are additional components; isomers are also separate components.

On the phase diagram, a horizontal line (constant T and varying composition) is termed an *isotherm*, and a vertical line (constant composition, varying T) is an *isopleth*. In a two-phase system at equilibrium interest lies not only in the compositions of the coexisting phases but also in the relative amounts of those phases, which can be obtained by use of the *lever rule*. The cloud point curve in systems with the nonionic surfactant C_8E_4 and water is shown as an example in Fig. 9.33.[9] Systems above the curve are cloudy in appearance and consist of two coexisting phases, one dilute phase at the *cmc* and the other rich in surfactant and consisting of aggregated micelles and water. In Fig. 9.33 the composition is expressed as wt % of surfactant, but this is readily converted to mole fraction x of surfactant if required. A system with an overall composition and temperature represented by point X separates, at equilibrium, to give systems represented by points A and B. The isotherm AB is the *tie line* and the compositions of the two phases are known from the diagram. Imagine now a two-phase system, at the same temperature but more concentrated in surfactant. This system still separates to give coexisting phases at A and B but obviously there will now be more of the surfactant-rich phase present. The lever rule allows the calculation, from the phase diagram, of the fraction of the total system present in the two phases. If the fraction of the total weight that is present in the phase with composition at A is denoted f_A, and similarly the fraction in the phase at B, f_B, the lever rule states that

$$\frac{f_A}{f_B} = \frac{XB}{XA} \qquad (9.11.1)$$

[8] *Lyotropic* liquid crystals are formed by addition of solvent (to surfactant); the term *mesophase* indicates that the phase lies in between liquid and crystalline states.

[9] In §9.8 and Fig. 9.19 it was seen how the *CPt* of a surfactant solution with fixed overall surfactant concentration is affected by *solubilization*. The curve in Fig. 9.33 gives the variation of the *CPt* with surfactant concentration in the absence of solubilization. Note that as T falls the compositions of the coexisting phases become closer (tie lines shorter). The difference vanishes at the *critical point*.

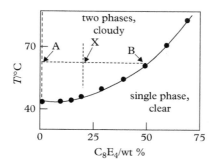

Figure 9.33 *Cloud point curve of C_8E_4 in aqueous solution. Systems above the curve are two-phase systems, with a surfactant-lean and a surfactant-rich phase. Redrawn from Mitchell et al. (1983).*

where XA and XB denote lengths along the tie line. If mole fractions are used in the phase diagram rather than weight fractions, then f_A and f_B are fractions of the total number of moles present.

As mentioned, increasing the concentration of a micellar solution brings micelles mutually closer; micelles can then aggregate to give mesophases with a variety of structures. Suppose, as an example, micelles in dilute solution are initially spherical. At some concentration the micelles pack into an ordered array of spheres giving rise to a normal micellar cubic phase (designated I_1 and sometimes referred to as a 'middle' phase-not to be confused with the term middle phase in Winsor systems to be discussed in Chapter 10) in which water fills the spaces (Fig. 9.34a). Upon further concentration, micelles will experience increased mutual repulsion which results in a transition from cubic to a hexagonal phase (H_1) (Fig. 9.34c). In the latter the micelles are cylindrical, and a smaller volume of water is needed to fill the intermicellar space. In addition, the average separation between micelle surfaces is increased, which leads to a reduction in micellar repulsion. The energetic advantages of these changes are sufficient to overcome any unfavourable curvature change in converting spheres into cylinders.

Reverse cubic (I_2) and hexagonal (H_2) phases are also observed. The I_2 phase has a similar structure to I_1 except that the individual micelles are reverse micelles, with water forming the cores with the surfactant tails on the outside. In H_2 phases, water forms the core of the cylinders and again the tails are on the outside. Low curvature *bicontinuous* cubic phases (V_1) are also possible in which the structure is formed from interconnecting surfactant films with curvatures of opposite sign, as shown in Fig. 9.34b. Similar bicontinuous but inverted structures can also occur (V_2). Lamellar phases (L_α), (also called 'neat' phases) have alternating surfactant bilayers and water layers, as shown in Fig. 9.34d; the layers can be more or less flexible but are liquid-like. If the chains are solid-like, lamellar *gel phases* (L_β) result, in which the hydrocarbon chains can be normal to the layers, tilted or interdigitated, as illustrated in Fig. 9.35. Micellar and reverse micellar solutions are denoted L_1 and L_2 and have, respectively, the largest positive and largest negative surface curvatures.

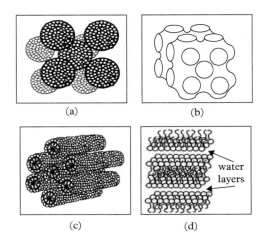

Figure 9.34 *Representation of some surfactant mesophase structures. (a) A cubic phase (I_1) of spherical micelles. (b) A bicontinuous cubic phase (V_1) in which connecting surfactant films possess curvatures of opposite sign, the net film curvature being low. (c) A hexagonal phase (H_1) in which rod-like micelles are arranged hexagonally. (d) A lamellar phase (L_α) made up of surfactant bilayers with water layers between the head groups.*

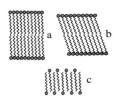

Figure 9.35 *Representation of L_α gel phases. In all cases the chains are solid-like in all trans configurations. (a) Normal chains, (b) tilted chains, (c) interdigitated chains.*

A further type of liquid crystalline phase, not referred to above, has a *nematic*[10] structure containing rod or disc-shaped micelles in a loosely-ordered linear arrangement, as illustrated in Fig. 9.36. The aggregates are roughly parallel but are not organized into layers or rows.

Some of the factors which influence the structure of lyotropic liquid crystalline phases have been usefully summarized by Huang and Gui (2018); see also Box 9.5.

---

[10] *Nematic* refers to threadlike chains (from Ancient Greek *nema* meaning thread, as in nematode, a worm).

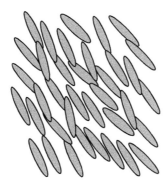

Figure 9.36 *Representation of a nematic liquid crystalline phase consisting of loosely ordered rod or disc-shaped micelles.*

Box 9.5 Liquid crystals without a solvent—thermotropic liquid crystals

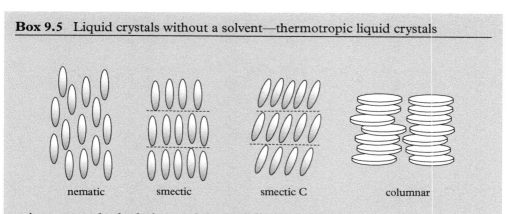

nematic   smectic   smectic C   columnar

Arrangement of molecules in some thermotropic liquid crystalline phases.

There are two types of liquid crystals: lyotropic and thermotropic. Lyotropic liquid crystals, the subject of §9.11, contain both surfactant and a solvent (e.g. water), whereas thermotropic liquid crystals do not contain solvent but are usually just a single chemical component, frequently organic molecules that tend to be elongated or rod-shaped. As the term implies, liquid crystals have some properties typical of liquids (e.g. fluidity) and also properties associated with crystallinity such as periodic arrangement of molecules in one or more directions. Structures of thermotropic liquid crystals respond to changes in temperature, whereas the structures of lyotropic liquid crystals respond to both solution concentration and to temperature. Depending on the arrangement of the molecules within the phase, thermotropic liquid crystals are classified as nematics, smectics, cholesterics, or columnar mesophases. *Nematic* liquid crystals are the ones most commonly employed in liquid crystal displays. The elongated molecules possess a long-range orientational order with molecular long axes aligned along a preferred direction (the *director*). There is no long-range order in the positions of centres of mass of molecules, and the director can vary throughout the

---

Box 9.5 *Continued*

---

phase. (Rod-shaped *micelles* can also form nematic phases as mentioned in the main text). In a nematic phase the molecules are able to rotate around their long axes. *Cholesteric* phases are similar to nematic phases in that there is no positional order but long-range orientational order. However, the director now varies regularly throughout the medium, as if a nematic phase with molecules aligned along the *y*-axis were twisted about the *x*-axis (Andrienko, 2018). A *smectic* mesophase differs from a nematic or a cholesteric phase, in that the molecules are arranged in layers in addition to having orientational order. The layers can slide over one another, and depending on the molecular order *within* layers, various types of smectic phases can be identified. In a smectic A phase the molecules are arranged perpendicularly to the planes, but are otherwise disordered within the planes whilst in smectic B phases there is hexagonal packing within the planes. The director in smectic C phases is at an angle to the planes. *Columnar* mesophases are often formed by disc-shaped molecules, and there can be various two-dimensional arrangements of the columns, for example, rectangular and hexagonal. Some materials that form the various types of thermotropic liquid crystalline phases are given in the article by Andrienko (2018).

Supposing that for a given system micelles in dilute solution are spherical, then the sequence of formation of phases as the surfactant concentration is increased across a phase diagram is expected to be in the order

$$L_1 \rightarrow I_1 \rightarrow H_1 \rightarrow V_1/L_\alpha/V_2 \rightarrow H_2 \rightarrow I_2 \rightarrow L_2$$

high + ve curvature   zero   high − ve curvature

although of course not all the phases are necessarily expected to form at a given temperature. If micelles in dilute solution are cylindrical rather than spherical then the cubic phase I_1 is not expected to form. Examples of phase diagrams for an aqueous cationic surfactant dodecyltrimethylammonium chloride (C_{12}TMACl) and for water and nonionic surfactant ($C_{12}E_6$) are shown in Fig. 9.37. For C_{12}TMACl single phase micellar solutions exist up to very high concentrations (ca. 40 wt %). Note that the concentration axis starts at 40 wt %. At (say) 50°C, the first mesophase to appear is a micellar cubic phase. As the surfactant concentration is increased further the phases form in the order H_1, V_1, and L_α. The dark grey shaded areas on the phase diagram are two-phase regions consisting of mixtures of the phases occurring on either side. The phase diagram for the nonionic surfactant $C_{12}E_6$ (Fig. 9.35b) is remarkably similar to that for C_{12}TACl, an obvious difference however being the absence of a micellar cubic phase I_1 at lower concentrations. S denotes the presence of solid surfactant.

The phase rule is

$$F = C - P + 2 \tag{9.11.2}$$

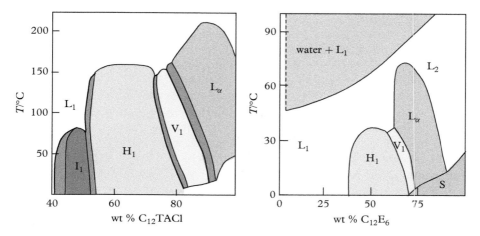

Figure 9.37 *(a) Phase diagram for $C_{12}TACl$ + water. Single phase micellar solutions exist up to about 40 wt % at room temperature. Shaded grey areas are two-phase regions. Redrawn from Balmbra et al. (1969), with low concentration (<40 wt %) and low temperature (T < 25°C) regions omitted for simplicity. (b) Phase diagram for $C_{12}E_6$ + water redrawn from Mitchell et al. (1983).*

in which C is the number of components (2 in the present case), P the number of coexisting phases at equilibrium, and F is the variance (i.e. number of degrees of freedom). The variance is the number of intensive thermodynamic variables that must be specified in order fully to define the system at equilibrium. Intensive variables include T, p, and chemical potential (hence composition). For the two-phase (grey shaded) regions in Fig. 9.37a, for which pressure has been fixed, $F = 2 - 2 + 1 = 1$. Thus to define a system lying in a two-phase region of the diagram it is only necessary to specify T. At this temperature the compositions of the two phases are defined since they lie at the ends of the tie line. The phase rule does not relate to extensive properties, and as discussed above, the relative *amounts* of the two phases are given by application of the lever rule. In a single phase region of the (isobaric) diagram, $F = 2 - 1 + 1 = 2$ and so to define the system both T and composition must be specified.

The rigorous determination of equilibrium phase diagrams for surfactant systems (such as those in Fig. 9.37) requires both patience and skill. Light scattering can often give useful information, since isotropic phases[11] (micellar solutions and cubic phases) are optically clear whereas anisotropic phases (lamellar L_α and L_β, hexagonal and nematic) scatter light giving them a cloudy appearance. Cubic phases are very viscous and so, very often a clear viscous sample is a cubic phase. Hexagonal phases also tend to be very viscous, much more so than lamellar phases. Isotropic and anisotropic phases can be distinguished by observing samples with a polarizing microscope. Anisotropic phases,

---

[11] *Isotropic* phases have the same properties in all directions, whereas *anisotropic* phases have different properties (e.g. refractive index) in different directions.

which are birefringent (i.e. have two refractive indices depending of the direction of the light relative to the liquid crystal alignment), give bright images rather than the black appearance produced by isotropic phases. Further, different anisotropic structures can often be distinguished by the quite distinctive patterns they produce through the polarizing microscope (see e.g. Oyafuso et al., 2017).

A very useful indication of the sequence of phases that is formed as surfactant concentration is changed at constant temperature (i.e. along an isotherm in a phase diagram) can be obtained by placing, say, a dry surfactant crystal on the stage of a polarizing microscope and then covering the sample with water. The water penetrates into the surfactant and gives a concentration gradient from the outside to the centre of the surfactant sample (see e.g. Häntzschel et al., 1999). Layers of the different liquid crystalline structures corresponding to the range of concentrations are revealed from the textures observed.

Diffraction of neutrons, light, or X-rays (SAXS—see e.g. Mezzenga et al., 2005) can be employed to detect the long range order present in liquid crystalline phases. By way of example, X-ray diffraction by a hexagonal lyotropic liquid crystal phase is considered here. A schematic representation of the X-ray experiment is shown in Fig. 9.38. The Bragg diffraction law is

$$n\lambda = 2d \sin \theta \qquad (9.11.3)$$

in which λ is the X-ray wavelength, d a fundamental crystal spacing, n is the diffraction order (integer 1, 2, 3. . ..), and here the angle θ is the diffraction angle shown in Fig. 9.38. For normal or reverse hexagonal phases the repeat distance d and cylinder diameter D (shown in Fig. 9.39) can be obtained. A schematic X-ray diffraction pattern for H_1 phases is shown in Fig. 9.40 in which the abscissa $q = 4\pi\sin \theta/\lambda$. Between two and five peaks are usually observed, with peak positions (relative to the most intense peak) being (see Fig. 9.40)

$$\sqrt{1} : \sqrt{3} : \sqrt{4} : \sqrt{7}......$$

The spacing d ($= 2\pi/q$) obtained from the diffraction pattern is related to D, the cylinder diameter by

$$D = \frac{2d}{\sqrt{\frac{2\pi(1+\{\phi_{water}/\phi_{surfactant}\})}{3}}} \qquad (9.11.4)$$

in which ϕ are volume fractions of the subscripted species.

Another technique that has been used to elucidate liquid crystalline structures is ^2H NMR (Holmberg et al., 2003). Isotropic phases (micellar solutions and cubic phases, as well as microemulsions—see Chapter 10) containing heavy water give a narrow singlet in the ^2H NMR spectrum. A single phase system of anisotropic mesophase on the other hand yields a doublet, in which the splitting reflects the degree of anisotropy present.

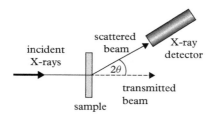

Figure 9.38 *The X-ray diffraction setup.*

Figure 9.39 *Repeat distance d and cylinder diameter D in a hexagonal lyotropic liquid crystalline phase.*

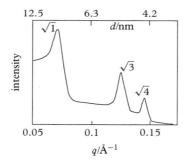

Figure 9.40 *Sketch of X-ray diffraction pattern for an hexagonal H_1 phase. The abscissa $q = 4\pi sin$ θ/λ, and corresponding values of d ($=2\pi/q$) are also shown.*

If two phases coexist, the spectra of the different phases are superimposed. So, for a mixture of an isotropic and an anisotropic phase, the spectrum shows a singlet with a superimposed doublet, and for two anisotropic phases (hexagonal and lamellar) two doublets are observed.

There has been a great deal of effort directed towards the determination of *three-component (ternary)* phase diagrams where, for example, the third component is an alkyl chain polar molecule such as an alkanol or, in microemulsion systems, a hydrocarbon (see §10.4). It is beyond the scope of this presentation to cover such diagrams except to say that phase information, obtained at constant T and p, can be represented on a triangular diagram, the corners of which represent pure (100 wt %) components, A, B, and C (Fig. 9.41a). Points for a given concentration of say A, lie on a line drawn parallel to the opposite side (BC) of the triangle. A similar situation holds for the other two

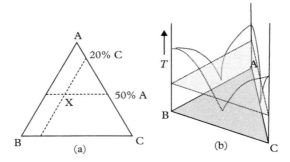

Figure 9.41 *(a) Triangular diagram for the representation of phase behaviour in three-component systems at constant T and p. (b) Representation of phase behaviour in three-component systems as a function of T.*

components. Thus, the phase at point X in Fig. 9.41a contains 50 wt % A and 20 wt % C, and hence 30 wt % B.

The effects of temperature on the phase behaviour of ternary surfactant systems is often of interest, and can be recorded on a triangular prism (Fig. 9.41b) with temperature represented vertically. Such a prism is formed by stacking triangular diagrams for a range of temperatures. Each of the three vertical faces of the prism is a binary phase diagram for a pair of the components. The ternary phase behaviour is represented within the body of the prism. There are however two commonly encountered types of *pseudobinary* diagram which represent effects of T on phase formation, and both are based on planes taken through the triangular prism normal to the triangle, as described in §10.4.2.

∙∙∙

REFERENCES

D.P. Acharya and H. Kunieda, Wormlike micelles in mixed surfactant solutions. *Adv. Colloid Interface Sci.*, 2006, **123–126**, 401–413.

B. Andersson and G. Olofsson, Calorimetric study of non-ionic surfactants. Enthalpies and heat-capacity changes for micelle formation in water of C_8E_4 and Triton X-100 and micelle size of C_8E_4. *J. Chem. Soc., Faraday Trans 1*, 1988, **84**, 4087–4095.

D. Andrienko, Introduction to liquid crystals. *J. Mol. Liquids*, 2018, **267**, 520–541.

E.A.G. Aniansson, The mean lifetime of a micelle. *Progr. Colloid Polym. Sci.*, 1985, **70**, 2–5.

E.A.G. Aniansson, S.N. Wall, M. Almgren, H. Hoffmann, I. Kielmann, W. Ulbricht, R. Zana, J. Lang, and C. Tondre, Theory of the kinetics of micellar equilibria and quantitative interpretation of chemical relaxation studies of micellar solutions of ionic surfactants. *J. Phys. Chem.*, 1976, **80**, 905–922.

R. Aveyard, B.P. Binks, P.D.I. Fletcher, C.E. Rutherford, P.J. Dowding, and B. Vincent, Dissociation of AOT monolayers stabilising oil-in-water microemulsions in Winsor I systems. *Phys. Chem. Chem. Phys.*, 1999, **1**, 1971–1978.

R.B. Balmbra, J.S. Clunie, and J.F. Goodman, Cubic mesomorphic phases. *Nature*, 1969, **222**, 1159.

A.D. Bangham, M.M. Standish, and J.C. Watkins, Diffusion of univalent ions across the lamellae of swollen phospholipids. *J. Mol. Biol.*, 1965, **23**, 238–252.

B.W. Barry and D.I.D. El Eini, Solubilization of hydrocortisone, dexamethasone, testosterone and progesterone by long-chain polyoxyethylene surfactants. *J. Pharm. Pharmacol.*, 1976, **28**, 210–218.

S.D. Christian and J.F. Scamehorn (eds.), *Solubilization in Surfactant Aggregates*, Surfactant Science Series, vol. 55. Marcel Dekker, New York, 1995.

J.H. Clint, Micellization of mixed nonionic surface active agents. *J. Chem. Soc., Faraday Trans. 1*, 1975, **71**, 1327–1334.

T.S. Davies, A.M. Ketner, and S.R. Raghaven, Self-assembly of surfactant vesicles that transform into viscoelastic wormlike micelles upon heating. *J. Am. Chem. Soc.*, 2006, **128**, 6669–6675.

S. Ezrahi, E. Tuval, and A. Aserin, Properties, main applications and perspectives of worm micelles. *Adv. Colloid Interface Sci.*, 2006, **128–130**, 77–102.

A.A. Gabizon, Stealth liposomes and tumor targeting: one step further in the quest for the magic bullet. *Clin. Cancer Res.*, 2001, 7, 223–225.

H. Gharibi, N. Takisawa, P. Brown, M.A. Thomason, D.M. Painter, D.M. Bloor, D.G. Hall, and E. Wyn-Jones, Analysis of the fast relaxation times for micelle kinetics taking into account new EMF data. *J. Chem. Soc., Faraday Trans.*, 1991, **87**, 707–710.

T. Gilanyi and E. Wolfram, Interaction of ionic surfactants with polymers in acqueous solution. *Colloids Surf.*, 1981, **3**, 181–198.

E.D. Goddard, Polymer–surfactant interaction. Part II. Polymer and surfactant of opposite charge. *Colloids Surf.*, 1986, **19**, 301–329.

D.G. Hall, in *Aggregation Processes in Solution* (ed. E. Wyn-Jones and J. Gormally), Chapter 2. Elsevier, Amsterdam, 1983.

D. Häntzschel, J. Schulte, S. Enders, and K. Quitzsch, Thermotropic and lyotropic properties of *n*-alkyl-β-D-glucopyranoside surfactants. *Phys. Chem. Chem. Phys.*, 1999, **1**, 895–904.

P.M. Holland and D.N. Rubingh, Nonideal multicomponent mixed micelle model. *J. Phys. Chem.*, 1983, **87**, 1984–1990.

K. Holmberg, B. Jónsson, B. Kronberg, and B. Lindman, *Surfactant and Polymers in Aqueous Solution*, 2nd edition. John Wiley and Sons Ltd, Chichester, UK, 2003.

Y. Huang and S. Gui, Factors affecting the structure of lyotropic liquid crystals and the correlation between structure and drug diffusion. *RSC Adv.*, 2018, **8**, 6978–6987.

I. Iliopoulos, T.K. Wang, and R. Audebert, Viscometric evidence of interactions between hydrophobically modified poly (sodium acrylate) and sodium dodecyl sulfate. *Langmuir*, 1991, 7, 617–619.

J.N. Israelachvili, D.J. Mitchell, and B.W. Ninham, Theory of self-assembly of hydrocarbon amphiphiles into micelles and bilayers. *J. Chem. Soc., Faraday Trans. 2*, 1976, **72**, 1525–1568.

M. Kahlweit, Kinetics of formation of association colloids. *J. Colloid Interface Sci.*, 1982, **90**, 92–99.

E.W. Kaler, A.K. Murthy, S.Z. Rodriguez, and J.A.N. Zasadzinski, Spontaneous vesicle formation in aqueous mixtures of single-tailed surfactants. *Science*, 1989, **145**, 1371–1374.

H.B. Klevens, Solubilization. *Chem. Rev.*, 1950, **47**, 1–74.

G. Kocak, C. Tuncer, and V. Bütün, pH-Responsive polymers. *Polym. Chem.*, 2017, **8**, 144–176.

H. Lange, Wechselwirkung zwischen Natriumalkylsulfaten und Polyvinylpyrrolidon in wäßrigen Lösungen. *Kolloid Z. Z. Polym.*, 1971, **243**, 101–109.

D.J. Lunn, J.R. Finnegan, and I. Manners, Self-assembly of "patchy" nanoparticles: a versatile approach to functional hierarchical materials. *Chem. Sci.*, 2015, **6**, 3663–3673.

J.W. McBain and H.E. Martin, Studies of the constitution of soap solutions: the alkalinity and degree of hydrolysis of soap solutions. *J. Chem. Soc., Trans.*, 1914, **105**, 957–977.

J.W. McBain and P.H. Richards, Solubilization of insoluble organic liquids by detergents. *Ind. Eng. Chem.*, 1946, **38**, 642–646.

U. Menge, P. Lang, and G. H. Findenegg, From oil-swollen wormlike micelles to microemulsion droplets: a static light scattering study of the L1 phase of the system water + $C_{12}E_5$ + decane. *J. Phys. Chem. B*, 1999, **103**, 5768–5774.

F.M. Menger, The structure of micelles. *Acc. Chem. Res.*, 1979, **12**, 111–117.

R. Mezzenga, C. Meyer, C. Servais, A.I. Romoscanu, L. Sagalowicz, and R.C. Hayward, Shear rheology of lyotropic liquid crystals: a case study. *Langmuir*, 2005, **21**, 3322–3333.

D.J. Mitchell and B.W. Ninham, Micelles, vesicles and microemulsions. *J. Chem. Soc., Faraday Trans. 2*, 1981, **77**, 601–629.

D.J. Mitchell, G.J.T. Tiddy, L. Waring, T. Bostock, and M.P. McDonald, Phase behaviour of polyoxyethylene surfactants with water. Mesophase structures and partial miscibility (cloud points). *J. Chem. Soc., Faraday Trans 1*, 1983, **79**, 975–1000.

R. Nagarajan, Molecular packing parameter and surfactant self-assembly: the neglected role of the surfactant tail. *Langmuir*, 2002, **18**, 31–38.

M.H. Oyafuso, F.C. Carvalho, T.M. Takeshita, A.L. Ribeiro de Souza, D.R. Araújo, V. Merino, M. Palmira, D. Gremião, and M. Chorilli, Development and in vitro evaluation of lyotropic liquid crystals for the controlled release of dexamethasone. *Polymers*, 2017, **9**, 330.

A. Patist, J.R. Kanicky, P.K. Shukla, and D.O. Shah, Importance of micellar kinetics in relation to technological processes. *J. Colloid Interface Sci.*, 2002, **245**, 1–15.

H. Rehage and H. Hoffmann, Viscoelastic surfactant solutions: model systems for rheological research. *Mol. Phys.*, 1991, **74**, 933–973.

M.J. Rosen and J.T. Kunjappu, *Surfactants and Interfacial Phenomena*. Wiley, Hoboken, NJ, 2012.

S.D. Stoyanov, V.N. Paunov, H. Rehage, and H. Kuhn, A new class of interfacial tension isotherms for nonionic surfactants based on local self-consistent mean field theory: classical isotherms revisited. *Phys. Chem. Chem. Phys.*, 2004, **6**, 596–603.

C. Tanford, *The Hydrophobic Effect*. Wiley-Interscience, New York, 1980.

J. Tolls, J. van Dijk, E.J.M. Verbruggen, J.L.M. Hermens, B. Loeprecht, and G. Schüürmann, Aqueous solubility–molecular size relationships: a mechanistic case study using C_{10}- to C_{19}-alkanes. *J. Phys. Chem. A*, 2002, **106**, 2760–2765.

S.-H. Tung, Y.-E. Huang, and S.R. Raghaven, A new reverse wormlike micellar system: mixtures of bile salt and lecithin in organic liquids. *J. Am. Chem. Soc.*, 2006, **128**, 5751–5756.

M. Velinova, D. Sengupta, A.V. Tadjer, and S-J. Marrink, Sphere-to-rod transitions of nonionic surfactant micelles in aqueous solution modeled by molecular dynamics simulations. *Langmuir*, 2011, **27**, 14071–14077.

B. Vincent, McBain and the centenary of the micelle. *Adv. Colloid Interface Sci.*, 2014, **203**, 51–54.

Y.-Y. Won, H.T. Davis, and F.S. Bates, Giant wormlike rubber micelles. *Science*, 1999, **283**, 960–963.

10

Surfactants in systems with oil and water, including microemulsions

10.1 Dilute surfactant systems: some experimental observations

In the last chapter it was seen that micelles in aqueous solution can solubilize relatively small amounts of say nonpolar oil before phase separation occurs. Equally, surfactant dissolved in oil can solubilize water. The solubilization behaviour in such systems is encapsulated in phase diagrams such as those in Fig. 9.20. In the present chapter, focus will be mainly on systems in which surfactant is present together with *comparable amounts* of oil and water.

10.1.1 Surfactant distribution between nonpolar oil and water

Suppose that surfactant is added to a system comprised of roughly equal volumes of water and a hydrocarbon oil in mutual contact. Many surfactants, for example, Aerosol OT (sodium di(2-ethylhexyl)sulfosuccinate) and some nonionic surfactants of the type C_nE_m, are soluble to some extent in hydrocarbons in the absence of water. When water and oil phases are in contact, with surfactant concentrations below an aggregation concentration, surfactant will distribute as monomers between the phases. Say the oil is an alkane such as octane, then AOT monomers will distribute overwhelmingly in favour of the aqueous phase, whereas say $C_{12}E_5$ monomers distribute strongly into the oil phase. As the surfactant concentration is increased in the two-phase system it will eventually reach a critical aggregation concentration (akin to the *cmc* in an aqueous system but now designated a *cac*). The interesting questions are: in which of the phases will aggregation occur and why, and under what conditions? Do the aggregates form in the phase favoured by monomers or are monomer and aggregate distributions unrelated?

To address these questions, the way in which AOT and $C_{12}E_5$ distribute between water (or dilute aqueous NaCl) and heptane at 293 K are taken as examples. The monomer molar distribution ratio [surfactant in water]/[surfactant in oil], measured up to the *cac* for $C_{12}E_5$, is 4×10^{-3}, that is, the surfactant monomers distribute about 250 in favour of the oil phase. On the other hand, as mentioned, AOT monomers distribute

Surfactants: In Solution, at Interfaces and in Colloidal Dispersions. Bob Aveyard. © Bob Aveyard 2019.
Published in 2019 by Oxford University Press. DOI: 10.1093/oso/9780198828600.001.0001

virtually completely in favour of water. At the *cac* aggregates begin to form in one of the phases. The distribution of AOT, both below and above the *cac*, between heptane and aqueous NaCl solutions is shown in Fig. 10.1. The abscissa is the initial concentration of surfactant in the aqueous phase, that is, before contacting with the oil phase. The ordinate is the AOT concentration in the aqueous phase at distribution equilibrium. For water or low concentrations of NaCl (0.0171 mol dm^{-3} in the figure) all the AOT remains in the aqueous phase both below and above the *cac*; the dashed line has been drawn with unit slope. For higher salt concentration (e.g. 0.1027 mol dm^{-3}) monomers again distribute almost completely into the aqueous phase; the plot is linear with unit slope up to the *cac* (at point *A*). Above the *cac* however the plot is horizontal indicating that, although at this salt concentration AOT monomers reside in the aqueous phase, the surfactant aggregates form in the oil phase. For C$_{12}$E$_5$ in the water/heptane system at 293K, although the surfactant monomers reside largely in the oil phase, the aggregates form in the aqueous phase.

10.1.2 Formation of microemulsion droplets: Winsor systems

When aggregates form in the oil, it is found that water is also present. The ratio [mol water/mol aggregated surfactant] in the oil phase is designated $\mathcal{R}$, and values for AOT in heptane at 298 K are plotted against the aqueous phase NaCl concentrations in Fig. 10.2. The ratio is independent of surfactant concentration in the oil and it falls with increasing aqueous phase concentration of NaCl. The oil phase is optically clear so the water is not present as large drops, which would scatter light strongly and cause cloudiness. Intuitively therefore it can be supposed that the aggregates consist of very small water droplets coated with surfactant monolayers. It is simple to show that, assuming all surfactant is present on the drops, the radius r_c of the water cores (see Fig. 10.3b) is given by

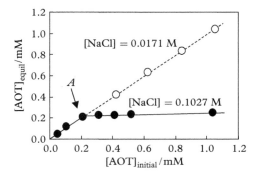

Figure 10.1 *Distribution of AOT between heptane and aqueous NaCl at 298 K. Abscissa is AOT concentration in the aqueous phase before equilibration between heptane and aqueous phases and the ordinate is the aqueous phase concentration after equilibration. Open circles for [NaCl] = 0.0171 mol dm^{-3} and filled points for [NaCl] = 0.1027 mol dm^{-3}. Redrawn from Aveyard et al. (1986).*

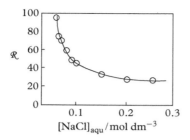

Figure 10.2 $\mathcal{R}$ *values as a function of aqueous phase NaCl concentration for AOT in heptane in equilibrium with aqueous NaCl at 298 K. Redrawn from Aveyard et al. (1986).*

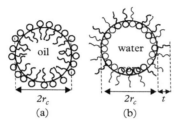

Figure 10.3 *Representation of microemulsion droplets. (a) Oil-in-water (O/W) droplet; (b) water-in-oil (W/O) droplet. r_c is the core radius and t the thickness of the surfactant layer.*

$$r_c = 3\mathcal{R}v_w/ar_c \qquad (10.1.1)$$

where v_w is the volume of a water molecule in liquid water (0.03 nm^3) and a is the area per surfactant molecule at the drop surface. From (10.1.1) it is seen that the droplet core radius is directly proportional to $\mathcal{R}$, and the size (Box 10.1) of the water droplets in oil therefore falls as the salt concentration in water rises. The aggregates in oil are termed water-in-oil (W/O) *microemulsion* droplets and the droplet radius, r_h, is equal to the core radius r_c plus the thickness t of the surfactant layer coating the drops (which is of the order of 1 nm).

Box 10.1   A simple way to estimate drop size

If the thickness of the surfactant layer is ignored an estimate of drop size can be obtained as follows (De Gennes and Taupin, 1982). Assume that unit volume of microemulsion phase contains n surfactant molecules all present at droplet surfaces. The volume fraction of droplets is φ and the area per surfactant at a drop surface is a. The number of drops present with radius r is $3\varphi/4\pi r^3$ and the total area of the drops is $3\varphi/r$. This area is also equal to na so that $r = 3\varphi/na$

To calculate r_c and hence r_h (the hydrodynamic radius) using (10.1.1) the value of a is required. The close-packed areas of the surfactant molecules adsorbed at the *planar* oil/water interface (§5.2) could be used. Hydrodynamic radii can however be determined independently, by say dynamic light scattering (DLS) or small angle neutron scattering (SANS) (see §A10.2 and §A10.3), and a then extracted from such data. A plot of r_h (obtained by DLS) against $\mathcal{R}$ for the AOT/heptane systems is shown in Fig. 10.4, and it is seen that the droplet radii lie between about 2 and 20 nm, much smaller than emulsion drops (see Chapter 14). Microemulsions obtained in two different ways are represented in the figure; the open symbols are for systems in which the oil phase is in equilibrium with aqueous solutions. It is also possible however to prepare the oil-in-water (O/W) microemulsions simply by adding the required amounts of surfactant and water to heptane, and systems prepared in this way are represented by the filled points. The relationship between r_h and $\mathcal{R}$ is seen to be linear, the intercept at $\mathcal{R} = 0$ giving the surfactant monolayer thickness t as 1.4 nm, which is reasonable for the AOT molecule. From the linearity of the plot, the value of a, obtained from the slope appears to be constant with a value of 0.5 nm^2. This is a little higher than, but similar to, the 'geometrical' area of the AOT head group (estimated to be 0.45 nm^2).

In Fig. 10.2 the $\mathcal{R}$ values for W/O microemulsion drops were shown for salt concentrations of about 0.07 mol dm^{-3} and upwards. For salt concentrations less than about 0.04 mol dm^{-3} however, it is found that the AOT forms aggregates in the (clear) aqueous phase rather than in heptane, and that the aggregates contain heptane. The size of these aggregates, which are O/W microemulsion drops, rises with increasing NaCl concentration.

At intermediate salt concentrations, centred on about 0.05 mol dm^{-3} in the AOT/heptane/aqueous NaCl system at 298 K, a third surfactant-rich phase is formed which rests between the oil and aqueous phases (and is consequently called a *middle*

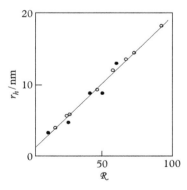

Figure 10.4 *Plot of r_h versus $\mathcal{R}$ for W/O microemulsion drops in heptane stabilized by AOT at 298 K. Filled points are for 'made up' microemulsions and open symbols are for W/O microemulsions in equilibrium with aqueous NaCl phases. Redrawn from Aveyard et al. (1986).*

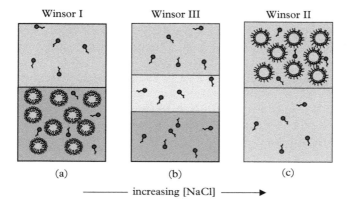

Winsor I   Winsor III   Winsor II

(a)   (b)   (c)

—————— increasing [NaCl] ——————→

Figure 10.5 *Representation of Winsor systems. (a) Winsor I (WI) systems are two-phase and consist of an oil-in-water microemulsion in equilibrium with an excess oil phase, (c) Winsor II (WII) systems consist of water-in-oil microemulsions in equilibrium with an excess aqueous phase. (b) Winsor III (WIII) systems have three phases in equilibrium: excess oil and water phases and a surfactant-rich third (middle) phase which contains oil, water and surfactant. Note, all phases contain some monomeric surfactant.*

phase[1]), as shown schematically in Fig. 10.5. The middle phase contains both oil and water and often much of the surfactant present in the system.

Systems with microemulsion droplets in an aqueous phase in equilibrium with excess oil (possibly containing significant amounts of surfactant monomer, depending on surfactant type) is termed a Winsor I (WI) system and is illustrated in Fig. 10.5a. A Winsor II (WII) system has microemulsion drops in the oil phase which is in equilibrium with excess aqueous phase, containing surfactant monomers. The three-phase system is referred to as Winsor III (WIII). The surfactant monolayers coating microemulsion drops are obviously curved, and the curvature (defined as positive for O/W droplets) in the WI region falls (oil droplets become larger) as the salt concentration is increased. In the inverted WII systems the monolayer curvature has changed sign, and the negative curvature increases (water drop size decreases) as the salt concentration (in the aqueous phase) is increased. The middle phase in WIII systems has a structure such that the (net) monolayer curvature is close to zero. The nature of the phase depends on how flexible the surfactant monolayers are, as discussed below.

The Winsor sequence brought about by changes in electrolyte concentration in systems with ionic surfactants results (at least in part) from the screening of electrical repulsion between surfactant head groups in surfactant monolayers. Inert electrolyte can also give Winsor transitions in systems with nonionic surfactants such as $C_{12}E_5$. The mode of action of the electrolyte however is quite different (see §5.4.2) and much higher

---

[1] The order in which the phases form depends upon their densities. A surfactant-rich phase containing only little oil (density < 1 g cm^{-3}) and an appreciable amount of surfactant can appear as the bottom phase.

salt concentrations are required to bring about the changes. For example, for systems containing $C_{12}E_5$, nonane, and aqueous NaCl at 304 K, the WIII systems are centred on a salt concentration of 0.5 mol dm^{-3}, an order of magnitude greater than that for the AOT system.

Winsor transitions can be effected by variables other than electrolyte concentration, including temperature, oil molecular structure (chain length in the case of normal alkanes) and addition of a cosurfactant (e.g. medium chain length alkanol, say butanol and upwards) or a second surfactant. The situation is summarized in Table 10.1, and the effects are explained in §10.2.1.

10.1.3 Interfacial tensions in Winsor systems

An important observation is the way in which the interfacial tensions vary with changes in the variables mentioned in Table 10.1. An example is shown in Fig. 10.6 for systems with $C_{12}E_5$, 0.085 mol dm^{-3} aqueous NaCl and alkanes with a range of chain length, N_c, at 40°C; log(tension) is plotted against N_c.

The tension, denoted γ_c, in a given two-phase system is that attained at and beyond the *cac* and it remains fairly constant with respect to surfactant concentration close to the *cac*. In WIII systems there are three tensions of interest. The excess oil and aqueous phases can be taken from either side of the middle phase and the tension between them determined; this is the tension plotted in the WIII region (denoted schematically as the region between the dashed lines) in Fig. 10.6. Tensions between the middle phase and each of the excess phases can also be measured.[2] It is seen that γ_c becomes ultralow in the WIII region. The longer chain alkanes produce WI systems and shorter chains favour WII systems, in line with the information given in Table 10.1. Very similar tension

Table 10.1 *Effects of variables on Winsor transitions**

Surfactant type	Increase in variable	WI↔WIII↔WII
ionic and nonionic	salt concentration	⟶
ionic	temperature	⟵
nonionic	temperature	⟶
ionic and nonionic	alkane chain length	⟵
ionic and nonionic	alkanol (cosurfactant) concentration	⟶

*Surfactant head and tail group structures are obviously also important variables, as reflected in the packing factor, P, for the surfactant. Other things being equal, surfactants with a bulky hydrophobic tail (relative to the head group) favour WII systems, whereas a bulky head group favours the formation of WI systems.

---

[2] It is shown elsewhere that in any three-phase system at equilibrium, the largest of the three interfacial tensions (that plotted in Fig. 10.6 in the WIII region) must be less than or equal to the sum of the other two tensions.

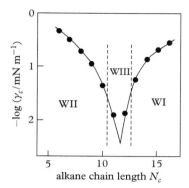

Figure 10.6 *Interfacial tensions as a function of alkane chain length in systems with $C_{12}E_5$, 0.085 mol dm^{-3} aqueous NaCl and normal alkanes at 40°C (see text). Redrawn from Aveyard and Lawless (1986).*

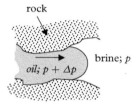

Figure 10.7 *Representation of the curved interface between crude oil and brine passing through a pore in the rock.*

behaviour can be observed with T, salt concentration and cosurfactant concentration as the variable, and these observations are rationalized later.

Ultralow oil/water tensions (down to 10^{-3} or 10^{-4} mN m^{-1}) and the concomitant Winsor phase behaviour generated considerable interest towards the end of last century in connection with enhanced oil recovery (EOR). One of the goals was to produce very low interfacial tensions between crude oil and surrounding brine in order to mobilize the oil through fine pores in the rock, as illustrated in Fig. 10.7. The interfacial tension between crude oil and brine is about 20 mN m^{-1}. To force the oil/water interface into and through a pore, and thereby cause the interface to curve as shown, a pressure equal to the pressure drop, Δp across the curved interface must be provided, which according to the Laplace equation (equation (3.4.2)) is $2\gamma/r$, where r is the radius of curvature of the (assumed spherically) curved interface. Clearly if the interfacial tension can be reduced by a factor of say 10^4 (to ca. 2×10^{-3} mN m^{-1}) the pressure required to mobilize the oil is also reduced by the same factor. Surfactant systems have been formulated that readily achieve such low interfacial tensions between crude oil and brine and which give good improvements in oil displacement in laboratory experiments. However EOR has not been very widely used partly because the economics have not been favourable.

10.2   Curvature properties of surfactant layers

10.2.1   Spontaneous or preferred curvature and why it changes

In a Winsor system, a microemulsion drop can spontaneously achieve an equilibrium (largest) size for the prevailing conditions. If a microemulsion oil drop in an aqueous phase is smaller than the 'preferred' size, it can readily swell by taking up some oil from the contacting excess oil phase. If a drop size exceeds the equilibrium size then core liquid is expelled to the excess phase in contact with the microemulsion. The drop radius is (neglecting monolayer thickness) the radius of the curved surfactant layer coating the core and so, from what has been seen in §10.1, the preferred curvature (Box 10.2) of surfactant monolayers in Winsor systems is inversely related to the interfacial tension γ_c. The tension is measured for a plane surface but in Winsor systems surfactant molecules in planar monolayers are in equilibrium with those in curved monolayers at drop surfaces.

Box 10.2 'Made-up' microemulsion droplets

Although microemulsion drops in equilibrium Winsor systems have their preferred curvature, it is easy to make up microemulsions simply by mixing the ingredients, in which case a range of drop sizes up to (but not beyond) the preferred size in the corresponding Winsor system is possible. The systems are at equilibrium and do not phase separate, but of course they are not in equilibrium with an excess phase as in Winsor systems. The solubilized systems represented in Fig. 9.20 are such made-up single phase microemulsions.

As seen, in the progression WI → WIII → WII, the preferred curvature of surfactant monolayers at oil/water interfaces changes from large positive, through zero to large negative values (Fig. 10.8), and these changes can be brought about by changes in salt and cosurfactant concentration, oil molecular structure and temperature. The plane oil/water interfacial tension falls as microemulsion drop size rises in WI systems and vice versa in WII systems. In the WIII regions the tension becomes extremely low and the 'drop' size is effectively infinite.

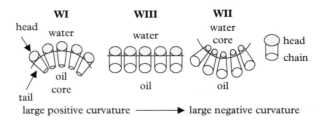

Figure 10.8 *Monolayer curvature in Winsor systems.*

The concepts of the surfactant packing factor P and effective molecular shape were introduced in §9.5, and it was shown how changes in P bring about changes in surfactant aggregate shape and structure in surfactant + water systems. The same ideas can be applied to microemulsion properties. Recall that P is defined as $v/a_o l_c$ where v is the effective chain volume, and l_c the critical chain length (say 80% of the fully extended length); a_o is the optimal head group area. When $v/l_c = a_o$, $P = 1$, the monolayer does not have a tendency to curve and the radius of curvature (and drop size) is infinite. In WI systems, $P < 1$, and in WII systems, $P > 1$.

Inorganic electrolytes screen the electrostatic repulsion between head groups of ionic surfactants so the heads are able to approach more closely and a_o falls, P rises and oil drop size increases. When $P \geq 1$, the drops invert and W/O drops form in the oil phase. The salt affects nonionic surfactant film curvature by a different mechanism and much higher salt concentrations are required as mentioned. It could be that the salt 'dehydrates' the polar head groups, reducing a_o. An increase in temperature is also expected to cause head group dehydration of nonionic surfactants and drive the WI to WII transition. Interestingly T has the opposite effect in systems with ionic surfactant (Table 10.1), consistent with an increase in T causing an increase in head group dissociation (and hence in repulsion and a_o) and a decrease in P.

In systems with alkane as the oil phase, increasing the chain length can cause a WII → WI transition (Fig. 10.6). This arises since longer alkanes penetrate into the monolayer chain region less (in volume terms) than do short alkanes. Penetration causes an increase in the effective chain volume and an increase in P. A similar chain length effect on alkane penetration is seen with planar monolayers at the air/water interface (§5.6).

Many surfactants are incapable of forming Winsor systems (i.e. microemulsions) or producing the associated ultralow oil/water interfacial tensions. For example it is not possible in oil/water systems to achieve packing factors of unity or greater with surfactants such as sodium alkylsulfates and alkyltrimethylammonium bromides, which have large head groups (due in part to electrical repulsion) and single alkyl chains. The Winsor systems already discussed involving AOT occur because this surfactant has two branched hydrocarbon chains. Nonetheless, it is possible to mix a cosurfactant, with a packing factor greater than unity, with say a single chain ionic surfactant, for which $P < 1$ and fine tune P so that it can pass through unity in the mixed system. For example, sodium dodecylbenzenesulfonate (SDBS) in systems with aqueous NaCl and heptane will neither give ultralow interfacial tensions (the lowest at 25°C being around 1.5 mN m^{-1}) or produce Winsor systems by varying the salt concentration alone. However, in the presence of varying amounts of an alkanol as cosurfactant (butanol in this case) Winsor behaviour becomes evident. Interfacial tensions and surfactant distribution between aqueous phase and heptane are illustrated in Fig. 10.9. There is an inversion from WI to WII as the total concentration of butanol is increased, centred on a butanol concentration of about 0.5 mol dm^{-3}. The oil/water tension is also a minimum close to this concentration, falling to between 10^{-2} and 10^{-3} mN m^{-1}. The change in P brought about by the cosurfactant occurs not only because P of the alkanol alone is greater than unity, but also because the alkanol in the mixed curved monolayers screens the repulsion between the charged surfactant head groups.

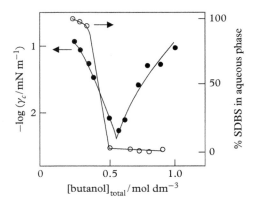

Figure 10.9 *Interfacial tensions (filled symbols) and surfactant distribution (open circles) in systems with SDBS, 0.3 mol dm^{-3} aqueous NaCl and heptane at 25°C as a function of the total molar concentration of butanol in the system. Redrawn from Ahsan et al. (1991).*

10.2.2   The relation between microemulsion drop size and γ_c: bending elastic moduli

Since equilibrium drop size in Winsor systems varies inversely with the interfacial tension γ_c, it can be anticipated that phase behaviour and the curvature properties of surfactant monolayers are interrelated. The latter can be characterized by the local principal radii of curvature r_1 and r_2, (see §3.4.1) the spontaneous radius of curvature r_0 and the bending elastic (or rigidity) moduli κ and $\overline{\kappa}$. Curvatures, c, are the inverse of the radii of curvature, for example, $c_1 = 1/r_1$ etc. The bending elastic modulus κ is always positive. The constant $\overline{\kappa}$ is termed the Gaussian or saddle-splay elastic modulus and is associated with the surface topology. It can be positive or negative, depending on the surface topology.

From the seminal work of Helfrich (1973) the curvature energy g_c per unit area of surface of spherical droplets, curvature c and spontaneous curvature c_o, can be written

$$g_c = 2\kappa(c - c_o)^2 + \overline{\kappa}c^2 \qquad (10.2.1)$$

It is assumed in writing (10.2.1) that the adsorbed film has zero thickness (see Box 10.3). The equation expresses the fact that since a monolayer has a spontaneous curvature (at which curvature energy is minimum), work is required to change the curvature away from c_o; κ and $\overline{\kappa}$ have the dimensions of energy. The work of formation of unit area of curved interface therefore is made up of the work of *extending* the surface by unit amount (see e.g. §4.2), together with the work of *bending* the unit area to the required curvature. In WIII systems, where the preferred curvature is around zero, γ_c is extremely small, ca. 10^{-2} to 10^{-3} mN m^{-1} or less as seen, so not only is the curvature energy very low, but so also is the free energy of extension of the surface. Thus, in WI and WII systems it is supposed that the measured tensions γ_c arise largely from curvature effects. There is, however, another factor that must be considered if the magnitudes of γ_c are to be accounted for fully (Box 10.4).

Box 10.3 The neutral surface

The surfactant monolayers of course have finite thickness, of the order of the length of a surfactant molecule. The curvatures (radii) in (10.2.1) are for surfaces that do not change area on bending, so-called *neutral* surfaces. A neutral surface can reasonably be supposed to be the one passing through the region where the head and tail groups of the surfactant are linked. If this is done then radii can be taken to be drop 'core' radii (see Fig. 10.3). A surface outside the neutral surface expands when the surface is bent and a surface on the inside contracts.

Box 10.4 Interfacial tension of microemulsion drop surfaces

The fact that microemulsion droplets form spontaneously in WI and WII systems indicates that the interfacial tension at *droplet* surfaces must be very low. The negative free energy change accompanying the spontaneous formation of a microemulsion is the sum of the positive free energy change of forming the droplet surfaces ($\gamma_d \mathcal{A}$, where γ_d is the droplet interfacial tension and $\mathcal{A}$ the surface area of the droplets) and the negative free energy change arising from the positive entropy of droplet dispersion in the medium.

Imagine that a plane oil/water interface is formed by bringing microemulsion drops to the interface and 'unbending' them. The drops must be de-mixed from the microemulsion phase and this is associated with a negative change in entropy and hence a positive change in free energy. Conversely, when a microemulsion phase is formed in a WI or WII system, the entropy of mixing favours the creation of a larger number of small drops over a smaller number of large drops. This means that at equilibrium the (measured) drop radius r_m is smaller than that, r_o, favoured by curvature energies alone. There is an energy cost in reducing the drop size which is offset by the extra entropy of dispersion resulting from the larger number of drops.

The curvature energy per unit area of drops with radius r_m is, from (10.2.1),

$$2\kappa(1/r_m - 1/r_o)^2 + \overline{\kappa}/r_m^2 \tag{10.2.2}$$

and for unit area of plane surface ($c = 1/r = 0$) it is

$$2\kappa/r_o^2 \tag{10.2.3}$$

The change in curvature energy associated with the formation of unit area of plane surface from the surfaces of drops, radius r_m, is therefore

$$- (2\kappa + \overline{\kappa}) / r_m^2 + 4\kappa / r_m r_o \tag{10.2.4}$$

To allow for the change in the entropy of dispersion, a term in the volume fraction φ of droplets in the microemulsion is added and the free energy of forming unit area of plane surface (F_u) from droplets becomes

$$F_u = - (2\kappa + \overline{\kappa}) / r_m^2 + 4\kappa / r_m r_o - \left(kT / 4\pi r_m^2 \right) f(\varphi) \tag{10.2.5}$$

The relationship between r_m and r_o can be obtained by noting that at equilibrium $\partial F_u / \partial r = 0$ which yields

$$\frac{r_m}{r_o} = \frac{2\kappa + \overline{\kappa}}{2\kappa} + \frac{kT}{8\pi\kappa} f(\varphi) \tag{10.2.6}$$

If the microemulsion droplets are considered to behave as a 'gas' of non-interacting hard spheres then, for low φ, as an approximation $f(\varphi) = \ln \varphi - 1$, and then combination of (10.2.5) with (10.2.6) gives for F_u (which is equivalent to the measured tension γ_c if the extensional work is neglected as discussed earlier)

$$F_u = \gamma_c = \frac{2\kappa + \overline{\kappa}}{r_m^2} + \frac{kT}{4\pi r_m^2} (\ln \varphi - 1) \tag{10.2.7}$$

In (10.2.6) and (10.2.7) the radius r_m relates to the neutral surface of the equilibrium droplets as explained; the *external* drop radius is $r_{ex} = r_m + h$ where h depends on the surfactant molecular structure and is of the order of the surfactant head group diameter. On this basis, Kellay et al. (1993) replace the term in brackets in (10.2.7) by $\{\ln \varphi - [3r_m / (r_m + h)]\}$.

Equation (10.2.7) interrelates γ_c, r_m and $2\kappa + \overline{\kappa}$; the measurement of ultralow interfacial tensions was discussed in §A3.5, and the determination of drop radii is described in the Appendix to this chapter. From knowledge of interfacial tensions and drop sizes, $(2\kappa + \overline{\kappa})$ can be calculated by use of (10.2.7). Then, if κ can be measured independently, $\overline{\kappa}$ is also known. Up to now microemulsion droplets have been considered to be monodisperse spheres, but the droplets undergo thermal fluctuations leading to polydispersity in size and also to shape fluctuations. Size distributions are influenced by the bending elastic moduli as might be expected. It can be shown that the polydispersity index p, which can be determined using SANS (see Appendix A10.3), is related to $(2\kappa + \overline{\kappa})$ by (see e.g. Gradzielski et al., 1996)

$$p^2 = \frac{kT}{8\pi (2\kappa + \overline{\kappa}) + 2kTf(\varphi)} \tag{10.2.8}$$

The bending elastic modulus κ alone can be determined using ellipsometric measurements on the plane oil/water interface in Winsor systems (see e.g. Meunier and Lee, 1991). For a sinusoidal deformation of the interface there are three restoring forces operating, on different length scales.

On the macroscopic scale (long wavelength, small wave vector q) gravity dominates. At larger q effects due to interfacial tension are the most important, and so ellipsometry can be used to measure low interfacial tensions as mentioned in §A3.5. On the very small length scale however (say around 10 nm) the curvature energy becomes the most important restoring factor and the measurement of the amplitude of thermally excited waves allows the determination of κ.

Some values of $(2\kappa + \overline{\kappa})$, reported by Gradzielski et al. (1996) for nonionic surfactants in WI systems containing D_2O and alkanes, are shown in Table 10.2. Values were obtained by the two methods described above (using (10.2.7) and (10.2.8)) as indicated in the table. Increase in the chain length of the surfactants C_nE_3 from $N_c = 8$ to 12, keeping the alkane (decane) chain length the same, causes a significant increase in $(2\kappa + \overline{\kappa})/kT$ from around 1 to 3.5. Values obtained using the two methods are in remarkably good agreement. For the surfactant $C_{10}E_4$, $(2\kappa + \overline{\kappa})/kT$ remains unchanged as the alkane chain length is increased. In Winsor systems with alkanes and AOT, the value of κ alone (obtained using ellipsometry) is found to depend on the oil chain length with κ/kT being about unity for alkanes heptane to decane but falling progressively with alkane chain length thereafter up to tetradecane, for which κ is almost zero. This fall could be a result of the smaller penetration of the longer alkanes into the chain regions of the surfactant monolayers. The sum $(2\kappa + \overline{\kappa})$ for the same systems is reported by Kellay et al. (1994) and so $\overline{\kappa}$ can be obtained. It has been shown for example that, close to the WII–WIII boundary, $\overline{\kappa}$ changes sign with alkane chain length; it is negative for octane $(-1.2kT)$, close to zero in the case of decane and positive $(+1kT)$ for dodecane, and these values are referred to below.

10.2.3   The structure of surfactant-rich phases in WIII systems in relation to elastic moduli

From (10.2.1) the curvature energy g_c of a surface is seen to be made up from two contributions, one involving κ and the other $\overline{\kappa}$, and it depends on the magnitude and sign of the two principal radii of curvature, r_1 and r_2, which describe the local interface curvature (as illustrated in Fig. 10.10). For a spherically curved surface the radii of curvature are equal in magnitude and have the same sign. A cylindrically curved interface has one radius that is infinite (zero curvature) and the other finite. In the case of saddle-shaped surfaces of the kind illustrated, r_1 and r_2 are of opposite sign. Conventionally r is taken to be positive for interfaces which have oil on the 'inside', such as O/W microemulsion droplets. The *Gaussian curvature* is the product of the two curvatures, that is, $(1/r_1)(1/r_2) = c_1c_2$ (see §3.4.1).

Table 10.2 *Bending elastic moduli in oil–water surfactant systems. (a) Some values of $(2\kappa + \bar{\kappa})/kT$ in systems with decane and D_2O at $10°C$ containing nonionic surfactants C_nE_3 and for $C_{10}E_4$ in a series of n-alkanes and D_2O at $10°C$. Results from Gradzielski et al. (1996). (b) Some approximate values of κ/kT in systems with alkanes and water at $20°C$ containing the anionic surfactant AOT. Results taken from graphical data in Kellay et al. (1994).*

(a)

Surfactant	C_8E_3	$C_{10}E_3$	$C_{12}E_3$
$(2\kappa + \bar{\kappa})/kT$ from polydispersity	1.2	2.28	3.42
$(2\kappa + \bar{\kappa})/kT$ from γ_c and r	0.9	2.34	3.5
Alkane	**heptane**	**octane**	**decane**
$(2\kappa + \bar{\kappa})/kT$ from polydispersity	2.3	2.32	2.28
$(2\kappa + \bar{\kappa})/kT$ from γ_c and r	2.26	2.75	2.34

(b)

octane	decane	dodecane	tetradecane
0.85	0.95	0.12	0.05

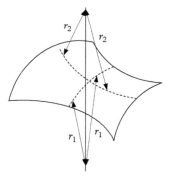

Figure 10.10 *A curved surface showing two principal radii of curvature r_1 and r_2 corresponding to curvatures c_1 and c_2 (see §3.4.1 for the definition of principal radii of curvature).*

In general the curvature energy E_c of a surface with area $\mathcal{A}$ can be expressed

$$dE_c = \left[\frac{\kappa}{2}\left(\frac{1}{r_1} + \frac{1}{r_2} - \frac{2}{r_o}\right)^2 + \frac{\bar{\kappa}}{r_1 r_2} \right] d\mathcal{A} \tag{10.2.9}$$

For a spherically curved surface (10.2.9) reduces to (10.2.1) where $g_c = dE_c/d\mathcal{A}$ and $r_1 = r_2 = r$. The curvature energy $E_{\bar{\kappa}}$ arising from $\bar{\kappa}$ is, for the whole surface S (Mauritsen, 2005),

$$E_{\bar{\kappa}} = \bar{\kappa} \oint_S \frac{1}{r_1 r_2} \, d\mathcal{A} = 4\pi\bar{\kappa}\,(1 - g_n) \tag{10.2.10}$$

where g_n is termed the *genus number*. For a sphere $g_n = 0$ and so for unit area of spherically curved surface, $E_{\bar{\kappa}}/\mathcal{A} = g_{\bar{\kappa}} = \bar{\kappa}/r^2$, as can also be seen from (10.2.1).

A *minimal* surface is one where $r_1 = -r_2$ everywhere so that the mean curvature H is always zero, that is,

$$H = \frac{1}{2}\left(\frac{1}{r_1} + \frac{1}{r_2}\right) = 0 \tag{10.2.11}$$

The surfactant-rich phases formed in WIII systems (which although not droplet dispersions are also termed microemulsions) have close to minimal surfaces. Some of the phases formed are similar to those in surfactant + water systems with lamellar and cubic structures (see Fig. 9.34b and d). In microemulsion systems however oil is present in which the surfactant chains reside, as shown in Fig. 10.11 for the simple case of a lamellar (so-called L_α) phase. Although these structures have close to zero net curvature, they are nonetheless susceptible to thermal fluctuations which are controlled by the bending elastic moduli of the surfactant layers within the structure. The phases have associated with them a persistence length, ξ, which is analogous to droplet sizes in droplet microemulsions (Kellay et al., 1994):

$$\xi = a\exp\left(\frac{2\pi\kappa}{kT}\right) \tag{10.2.12}$$

In (10.2.12), a is a molecular length. For large values of κ (say $10kT$ and upwards) the layers are rigid and hence flat over large distances relative to molecular size. There is long-range order in such structures. Lipid bilayers, for example, tend to have such large bending elastic moduli. Surfactant monolayers in some of the systems of present interest, however, have much smaller values of κ, of the order of kT or less (Table 10.2). In this

Figure 10.11 *A lamellar (L_α) phase containing both oil and water.*

case ξ becomes much smaller (e.g. around 20 nm for $a = 1$ nm and $\kappa/kT = 0.5$) and long-range order is lost.

Various structures have been observed in WIII systems, depending on the volume fractions of oil and water present in the third phase, such as those with local structures illustrated in Fig. 10.12. The surfactant monolayers are 'wavy' as a result of thermal fluctuations. If the amounts of oil and water are comparable, a bicontinuous sponge-like structure is given in which the surfactant monolayers are similar to that illustrated in Fig. 10.12a. If however the phase is largely surfactant and water with only a small amount of oil an L_3 phase is given in which surfactant bilayers swollen with some oil separate large water domains (Fig. 10.12b). Phases which contain much more oil than water (designated L_3') have surfactant bilayers swollen by water separating large oil domains as, shown in Fig. 10.12c. A representation of a bicontinuous cubic phase with surfaces of zero net curvature is given in Fig. 10.13; the structure is made up of a collection of saddle shapes. In the case of a bicontinuous sponge-like phase for example, the surfaces consist of surfactant monolayers separating oil and water regions (see Fig. 10.12a). For an L_3 phase the surfaces are oil-swollen surfactant bilayers separating water regions (Fig. 10.12b).

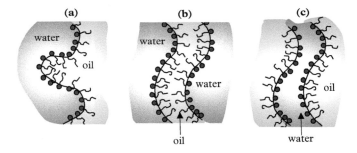

Figure 10.12 *Representations of surfactant layers in surfactant-rich third phases in WIII systems. (a) Monolayer in a bicontinuous sponge-like microemulsion phase with roughly equal volumes of oil and water. (b) Surfactant bilayer swollen by oil in water in an L_3 phase. (c) Surfactant bilayer swollen by water in oil in an L_3' phase.*

Figure 10.13 *A cubic bicontinuous structure with zero net curvature.*

Finally, the role of $\bar{\kappa}$ in the determination of phase structures is mentioned; some values of $\bar{\kappa}$ were given in the previous section for systems with AOT, aqueous NaCl, and various alkanes. The term in $\bar{\kappa}$ in (10.2.9) is $\bar{\kappa}/r_1r_2$, and so when the signs of r_1 and r_2 are different, positive $\bar{\kappa}$ gives a more negative curvature energy than does a negative modulus. This means that bicontinuous structures exhibiting saddle-splay surfaces are favoured by positive Gaussian elastic moduli. When the two radii of curvature have the same sign however, the more negative curvature energy is given for negative $\bar{\kappa}$. Thus spherical droplets and lamellar structures are favoured. The signs of $\bar{\kappa}$ found by Kellay et al. (1994) for third phases in systems with AOT, aqueous NaCl, and alkanes were in broad agreement with the phase structures observed.

10.2.4 What is the thermodynamic significance of tension minima in Winsor systems?

It was seen that the tension, γ_c, of the plane oil/water interfaces in Winsor microemulsion systems is inversely related to the equilibrium droplet size. Changes in droplet size and γ_c can be brought about by changing the concentration of an added inert electrolyte, the concentration of a cosurfactant (such as an alkanol) and temperature. The thermodynamic significance of the changes in tensions with respect to these variables is now considered in terms of the properties of the plane and the droplet interfaces.

First, the effects of inert electrolyte S on the interfacial tensions in Winsor systems containing a nonionic surfactant D (as exemplified in Fig. 10.14 for systems containing $C_{12}E_5$, NaCl and nonane) is examined. In an oil/water system the interfacial tension falls as surfactant is progressively added up to the concentration (*cac*) where aggregation begins, and thereafter it remains essentially constant (at γ_c) close to the *cac*, as shown schematically in Fig. 10.15. Tension values at points A, B, and C in Fig. 10.15 lie on a curve of the type shown in Fig. 10.14. An expression is now derived for the way in which γ_c varies with the concentration m_s of added electrolyte.

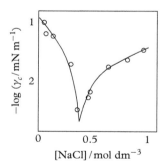

Figure 10.14 *Variation of γ_c with NaCl concentration in Winsor systems with $C_{12}E_5$, water, and nonane at 31°C. Redrawn from Aveyard et al. (1986).*

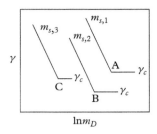

Figure 10.15 *Schematic representation of the variation of interfacial tension with ln m_D for different concentrations m_S of added electrolyte.*

Suppose a microemulsion drop containing n_D nonionic surfactant molecules and n_s^+ adsorbed/desorbed electroneutral salt molecules is formed at the *cac* in a Winsor system. For variations at the aggregation point, Hall (1983) has shown that

$$\sum_i \left(n_i + n_i^+ \right) d\mu_i = 0 = n_D d\mu_D + n_s^+ d\mu_s \tag{10.2.13}$$

For the plane interface in equilibrium with the aggregates in solution the Gibbs equation is

$$-d\gamma = \Gamma_D^\sigma d\mu_D + \Gamma_S^\sigma d\mu_S \tag{10.2.14}$$

Here Γ_D^σ is the surface excess of surfactant, which is effectively equal to the surface concentration (§4.1), and Γ_S^σ is the surface excess of the salt (§5.4.2). Noting that for a 1:1 electrolyte $d\mu_S = 2RT d\ln m_S f_\pm$, combination of (10.2.13) and (10.2.14) at the aggregation point (where $\gamma = \gamma_c$), gives

$$\frac{d\gamma_c}{d\ln m_s} = RT\Gamma_D \left[\left(\frac{2n_S}{n_D} - \frac{2\Gamma_S^\sigma}{\Gamma_D} \right) \left(1 + \frac{d\ln f_\pm}{d\ln m_S} \right) \right] \tag{10.2.15}$$

For say a 1:1 anionic surfactant in the presence of a 1:1 inert electrolyte with common cation Na$^+$ it can be shown that the equivalent expression is

$$\frac{d\gamma_c}{d\ln m_{Na^+}} = RT\,\Gamma_D \left[(\alpha_p - \alpha_m) \left(1 + \frac{d\ln f_\pm}{d\ln m_S} \right) \right] \tag{10.2.16}$$

The degrees of surfactant dissociation, α_p and α_m, for surfactant in the planar and aggregate surfaces, are defined as (see §9.6) minus twice the (negative) surface excess of coions per surfactant. The equations for ionic and nonionic surfactants are obviously similar; in the case of nonionic surfactants concern is with desorption of electroneutral salt at the surfaces whereas for ionic surfactants it is the desorption of coions that is considered. In both cases however the rate of change of interfacial tension with salt concentration is associated with the difference between desorption (expressed per surfactant) at aggregate and planar surfactant-covered interfaces. The minimum in tension occurs when desorption is the same at planar and aggregate surfaces as expected, since the net curvature of surfaces in surfactant phases at the condition for minimum γ_c is zero, as discussed earlier. What is clear is that the difference in ion or salt desorption (hence in the degree of dissociation of ionic surfactant) is dependent on the surface curvature.

It can be seen crudely why salt desorption from curved aggregates of nonionic surfactant is curvature dependent on simple geometrical grounds. Desorption is proportional to the volume from which desorption occurs. If it is supposed that there is a salt-free layer (§5.4.2) around the aggregate (see Fig. 10.16) then for oil drops in water the volume from which salt desorption occurs is greater (roughly as represented by the two black triangles) than that (light grey shaded area) at a planar interface. This means that for O/W microemulsion drops (in WI systems) salt desorption is greater than from plane surfactant monolayers. It follows then from (10.2.15) that in the WI region γ_c falls with increasing salt concentration, that is, n_S/n_D is more negative than Γ_S^σ/Γ_D so $d\gamma_c/d\ln m_s$ is negative.

The changes in tensions in Winsor progressions brought about by changes in T and by cosurfactant additions are also associated with differences in the properties of curved and planar interfaces, as anticipated. In the case of changes in T it transpires that

$$\frac{d\gamma_c}{dT} = -\left(S_u^s - \sum_i \Gamma_i^s S_i\right) + \Gamma_D^s \Delta S_m \tag{10.2.17}$$

Here S_u^s is the entropy of unit area of plane oil/water interface, S_i are partial molar entropies of components i in bulk solution, and ΔS_m is the molar entropy change on forming aggregates. The superscript s denotes total (rather than excess) quantities for the surface so that Γ_i^s is the total surface concentration of species i. The term in brackets in (10.2.17) is the entropy of formation of unit area of plane interface from species

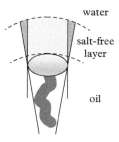

Figure 10.16 *Representation of the volume from which salt desorption occurs from the curved surface of a nonionic surfactant aggregate in water. Desorption occurs from a volume represented by the sum of the light and dark shaded areas.*

originally in bulk solution. Thus the temperature gradient of γ_c reflects the difference in entropy of forming aggregates containing Γ_D mol surfactant and unit area of plane interface (which also contains Γ_D mol surfactant).

Finally, the effects of addition of cosurfactant on γ_c in Winsor systems are described by

$$\frac{d\gamma_c}{d \ln a_{co}} = RT \, \Gamma_D \left(\frac{n_{co}}{n_D} - \frac{\Gamma_{co}}{\Gamma_D} \right) \tag{10.2.18}$$

Subscript *co* refers to the cosurfactant, often a medium chain length alkanol. Such materials tend to preferentially distribute into the oil phase, and if so they often partially associate to form dimers and higher associates. This leads to nonideality of the solutions, often at low concentrations in the oil, so that activities (a_{co}) must be used in place of concentrations. It is seen from (10.2.18) that the variation of γ_c with cosurfactant activity is determined by the difference in the molar ratio (cosurfactant/surfactant) in the aggregates (n_{co}/n_D) and at the plane interface (Γ_{co}/Γ_D).

10.3   Microemulsion droplet dynamics—what happens when droplets collide?

It has been seen in §9.10 that micelles, which are thermodynamically stable, are also dynamic structures; surfactant entities rapidly exchange between aggregates and solution and, on a longer timescale micelles disintegrate. It was also explained above that microemulsion droplets, although thermodynamically stable, are subject to thermal fluctuations leading to shape changes and polydispersity (§10.2.2). It is known that solutes in, say, W/O microemulsion droplets can exchange between droplets. Also, when microemulsions with droplets of different size are mixed, droplets of intermediate size are formed within a few seconds.

Droplet coalescence and breakup can be probed by studying chemical reactions in microemulsions (Fletcher et al., 1987). The central question is what happens when two droplets come into contact under normal diffusion to form (instantaneously at least) an encounter pair. Do the droplets simply part again or can they stick together and if so possibly coalesce to form a larger drop ('fused dimer') allowing the contents to mix? What are the timescales involved?

Very fast (effectively diffusion controlled) chemical reactions have been carried out in microemulsion systems. Fletcher et al. (1987) studied electron transfer reactions such as

$$\text{Ir(Cl)}_6^{2-} + \text{Fe(CN)}_6^{4-} \xrightarrow{k_f} \text{Ir(Cl)}_6^{3-} + \text{Fe(CN)}_6^{3-} \tag{i}$$

The two reactants were contained in similar but separate water-in-heptane microemulsions stabilized by AOT and then rapidly mixed using a stopped-flow technique, and the chemical reaction followed spectrophotometrically. The reactants are confined to the water droplets and cannot transfer between droplets through the oil phase, in which they

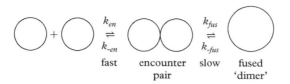

Figure 10.17 *Representation of the encounter and coalescence of two microemulsion droplets.*

are insoluble. Since the chemical reaction is so fast, the rate-determining step in kinetic experiments is the exchange of the reactants between droplets.

Consider the approach, encounter and coalescence of two drops, as represented in Fig. 10.17. The fused dimer is unstable and will break up but it has a lifetime which is sufficiently long for the reactants to mix randomly before breaking up to give two smaller droplets again. The experimentally determined second order rate constant can be equated with that for the exchange process, k_{ex}. It was found that, for a range of chemical reactions k_{ex} values were remarkably similar, supporting the claim that it is the droplet fusion process that is the rate limiting step and not the chemical reaction itself. For droplets with diameter 10 nm at 20°C the rate constant observed using reaction (i) was $k_{ex} = 7.5 \pm 1.5 \times 10^6$ dm^3 mol^{-1} s^{-1}. If every encounter between a pair of droplets were to lead to reactant exchange, k_{ex} would be expected to be close to the collision frequency of droplets. The diffusion controlled rate constant k_{dc} for droplet encounters can be calculated from the Smoluchowski equation; Fletcher et al. give a value for $k_{dc} = 1.7 \times 10^{10}$ dm^3 mol^{-1} s^{-1} for water droplets in heptane at 25°C. This means that, for say a droplet concentration of 1 mmol dm^{-3}, the time between encounters is about 60 ns. Comparison of k_{ex} obtained using reaction (i) with k_{dc} shows that only 1 in about 10^3–10^4 droplet collisions results in coalescence so there is a significant barrier (energy W) to droplet fusion.

The mechanism of droplet coalescence and the nature of the energy barrier involved are interesting, and probably relate to the way in which droplets mutually interact (attract) and also to the curvature properties of the surfactant layers coating microemulsion droplets. It was seen elsewhere (§9.8.3) that water droplets stabilized by nonionic surfactants (of the type C_nE_m) in nonpolar oil become more mutually attractive ('sticky') as the temperature is lowered towards the haze boundary. Conversely, these attractions gradually disappear with an increase in temperature as the solubilization boundary is approached, when the droplets attain their preferred curvature and excess aqueous phase begins to separate.

Fletcher and Horsup (1992) explored the effects of (i) proximity to phase boundaries and (ii) the magnitude of bending elastic moduli of surfactant monolayers at the oil/water interface (§10.2.2), on droplet exchange in W/O microemulsions stabilized by the nonionic surfactants C_8E_3, $C_{10}E_4$, and $C_{12}E_5$. Comparisons were made in 'equivalent' systems, that is, using droplets of the same size close to the solubilization phase boundaries, ensuring that effects due to droplet attractions were minimized. This entailed using different alkanes as oils with the different surfactants. Effects of moving away from solubilization boundaries (hence increasing inter-droplet attractions) were also investigated. Droplet exchange rates were measured by following the quenching of fluorescence of tris(2, 2′-bipyridyl)ruthenium(II) (RB) by methylviologen (MV). Droplets containing 1 RB molecule were mixed with droplets with 1 MV molecule and the fluorescence quenching was followed using a time-resolved fluorescence (TRF) technique. The quenching in bulk solution is close to the diffusion-controlled limit, and it is reasonable to suppose diffusion within a droplet is the same. The second order rate constant for the exchange process between droplets (k_{ex}) was extracted from the TRF data, and it was found that for a given system the droplet exchange was slowest at the solubilization (i.e. high temperature) boundary, and increased as the temperature was lowered. As T is lowered, inter-droplet attractions increase and it is to be expected on this count that the rate of exchange between droplets will increase.

To relate droplet fusion rates to bending elastic moduli, κ, of the stabilizing surfactant monolayers on the drops, data determined at solubilization boundaries are used, to minimize effects of droplet attractions. Suppose the fusion process can reasonably be represented by the scheme given in Fig. 10.17. Then the energy barrier (W) to droplet fusion is associated with a transition structure which is formed as the encounter pair transforms to a fused dimer. If the transition state involves a neck, as shown in Fig. 10.18, then it is necessary to bend the surfactant monolayers around the drops away from their natural curvature.[3] This requires an input of energy (W), which is expected to be related in some way to the bending elastic modulus of the monolayer.

In line with these ideas, Fletcher and Horsup (1992) report values of W and κ shown in Table 10.3 for the systems they studied. The energy barrier was estimated from the diffusion controlled rate constant k_{dc} and k_{ex} according to

$$W = RT \ln (k_{dc}/k_{ex}) \tag{10.3.1}$$

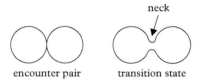

encounter pair   transition state

Figure 10.18 *Formation of transition state droplet pair showing neck with net curvature of opposite sign to that of precursor droplets.*

---

[3] Note that, with reference to Fig. 10.18, the monolayer in the bridge has two curvatures of opposite sign, one in the plane of the paper and one at right angles to the paper. It is the *net* curvature in the bridge that is associated with the energy required to form the bridge.

Table 10.3 *Values of bending elastic moduli and activation energies for droplet exchange in the systems indicated. Results from Fletcher and Horsup (1992).*

System	$T/^{\circ}C$	κ/kT	W/RT
C_8E_3–decane	37	0.35	0.8
$C_{10}E_4$–octane	49	0.76	2.4
$C_{12}E_5$–hexane	60	1.25	5.1

10.4   Representation of phase behaviour

Ways of representing phase behaviour in systems with surfactant and water were presented in §9.11, where systems with three components and the use of triangular phase diagrams were only briefly touched upon. The microemulsion systems already considered occupy regions in triangular phase diagrams in which the apices represent pure surfactant, water and 'oil' (such as an alkane). The application of the phase rule,

$$F = C - P + 2 \tag{9.11.2}$$

to three-component systems shows that at constant T and p, $F = 3 - P$. For a single-phase system this means the composition with respect to two of the components must be specified to determine the position in the phase diagram. If two phases are present at equilibrium then composition with respect to one of the components only need be specified. When 3 phases are present the system is invariant.

10.4.1   Systems at constant temperature

The Winsor systems described in §10.1.2 are considered now. As with surfactant + water systems already discussed, a range of surfactant structures can exist for a ternary system (e.g. hexagonal, lamellar, etc., including the 'reverse' forms), depending on composition and temperature, and on the molecular structure of the surfactant and oil present. Here, in Fig. 10.19, the regions where microemulsions of the type described are indicated, with other phases omitted for clarity. Figure 10.19a represents a WI system. The dashed line in the two-phase (2ϕ) region is a tie line, and a system with overall composition at point A (*anywhere* on the tie line) will separate into two phases with compositions corresponding to the two ends of the tie line. In this case one of the phases is almost pure oil (containing surfactant monomers at the *cac* of the system) and the other (point X) is largely water containing O/W microemulsion droplets. A mixture with composition represented by point B in a WII system (Fig. 10.19c) has one phase which is largely water, which contains surfactant monomers, and the other (at point Y) which is a W/O microemulsion. Any point in a 2ϕ region must lie on a tie line so the composition with

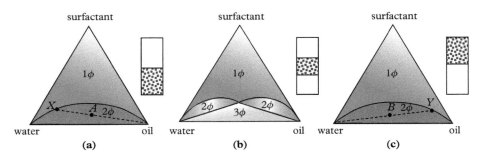

Figure 10.19 *Schematic triangular phase diagrams representing Winsor systems (other phases omitted): (a) Winsor I, (b) Winsor III, and (c) Winsor II. The shaded phases in the tubes are microemulsions.*

respect to one component need only be specified ($F = 1$ at constant T and p) to define the overall composition.

A WIII system is represented in Fig. 10.19b; the system is invariant ($F = 0$) in the 3ϕ region. The physical meaning of this is that a system with overall composition falling anywhere within this region separates out to three phases whose compositions are independent of the overall composition. As the latter is changed, the *amounts* of the three equilibrium phases change, but their compositions do not. The equilibrium phase compositions are those corresponding to the apices of the three-phase triangle. One of the phases is almost pure water and another is almost pure oil. The third (surfactant-rich) phase contains both oil and water and most of the surfactant present in the system. This phase is represented as the middle phase in Fig. 10.19b.

10.4.2 Simple representations of the effects of temperature on phase behaviour

Temperature is a variable of great importance when dealing with phase behaviour, both for practical and theoretical reasons. A full representation can be given in terms of a triangular prism (Fig. 9.41b) in which the axis normal to the triangle represents temperature. Determination of extensive phase information over a range of temperature is very laborious and very often a knowledge of partial phase behaviour is sufficient. There are two commonly encountered types of *pseudobinary* diagram which represent effects of T on phase formation, and both are based on planes taken through the triangular phase prism normal to the triangle. In one case (Fig. 10.20a), the (shaded) plane represents a system with a constant ratio of oil (O) to water (W) and varying amounts of surfactant (S). The plane in Fig. 10.20b is for a constant overall amount of S but varying amounts of O and W.

A schematic diagram representing the effects of temperature on phase behaviour in systems with a fixed oil to water ratio, say 50 wt %, and varying amounts δ (wt % say) of nonionic surfactant is shown in Fig. 10.21 (Kahlweit, 1999). The lines denoting the phase boundaries resemble the shape of a fish. The body of the fish is a three-phase region containing oil, water and microemulsion. The two-phase region at higher temperatures

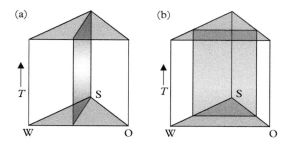

Figure 10.20 *Two possible 'cuts' through a triangular prism. (a) The shaded plane corresponds to a fixed ratio of oil (O) to water (W) with varying amounts of surfactant (S). (b) This plane is for a given overall amount of S with varying amounts of O and W.*

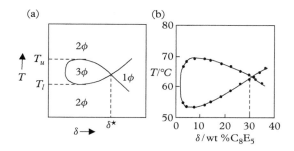

Figure 10.21 *(a) A schematic 'fish' phase diagram for nonionic surfactant in oil + water systems showing regions of one, two, and three phases. (b) Diagram for equal masses of water and octane with added C_8E_5. Redrawn from Kahlweit (1999).*

consists of a W/O microemulsion and an aqueous phase whereas the phases in the lower 2ϕ region are an O/W microemulsion and an oil phase. The surfactant concentration δ^* at the beginning of the tail of the fish marks the onset of the single phase regime and represents the minimum amount of surfactant required to completely solubilize all the oil and water present.

An experimentally determined phase diagram for systems with equal masses of octane and water with added C_8E_5 is shown in Fig. 10.21b. The phase diagrams for C_8E_4 and C_8E_3 (also with octane and water) are similar although shifted to lower temperatures. The mean temperature $T_m = (T_l + T_u)/2$ (see Fig. 10.21a) and δ^* for the various systems are given in Table 10.4. The greater the hydrophilicity of the surfactant, the greater is the temperature that is needed to force aggregation to occur in the oil phase, as might be expected.

Phase diagrams are now considered for systems with varying proportions of oil and water but with a fixed overall amount of surfactant, as represented by the shaded plane in Fig. 10.20b. Typically, the amount of surfactant used is less than 10 wt %. Parts of such diagrams have already been discussed in §9.8.3 in connection with solubilization of oil and of water by surfactants. A schematic representation is shown in Fig. 10.22; such figures are sometimes referred to as 'Shinoda diagrams' since Shinoda and his

Table 10.4   *Values of T_m and δ^* for systems with C_8E_m and equal weights of octane and water. Results taken from Kahlweit (1999).*

Surfactant	$T_m/°C$	$\delta^*/wt \%$
C_8E_3	16	20
C_8E_4	42	25
C_8E_5	61	32

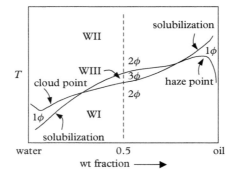

Figure 10.22 *Schematic 'Shinoda' phase diagram for a nonpolar oil + water + nonionic surfactant. The overall amount of surfactant is constant over the whole diagram, and is typically <10 wt %.*

co-workers have used such phase representations over a number of years (see e.g. Shinoda and Friberg, 1986). Phases other than those shown often appear but are omitted here for simplicity. Single phase areas of solubilization of oil and of water are shown at the water-rich and oil-rich ends of the diagram, respectively, and the solubilization, cloud point, and haze-point curves are indicated (see §9.8.3 for a discussion of the significance of these curves).

Consider systems consisting of roughly equal amounts of water and oil with nonionic surfactant present. At lower temperatures the systems are OW microemulsions in equilibrium with excess oil containing monomeric surfactant. At intermediate temperatures a surfactant- rich phase in equilibrium with both excess oil and water is formed. The systems again become two-phase at higher temperatures, with a W/O microemulsion in equilibrium with excess water containing surfactant monomers at the *cac*. This progression with T is the Winsor progression previously discussed.

This kind of phase diagram can also be constructed with parameters (termed HLB variables) other than temperature as ordinate. The term HLB is shorthand for hydrophile–lipophile balance, and the HLB concept is discussed in connection with emulsion formation in Chapter 14. Broadly, in the present context, an HLB parameter is one which can bring about phase inversion (such as a Winsor progression—see Table 10.1) in an oil + water + surfactant system. Clearly temperature is one such variable; some others are concentration of ionic surfactant added to nonionic surfactant (or vice

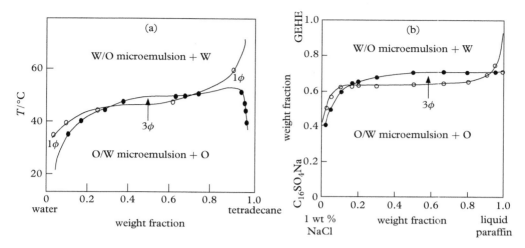

Figure 10.23 *Phase diagrams for (a) 5 wt % $C_{12}E_5$ in the tetradecane + water system. (b) Mixtures of glycerol mono(2-ethylhexyl)ether (GEHE) and sodium hexadecylsulfate (total 5 wt %) in systems with 1 wt % aqueous NaCl and liquid paraffin at 25° C. Redrawn from Shinoda et al. (1984). Some phases omitted for clarity.*

versa), concentration of cosurfactant added to surfactant, alkane chain length when the oil is alkane or composition with respect to two similar C_nE_m surfactants with different values of *m*.

Experimental examples of phase diagrams (taken from Shinoda et al. 1984) are shown in Fig. 10.23; in (a) the effects of *T* on systems containing water and tetradecane with added $C_{12}E_5$ are shown. The phases obtained are as illustrated in Fig. 10.22. The ordinate in Fig. 10.23(b) is the weight fraction of glycerol mono(2-ethylhexyl)ether (GEHE) in mixtures with sodium hexadecylsulfate ($C_{16}SO_4Na$) in systems with 1 wt% aqueous NaCl and liquid paraffin as the oil phase. Ionic surfactants such as $C_{16}SO_4Na$ favour the formation of OW microemulsions (or swollen micelles); the packing factor in this kind of system is less than unity as discussed in §10.2.1. Addition of GEHE (a cosurfactant), however, serves to increase the effective value of *P*, and the system passes through a three-phase regime and then transforms into a two-phase system consisting of W/O microemulsion and excess aqueous phase. Some details of the diagrams in Fig. 10.23 have been omitted.

Appendix A10:   Scattering methods to determine microemulsion droplet size

The theory and practice of the measurement of 'particle' sizes and size distributions has been central to the development of colloid science, and knowledge of particle sizes is crucial to quality control in a range of industries. Particles can be solid, liquid (e.g. emulsion or microemulsion droplets) or gas (bubbles). A number of methods are used and include light, X-ray, and neutron scattering techniques, acoustic spectroscopy (attenuation of ultrasound), centrifuge sedimentation, electron microscopy, and microscopic image analysis, depending on the systems of interest. The scattering techniques and acoustic spectroscopy involve measuring a system property and then using appropriate theory to *calculate* particle size. Acoustic spectroscopy (Dukhin and Goetz, 2010) has the advantage over light scattering since, like neutron scattering, it can be used on optically opaque concentrated dispersions, including emulsions, without sample dilution.

In what follows discussion is restricted to light and neutron scattering methods that have commonly been used to measure sizes of micelles and microemulsion droplets.

A10.1   Conventional light scattering

Light can be described in terms of a fluctuating electrical field E at right angles to an associated fluctuating magnetic field, H, as represented in Fig. A10.1. For *plane polarized* light (see also §8.3) the plane which includes the electric vector, and the direction of propagation, are each fixed. *Unpolarized* light (intensity I_o) can be represented as the combination of two plane polarized rays, each of intensity $I_o/2$, at right angles to each other. When light impinges on a molecule, the *electric* field interacts with the *electrons* in the molecule. Consequently, a small particle (Box 10.5) in a light beam is polarized by the light and a dipole moment is created in the particle. Some of the energy of the light is used in creating the fluctuating dipole, which in turn acts as a source of radiated light of the same frequency. Since the frequency of the light is unchanged, the scattering is said to be *elastic*.

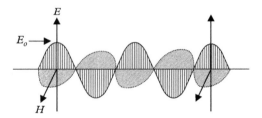

Figure A.10.1 *Electromagnetic radiation with a fluctuating electric field, E, at right angles to a fluctuating magnetic field, H.*

Box 10.5   How small is small?

Small in the context of the Rayleigh theory of light scattering means the particle dimensions are less than about $\lambda/20$, where λ is the wavelength of the light being scattered. When light falls on such a particle, in effect the entire particle is instantaneously subjected to the same electric field. For a large particle not all parts are subjected to the same electric field at a particular instant; the Rayleigh approach is then no longer valid and theories developed by Gans, Debye, and Mie must be deployed, depending on particle size. Everett (1988) gives a clear and simple exposition of the elements of these theories. Further references to classic works on the subject are also cited by Everett.

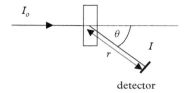

detector

Figure A.10.2 *Schematic representation of light scattering experiment. The incident beam has intensity I_o and the scattered beam, observed in the plane of the paper at an angle θ to the direction of the incident light, has intensity I.*

A simple scattering experiment is represented schematically in Fig. A10.2. It is supposed that incident light (in the plane of the paper), intensity I_o, impinges on a dilute dispersion of small spherical particles with radius R. The intensity of scattered light (observed in the plane of the paper) at an angle θ from the direction of the incident beam and a distance r from the sample, is I. It is noted that the intensity of light is proportional to the square of the amplitude E_o indicated in Fig. A10.1. For scattering of unpolarized light by unit volume of a dispersion of small spherical particles, concentration c, the Rayleigh theory gives (Box 10.6)

$$\frac{I}{I_o} = \frac{c}{r^2}\left[\frac{16\pi^4 R^6}{2\lambda^4}\left(\frac{n^2-1}{n^2+2}\right)^2\left(1+\cos^2\theta\right)\right] = \frac{R_\theta c}{r^2} \tag{A10.1}$$

Here λ is the wavelength of the light (incident and scattered) and n the relative refractive index, which is the ratio of the refractive indices of the particle and the dispersion medium (n_1/n_0). The term in square brackets is referred to as the Rayleigh ratio (R_θ).

Box 10.6 Now you see it, now you don't

Equation (A10.1) demonstrates that $I = 0$ if $n_1 = n_0$, that is, that if the refractive indices of particle and medium are matched, the particles are effectively 'invisible' and do not scatter light. Further, the intensity of scattered light is seen to be proportional to $1/\lambda^4$ so that blue light (short wavelength) is scattered much more strongly than red light (long wavelength). This accounts for the blue appearance of the sky and red/orange colour of sunsets, both due to light scattering by molecules. In the former case, light is viewed say at right angles to the sun's rays (equivalent to $\theta = 90°$ in Fig. A10.2) whereas in the latter case the light is viewed through the scattering molecules ($\theta = 0$). In the present context, it is seen that in principle the radius R of spherical scattering particles can be obtained from measurements of I and I_o if all the other parameters in (A10.1) are known.

As mentioned, for a sufficiently small particle, the entire particle is instantaneously subjected to the same electric field. The radiation envelope (intensity as a function of angle θ) is symmetrical, as seen schematically in cross-section in Fig. A10.3 (top). It can be appreciated from (A10.1) that

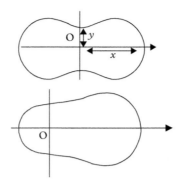

Figure A.10.3 *Top: Symmetrical cross-section of radiation intensity envelope for scattering of non-polarized light by a small spherical particle at the origin O (Rayleigh theory). The ratio x:y is 2:1. Bottom: Corresponding radiation envelope (schematic) for a large spherical particle according to Rayleigh–Gans–Debye theory.*

the ratio of scattered intensity at $0°$ $(180°)$ to that at $90°$ $(270°)$ is 2:1 since $(1 + \cos^2\theta)$ is 2 for $\theta = 0°$ $(180°)$ and 1 for $\theta = 90°$.

For larger particles of intermediate size however, this is no longer the case and light scattered from different parts of a particle are out of phase and can interfere, thus reducing the intensity of scattered light (Rayleigh–Gans–Debye theory). Backscattering intensity is reduced and the radiation envelope becomes asymmetric. This is illustrated schematically in Fig. A10.3 (bottom). For particles in this larger size range the expression for Rayleigh scattering intensity (I_R) is modified to

$$I = I_R P(Q) \tag{A10.2}$$

in which the *form factor* $P(Q)$ is given by

$$P(Q) = 1 - \frac{1}{3}(QR_G)^2 + \cdots \tag{A10.3}$$

where R_G is the radius of gyration of a particle with arbitrary shape.[4] The *wave vector* Q is related to θ and λ by (see also (8.1.1))

$$Q = \left(\frac{4\pi}{\lambda}\right) \sin\left(\frac{\theta}{2}\right) \tag{A10.4}$$

From (A10.3) and (A10.4) it is seen that when $\theta = 0$, $P(\theta) = 1$, and when $\theta = 180°$

$$P(Q) = 1 - \frac{1}{3}\left(\frac{4\pi R_G}{\lambda}\right)^2 + \cdots \tag{A10.5}$$

---

[4] The radius of gyration for a (homogeneous) sphere with radius R is $(3/5)^{0.5} R$.

which shows that the intensity of backscattering is reduced relative to Rayleigh scattering, as illustrated in Fig. A10.3. If only the first term of the series in Q in (A10.3) is important (i.e. when $QR_G << 1$), the ratio of scattered intensity at 45° and 135°, I_{45}/I_{135}, is

$$\frac{I_{45}}{I_{135}} = \frac{1 - 0.049(4\pi R_G/\lambda)^2}{1 - 0.285(4\pi R_G/\lambda)^2} \tag{A10.6}$$

It follows that from measurements of scattering intensity at 45° and 135°, R_G can be extracted and if the particles are spherical, the radius R can be obtained. As will be seen, the treatment of neutron scattering has parallels with this treatment of light scattering.

For even larger particles, a theory of scattering has been presented by Mie, and the scattering pattern is more complex and detailed; there are a number of lobes of intensity, large ones for forward scattering and smaller ones for backscattering.

A10.2   Dynamic light scattering (DLS)

In conventional light scattering described above the intensity of light is the average observed over a 'long' period of time. In DLS (which, for reasons that will become obvious, is also referred to as photon correlation spectroscopy, PCS), laser light is used which is monochromatic and coherent.[5] When a particle undergoing random Brownian motion scatters coherent light, the movement imparts a randomness to the phase of the scattered light. When two or more moving particles are present the scattered light from different particles undergoes time-dependent constructive or destructive interference. This produces fluctuations in the scattered light intensity (twinkling) which, in DLS experiments, are observed using a fast photon counter. A schematic depiction of the variation of intensity of scattered light with time is shown in Fig. A10.4. It can be appreciated that the intensity fluctuations are related to the diffusion rate of particles in the medium. Further, for monodisperse spheres the diffusion coefficient D is related to the hydrodynamic radius, R_H, of the particles by the Stokes–Einstein equation:

$$R_H = \frac{kT}{6\pi \eta D} \tag{A10.7}$$

where η is the viscosity of the medium in which the particles are dispersed.

In a DLS experiment scattered intensity at time t is compared with that at time $t + \tau$ (see Fig. A10.4). When the delay time τ is short the correlation is high since particles have not moved far in that time. As τ becomes larger however the correlation becomes less, and decays exponentially. Information about the motion is extracted from the intensity fluctuations in terms of a *correlation function* $g(\tau)$, which is a measure of the probability of a particle moving a given distance in time τ. For monodisperse spherical particles (e.g. microemulsion droplets),

$$g(\tau) = \exp(-\Gamma\tau) \tag{A10.8}$$

---

[5] In coherent light all the waves in a beam are in phase along a line normal to the direction of the beam.

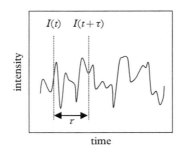

Figure A.10.4 *Schematic representation of the variation of scattered light intensity with time in a dynamic light scattering experiment. Fluctuations occur on the timescale that particles move about one wavelength of light.*

where Γ, a decay constant, is given as

$$\Gamma = Q^2 D \tag{A10.9}$$

From a plot of $\ln(g(\tau))$ versus τ, Γ can be obtained and since Q, the wave vector (see also A10.4), is known, D and hence R_H can be calculated by use of (A10.7).

Although DLS has been discussed here in connection with microemulsion droplet sizes, it is a versatile technique and can be used for the determination of particle sizes of, for example, macromolecules of various kinds (carbohydrates, proteins, synthetic polymers), solid particles, micelles, etc. A number of commercial instruments are readily available for laboratory use.

A10.3   Small angle neutron scattering (SANS)[6]

Light interacting with matter causes polarization of the *outer electrons* of atoms which is related to the refractive index of the scattering material. As mentioned in §8.1, neutrons, which interact with the *nuclei* of atoms, have wavelengths λ of the order 10^{-3} the wavelength of visible light and similar to atomic sizes and spacings. Neutron wavelength is related to the velocity v of the neutrons by the de Broglie relationship

$$\lambda = \frac{h}{mv} \tag{A10.10}$$

where h is the Planck constant and m the mass of a neutron. The velocity of neutrons can be moderated to give various wavelengths, with thermal neutrons having $\lambda \approx 0.1\,\text{nm}$ while cold neutrons have λ between about 0.4 to 2 nm. The angles observed for the elastic scattering of neutrons are small (a few degrees) and so the technique is referred to as small-angle neutron scattering or SANS.

---

[6] A useful and simple account of the technique of SANS by colloidal dispersions is given by Ottewill (1982), and the present exposition has drawn upon this.

Since the wavelength of the neutrons used in SANS is very small relative to (say) the radius R of a spherical colloidal particle, $R/\lambda >> 1$, and for a dispersion of non-interacting[7] monodisperse homogeneous spherical particles (at volume fraction φ) the intensity $I(Q)$ of scattered neutrons is given by an expression similar in form to (A10.2):

$$I(Q) = A(\rho_p - \rho_m)^2 N_p v_p^2 P(Q) \tag{A10.11}$$

where A is an instrument constant, involving λ, the distance from sample to detector (equivalent of r in Fig. A10.2) and I_o, and v_p is the volume of the particle ($= (4/3)\pi R^3$). The number concentration of particles in the dispersion is denoted N_p and is equal to φ/v_p. The quantities ρ_p and ρ_m are the neutron scattering length densities (see §8.1) of the particle and the medium respectively, and are related to the neutron refractive indices of the materials as shown in (8.1.2). The expression for $P(Q)$ (the form factor) given in (A.10.5) for light scattering no longer holds and is now given, for example, for a homogeneous sphere with radius R, as

$$P(Q) = \left[\frac{3 \, (\sin QR - QR\cos QR)}{Q^3 R^3} \right]^2 \tag{A10.12}$$

For a given system, the shape of a plot of $I(Q)$ against Q is determined by the form factor.

The scattered intensity at zero Q (hence zero θ), for which $P(Q) = 1$, is seen from (A10.11) to be

$$I(0) = A(\rho_p - \rho_m)^2 N_p v_p^2 = A(\rho_p - \rho_m)^2 \varphi v_p \tag{A.10.13}$$

and so

$$\ln \left(I(Q)/I(0) \right) = \ln P(Q) \tag{A10.14}$$

A plot of $\ln\{(I(Q)/I(0)\}$ against Q is shown in Fig. A10.5 for homogeneous monodisperse spherical particles with radius 35 nm in the range of Q from zero to 0.4 nm^{-1}. If the wavelength of the neutrons is 1 nm then $Q = 0.4$ nm^{-1} corresponds to $\theta \approx 2.2°$. The plot is an ideal curve, and in practice due to particle polydispersity and a spread of wavelengths, λ, the deep minima can be much less marked than shown.

From (A10.13) it is seen that

$$\pm\sqrt{I(0)} = (\rho_p - \rho_m)\left(A\varphi v_p\right)^{1/2} = \rho_p\left(A\varphi v_p\right)^{1/2} - \rho_m\left(A\varphi v_p\right)^{1/2} \tag{A10.15}$$

A value of $\sqrt{I(0)}$ can be obtained by extrapolation of a plot of $I(Q)$ versus Q to zero Q. The scattering length density of the medium in aqueous systems can be varied by using mixtures of H_2O and D_2O as explained in §8.1. When $\rho_p = \rho_m$, $\sqrt{I(0)} = 0$ and so a plot of $\sqrt{I(0)}$ against ρ_m can yield a value of ρ_p. The slope of such a plot is, according to (A10.15),

---

[7] If the colloidal particles interact mutually, as they do in concentrated dispersions, then the right hand side of (A.10.11) must be multiplied by a *structure factor*, $S(Q)$. Importantly, neutron scattering can be used to probe structuring (e.g. obtain radial distribution functions) in concentrated dispersions, for which the use of light scattering is prohibited by the opacity of the systems.

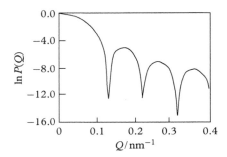

Figure A.10.5 *Plot of ln P(Q) versus Q calculated using (A10.12) for monodisperse homogeneous spherical particles with radius R = 35 nm.*

$$-\left[A\varphi v_p\right]^{1/2}$$
(A10.16)

and the intercept is

$$\rho_p\left[A\varphi v_p\right]^{1/2}$$
(A10.17)

Clearly, if A and φ are known the particle volume and hence radius R can be determined.

Particle size can also be obtained from scattering data at very small angles such that $QR << 1$. The form factor can be expanded to give

$$P(Q) = \left(1 - \frac{Q^2 R^2}{10} \cdots\right)^2 = \left(1 - \frac{Q^2 R_G^2}{6} \cdots\right)^2$$
(A10.18)

where R_G is the radius of gyration (see footnote 4). Expansion of (A10.18) then gives

$$P(Q) = \left(1 - \frac{Q^2 R_G^2}{3} \cdots\right) \approx \exp\left(-\frac{Q^2 R_G^2}{3}\right)$$
(A10.19)

where, for $QR << 1$ terms beyond those shown in (A10.19) can be neglected. So, from (A10.14) and (A10.19),

$$\ln\left(\frac{I(Q)}{I(0)}\right) = \ln P(Q) = -\frac{Q^2 R_G^2}{3}$$
(A10.20)

Equation (A10.20) is referred to as Guinier's law and shows that for very low scattering angles $\ln I(Q)$ and $\ln P(Q)$ are linear functions of Q^2. Thus, R_G and hence R can be obtained from the slope of a plot of $\ln P(Q)$ versus Q^2. Such a plot is shown in Fig. A10.6 for the same system as represented in Fig. A10.5, and covers values of θ from zero to 0.2°.

The ability to change the scattering length density of an aqueous medium by mixing H_2O and D_2O allows particles other than homogeneous entities to be probed, in a way that is not possible with light or X-rays. An obvious example is that of core-shell particles (e.g. particles with attached

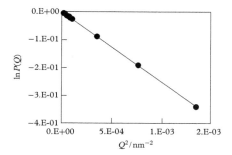

Figure A.10.6 *Guinier plot, ln P(Q) versus Q^2, for monodisperse homogeneous spherical particles with radius R = 35 nm. The abscissa corresponds to a range of θ from zero to 0.2°.*

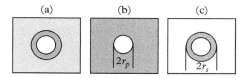

Figure A.10.7 *Use of contrast variation to probe the structure and size of core–shell particles by varying ρ_m. (a) All ρ are different; (b) $\rho_m = \rho_{shell}$; (c) $\rho_m = \rho_{core}$.*

layers acting as steric stabilizer) where ρ_m can be matched to that of the core ρ_{core} and then to that of the shell ρ_{shell}, as illustrated in Fig. A10.7. When $\rho_m = \rho_{shell}$ (case (b)) the neutrons 'see' only the particle core, and the particle radius r_p is obtained. When $\rho_m = \rho_{core}$ (case (c)) scattering is given by the concentric shell only. Using the known value of r_p, scattering data for the case where $\rho_m = \rho_{core}$ is fitted by adjusting the value of r_s and hence the shell thickness $r_s - r_p$ is obtained.

• •

REFERENCES

T. Ahsan, R. Aveyard, and B.P. Binks, Winsor transitions and interfacial film compositions in systems containing sodium dodecylbenzene sulphonate and alkanols. *Colloids Surf.,* 1991, **52**, 339–352.

R. Aveyard, B.P. Binks, S. Clark, and J. Mead, Interfacial tension minima in oil–water–surfactant systems. Behaviour of alkane–aqueous NaCl systems containing aerosol OT. *J. Chem. Soc., Faraday Trans. 1,* 1986, **82**, 125–142.

R. Aveyard and T.A. Lawless, Interfacial tension minima in oil–water–surfactant systems. Systems containing pure non-ionic surfactants, alkanes and inorganic salts. *J. Chem. Soc., Faraday Trans. 1,* 1986, **82**, 2951–2963.

A.S. Dukhin and J.P. Goetz, *Characterization of Liquids, Nano- and Microparticulates, and Porous Bodies using Ultrasound.* Elsevier, 2010.

D.H. Everett, *Basic Principles of Colloid Science.* Royal Society of Chemistry, London, 1988, Chapter 7.

P.D.I. Fletcher and D.I. Horsup, Droplet dynamics in water-in-oil microemulsions and macroemulsions stabilised by non-ionic surfactants. Correlation of measured rates with monolayer bending elasticity. *J. Chem. Soc., Faraday Trans.*, 1992, **88**, 855–864.

P.D.I. Fletcher, A.M. Howe, and B.H. Robinson, The kinetics of solubilisate exchange between water droplets of a water-in-oil microemulsion. *J. Chem. Soc., Faraday Trans. 1*, 1987, **83**, 985–1006.

P.D. De Gennes and C. Taupin, Microemulsions and the flexibility of oil/water interfaces. *J. Phys. Chem.*, 1982, **86**, 2294–2304.

M. Gradzielski, D. Langevin, and B. Farago, Experimental investigation of the structure of nonionic microemulsions and their relation to the bending elasticity of the amphiphilic film. *Phys. Rev. E*, 1996, **53**, 3900.

D.G. Hall, in *Aggregation Processes in Solution* (ed. E. Wyn-Jones and J. Gormally), Chapter 2. Elsevier, Amsterdam, 1983.

W. Helfrich, Elastic properties of lipid bilayers: theory and possible experiments. *Z. Naturforsch. C*, 1973, **28**, 693–703.

M. Kahlweit, Microemulsions. *Annu. Rep. Prog. Chem., Sect. C*, 1999, **95**, 89–116.

H. Kellay, B.P. Binks, Y. Hendrikx, L.T. Lee, and J. Meunier, Properties of surfactant monolayers in relation to microemulsion phase behaviour. *Adv. Colloid Interface Sci.*, 1994, **49**, 85–112.

H. Kellay, J. Meunier, and B.P. Binks, Saddle-splay modulus of the AOT monolayer in the AOT-brine–oil system. *Phys Rev Lett.*, 1993, **70**, 1485.

O.G. Mauritsen, *Life – As a Matter of Fat. The Emerging Science of Lipidomics*, Springer, Berlin, 2005, Chapter 6.

J. Meunier and L.T. Lee, Bending elasticity measurements of a surfactant monolayer by ellipsometry and x-ray reflectivity. *Langmuir*, 1991, **7**, 1855–1860.

R.H. Ottewill, in *Colloidal Dispersions* (ed. J.W. Goodwin), Chapter 7. Royal Society of Chemistry, London, 1982.

K. Shinoda and S. Friberg, *Emulsions and Solubilization*. Wiley-Interscience, New York, 1986.

K. Shinoda, H. Kunieda, T. Arai, and H. Saijo, Principles of attaining very large solubilization (microemulsion): inclusive understanding of the solubilization of oil and water in aqueous and hydrocarbon media. *J. Phys. Chem.*, 1984, **88**, 5126–5129.

Section IV

Surface forces and thin liquid films

The way in which *lateral* forces acting between adsorbed surfactant *molecules* affect the behaviour within monolayers (as reflected in surface equations of state and adsorption isotherms) has been explored in Chapters 5 and 7. Interest in the present section (Chapters 11 and 12) is focused on forces that act (say) normally between the surfaces of macroscopic bodies in close proximity, that is, so-called *surface forces*. Such forces, which include amongst others van der Waals and electrical (Coulombic) forces, are responsible for the stability or otherwise of various kinds of systems in which thin liquid films are present. Films can be those between approaching particles in colloidal dispersions (Chapters 11 and 13) or droplets in emulsions (Chapter 14). Thin liquid films are also present in a range of wetting processes (Chapters 12 and 16).

11

Surface forces and colloidal behaviour

11.1   Introductory remarks about colloidal systems

Earlier, some of the ways in which molecules can interact laterally within an adsorbed layer, and how this can be described in terms of surface equations of state and adsorption isotherms, were described (see e.g. §5.3.2). In many systems of interest, however, there are two or more surfaces in close proximity and these can interact with each other across an intervening phase. The interaction between surfaces (area A) results in a force, $F(h)$, normal to the surfaces, the magnitude and sign of which depend on the normal separation h of the surfaces. It is common to refer to forces of this kind as *surface forces* or, when dealing with *colloids* (Box 11.1), *colloidal interactions*. Interactions can equally be expressed in terms of *interaction (free) energies* (per unit area), often designated $V(h)$. In the context, the distinction between free energy and potential energy of interaction is often ignored. In the case of plain parallel liquid films (discussed in Chapter 12) it will also be useful to consider the so-called *disjoining pressure* $\Pi_D(h)$ generated by the interactions.[1] The importance of surfactants in conferring stability on thin films (in foams, for example) and on dispersions of solid particles and liquid droplets (in emulsions) arises, in part, from their ability to modify surface forces.

A colloid can be defined as a dispersion of particles (liquid, solid, or gas) in a medium, the particles being larger than ordinary molecules yet small enough (when the medium is a fluid) not to settle out under gravity. They are held in suspension by the motion of the molecules of the surrounding medium, and undergo Brownian motion. Other systems, for example, soap films and fibres, which have one of their dimensions in the colloid size range, can also be classed as colloids. Systems with larger 'particles', such as emulsions,

---

[1] $F(h)$, $V(h)$, and $\Pi(h)$ are related through the usual definitions. Force acting over area A is obtained by differentiating energy of interaction per unit area with respect to separation, that is $F(h) = -A\,\mathrm{d}V(h)/\mathrm{d}h$. The disjoining pressure is the rate of change of the interaction energy per unit area with separation, that is, $\Pi_D(h) = -\mathrm{d}V(h)/\mathrm{d}h$.

Surfactants: In Solution, at Interfaces and in Colloidal Dispersions. Bob Aveyard. © Bob Aveyard 2019.
Published in 2019 by Oxford University Press. DOI: 10.1093/oso/9780198828600.001.0001

Box 11.1   The early years of colloid science

Michael Faraday, 1791–1867. GL Archive/Alamy Stock Photo.

Faraday's work marks the beginning of modern colloid science (Edwards and Thomas, 2007). In 1857 Faraday presented his work, in the Bakerian Lecture to the Royal Society of London, on the interaction of light with metal (mainly gold) particles which were 'very minute in their dimensions'. He realized that in his preparations the metals were not dissolved in water but were, rather, very small dispersed solid particles. Different particle sizes produced different coloured clear dispersions, the famous ruby coloured gold dispersions containing particles with diameters of around 6 nm. The gold colloidal dispersions were very stable and indeed at least one sample survived up to the Second World War when it was destroyed accidentally by enemy action (Thompson, 2007). However, the term 'colloid' was not coined until 1861 by Thomas Graham and derives from the Greek word for glue: *kolla*. Graham discovered that 'glue-like' substances, such as glue, gelatine, and starch, could be separated from small chemical species, such as sugar or salt, by dialysis. The low rates of diffusion of the 'particles' led Graham to the conclusion that they were large relative to ordinary molecules; he suggested a range of sizes (1–1000 nm 'diameter') which is still used today. Of course colloids range far beyond glue and similar materials.[†] Although colloids generated early scientific interest from eminent scientists including Faraday, Tyndall, Ostwald, van't Hoff, and Nernst, Calvert (2002) has succinctly observed that chemists in general 'could not get excited over glue'. Hodges wrote in 1931 'To some the word 'colloidal' conjures up visions of things indefinite in shape, indefinite in chemical composition and physical properties, fickle in chemical deportment, things infilterable and generally unmanageable'. Scientific progress in the first half of last century was relatively slow because of the complexity and breadth of the kind of systems and behaviour encountered. Happily, modern colloid science is flourishing, but it is very much an interdisciplinary subject, hence colloid *science*. Crucially, it has benefitted greatly from the advent of modern instrumentation, much of which is now laboratory-based and available commercially.

[†] Two types of colloid are often referred to: *lyophilic* and *lyophobic*. Lyophobic (hydrophobic in water) colloids have particles that are solvent 'hating', that is, insoluble, such as metal and mineral particles in water. Micelles and microemulsion droplets are examples of lyophilic (hydrophilic) colloids.

are often grouped together with colloids because aspects of their behaviour resemble the behaviour of colloids. Colloidal systems include, for example, paints, inks, cosmetics, biological membranes, and many food, pharmaceutical, and agrochemical formulations.

As seen in Chapters 9 and 10, small surfactant molecules can form colloidal entities such as micelles and microemulsion droplets. They can also adsorb at particle and droplet interfaces, so that colloid science and the physical chemistry of surfactant systems are inextricably linked. Since colloidal particles are small, the proportion of constituent molecules/ions that are in the surface is large, as is the surface area. It is estimated for example that the surface area of the fat globules in a pint of homogenized milk (which is a colloidal suspension) is about 200m², the floor area of an average house (Fisher, 2003).Understanding colloidal interactions, that is, the way in which surface forces operate between colloid particles, is at the core of colloid science. Usually, there are competing attractive and repulsive forces between particle surfaces in a dispersion or between the surfaces of a thin film. Since the distance dependence of these forces differ however, it is possible to have maxima and minima in, say, plots of $V(h)$ against h. The stabilization (Box 11.2) of dispersions by adsorption of surfactants onto the particles relies on the adsorbed layer of surfactant providing a repulsive energy barrier of some kind to particle contact and aggregation. In the case of emulsions, surfactant can provide a barrier to *contact* of emulsion droplets and also (by a different mechanism) to the *coalescence* of contacting droplets (Chapter 14).

Box 11.2 What is stable?

Colloidal dispersions are often not stable in the *thermodynamic* sense. For example, although liquid droplets may remain dispersed throughout a 'stable' emulsion (Chapter 14), and rarely coalesce on close approach, the free energy of the coalesced state is nonetheless more negative than that of the dispersed state. It is often said therefore that such dispersions are only *kinetically stable*. On the other hand microemulsion droplets and micelles (which are also colloidal dispersion), form spontaneously and are thermodynamically stable. In the case of microemulsions for example, the interfacial tension of droplet interfaces is close to zero, and the favourable (positive) entropy of droplet dispersion in the medium outweighs the work of forming the droplet interfaces (§10.2.2.). The same is not true for emulsions, which for the same volume of dispersed phase, have a lower entropy of dispersion (fewer droplets) and higher droplet interfacial tensions.

The landmark theory of colloid stability in systems with charged particles took account of the attractive interactions between particles due to van der Waals forces and the repulsion resulting from electrostatic (Coulombic) interactions. It was independently devised in the 1940s by Derjaguin and Landau (1941) in Russia and Verwey and Overbeek (1948) in Holland. The theory is now referred to as the Derjaguin–Landau–Verwey–Overbeek (DLVO) theory (Box 11.3). Below, a rudimentary account of the theory is given. Following this the effects of other surface forces, that are not taken into account in the DLVO theory but that can have a crucial influence on the stability of dispersions and thin films, are introduced.

Box 11.3 D, L, V, and O—who were they?

Boris Derjaguin who died in 1994 was a distinguished and prolific physical chemist in the area of colloids and surface phenomena. He collaborated with Lev Landau (died 1968) in the study of colloid stability, and their work was published in 1941. Landau, a physicist and Nobel Laureate, contributed widely to physics. His most famous book is the 10-volume treatise with E.M. Lifshitz (who had been Landau's research student) entitled a *Course of Theoretical Physics*. The eminent Dutch contingent of the DLVO quartet worked independently of the Russians during the Second World War and published their work in 1948. Evert Verwey (died 1981) was a chemist who produced a doctoral thesis on the stability of lyophobic colloids. He subsequently worked at the laboratories of Philips in Eindhoven where he had a distinguished career. Theo Overbeek was the youngest of the quartet and died most recently, in 2007 at the age of 96. He was not only productive in the area of lyophobic colloid stability, but also worked, for example, on irreversible thermodynamics, polyelectrolytes, wetting, and biochemical problems. He was known and respected by many current practitioners of colloid science around the world.

11.2   A simple account of the DLVO theory

11.2.1   Dispersion forces between molecules and between macroscopic bodies

Attractive forces between molecules in a gas were postulated by van der Waals and said to be responsible (together with the finite size of the molecules) for deviations from ideal gas behaviour. The origin and the range of these attractive forces were not however understood until much later, following the quantum-mechanical theory given by Eisenschitz and London (1930). This discussion below is restricted to dispersion forces (often referred to as London-van der Waals forces) between molecules and between particles. Dipole–dipole (Keesom) and dipole–induced dipole (Debye) interactions, present in systems with polar molecules, are not considered here.

The fact that gases such as hydrogen and nitrogen can be liquefied indicates that attractive forces exist between nonpolar molecules. If the molecules are slowed down sufficiently (cooled) and brought near enough to each other (subjected to pressure), the London–van der Waals forces overcome thermal energy and condensation occurs. Dispersion forces can be thought of as resulting from the instantaneous polarization of one molecule by the motions of the outer (dispersion) electrons of the other molecule, and *vice versa*. The energy of attraction U_a falls off as r^{-6} where r is the separation between the nuclei (or centres of mass) of two molecules. The (free) energy U_r at separation r is equal to minus the work of parting the molecules from separation r to infinity.

At very close separation the electron clouds of the molecules begin to repel giving rise to Born repulsion, which rises very steeply with decreasing separation. The total energy of interaction, $U(r)$, can be expressed by, for example, the Lennard-Jones 6-12 potential,

$$U(r) = -Cr^{-6} + Br^{-12} \tag{11.2.1}$$

where C and B are constants for a given material. For the interaction of like molecules, C is given by

$$C = (3/4) \, h\nu\alpha^2/(4\pi\varepsilon_o)^2 \tag{11.2.2}$$

where ε_o is the permittivity of free space, h is Planck's constant, and α the polarizability of the molecules. The larger and more polarizable the molecules, the greater the dispersion forces between them. The quantity ν is a characteristic frequency in the ultraviolet corresponding to the first ionization potential.

In the treatment of the interactions between macroscopic bodies in a vacuum it can be assumed, as an approximation, that the pairwise interaction between any two molecules is independent of the presence of other neighbouring molecules. In the so-called *microscopic approach* the interaction between the bodies is obtained by summing (integrating) the interactions between all molecular pairs between the two bodies. Hamaker (1937) first performed such a summation for interactions between two spheres and between a sphere and a flat wall. For two overlapping equal parallel flat plates with thickness and lateral extent greater than the separation h (Fig. 11.1), the interaction (free) energy *per unit area* is

$$V_A(h) = -A_H/12\pi h^2 \tag{11.2.3}$$

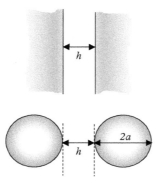

Figure 11.1 *Top: two semi-infinite plane parallel surfaces, separation h. Bottom: two spheres, radius a, separated by distance h.*

where the negative sign indicates attraction. The constant A_H, termed the *Hamaker constant*, is related to the constant C for the interaction between the molecules (11.2.2) by

$$A_H = (3/4)\,\pi^2 n^2 h v \alpha^2 / (4\pi\varepsilon_o)^2 = C\pi^2 n^2 \tag{11.2.4}$$

Here, n is the number of molecules per unit volume in the particle phase.

An approximate expression for the attraction of two equal spheres of radius a in vacuum when $h/a << 1$ is

$$V_A(h) = -A_H a / 12h \tag{11.2.5}$$

The general expression for spheres derived by Hamaker is more complicated but it shows that at small separations $V_A(h)$ varies with $1/h$ (as in (11.2.5)) and at very large separations it varies with h^{-6} (as for molecules). Importantly, the forms of (11.2.3) and (11.2.5) show that the London–van der Waals attraction between macroscopic bodies falls off much more slowly with separation than does the similar interaction between molecules. As will be seen, the range of this interaction is a very important factor in determining the stability of colloidal systems.

Interest usually centres on interactions between particles immersed in a liquid medium rather than in vacuum, and the presence of the medium lowers the inter-particle interactions. This is because the particles interact with the medium as well as with each other and so *effective* values of Hamaker constants must be used. A simple picture is employed to show how such constants can be obtained. Consider a particle of 1 interacting with a particle of 2 across an intervening medium 3. At large enough separation, the particles interact only with the medium, with Hamaker constants A_{13} and A_{23}. At small separations however the particle-medium interactions are replaced by particle–particle interactions (Hamaker constant A_{12}) and medium–medium interactions (Hamaker constant A_{33}). Thus for two particles interacting in the medium the effective Hamaker constant A_{132} is

$$A_{132} = A_{12} + A_{33} - A_{13} - A_{23} \tag{11.2.6}$$

Note that the Hamaker constants on the right-hand side of (11.2.6) relate to interactions in a vacuum. If it is assumed that the interactions between unlike phases is the geometric mean of the interactions of the 2 phases with themselves, then

$$A_{12} = (A_{11}A_{22})^{0.5}; \quad A_{13} = (A_{11}A_{33})^{0.5}; \quad A_{23} = (A_{22}A_{33})^{0.5} \tag{11.2.7}$$

Then, from (11.2.6) and (11.2.7), the expression for A_{132} is

$$A_{132} = \left(A_{11}^{0.5} - A_{33}^{0.5}\right)\left(A_{22}^{0.5} - A_{33}^{0.5}\right) \tag{11.2.8}$$

If the particles are of the same material, say 1, (11.2.8) gives

$$A_{131} = \left(A_{11}{}^{0.5} - A_{33}{}^{0.5}\right)^2 \tag{11.2.9}$$

From (11.2.8) it can be seen that, depending on the relative magnitudes of A_{11}, A_{22}, and A_{33}, A_{132} can be positive (attraction between particles 1 and 2) or negative (repulsion between particles). As Israelachvili (1974) has put it, Archimedes' principle also applies to van der Waals forces. In the Earth's gravitational field, a cork in air falls towards the Earth, but the cork when immersed in water rises, being apparently repelled by the Earth. However, (11.2.9) indicates that *like* particles in a medium always attract, that is, A_{131} must always be positive since the term on the right of (11.2.9) is squared. Repulsion between unlike particles occurs if A_{33} lies in *between* A_{11} and A_{22}. Otherwise (i.e. if A_{33} is greater than or less than each of A_{11} and A_{12}), the unlike particles attract in medium 3.

Pairwise additivity of interactions between molecules was assumed in the microscopic approach discussed above. However, amongst other problems arising from this assumption, polarizabilities of molecules are modified by the presence of other molecules in the vicinity, particularly in condensed phases. The *macroscopic* theory of Lifschitz and co-workers (1956, 1961) circumvents problems arising from the assumption of additivity (Israelachvili, 2011). Particles and medium are treated as continuous phases, and Hamaker constants are obtained in terms of the dielectric and optical properties of the media as a function of frequency. As an example, for dielectric materials, that is, non-metals, the non-retarded Hamaker constant (see *retardation effect* below) for the interaction between particles of 1 and 2 in medium 3 is given approximately by (Israelachvili, 2011)

$$A_{132} \approx \frac{3}{4}kT\left(\frac{\varepsilon_1 - \varepsilon_3}{\varepsilon_1 + \varepsilon_3}\right)\left(\frac{\varepsilon_2 - \varepsilon_3}{\varepsilon_2 + \varepsilon_3}\right)$$
$$+ \frac{3h\nu_e}{8\sqrt{2}} \frac{\left(n_1^2 - n_3^2\right)\left(n_2^2 - n_3^2\right)}{\left(n_1^2 + n_3^2\right)^{1/2}\left(n_2^2 + n_3^2\right)^{1/2}\left\{\left(n_1^2 + n_3^2\right)^{1/2} + \left(n_2^2 + n_3^2\right)^{1/2}\right\}} \tag{11.2.10}$$

Here n and ε are, respectively, the refractive indices (in the visible region) and static relative permittivities of the subscripted phases, h is Planck's constant, and ν_e is the main electronic absorption frequency in the UV which is approximately 3×10^{15} s^{-1}. For two like phases, 1, interacting across medium 3 (11.2.10) becomes

$$A_{131} \approx \frac{3}{4}kT\left(\frac{\varepsilon_1 - \varepsilon_3}{\varepsilon_1 + \varepsilon_3}\right)^2 + \frac{3h\nu_e}{16\sqrt{2}} \frac{\left(n_1^2 - n_3^2\right)^2}{\left(n_1^2 + n_3^2\right)^{3/2}} \tag{11.2.11}$$

It is seen from (11.2.11) that the Hamaker constant for the interaction of 1 with 1 over medium 3 is the same as that for interaction of 3 with 3 over medium 1. Some values of Hamaker constants, calculated from (11.2.11) for interaction of materials in a vacuum

Table 11.1 *Non-retarded Hamaker constants for interaction over vacuum obtained from (11.2.11) with $\varepsilon_3 = 1$ and $n_3 = 1$; values taken from Israelachvili (2011).*

Material	Relative permittivity	Refractive index	$\nu_e/10^{15}$ s^{-1}	$A_H/10^{-20}$ J for interaction over vacuum
Water	80	1.333	3.0	3.7
n-Dodecane	2.01	1.411	3.0	5.0
n-Hexadecane	2.05	1.423	2.9	5.1
Polystyrene	2.55	1.557	2.3	6.5
PTFE	2.1	1.359	2.9	3.8
Fused quartz	3.8	1.448	3.2	6.3
Mica	7.0	1.60	3.0	10
Alumina (Al$_2$O$_3$)	11.6	1.75	3.0 (estimated)	14

Table 11.2 *Non-retarded Hamaker constants for interactions between bodies (1 and 2) immersed in a medium (3) obtained using (11.2.10)*

Phase 1	Medium 3	Phase 3	$A_H/10^{-20}$ J from (11.2.10)	$A_H/10^{-20}$ J from (11.2.8) or (11.2.9)
n-Hexadecane	Water	n-Hexadecane	0.49	0.11
Water	n-Dodecane	Air	0.81	0.70
n-Dodecane	Water	Air	−0.47	−0.60
Air	Water	PTFE	0.04	−0.048
Air	Water	Fused quartz	−0.89	−1.13

(air), are given in Table 11.1. Using the Lifschitz theory, the Hamaker constants for interactions between bodies, like or unlike, in a medium, arise naturally from (11.2.10). Some values are given in Table 11.2 and compared with those obtained using (11.2.8) or (11.2.9) together with the Hamaker constants for phases interacting in a vacuum given in Table 11.1.

Constants obtained from (11.2.10) and (11.2.11) generally agree well with experimentally derived values; it can be seen from Table 11.2 however that Hamaker constants from (11.2.8) or (11.2.9) are only in moderate agreement with those from (11.2.10) and (11.2.11) and hence with those from experiment. Although in the microscopic and macroscopic methods Hamaker constants are obtained by different routes, nonetheless the expressions for the dependence of interaction energies on separation (e.g. that in (11.2.5)) remain the same in the two approaches. It is noted that the Hamaker constants so far referred to are the so-called *non-retarded* Hamaker constants; the retardation effect is discussed below. The relevance of Hamaker constants to wetting is discussed in §16.2.2.

Effect of adsorbed layers

In systems with surfactants, solids in solution will usually be coated with adsorbed surfactant layers, which can be expected to modify the London–van der Waals forces between surfaces. The symmetrical case, depicted in Fig. 11.2, has plane parallel solid surfaces of 1 coated with like surfactants layers (2) of thickness t, with the outer parallel surfaces separated by distance h in medium 3 (surfactant solution). An approximate expression for the energy of interaction between unit areas of the surfaces is (Ninham and Parsegian, 1970)

$$V_A(h) = -\frac{1}{12\pi}\left(\frac{A_{232}}{h^2} - \frac{2A_{123}}{(h+t)^2} + \frac{A_{121}}{(h+2t)^2}\right) \tag{11.2.12}$$

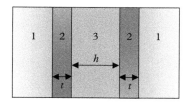

Figure 11.2 *Interaction of plane parallel solids 1 coated with layers of surfactant 2, thickness t, over a layers 3, thickness h.*

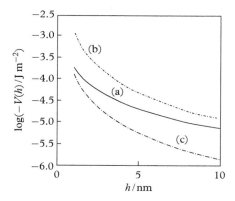

Figure 11.3 *(a) Interaction energy between alumina coated with surfactant layers (thickness t = 2 nm) over water, calculated according to (11.2.12). (b) Interaction energy in absence of surfactant layers and (c) interaction of surfactant with surfactant over water. Hamaker constants in units of 10^{-20} J were taken as (using numbering as in Fig. 11.2): $A_{131} = 4.4, A_{232} = 0.49, A_{123} = -0.53,$ and $A_{121} = 2.71.$*

In Fig. 11.3 (curve (a)) it is shown how the interaction energy between alumina sheets (phase 1) coated with surfactant layers (phase 2), interacting over water (phase 3) varies with separation h. In the absence of the layers (i.e. for alumina with alumina interacting over water) curve (b) is given. Curve (c) is for the interaction of the surfactant phase over water. The Hamaker constants used are those given in the legend to the figure and it has been supposed that the thickness of the layers is 2nm. The Hamaker constant for the surfactant is approximated by that for hexadecane. It can be seen that for small separations of the surfactant-coated surfaces, the interactions are close to those for the surfactant layers alone, whereas at high separations the interactions approach those between uncoated alumina surfaces.

Retardation effect

The interactions so far discussed relate to particles that are fairly close together, and as seen the dispersion forces arise from the polarization of one molecule by the motions of the outer (dispersion) electrons of the other molecule. For the interaction of two isolated molecules or atoms the time taken for the electromagnetic wave, produced by the moving electron on one molecule, to reach the second molecule and return to the first is very small compared to the period of the fluctuating dipoles, that is, the dipoles are induced instantaneously and the attracting dipoles 'flicker' in unison. The dispersion forces between, say, colloidal particles, however, have a much longer range than those between isolated molecules (Box 11.4). As the particle separation increases, the time taken for the electromagnetic waves from one particle (travelling at the speed of light) to reach the other particle and return to the first is such that the electronic state of molecules in the first particle have changed, and so the attraction begins to fall off more rapidly with distance. This effect is termed *retardation*, and the forces are referred to as *retarded* forces. For example, for the case of two parallel plates discussed above, the interaction energy, instead of falling with $1/h^2$, falls with $1/h^3$ in the retarded region. It follows from this that Hamaker constants are not strictly constant but vary with separation of bodies. As Israelachvili (2011) points out, for molecules in a vacuum the onset of retardation starts at separations of about 5 nm. For molecules in a medium however retardation occurs at smaller separations because the speed of light is reduced. Since dispersion forces (Box 11.5) between macroscopic bodies are much longer range, retardation of forces between particles immersed in liquid can, in some contexts, become important. In §16.2.2 the relevance of retardation to the wetting of water by liquid alkanes and of alkanes by water is discussed.

Box 11.4   Casimir forces

Overbeek noted, in applying the DLVO theory (as originally devised) to experimental results, that there were discrepancies between experiment and theory consistent with dispersion forces between colloidal particles falling off more rapidly with separation at relatively large separations than is the case at closer distances. He discussed this with Hendrik Casimir at the Philips Research Laboratories in Eindhoven, Holland, and this stimulated Casimir

Box 11.4 *Continued*

and Polder (1948) to carry out a theoretical study on what colloid scientists often refer to as *retarded* London–van der Waals forces. Casimir considered the attraction between two perfectly conducting plates (mirrors) in a vacuum. The so-called Casimir force arises as a result of vacuum electromagnetic fluctuations between the plates and is a quantum mechanical effect (Lambrecht, 2002). Lifshitz (1956) and co-workers later extended this work to include a wide range of real materials; the theory describes interactions arising from quantum fluctuations at all separations of surfaces, and so includes the van der Waals short-range forces and the longer range (retarded) Casimir forces. Lamoreaux (2007) discusses more recent developments of Casimir's work, which has proved to be of interest in areas beyond colloid science, especially in recent years (see e.g. Ball, 2007). Casimir (1909–2000) made other wide-ranging contributions to theoretical physics. He gained his PhD at Leiden University in 1931 in the area of quantum mechanics. He spent time with Niels Bohr in Copenhagen and after receiving his PhD he became an assistant to Wolfgang Pauli in Zurich. In 1938 he took up an academic post at the University of Leiden, later joining Philips Research Laboratories in 1942 where he continued to carry out his research investigations. He became a member of the board of directors of Philips in 1956, retiring in 1972.

Box 11.5 Feel the force

The magnitude of the force acting between two colloidal particles is of course very small. It is possible however get a good feel for what van der Waals forces are capable of by noting the cumulative effect of many millions of such pairwise interactions. The Gecko lizard can happily walk up walls and across ceilings. Geckos have about 2 million keratin hairs (*setae*) on each toe, each of which subdivides into hundreds or thousands of finer structures referred to as *spatulae*, of which the Gecko has over 1 billion. The total van der Waals dispersion forces between a surface and the closely situated spatulae are easily sufficient to hold a Gecko to a ceiling. Not surprisingly, the Gecko's foot is a good model for fabricating adhesive surfaces. For example, researchers at BAE Systems in the UK have produced a super-adhesive material coated with fine polymer hairs, and it has been claimed that 1 m^2 can suspend an average family car.

11.2.2 The electrical double layer at charged surfaces

The electrical double layer that exists at the interface between a charged solid and an aqueous solution has already been described (see §7.3.1 and Fig. 5.11). Here the focus will be on the way in which the electrical double layers on like surfaces interact repulsively as the surfaces approach in aqueous electrolyte solution. The electrical potential, ψ, at any point in a solution close to a charged surface will be of interest. Suppose two charges, q_1 and q_2, in solution are brought from infinity to a separation d; the force F between the charges is given by Coulomb's law as

$$F = q_1 q_2 / 4\pi\varepsilon\, d^2 \qquad (11.2.13)$$

where ε is the permittivity of the medium (solution). The electrical work, ΔG_{el}, done in this process is

$$\Delta G_{el} = -\int_{\infty}^{d} F \, dh = q_1 q_2 / 4\pi\varepsilon d \qquad (11.2.14)$$

Suppose now that q_2 is a unit charge, then the work done is $q_1/4\pi\varepsilon d$, which is equivalent to the electrical potential ψ at distance d from charge q_1.

First consider the way in which ψ falls off with distance from a charged surface into a solution of symmetrical electrolyte with ions of charge number z. The approach described below is referred to as the *Gouy–Chapman* theory. A plane, positively charged semi-infinite surface in electrolyte solution is illustrated in Fig. 11.4. For simplicity it is supposed that there are no anions specifically or non-specifically adsorbed in the inner or outer Helmholtz planes (see Fig. 7.1), that is, the solution side of the interface consists only of the diffuse part of the double layer resulting from a competition between electrical and thermal effects on the ions (assumed to be point charges). Hence, the concentration (number per unit volume) $n^-(x)$ and $n^+(x)$ of anions and cations respectively at a normal distance x from the surface, at which the potential is $\psi(x)$, is given by Boltzmann's law as

$$n^-(x) = n^o \exp\left(\frac{+ze\psi(x)}{kT}\right);$$

$$n^+(x) = n^o \exp\left(\frac{-ze\psi(x)}{kT}\right) \qquad (11.2.15)$$

in which n^o is the (equal) concentration of anions and cations in bulk solution, and e is the elementary (proton) charge. In (11.2.15) $\psi(x)$ is the electrical potential *relative* to that in bulk solution (often, as here, arbitrarily taken to be zero).

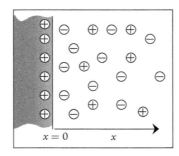

Figure 11.4 *A plane, semi-infinite solid surface carrying a net positive charge immersed in solution containing univalent positive and negative ions. The system of charges constitutes a diffuse double layer.*

The net volume charge density $\rho(x)$ at a point a normal distance x from the surface is[2]

$$\rho(x) = ze\left(n^+(x) - n^-(x)\right) = zen^o\left[\exp\left(\frac{-ze\psi(x)}{kT}\right) - \exp\left(\frac{+ze\psi(x)}{kT}\right)\right]$$

$$= -2zen^o\sinh\left(\frac{ze\psi(x)}{kT}\right) \tag{11.2.16}$$

The relationship between $\psi(x)$ and $\rho(x)$ is given by the Poisson equation, which for a single dimension, in this case in the direction normal to the charged surface, is

$$\frac{\partial^2\psi(x)}{\partial x^2} = -\frac{\rho(x)}{\varepsilon} \tag{11.2.17}$$

From (11.2.16) and (11.2.17),

$$\frac{\partial^2\psi(x)}{\partial x^2} = \frac{2zen^o}{\varepsilon}\sinh\left(\frac{ze\psi}{kT}\right) \tag{11.2.18}$$

which is the Poisson–Boltzmann equation. Using the boundary conditions $\psi(x) = \psi(0)$ at $x = 0$ and $\psi(x) = 0$ and $\partial\psi(x)/\partial x = 0$ at $x = \infty$, (11.2.18) can be solved to give

$$\psi = \frac{2kT}{ze}\ln\left(\frac{1 + B\exp[-\kappa x]}{1 - B\exp[-\kappa x]}\right) \tag{11.2.19}$$

in which B and κ are

$$B = \frac{\exp[ze\psi(0)/2kT] - 1}{\exp[ze\psi(0)/2kT] + 1} = \tanh\left(\frac{ze\psi(0)}{4kT}\right);$$

$$\kappa = \left(\frac{2z^2e^2n^o}{\varepsilon kT}\right)^{0.5} \tag{11.2.20}$$

Plots of $\psi(x)$ versus x for $\psi(0) = 50$ mV, obtained using (11.2.19) for a range of electrolyte concentrations, are shown in Fig. 11.5. The potential falls more rapidly from the surface the higher the electrolyte concentration. This effect is termed *compression* of the double layer. In effect, the surface charge is neutralized over a shorter distance as the concentration of electrolyte increases.

For $ze\psi \ll kT$, the exponential terms in (11.2.19) can be approximated by $\exp(ze\psi/2kT) \sim 1 + (ze\psi/2kT)$ and (11.12.19) then becomes

$$\psi(x) = \psi(0)\exp(-\kappa x) \tag{11.2.21}$$

---

[2] The hyperbolic sine function $\sinh(x) = 0.5(\exp(x) - \exp(-x))$.

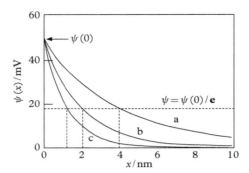

Figure 11.5 *Potential–distance curves calculated using (11.2.19) with $\psi(0) = 50$ mV and c for 1:1 electrolyte (a) 5 mM, (b) = 20 mM, and (c) 50 mM. The horizontal dashed line is drawn for $\psi(x) = \psi(0)/e$ (see text). The vertical dashed lines represent double layer thicknesses ($1/\kappa$) for the various values of c.*

This shows that for low potentials, $\psi(x)$ decays exponentially with distance from the surface. It is also apparent from (11.2.21) that the reciprocal of κ (defined in (11.2.20)) is the value of x for which $\psi(x) = \psi(0)/e$, where e is the mathematical constant (2.718...); $1/\kappa$ is termed the *double layer thickness*.[3] The horizontal dashed line in Fig. 11.5 is drawn at $\psi(x) = \psi(0)/e$ and the vertical lines represent the double layer thicknesses for the concentrations given in the legend. For, say, a positively charged surface, anions are attracted to the surface and cations repelled, and the ionic concentration profiles normal to the surface can be calculated using (11.2.16) together with (11.2.19). Such profiles, for a system in which the 1:1 electrolyte concentration $c^0 = 0.05$ mol dm^{-3} and $\psi(0) = 50$ mV, are shown in Fig. 11.6.

So far attention has been centred on the electrical potential at a charged interface, but obviously the electrical potential arises from the presence of the surface charge, and as might be expected, the surface charge and potential, $\psi(0)$, are related. The charge is neutralized by the net charge in the diffuse part of the double layer and so, noting (11.2.17),

$$\sigma(0) = -\int_0^\infty \rho(x)\mathrm{d}x = -\int_0^\infty \varepsilon \frac{\partial^2 \psi(x)}{\partial x^2}\mathrm{d}x \qquad (11.2.22)$$

where $\sigma(0)$ is the charge density (charges per unit area) at the surface and ρ is the net volume charge density in the diffuse part of the double layer. Using the boundary conditions given earlier ($\psi(x) = \psi(0)$ at $x = 0$ and $\psi(x) = 0$ and $\partial \psi(x)/\partial x = 0$ at $x = \infty$) it can be shown that $\sigma(0)$ and $\psi(0)$ are related (in the case of a single symmetrical electrolyte) through

---

[3]   The double layer thickness $1/\kappa$ is the same as the parameter in the Debye–Hückel theory of electrolyte solutions, where it is referred to as the radius of the ion atmosphere.

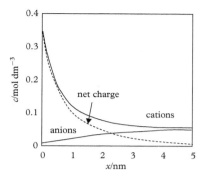

Figure 11.6 *Concentrations of anions and cations in the diffuse part of the double layer at a negatively charged surface at 298 K calculated using (11.2.19) and (11.2.16). The dashed curve represents the net positive charge. The potential at the surface $\psi(0) = 50\ mV$ and the concentration of 1:1 electrolyte in bulk solution is 50 mM.*

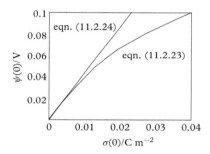

Figure 11.7 *Potential at surface as a function of surface charge density at 298 K in the presence of 10 mM 1:1 electrolyte, according to (11.2.23) and (11.2.24). The two equations give very similar results for $\psi(0)$ up to about 30 mV.*

$$\sigma(0) = \left(8n^{o}\varepsilon kT\right)^{0.5} \sinh\left(\frac{ze\psi(0)}{2kT}\right) \qquad (11.2.23)$$

which for $ze\psi \ll kT$ becomes

$$\sigma(0) = \left(\frac{2\varepsilon n^{o}}{kT}\right)^{0.5} ze\psi(o) = \varepsilon\kappa\psi(0) \qquad (11.2.24)$$

Plots of potential at a charged surface versus the surface charge density, according to (11.2.23) and (11.2.24), are shown in Fig. 11.7 for an electrolyte concentration of 10 mM. Equation (11.2.24) gives a linear relationship between $\psi(0)$ and $\sigma(0)$, and results given by the two equations are very similar up to potentials of around 30 mV. Beyond this however the line generated by (11.2.23) becomes more curved and falls progressively

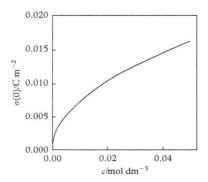

Figure 11.8 *Dependence of $\sigma(0)$, calculated from (11.2.23), for 1:1 electrolyte concentration c with $\psi(0) = 30$ mV.*

below the line from (11.2.24). The way in which $\sigma(0)$ (calculated using (11.2.23)) varies with electrolyte concentration for a fixed potential is shown in Fig. 11.8.

To summarize, the electrical potential in the diffuse double layer at a charged surface falls off with distance from the surface, the decay becoming exponential for low potentials. The double layer thickness $1/\kappa$ is the normal distance from the surface at which $\psi(x) = \psi(0)/e$, according to (11.2.21); the double layer is compressed by the addition of electrolyte. The distribution of anions and cations normal to the surface are predicted by the Boltzmann equation. Equations have been presented that relate $\sigma(0)$ and $\psi(0)$ showing that, at a fixed electrolyte concentration the potential is determined by the surface charge density, as shown in Fig. 11.7. At a fixed potential, $\sigma(0)$ rises with electrolyte concentration as illustrated in Fig. 11.8, whereas for a fixed $\sigma(0)$, $\psi(0)$ falls with electrolyte concentration as seen from (11.2.23) and (11.2.24).

The above account of the electrical double layer is presented to demonstrate some important behaviour of charged surfaces in solution, but it is rudimentary and a number of limiting assumptions are made in the treatment. It has been assumed that the permittivity of the medium (water) is the same in the double layer as in bulk solution, but the electric field strength in the double layer, $d\psi(x)/dx$, can be high; it varies with x and is likely to modify ε significantly. It can be shown, from the ideas presented above, that $d\psi(x)/dx$ is given by

$$\frac{\partial \psi(x)}{\partial x} = -\left(\frac{8n^0 kT}{\varepsilon}\right)^{0.5} \sinh\left(\frac{ze\psi(x)}{2kT}\right) \qquad (11.2.25)$$

So, for $\psi(x) = 100$ mV and $c = 0.01$ mol dm^{-3}, for example, the gradient is $\sim 2 \times 10^6$ V m^{-1}. A further assumption made, which can give rise to serious errors, is that ions behave as point charges, which they obviously are not. The number of point charges that can be accommodated in a given volume is infinite, and consequently the Boltzmann expression (11.2.16) can predict unreasonably high local ion concentrations. In addition, because point charges are assumed, the lower limit in the integral in (11.2.22) is set at zero, that is, the charge can come infinitely close to the surface. The integral limit is however more reasonably taken to be an ionic radius. Another limitation of the Gouy–Chapman treatment is the assumption that, although the surface charge arises from discreet ions, it is smeared out along the surface. A number of these limitations have been addressed, but for present purposes the simple approach gives the insight needed here.

Up to now no account has been taken of adsorbed ions in either the inner or outer Helmholtz planes; only the ions in the diffuse part of the double layer have been considered. As discussed in §7.3.1, if ions (say surfactant ions) are specifically adsorbed (as opposed to being adsorbed solely by electrostatic forces), they are likely to be only partially solvated or unsolvated and their centres occupy the inner Helmholtz plane (i.H.p.). If hydrated ions are present then their centres lie in the outer Helmholtz plane (Stern plane). In the *Stern model* specific adsorption is supposed to occur in the Stern *layer*, that is, the layer between the Stern plane and the surface. The potential ψ obviously changes through the Stern layer,[4] and then changes further in the diffuse part of the double layer as discussed above. Specific adsorption of ionic surfactant can lead to the charging of an originally neutral surface or to a change in charge or even charge reversal (super-equivalent adsorption) at an originally charged surface. Take as an example the adsorption of a cationic surfactant at a negatively charged solid surface in aqueous solution. At very low surfactant concentration Coulombic forces obviously favour adsorption. As the adsorbed monolayer becomes more concentrated however, and after the original surface charge has been neutralized, lateral attractive interactions between the adsorbed surfactant chains provide a strong driving force for further adsorption, leading to a build-up of positive charge. For this latter situation the change in potential can be depicted as in Fig. 11.9. The Stern–Langmuir isotherm equation covering adsorption in the Stern layer has been referred to in §7.4.

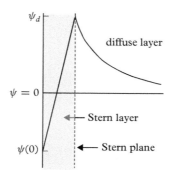

Figure 11.9 *Effect on potential, ψ, of the adsorption of a cationic surfactant in the Stern layer at the interface between a negatively charged solid and electrolyte solution. The potential changes sign when the original surface charge has just been neutralized by surfactant adsorption.*

---

[4] Taking the Stern layer to be a capacitor, thickness δ and with uniform permittivity ε_δ, the change in potential $\Delta\psi_{St}$ across the layer is $\sigma\delta/\varepsilon_\delta$ where σ is the charge density at the surface.

11.2.3   Interaction between like double layers

Here as before, semi-infinite plane surfaces are considered for simplicity but it is noted that equations for spherically curved surfaces, such as might be appropriate for solid particles or liquid droplets, differ in some detail from those for plane surfaces. Nevertheless, the general picture described below remains applicable to particles also. It will be assumed that the repulsion between charged surfaces arises only from the interaction of the diffuse parts of the double layers, that is, that Stern layers, if present, are not involved.

When two like double layers approach, at some distance a mutual repulsion occurs and becomes progressively larger as the separation h falls. As an approximation it is supposed that the potentials are additive in the region of overlap, as illustrated in Fig. 11.10. The elevated electrical potential indicates an increase in the electrical contribution to the free energy of the system, corresponding to repulsion. One way of viewing this repulsion (Everett, 1988) is that the diffuse part of the double layer screens the charge at the surface, and at large separations the two surfaces appear effectively neutral. At smaller separations however, in the overlap region, the charged surfaces are only partially screened and hence mutually repel.

Here, an approach is used in which repulsion between the surfaces is considered to arise from an osmotic pressure forcing the surfaces apart (Hunter, 1993; Israelachvili, 2011). Suppose an uncharged surface in water undergoes some dissociation, producing counterions. There will be an attractive electrostatic attraction between the surface and counterions but, nonetheless, the counterions leave the surface spontaneously and enter the solution to contribute to the formation of an equilibrium diffuse double layer. The reason is that by so doing they gain configurational entropy, and there is an osmotic pressure associated with this arising from the mutual repulsion between counterions and counterions and surface. When two double layers overlap there is a tendency of the increasing osmotic pressure to force counterions back onto the surface. Although the electrostatic part of the energy change when the ions go to the surface is attractive, this is outweighed by the repulsive entropic/osmotic effect.

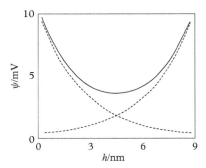

Figure 11.10 *Potentials in the overlap region of 2 plane double layers. Dashed curves for the individual double layers are generated using (11.2.21) with $c^0 = 10$ mM and $\psi(0) = 10$ mV. The net potential is shown by the full curve. It is assumed that the potentials are additive.*

In dilute solution osmotic pressure Π is related to the concentration c of a species in solution by

$$\Pi = nkT = cRT \qquad (11.2.26)$$

which is analogous to the ideal gas equation. The excess concentration of ions (over that in bulk) between the plates, for a symmetrical electrolyte, is $(n^+ + n^- - 2n^o)$ so the (excess) osmotic pressure Π_e is

$$\Pi_e = kT\left(n^+ + n^- - 2n^o\right) \qquad (11.2.27)$$

The total excess number of ions in the diffuse part of a double layer is shown as a function of x in Fig. 11.11 for conditions given in the legend. Note that, although (for a positively charged surface) $n^+ < n^o$, the total number of ions $n^+ + n^- > 2n^o$. With reference to Fig. 11.10, the relative osmotic pressure, Π_m, at the mid-point $h = H/2$ between equal parallel plates separated by distance H can be expressed

$$\Pi_m = kT\left(n^+ + n^- - 2n^o\right)_m \qquad (11.2.28)$$

which, noting (11.2.15), becomes[5]

$$\Pi_m = kTn^o\left[\exp\left(\frac{ze\psi_m}{kT}\right) + \exp\left(-\frac{ze\psi_m}{kT}\right) - 2n^o\right]$$
$$= 2n^okT\left(\cosh\left(\frac{ze\psi_m}{kT}\right) - 1\right) \qquad (11.2.29)$$

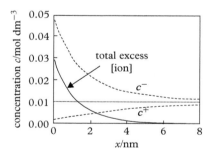

Figure 11.11 *Ion concentrations in the diffuse part of an aqueous plane double layer at a positively charged surface; electrolyte concentration $c^o = 10$ mM and $\psi(0) = 40$ mV. Top dashed line is for c^- and the bottom dashed for c^+. The full line is the total excess ion concentration in the double layer.*

---

[5] Note that $\cosh x = 0.5(\exp(x) + \exp(-x))$.

For small potentials the cosh term can be expanded,[6] and then

$$\Pi_m \approx \frac{\kappa^2 \varepsilon}{2} \psi_m^2 \tag{11.2.30}$$

When the potentials are additive in the overlay region ψ_m is $2\psi_{h=H/2}$.

The potential energy of interaction per unit area is obtained by integration of the osmotic pressure Π_m with respect to separation:

$$V_R(h) = -\int_\infty^H \Pi_m \mathrm{d}h \tag{11.2.31}$$

Results of the integration depend on the assumed conditions. For small enough overlap such that potentials are additive it can be shown that (Hunter, 1993), where B and κ are defined in (11.2.20),

$$V_R(h) = \frac{64n^o kTB^2}{\kappa} \exp\left(-\kappa H\right) = \frac{64n^o kT}{\kappa} \tanh^2\left(\frac{ze\psi(0)}{4kT}\right) \exp\left(-\kappa H\right) \tag{11.2.32}$$

In the above treatment, which is appropriate for relatively small overlap of potentials, it is assumed that the surface charge density $\sigma(0)$ and surface potential $\psi(0)$ (related through (11.2.24) for low potentials) are those for the isolated surfaces and remain unchanged during the approach of the surfaces. As the surfaces approach closely however there is the possibility that some of the counterions adsorb at the surfaces, hence reducing the net surface charge and lowering the repulsion. The repulsive interaction at small separations lies somewhere between (a) an upper limit corresponding to constant surface charge and (b) a lower limit for constant surface potential (equal to that of an isolated surface).

11.2.4 Net interaction energies arising from van der Waals attraction and electrical repulsion between plane surfaces

Expressions for the energies of attraction, $V_A(h)$, arising from van der Waals forces, and electrical repulsion $V_R(h)$ acting between plane surfaces have been presented above. The DLVO theory of colloid stability is based on the net effects of these interactions ($V_T(h)$) and how it depends on separation h between surfaces and on electrolyte concentration. Below, some of the features of the potential energy curves ($V_T(h)$ vs h) are explored. The effect of adding electrolyte on the form of potential energy curves is illustrated in Fig. 11.12. The curves have been generated using (11.2.32) for the repulsion, with $\psi(0) = 30$ mV, and (11.2.3) for the attraction, with a Hamaker constant $= 4.2 \times 10^{-20}$ J. Addition

---

[6] For low x ($= ze\psi(0)/kT$), $\cosh x \sim 1 + x^2/2$. For $\psi(0) = 40$ mV $\cosh x$ and $1 + x^2/2$ differ by about 10%; for $\psi(0) = 30$ mV the difference is about 5%.

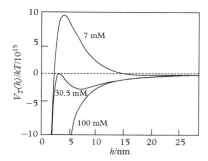

Figure 11.12 *$V_T(h)/kT$ vs h for the approach of two flat parallel charged plates in aqueous 1:1 electrolyte solution at 298 K. $V_A(h)$ has been calculated using (11.2.3) with $A_H = 4.2 \times 10^{-20}$ J and $V_R(h)$ using (11.2.32) with $\psi(0) = 30$ mV. Salt concentrations c^0 are as indicated on the curves.*

of electrolyte compresses the double layers at the surfaces and so for a given separation, the overlap potential ψ_m (see Fig. 11.10) falls, as therefore does the repulsion (see Box 11.6). The top curve (for $c^0 = 7$ mM) shows a marked maximum in $V_T(h)/kT$ of about 10^{16} m^{-2} at a separation of around 4.5 nm. At higher separations a shallow *secondary* minimum is evident (since the curves approach zero at high separations).

Box 11.6   The Schultze–Hardy rule

Lyophobic colloid stability (with respect to coagulation) is very sensitive to the presence of electrolyte as seen, and it is the *counterion charge* that is largely important in destabilizing colloids. It is found for example that divalent and trivalent cations are, respectively, about 60 and 700 times more efficient at causing precipitation of a negatively charged colloid than is a univalent cation. This kind of experimental information is embodied in the empirical *Schultze–Hardy rule*. The rule has been given some rational basis by supposing that the electrolyte concentration that just causes precipitation of the colloid (the critical coagulation concentration, *ccc*) can be taken to be the concentration that generates the curve for which simultaneously $dV_T(h)/dh = 0$ and $V_T(h) = 0$. In addition to the charge on an ion, however, its specific nature can also be very important and, for example, not all univalent cations act equivalently in destabilizing a negatively charged colloid.

When the plates approach very closely, they must of course begin to repel due to electron orbit overlapping (*Born* repulsion). Therefore at very small separations the potential energy curves rise very steeply (not shown in Figure 11.12), which means there is a *primary* minimum at small separations (also not shown). For higher salt concentration (30.5 mM) the maximum occurs at zero energy, and the secondary minimum is much more pronounced, occurring at a smaller separation (at about $h = 8$ nm). At $c^0 = 100$ mM the maximum has completely disappeared. Maxima in such curves, when they appear, are responsible for colloid stability if they are sufficiently high relative to thermal energy. This can be seen a little more clearly if particles rather than 'infinite' flat plates are envisaged. The units of $V_T(h)/kT$ for interacting plates are m^{-2}. Suppose that particles are cubic, of the same material as the plates, with sides 60 nm. If interactions occur in essentially the same way between the particle faces as between the surfaces of the 'infinite' plates, then for $V_T(h)/kT$ of 10^{16} m^{-2} for the plates (the maximum in the top curve of Fig. 11.12) the corresponding energy maximum for the particles is 35, that is, the energy of repulsion between two particles is $35kT$. The rate at which colliding particles can surmount such a barrier and adhere is expected to be low. On the same basis of calculation the secondary minimum in the middle curve in Fig. 11.12 corresponds to an *attraction* between a pair of particles of approximately $10kT$. The existence of a minimum of this magnitude or greater can lead to the formation of reversible flocs.

Hamaker constants A_H, and hence attraction between surfaces, are essentially unaffected by low concentrations of salt, and have been ignored above. Differences in A_H for different interacting materials however can affect the shape of the energy curves substantially. Energy curves for a range of values of A_H, generated by assuming constant salt concentration ($c^0 = 30.5$mM) and potential ($\psi(0) = 35$mV), are shown in Fig. 11.13. The series of curves are, from top to bottom, for increasing values of A_H as indicated in the legend. The Hamaker constants chosen are those for, in ascending order of A_H, quartz, polystyrene, mica, alumina, and rutile.

For some systems (e.g. mica surfaces in dilute aqueous KNO_3 and $Ca(NO_3)_2$) the DLVO theory predicts force-distance curves which are in remarkable agreement with direct force measurements (Israelachvili, 2011), and certainly the theory gives considerable insight into the origins of the forces operating between charged surfaces. It is important to stress however that although ionic charge is obviously important in the ions which affect the interactions between charged interfaces, it has long been known that the specific nature of the ions can also be very significant in a range of contexts. Not all the important properties of an ion are contained in its charge. For example the effectiveness within a series of, say, univalent cations in destabilizing a given negatively

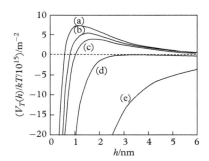

Figure 11.13 *Effect of the Hamaker constant on $V_T(h)/kT$ versus h curves for the approach of flat parallel plates in aqueous 1:1 electrolyte. $\psi(0)$ is taken as 35 mV and $c^0 = 30.5$ mM. The Hamaker constants for the curves (a) through to (e) are, in units of 10^{-20} J, 0.83, 1.4, 2.0, 5.3, and 26.*

charged lyophobic colloid, differs. The DLVO theory did not attempt to take account of specific ion effects. Ions experience van der Waals interactions in addition to Coulombic forces when close to an interface, especially the larger more polarizable species. One would not therefore expect a chloride ion to behave exactly like the larger and more polarizable bromide ion in causing the precipitation of a positively charged colloid. The specific nature of ions and the effects they exert in situations where charge is important, were long ago recognized by Hofmeister, and are embodied in the Hofmeister series (Boström et al., 2006).

11.3 Some other important forces acting between surfaces

The DLVO theory was concerned with lyophobic colloid stability and, as seen, the only forces considered were those arising from van der Waals and electrical interactions. Certainly, van der Waals forces are ubiquitous and electrical interactions between surfaces very common. However in practice, stabilization of a dispersion is not usually achieved by charge alone. In the first place to achieve and/or control surface charge in non-aqueous media can be difficult. Secondly, in aqueous media small amounts of multivalent counterions such as are found in hard water can have a marked effect on the stability of a dispersion. To understand and manipulate the behaviour of a wide range of colloidal systems account must be taken of other important surface forces, both repulsive and attractive, which can be capitalized upon.

11.3.1 Steric forces

It is well known that strongly adsorbed polymers,[7] or grafted polymer/oligomer chains, at a surface can act as very effective dispersion stabilizers. There is a repulsive force associated with the polymer layers that keeps the particles or surfaces far enough apart so that van der Waals attraction cannot cause instability. This steric force is insensitive to the presence of low concentrations of electrolyte and can operate in both aqueous and non-aqueous environments. The effect of polymer layers on the interaction energy between solid surfaces depends on the nature of the layers. If the layers, thickness δ, are compact so that they remain unaffected on collision of the surfaces then the surfaces are prevented from approaching to a separation $h < 2\delta$. Two extreme situations can then be envisaged corresponding to (a) the Hamaker constants for the layers and the medium being similar and (b) the solid and layers having similar Hamaker constants. In case (a) the layers and medium are indistinguishable so the interaction curve will be unchanged, except that the shortest distance of approach of surfaces will be 2δ rather than zero, so that the van der Waals attraction will be less on contact than if the layers were absent. For case (b) as far as the attraction is concerned the layers are simply an extension of

---

[7] Natural gums were used by the ancient Egyptians and Chinese to stabilize ink preparations. Faraday believed that the gum adsorbed and formed a 'protective' layer around the particles in the ink, preventing aggregation.

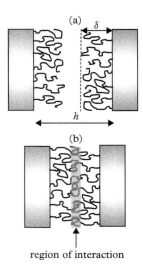

Figure 11.14 *(a) Two flat plates with grafted polymer layers, thickness δ. (b) The plates with the polymer layers contacting.*

the solid substrate. If the solid is in the form of spherical particles the effective particle radius is $(a + \delta)$ rather than simply a in the absence of the layer.

Of more interest here are the situations in which the (still concentrated) layers are more expanded and are perturbed in some way on impact. Adsorbed polymer chains that are present in a *good solvent* (Box 11.7) tend to extend into the solution, as illustrated in Fig. 11.14, and give effective steric barriers to the approach of surfaces. In practice, the thickness of the polymer layers can be large (several 10s of nm) depending on the molecular weight of the adsorbed or grafted polymer chains. The energy of steric interaction rises sharply when the polymer layers 'touch' ($h < 2\delta$), a situation illustrated schematically in Fig. 11.14b and can, a little artificially, be supposed to have contributions from two sources. After initial contact, as the surfaces approach further the chains begin to mix (shaded region in Fig. 11.14b) and the concentration of polymer segments increases in the overlap region. This generates an osmotic pressure, the solvent tending to be drawn into the region resulting in a repulsive pressure between the surfaces (*osmotic* or *mixing* effect, energy V_{Sm}). If the overlap of the layers becomes great, the conformations of the chains are substantially restricted, leading to a loss of configurational entropy and further repulsion, energy V_{Se} (*volume restriction* or *elastic* effect). Thus, the total steric energy of interaction is given as

$$V_S = V_{Sm} + V_{Se} \tag{11.3.1}$$

Often, effective steric stabilization of dispersions can be given without much overlap of the polymer layers in which case the osmotic term V_{Sm} dominates.

Box 11.7   Good and poor solvents for polymers

The quality of the solvent for a polymer molecule (hence polymer configuration in solution) is often discussed in terms of the Flory–Huggins χ-parameter $= [w_{12} - (w_{11} + w_{22})/2]z/kT$. Here w_{11}, w_{22}, and w_{12} are, respectively, interaction energies between solvent molecules, polymer segments, and polymer segments and solvent molecules; z is the number of nearest neighbours of a solvent molecule or segment in solution, the volumes of a segment and a solvent molecule being taken to be the same. For relatively weak interactions between solvent and segments ($\chi > 0.5$, 'poor' solvent) polymer configuration is dominated by segment-segment attractions, leading to compact coils and loss of solubility. Small χ (< 0.5) indicates strong interactions between segments and solvent and the polymer chains tend to be extended in solution ('good' solvent conditions). For $\chi = \frac{1}{2}$ the chain is a random coil, behaving as if it had no interactions with itself. This corresponds to the *theta*-condition which, for a given system, occurs (if at all) at only a single temperature, the theta (θ) temperature, which is analogous to the Boyle temperature for a real gas.

There have been a number of equations derived for V_S and its components based on various models and geometries (Napper, 1984; Vincent, 1987). The way in which V_S varies with separation h depends on the form of the polymer segment density distribution, $\rho(x)$, normal to the surfaces. A representation of how the segment density might vary for a polymer molecule terminally attached to a solid is shown in Fig. 11.15. It is useful, in order to bring out some salient features of steric repulsion, to consider a simple equation as an example. In (11.3.2) an approximate expression is reproduced for the mixing term V_{Sm} for the approach of flat surfaces (of unit area), or spheres radius a with $\delta << a$:

$$\frac{V_{Sm}}{kT} = \frac{2\varphi_{2s}^2}{v_1}(0.5 - \chi)X \qquad (11.3.2)$$

Here v_1 is the molecular volume of the solvent and φ_{2s} the volume fraction of polymer in the polymer layers. The repulsion depends on the degree of overlap and X is the overlap volume, which for plates of unit area (see Fig. 11.14) and spheres is given by (Goodwin, 2009)

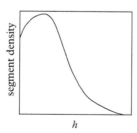

Figure 11.15 *Schematic representation of the segment density normal to a surface for a terminally attached polymer chain.*

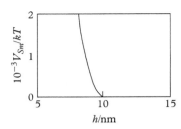

Figure 11.16 *Variation of V_{Sm} with separation for two spheres, radius 100 nm, coated with polymer layers thickness 5 nm. The polymer volume fraction in the layers is taken as 0.7 and χ is set at 0.4.*

$$X_{plates} = (2d - h)\,;\, X_{spheres} = (2\pi/3)\,(\delta - h/2)^2\,(3a + 2\delta + h/2) \qquad (11.3.3)$$

A plot of V_{Sm}/kT versus h according to (11.3.2) and (11.3.3) for two spheres is shown in Fig. 11.16, where it is seen that the energy rises steeply from zero at the point of contact of the polymer layers, that is, at $h = 2\delta$.

Inspection of (11.3.2) reveals that V_{Sm} is proportional to $(0.5 - \chi)$, which means that if $\chi > 0.5$ (i.e. the medium is a poor solvent for the polymer chains—see Box 11.7), V_S is negative and the steric interaction is attractive. For good solvents however ($\chi < 0.5$), V_S is positive giving rise to steric repulsion. Insofar as adsorbed polymer chains behave like their counterparts in bulk solution, it might be expected that steric interactions (hence particle flocculation in sterically stabilized particle systems) should be related to proximity of θ conditions (with respect to solvent and temperature), and this is often observed in practice.

Steric repulsion is expected to rise sharply with polymer concentration (volume fraction) in the surface, since V_{Sm} varies with the square of φ_{2s}. For effective stabilization of dispersions, the adsorption needs to be in or close to the plateau region of the adsorption isotherm (Chapter 7). Temperature has an effect on steric stabilization, both directly through $T (V_S \propto T)$ and through changes in χ with temperature. When polymer solubility rises with T, χ decreases giving rise to an increase in steric repulsion. The elastic term is not considered further here, but it is noted that it is always positive (repulsive).

The discussion so far has given a useful, but simple picture of steric repulsion, ignoring various important complicating factors. For example, the stabilizing polymer chains are likely to be polydisperse and this will affect the nature of the overlap region. There will also be local rearrangements of chains as the surfaces approach. Some possible effects that the presence of polymer layers can have on the van der Waals attraction between surfaces have already been touched upon. The polymer is also expected to influence electrical interactions between surfaces. A possible simple way around this (Vincent, 1987) is to replace the surface potential $\psi(0)$ in an expression for $V_R(h)$ (e.g. (11.2.32)) by the zeta (ζ)-potential and h by $(h - 2\delta)$. The ζ-potential is the potential at the plane of shear as a solid carrying a charge moves relative to the surrounding liquid, and is measured experimentally (Hunter, 1981). For a polymer-coated surface the plane of shear will be located in the liquid near the outer surface of the polymer layer.

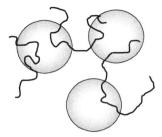

Figure 11.17 *Representation of flocculation of particles by polymer bridging.*

11.3.2 Bridging of surfaces

As seen, effective steric stabilization of particles needs high surface concentrations of polymer chains. At *low coverages* however, where uncovered surface is present, it is possible for a polymer molecule to adsorb on more than one particle simultaneously, forming *bridges* between the particles which can lead to flocculation (Gregory, 1987). Such bridging of particles, by polymer molecules of similar size to the particles, is illustrated in Fig. 11.17. An example of a system capable of forming bridges is one in which the particles are charged and the polymer is a polyelectrolyte with opposite charge. In such a case, not only can the polymer form bridges, but in so doing it can also (partially) neutralize the charge on the particles and thereby lower electrical repulsion. Some polymeric flocculants are obtained from natural materials, such as alginates and gelatine and these can be used for the purification/clarification of water or water-based systems such as wine and beer. Commercial flocculants are frequently high molecular weight polyacrylamide derivatives, which can be neutral or ionic in character. They are effective at low concentrations, in the region of a few parts per million.

11.3.3 Depletion forces

If, say, in a particulate system polymer molecules do not adsorb on the particle surfaces, they cannot give rise to either steric repulsion (or attraction) or form bridges. They are however capable of producing an *attraction* between the particles by *depletion forces* which, like electrical double layer and steric forces, arise from osmotic effects. Depletion forces are also apparent if the solid surfaces are saturated with polymer and there is an *excess* of polymer in solution. It is the presence of polymer molecules in solution and their absence (in non-adsorbed form) between the surfaces that gives rise to depletion forces.

There is, adjacent to a surface in solution, a layer from which the centres of small particles, surfactant micelles or (say globular) polymer molecules are excluded. In the case of polymer solutions, the thickness of this depletion layer is the radius of gyration, r_g, of the polymer molecule,[8] as illustrated in Fig. 11.18a. When two such surfaces approach

---

[8] The radius of gyration is the RMS distance of the atoms of the chain from the centre of gravity of the chain. For a large single freely jointed chain with n segments $r_g = n^{1/2}l/\sqrt{6}$, where l is the separation between two adjacent connected segments.

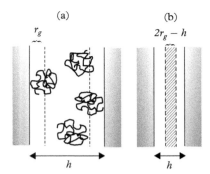

Figure 11.18 *(a) Non-adsorbing polymer molecules between flat surfaces (area $\mathcal{A}$) for $h > 2r_g$. There is a depletion layer, thickness r_g, next to the solid surfaces from which centres of polymer molecules are excluded. (b) Surfaces closer together giving overlap of depletion layers. Volume of overlap, v_{ex}, is $(2r_g - h)\mathcal{A}$ and is indicated by the hatched area.*

to separations $h < 2r_g$ polymer molecules are excluded from an overlap volume, v_{ex}, which increases with decreasing h (Fig. 11.18b). The thickness of the overlap region (shaded area in Fig. 11.18b) is $2r_g - h$, and the excluded or overlap volume, v_{ex} is $(2r_g - h)\mathcal{A}$, where, in the present context, $\mathcal{A}$ is the area of the plates. The depletion force arises from the exclusion of polymer from the volume v_{ex}, which leads to the tendency of solvent to diffuse osmotically from between the plates into the polymer solution.

The depletion attraction energy, V_{dep}, can be expressed in terms of the solution osmotic pressure Π and the excluded volume v_{ex} by the approximate relationship

$$V_{dep} = \Pi v_{ex} \approx (\rho kT)\, v_{ex} = \rho kT \left(2r_g - h\right) \mathcal{A} \tag{11.3.4}$$

in which ρ is the number density of polymer molecules in solution. The limiting (van't Hoff) expression for the osmotic pressure in dilute polymer solutions is used in (11.3.4) and is unlikely to be applicable in practice. This can be circumvented by using experimentally obtained osmotic pressures.

It is common to express polymer concentration in solution as a volume fraction, φ_2, of polymer. On the crude assumption that the polymer molecules are hard spheres with radius r_g,[9] φ_2 can be related to number density ρ by

$$\varphi_2 = \frac{4}{3}\pi r_g^3 \rho; \ \ \rho = \frac{3\varphi_2}{4\pi r_g^3} \tag{11.3.5}$$

[9] In practice, polymers can be expected to be polydisperse, that is, have a range of sizes. Further the radius of gyration will be influenced by the solvent quality, hence by the value of χ. In good solvents polymer molecules do not behave as hard spheres, and the segment density changes along a diameter as, therefore, does the solvent content.

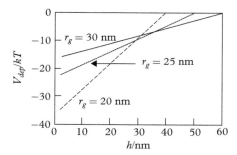

Figure 11.19 *Depletion energies, in units of kT, between flat plates (250×250 nm^2) as a function of separation at 298 K according to (11.3.6). The volume fraction of polymer in solution is 0.05 and r_g is taken as 20, 25, and 30 nm, as indicated.*

Combination of (11.3.4) and (11.3.5) gives the useful, but crude expression for the depletion energy of attraction, V_{dep}, between flat plates:

$$\frac{V_{dep}}{kT} = \frac{3\varphi_2 A}{4\pi} \left(\frac{2}{r_g^2} - \frac{h}{r_g^3} \right) \tag{11.3.6}$$

The energy is seen to be directly proportional to temperature and to the polymer volume fraction, φ_2. The way in which r_g affects V_{dep}, for a given φ_2 is illustrated in Fig. 11.19. The energy is zero when $h = 2r_g$, and the attraction increases more rapidly with falling h the smaller r_g. This is because for a given volume fraction, the *number* of polymer molecules is greater the smaller r_g, and osmotic pressure is a colligative property. Indeed osmotic pressure in (11.3.4) was expressed in terms of a *number* density of molecules.

To summarize, polymers at surfaces (grafted or adsorbed) and in bulk solution can give rise to a variety of forces (attractive or repulsive) between surfaces, which are additional to the van der Waals and electrical forces considered in the DLVO theory.

11.3.4   Some other surface forces

There are various other forces between surfaces that have been measured directly (see §11.4). Some come into play at very small distances between surfaces and others are evident at surprisingly large separations. The solvent can be directly implicated in some cases, and oscillatory forces have been observed. Whereas the origin of some of these forces is clear, there has been debate in the past about how some of the other forces arise. Here an outline is given of some of the forces that are possible in various types of system.

Oscillatory structural forces

Imagine two molecularly smooth plane parallel surfaces approaching in a liquid dispersion of small spherical colloidal particles or spherical micelles (radius a), present

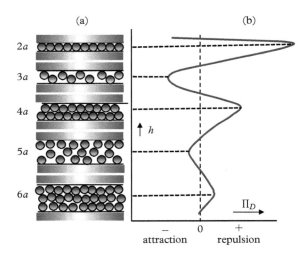

Figure 11.20 *(a) Packing of spherical particles, radius a, between smooth parallel plates. For h ∼ integral number of particle diameters, the packing is most efficient and the number density is larger than in bulk. For h ∼ odd numbers of particle radii the packing is poor and the number density is less than in bulk. (b) Disjoining pressure Π_D(h) as a function of separation h between the plates. There is a maximum in repulsion between the plates (for small h) when h is approximately an even number of particle radii.*

in high concentrations. The particles tend to line up into layers at the surfaces, and as two surfaces approach the particle number density, $\rho(h)$, between the plates oscillates. At some separations the particles fit efficiently between the plates ($\rho(h) > \rho(\infty)$, the number density in bulk) but at others they do not, and then $\rho(h) < \rho(\infty)$, as illustrated in Fig. 11.20a. The osmotic pressure between the plates at separation h, $\Pi(h)$, changes relative to that in solution, $\Pi(\infty)$, in an oscillatory fashion. The net pressure between the plates (the *disjoining pressure*, $\Pi_D(h)$, see Chapter 12) is

$$\Pi_D(h) = \Pi(h) - \Pi(\infty) \approx kT\left[\rho(h) - \rho(\infty)\right] \tag{11.3.7}$$

For poor packing (say at $h \sim 3a$, $5a$, etc.), solvent tends to diffuse from between the plates into solution, giving rise to attraction between plates (negative $\Pi_D(h)$). When the packing is efficient (e.g. for h of about $2a$, $4a$, etc.), $\rho(h) > \rho(\infty)$ and solvent tends to diffuse into the region between the plates giving maxima in $\Pi_D(h)$, as shown in Fig. 11.20b. The forces arising in this way are called colloid structural forces. The attractive regions are essentially the result of depletion forces.

Such oscillatory forces have been observed experimentally in a variety of systems, for example in thin liquid films (foam films) containing spherical surfactant micelles (see §12.2), and between smooth solid surfaces in liquids with small dispersed solid (e.g. latex) particles. Further, oscillatory forces are observed when smooth solid surfaces approach in pure liquids with large quasi spherical molecules such as silicone oil (Israelachvili, 2011).

Solvation (hydration) forces

In systems where a solvent has a high affinity for a surface, that is, the moieties at the surface are solvated, the surfaces have to be de-solvated in order to approach closely. This can give rise to a strong repulsive force at small separations (~5 nm and less). Adsorbed surfactant layers with polar head groups could for example give rise to hydration forces. Negatively charged solid surfaces (such as silica in aqueous electrolyte solution) which pick up strongly hydrated cations can also exhibit hydration forces. For this case, with 1:1 electrolytes, Pashley (1982) gives the empirical expression for the hydration interaction energy W per unit area

$$W = W_o \, exp \, (-h/\lambda_o) \tag{11.3.8}$$

in which W_o for a given system is a constant between 3 and 30 mJ m^{-2} and depends on the strength of hydration of the surface. The quantity λ_o is a decay length lying roughly between 0.6 and 1.1 nm. A calculated example is given in Fig. 11. 21 for the energy of interaction between charged parallel flat plates in 10^{-4} mol dm^{-3} aqueous 1:1 electrolyte. It has been supposed that the energy arises from van der Waals and Coulombic (i.e. DLVO) forces, together with hydration forces. Values taken for the various parameters are given in the figure legend. It is seen that at larger separations the DLVO forces dominate and the hydration interaction becomes important, and begins to rise steeply, only at separations of about 5 nm and less. In this range however it is the dominant interaction. Curves similar to that in Fig. 11.21 have been observed for the interaction of silica surfaces in aqueous electrolyte (Horn et al., 1989). Hydration forces are also observed in soap films and fluid bilayers.

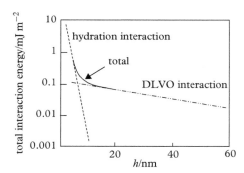

Figure 11.21 *Energy of interaction against separation for the approach of parallel charged plates in 10^{-4} mol dm^{-3} aqueous 1:1 electrolyte at 298 K. The energy is taken to have contributions from van der Waals, electrical, and hydration forces, the former two constituting the DLVO forces (dashed line as indicated). The steep dashed line represents hydration interaction. The net effect is shown as a full line close to where the lines for DLVO and hydration interactions cross. The parameters used are: Hamaker constant $A_H = 1 \times 10^{-20}$ J, $\psi(0) = 100$ mV, $W_0 = 20$ mJ m^{-2}, and $\lambda_0 = 1$ nm.*

Hydrophobic forces

It was seen in §2.2 that nonpolar molecules or groups in water undergo a solvent-mediated attraction ('hydrophobic bonding') when sufficiently close. It has been observed that *macroscopic* hydrophobic surfaces in water can also attract mutually, often over very large distances, up to 200–300 nm. At such separations, the attractions are much larger than can be accounted for by van der Waals forces. The origin of these hydrophobic forces has been the subject of much debate (see e.g. Meyer et al., 2006), and there may be no single explanation for the forces over the whole range of their action. Further, the magnitude and range of the forces are dependent on the detailed nature of the hydrophobic surfaces and the mode of their preparation. An example of a long-range attractive force is shown in Fig. 11.22; the results were obtained for mica surfaces rendered hydrophobic by coating with dioctadecyldimethylammonium bromide (Christenson and Claesson, 2001). Also shown for comparison is the curve expected if only van der Waals forces were acting between the surfaces. The ordinate is the normalized force which is simply related to the energy of interaction of unit areas of plane parallel surfaces (see §11.5).

Attard and co-workers (Parker et al., 1994; Tyrell and Attard, 2001) proposed that forces at long ranges could arise due to the presence of bridging sub-microscopic 'nanobubbles' (Box 11.8) between the surfaces. An AFM image of a smooth polystyrene surface in water putatively covered with very small bubbles is shown in Fig. 11.23. If two hydrophobic surfaces in water, each carrying such nanobubbles, come together and bubbles on approaching surfaces coalesce, vapour/gas bridges are formed, as illustrated in Fig. 11.24. Since the contact angle θ which water makes with the surface is greater than 90°, the bridge walls are concave as shown; from the Laplace equation (see §3.4.1), and depending on the geometry of the bridge, the pressure inside the bridge is less than that in the water and so the bridge tends to pull the surfaces together.[10] It is worth noting

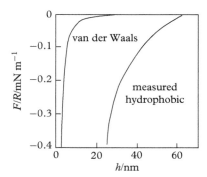

Figure 11.22 *Normalized force between mica surfaces coated with dioctadecyldimethylammonium bromide. The calculated van der Waals attraction is shown for comparison. Redrawn from Christenson and Claesson (2001).*

---

[10] The bridge in Fig. 11.24 has two principal radii of curvature r_1 and r_2 at right angles which have opposite sign. It is the *mean* curvature $(1/r_1) + (1/r_2)$ that determines the difference in pressure on the two sides of the curved liquid/vapour surface, as seen from equation (3.4.3).

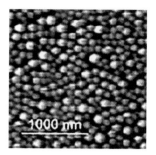

Figure 11.23 *AFM image of nanobubbles formed spontaneously on the surface of polystyrene coating a silicon wafer under water. Reprinted with permission from A.C. Simonsen, P.L. Hansen, and B. Klösgen, J. Colloid Interface Sci., 2004, 273, 291.*

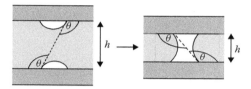

Figure 11.24 *Two nanobubbles at hydrophobic surfaces in water coalesce on the approach of the surfaces to form a gaseous bridge which tends to draw the surfaces together by capillary action. Note that the contact angle θ of water with the surfaces is greater than 90°.*

that degassing the liquid generally reduces the range and magnitude of the attraction, so that there is the possibility that often the nanobubbles contain air.

Box 11.8   Troubles with bubbles

The Laplace pressure inside a nanobubble attached to a surface is so high that it might be expected to disappear within seconds. For example in a bubble with radius 10 nm present in water at room temperature the pressure is about 150 atm. Such nanobubbles in water, attached to a solid surface, however can be long-lived, which requires explanation. It has been suggested by Lohse and Zhang (see Ball, 2014) that the stability could possibly arise as follows. Gas (air) contained in a nanobubble at high pressure has (by Henry's law) a higher concentration in water immediately surrounding the bubble than in bulk water far removed from the bubble. It is supposed that the 3-phase contact line around the bubble is pinned (see Chapter 16). Then for gas to leave the bubble, dissolved gas must diffuse away from the bubble interface into bulk solution at a finite rate. As it does so and more gas leaves the bubble, the curvature of the bubble falls (bubble becomes flatter) and the excess pressure and concomitant rate of gas loss is reduced. Such effects taken together could possibly explain the longevity of nanobubbles.

The long-range attraction, when it is observed, is related to the hydrophobicity of the solid surface through the contact angle θ as explained. However, water very close to a hydrophobic surface has different properties to bulk water and is involved in a 'partial drying' process. Both neutron and X-ray reflectivity measurements indicate that the density of water close to a hydrophobic surface is lower than that of bulk water over thicknesses from a few Å to a few nm (Ball, 2003; Poynor et al., 2006). Further, it is expected that any dissolved gas will accumulate (adsorb) at the solid/liquid interface (Dammer and Lohse, 2006). It is possible that the low density interfacial layer is involved in the nucleation of nanobubbles in some way.

In summary, the story of hydrophobic interactions between surfaces is probably not complete, and the magnitude and range of the forces is dependent on how the surfaces are prepared. Very long range forces can arise from the formation of gas/vapour bridges between surfaces, but the hydrophobic interactions at closer separations have different origins.

11.4   Influence of surfactant adsorption on the interaction between surfaces in solution

Up to now no account has been taken of the possibility that adsorption at surfaces might change as the surfaces approach. Imagine however that the system remains at *adsorption equilibrium* as the surfaces are brought together. If the adsorption (i.e. surface excess $\Gamma_2^{(1)}$ —see §4.3.1) of surfactant varies with the separation h of the surfaces, this will influence the interaction energy (Everett, 1988). For example, if surfactant desorbs as the surfaces approach, the energy of desorption must be supplied, which means that there is a repulsive component to the interaction energy arising from desorption of surfactant. The effect of adsorption on the force between surfaces can be formulated as follows.

Consider unit area of two overlapping flat parallel plates with separation h in surfactant solution; the adsorption (surface excess) at this separation is denoted $\Gamma_2^\sigma(h)$. The Gibbs adsorption equation is

$$d\gamma/d\mu_2 = -\Gamma_2^\sigma(h) \tag{4.3.5}$$

in which μ_2 is the chemical potential of surfactant in solution. An expression is required for the way in which the force between the plates is affected by the presence of surfactant. As the surfaces approach, if a force $F(h)$ acts between them it is manifested in a change in surface tension of the two surfaces (see also §12.1). The product of $F(h)$ and the change in separation dh is equal to the change in the sum of the tensions, $2d\gamma$, and so

$$F(h) = -2d\gamma/dh \tag{11.4.1}$$

Differentiation of the Gibbs equation with respect to separation gives

$$\frac{d^2\gamma}{d\mu_2 dh} = -\frac{d\Gamma_2^\sigma(h)}{dh} = \frac{d(d\gamma/dh)}{d\mu_2} \tag{11.4.2}$$

Then, combination of (11.4.1) and (11.4.2) results in an expression for the change in force with μ_2 and hence with surfactant concentration

$$\frac{dF(h)}{d\mu_2} = \frac{2d\Gamma_2^\sigma(h)}{dh} \tag{11.4.3}$$

The effect of surfactant adsorption alone on the interaction force can be seen by integration of (11.4.3). Note that the chemical potential of surfactant is related to its concentration c_2 (in ideal dilute solution) by $\mu_2 = \mu_2^o + RT \ln c_2$ so in the presence of solvent alone $\mu_2 = -\infty$. The integration gives

$$F(h) - F(h)^o = 2 \int_{-\infty}^{\mu_2} \left(\partial \Gamma_2^\sigma(h)/\partial h \right) d\mu_2 = 2RT \int_{c_2=o}^{c_2} \left(\partial \Gamma_2^\sigma(h)/\partial h \right) d\ln c_2 \tag{11.4.4}$$

where $F(h)^o$ is the force in solvent alone, and $F(h)$ is the force in the presence of surfactant. The right-hand side of (11.4.4) gives the contribution of surfactant adsorption to the force between unit areas of the surfaces.

Inspection of (11.4.4) indicates that if $\partial \Gamma_2^\sigma(h)/\partial h$ is positive, that is, adsorption decreases as the plates approach, the force due to surfactant adsorption is repulsive (positive). Conversely when adsorption increases as the surfaces approach (i.e. $\partial \Gamma_2^\sigma(h)/\partial h$ is negative), the force arising from surfactant adsorption is attractive (negative). (In this connection, see also §12.1.) Obviously, the detailed effect of surfactant adsorption on the force-distance relationship depends on how adsorption varies with separation.

11.5   Direct measurement of forces between solid surfaces

Two of the most powerful methods for the measurement of surface forces between solid surfaces involve the use of the surface force apparatus (SFA) and the atomic force microscope (AFM) (Claesson et al., 1996; Israelachvili, 2011). The application of AFM to the imaging of the topography of surfaces has been touched upon in Chapter 8, and the measurement of forces between fluid/fluid interfaces in thin liquid films (in the form of disjoining pressure isotherms) is described in Chapter 12.

11.5.1   Surface force apparatus

Development of the SFA for the direct measurement of forces between molecularly smooth solid surfaces in air or vacuum as a function of surface separation, was pioneered by Tabor and Winterton (1969) and by Israelachvili and Tabor (1972, 1973). Later it was developed for force measurements across intervening liquids. The SFA has been successfully used to verify predictions of the theory of van der Waals forces and also of the DLVO theory in appropriate systems. Other forces mentioned earlier, such as hydrophobic, hydration, steric and oscillatory structural forces have all been probed

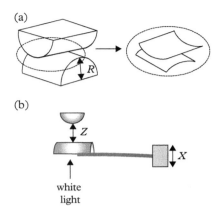

Figure 11.25 *(a) Half-cylinders crossed at right angles, which is locally equivalent to a sphere and a flat surface. (b) Schematic setup in a surface force apparatus (see text). The light beam entering from below is used for the interferometric determination of separation Z between the surfaces.*

using the SFA, which has proved to be enormously useful over a very wide area of surface and biological sciences.

The surfaces used have usually been muscovite mica (an aluminosilicate mineral), which can be cleaved to give molecularly smooth surfaces. The mica can be coated with surfactant or polymer layers for example, and electrolytes or other materials can be added to the liquid phase. The distance resolution is about 1 Å and forces down to about 10^{-8} N can be measured.

Cleaved mica sheets, a few μm thick, are silvered on one side to give semi-reflecting surfaces. The sheets are then glued (silver down) to half-cylindrical silica discs, radius R about 1 cm, in a crossed cylinder configuration (which is geometrically equivalent to a sphere near a plane surface), as illustrated in Fig. 11.25a. The shortest distance Z between the surfaces (Fig. 11.25b) is determined interferometrically; white light enters from below and is multiply reflected between the silver coatings on the (transparent) mica sheets. The beam passes to a spectrometer and analysis of the resulting fringes gives the separation Z (see e.g. Claesson et al., 1996). The lower disc is mounted on a double cantilever spring and can be moved vertically, and the upper disc is attached to a piezoelectric crystal which can give very fine vertical movement. In this way the separation Z can be altered in a very fine, controlled and reproducible way. With reference to Figure 11.25b, if X is changed (by moving either the upper or lower surface) and there is no interaction between the surfaces, Z changes by an equal amount. When interactions do occur, for repulsion between the surfaces Z changes less than X, and for attraction changes in Z are greater than those in X. From the spring constant, which is readily determined, and Hooke's law the force can be calculated.

In the discussion of surface forces, for the most part the energy of interaction between unit areas of plane parallel surfaces was considered since this is often what theoretical formulations give. The force or energy of interaction between two spherical particles is also often of interest, and the SFA yields forces between crossed cylinders. It is important therefore to know how *forces* measured in systems involving curved surfaces can be converted to the corresponding *energies per unit area* of interaction between plane parallel surfaces. The *Derjaguin approximation* allows such conversions to be made. The interaction energy between unit areas of plane parallel surfaces is denoted $V_{pp}(h)$, the force between two spheres with radii R_1 and R_2 as $F_{ss}(h)$ and the force between cylinders crossed at $90°$ as $F_{cc}(h)$; h is the closest distance between surfaces in all cases. It has already been noted that there is local geometrical equivalence of cylinders (radius R) crossed at $90°$ and a sphere, radius R, close to a plane surface. The Derjaguin approximation for the interaction of two spheres is

$$V_{pp}(h) = \frac{F_{ss}(h)\,(R_1 + R_2)}{2\pi R_1 R_2} \qquad (11.5.1)$$

and for the case of crossed cylinders (or sphere near a flat surface):

$$V_{pp}(h) = \frac{F_{cc}(h)}{2\pi R} \qquad (11.5.2)$$

The approximations hold for systems where radii are much greater than h. Data obtained using the SFA are plotted with F/R (referred to as the normalized force) or $F/2\pi R$ as the ordinate (see e.g. Fig. 11.22), which according to the Derjaguin approximation is equal to $2\pi V_{pp}(h)$ or $V_{pp}(h)$, respectively.

There have been various modifications to the SFA, which can be used for example to make dynamic measurements of viscous and viscoelastic properties of liquids as well as time-dependent interactions between surfaces.

11.5.2 Atomic force microscope

Since the AFM utilizes the force which exists between a probe and a surface to determine surface topology (§8.2.2), it can also be used to measure force-distance curves. With reference to Fig. 8.5, the AFM has a sensitive cantilever spring for force determination, and a piezoelectric positioning device with vertical movement (horizontal movement can be inactivated) so that the separation between tip and surface can be adjusted and determined. For the measurement of force-distance curves the (Hookean) cantilever spring constant has to be determined in order that deflection can be converted into force. There are various methods of calibration (Ralston et al., 2005). An obvious way is to note the deflection of the cantilever with a range of spherical particles of known mass (size and density) on the end. Another way is to determine the change in the resonant frequency of the cantilever with different masses on the end. Cantilevers with a wide range of spring constants are available, say from 0.01 to 100 N m^{-1}. The piezoelectric device, which expands and contracts with applied voltage, must also be calibrated in the z-direction.

The AFM method is more versatile than the SFA in terms of the systems that can be investigated. In an AFM experiment a (spherical) probe particle of interest (e.g. silica,

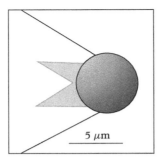

Figure 11.26 *Representation of an SEM image of a 6-μm diameter silica particle attached to a facet of the tip of an AFM cantilever. After an image reported by Ralston et al. (2005).*

silicon, polystyrene, or various metals) is glued to the end of the cantilever (Fig. 11.26). The particle is usually a few microns in diameter, and although this is outside the colloidal size range the AFM method is often referred to as *colloid probe microscopy* since it has been widely used to measure forces of direct interest to colloidal behaviour. The interaction of the probe with a plane surface can be measured through intervening liquid. Also after contact, the tip can be removed from the surface to give a pull-off force which is related to the work of adhesion of the two solids immersed in the liquid. The principle of the determination of force-distance curves using AFM is very similar to that described above for the SFA. When a force is exerted between probe particle and plane surface the cantilever spring is deformed and the extent of movement of the probe and the positioning device differ.

When experiments are performed in air containing water vapour, capillary condensation can occur between the tip and surface (see Fig. 8.6) and if so this will give rise to an attractive force between tip and surface. Obviously this kind of capillary force is absent when measurements are carried out under water.

11.6   Direct measurement of forces between fluid interfaces

Direct measurement of forces between fluid/fluid interfaces can be made using a similar principle to that employed in the SFA apparatus described in §11.5.1. Forces between two droplets in an immiscible liquid, or between gas bubbles in liquid, have also been measured directly using the AFM technique.

11.6.1   Liquid surface force apparatus

A schematic representation of a liquid surface force apparatus (LSFA) (for measurement of forces between two oil/water interfaces) is given in Fig. 11.27(a) (Aveyard et al., 1996). With reference to the figure, a thin flexible glass micropipette is attached to a piezo translator (piezo 1) which moves vertically (covering approximately 40 μm). An

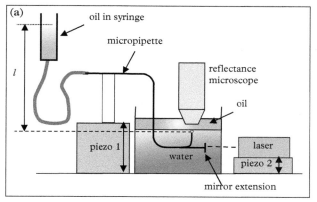

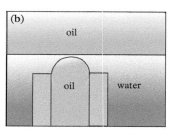

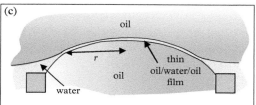

Figure 11.27 *(a) Schematic representation of an experimental arrangement for the determination of the force–distance curve for a small oil droplet in water on the end of a flexible micropipette approaching an oil/water interface (see text). (b) Oil drop on the end of the flexible pipette close to a plane oil/water interface. (c) Oil drop is pressed against the originally plane oil/water interface to form a thin circular and spherically curved oil/water/oil film.*

oil droplet is formed in aqueous solution on the internal circumference of the end of the pipette (internal diameter of 5 μm and upwards) (Fig. 11.27b) from oil contained in a syringe barrel, height l above the drop. The droplet is centred in the field of view of the objective of a reflectance microscope (Fig. 11.27a) used to determine the film thickness, h, of the oil/water/oil 'emulsion' film by optical interferometry as the droplet is forced upwards against the originally planar oil/water interface (Fig. 11.27c). The circular thin liquid film (radius r as shown in Fig. 11.27c) is spherically curved; r is readily measured and is typically a few microns, similar to the radii of emulsion films formed on droplet collisions in emulsions (see e.g. Fig. 14.9).

An equilibrium force–distance curve (i.e. interaction force acting normal to the film interfaces as a function of the film thickness) can be determined in the following way. As the pipette with attached drop is moved vertically towards the oil/water interface using piezo 1, when no force exists between the film surfaces (film is a *duplex* film), the mirror attached to the pipette (Fig. 11.27a) moves the same vertical distance z as piezo 1. This is detected by the movement of the reflected beam from a laser mounted on piezo translator, piezo 2. When a (say repulsive) force exists between the film surfaces however, the flexible pipette is deflected and the tip moves through a shorter distance than the translator piezo 1. The difference in distance (i.e. the *tip deflection* as measured using the reflected laser beam) is a direct measure of the force acting between the film surfaces. The force constant f of the pipette is readily determined by placing known masses on the end of the pipette and measuring the resulting deflection. The force acting across the film is then the product of f and the deflection. For a given hydrostatic head l of oil (Fig. 11.27a), the film thickness h is effectively unchanged by the vertical position (z) of the end of the pipette, but varies with l.

The force $F(h)$ acting across a film of area A is related to the interaction energy per unit area $V(h)$ between the film surfaces by

$$F(h) = -A dV(h)/dh \qquad (11.6.1)$$

and the disjoining pressure $\Pi_D(h)$, which is equal to the force per unit area, is

$$\Pi_D(h) = -dV(h)/dh \qquad (11.6.2)$$

The experimental setup described above can be used to determine disjoining pressures in thin films with areas comparable to those between colliding emulsion droplets (say ca. $10 \ \mu m^2$). With reference to Fig. 11.28, the radius of curvature of the thin oil/water/oil film is denoted r_f; neglecting the very small excess film tension resulting from the action of surface forces, the film has tension 2γ. The spherically curved oil/water interface of the oil drop adjacent to the film has a radius of curvature r_o and tension γ. The hydrostatic pressure in the oil drop on the tip (see Fig. 11.27a) is $l\rho g$ and is balanced by the Laplace pressure across both the curved film and oil/water surfaces. It follows therefore that $r_f = 2r_o$.

Water in the meniscus A around the film tends to suck liquid from the film and when the film is an equilibrium thickness h the (positive) disjoining pressure $\Pi_D(h)$ acting normal to the film surfaces just balances this capillary suction. Thus, the pressure p_f in the thin film less that, p_m, in the aqueous meniscus A is equal to the disjoining pressure, that is,

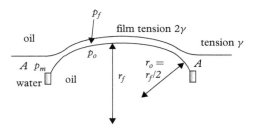

Figure 11.28 *Radius of curvature of the thin film with tension 2γ, is r_f and that of the oil/water interface, with tension γ, is denoted r_o. The radius $r_f = 2r_o$ (see text). The meniscus around the film is denoted A. Pressures p_m, p_o, and p_f are, respectively, pressures in the meniscus A, the oil drop, and thin aqueous film.*

$$\Pi_D(h) = p_f - p_m \tag{11.6.3}$$

The pressure in the film, p_f, is given by

$$p_f = p_o - \frac{2\gamma}{r_f} \tag{11.6.4}$$

and the pressure in the meniscus A is

$$p_m = p_o - \frac{2\gamma}{r_o} \tag{11.6.5}$$

Therefore from (11.6.3), (11.6.4), and (11.6.5), noting that $r_f = 2r_o$, the disjoining pressure in the film is

$$\Pi_D(h) = \frac{\gamma}{r_o} \tag{11.6.6}$$

With reference to Fig. 11.27, since the pressure drop $2\gamma / r_o$ across the undistorted oil drop surface is equal to $l\Delta\rho g$, where $\Delta\rho$ is the difference in density between the oil and surrounding air and g is the acceleration due to gravity, it follows that

$$\Pi_D(h) = l\Delta\rho g/2 \tag{11.6.7}$$

As mentioned, the film thickness h varies with l and so the disjoining pressure isotherm is readily determined.

Disjoining pressure isotherms obtained by use of the LSFA, for thin aqueous films in dodecane stabilized by the anionic surfactant AOT are shown in Fig. 11.29. Film thicknesses are in the range 30–80 nm and disjoining pressures are as high as 1500 Pa. The AOT concentrations used span the critical aggregation concentration of AOT which is 2.3 mM. Since the films are stabilized by an ionic surfactant in the absence of any further added electrolyte it is to be expected that the predominant surface forces arise from electrical repulsion.

An expression for the potential energy of repulsion, $V_R(h)$, arising from the overlap of like (plane) double layers in the presence of 1:1 electrolyte is given in (11.2.32). The energy is related to the disjoining pressure by (11.6.2), which yields for the electrical component of the disjoining pressure:

$$\Pi_{el} = 64 c k T \tanh^2 \left(\frac{e\psi(o)}{4kT} \right) \exp(-\kappa h) \tag{11.6.8}$$

The calculated curves in Fig. 11.29 were generated, using (11.6.8), by assuming values of $\psi(o)$ to give the best fit of the experimental data (points). Values of $\psi(o)$ taken for curves

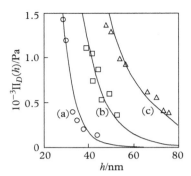

Figure 11.29 *Disjoining pressure isotherms for aqueous films in dodecane, stabilized by the twin-tail anionic surfactant AOT. Curves (a), (b), and (c) are for aqueous phase concentrations of AOT of, respectively, 3.0, 1.0, and 0.3 mM. Points are measured values and full curves are generated by use of (11.6.8)—see text. Redrawn from Aveyard et al. (1996).*

(a) to (c) were 90, 85, and 90 mV, which are very similar to measured zeta potentials of -100 mV in heptane-in-water emulsions stabilized by AOT.

The measurement of disjoining pressure isotherms in much larger, planar films, both symmetric and asymmetric, is covered in §12.2.2.

11.6.2 Forces between drops and bubbles using AFM

As two fluid interfaces approach, the interfaces can become deformed and fluid drains from the intervening phase. In the LSFA experiments described above force measurements are made only after equilibrium has been attained. However, information on *dynamic* forces is of great importance in gaining insights into the behaviour of various colloidal processes such as, for example, the approach and coalescence of droplets undergoing Brownian motion in emulsions. Dynamic forces have been probed using AFM as described below.

The AFM technique (see §11.5.2) has been used to investigate the time-dependent (i.e. *dynamic*) forces acting between approaching bubbles in water and between deformable oil drops in aqueous surfactant solutions. In the latter case (Dagastine et al., 2006) one oil droplet is immobilized on a plane substrate in the AFM and the other drop is attached to the cantilever, as illustrated schematically in Fig. 11.30. The relative vertical position of the drops (i.e. *h*) is varied using a piezo drive under the substrate. Drop radii are of the order of tens of microns and the oil/water interfacial tension is quite low due to surfactant adsorption, so the drops are expected to be deformable (see footnote 2, Chapter 14). Rates of approach of the drops used is of the order of the root mean square velocity due to Brownian motion of emulsion drops of this size. For example for oil drops with radius 40 μm the root mean square velocity is 7 μm s^{-1}.

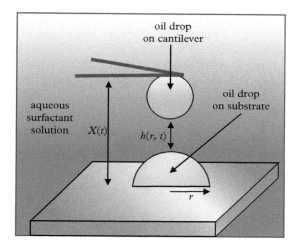

Figure 11.30 *Schematic representation of AFM setup for the determination of dynamic force-distance curves for the approach and separation of two oil droplets immersed in aqueous solution.*

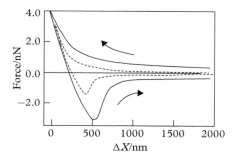

Figure 11.31 *Force against piezo drive motion ΔX for the interaction of two decane drops in 3mM aqueous SDS. Full curves are for velocity 28 $\mu m\ s^{-1}$, and dashed curves for 9.3 $\mu m\ s^{-1}$. The upper curve of a pair is for approaching drops and the lower curve is for separating drops, as indicated by the arrows. Zero ΔX is arbitrarily taken to correspond with the highest measured force. Redrawn from Dagastine et al. (2006).*

Some results for forces between two decane droplets (radii 43 and 90 μm) immersed in aqueous sodium dodecylsulfate (SDS) solutions, obtained by Dagastine et al. (2006), are reproduced in Fig. 11. 31. For a given velocity the force rises monotonously as the drops approach, repulsion arising from hydrodynamic effects of drainage of the film between the drops. On separation of the drops the force passes through an attractive (negative) minimum; the timescale of the fluid flow into the film between droplets is longer than the timescale of the measurements, resulting in attraction between the drops.

Vakarelski et al. (2010) have used a similar AFM technique to probe the dynamic interactions between ultrasonically generated microbubbles in water. Bubbles are of the order of 100 μm in diameter, and one bubble is attached to the AFM cantilever and the other to a plane substrate. Experimental results have been used to elucidate the coupling of hydrodynamic flow in the region between bubbles, van der Waals attractive forces and bubble deformation leading to bubble coalescence. It is interesting that, although capillary waves in thin liquid films are usually believed to be implicated in film rupture (§12.4), there was no evidence that this was the case in film rupture and bubble coalescence in the AFM study.

· ·

REFERENCES

R. Aveyard, B.P. Binks, W.-G, Cho, L.R. Fischer, P.D.I. Fletcher, and F. Klinkhammer, Investigation of the force−distance relationship for a small liquid drop approaching a liquid−liquid interface. *Langmuir*, 1996, **12**, 6561−6569.

P. Ball, Chemical physics: how to keep dry in water. *Nature*, 2003, **423**, 25.

P. Ball, Fundamental physics: feel the force. *Nature*, 2007, **447**, 772.

P. Ball, Why don't nanobubbles go pop? *Chem. World*, December 2014, 39.

M. Boström, V. Deniz, G.V. Franks, and B.W. Ninham, Extended DLVO theory: electrostatic and non-electrostatic forces in oxide suspensions. *Adv. Colloid Interface Sci.*, 2006, **123–126**, 5–15.

J.B. Calvert, Colloids, 2002. Available at: http://mysite.du.edu/~jcalvert/phys/colloid.htm.

H.B.G. Casimir and D. Polder, The influence of retardation on the London–van der Waals forces. *Phys. Rev.*, 1948, **73**, 360.

H.K. Christenson and P.M. Claesson, Direct measurements of the force between hydrophobic surfaces in water. *Adv. Colloid Interface Sci.*, 2001, **91**, 391–436.

P.M. Claesson, T. Ederth, V. Bergeron, and M.W. Rutland, Techniques for measuring surface forces. *Adv. Colloid Interface Sci.*, 1996, **67**, 119–183.

R.R. Dagastine, R. Manica, S.L. Carnie, D.Y.C. Chan, G.W. Stevens, and F. Grieser, Dynamic forces between two deformable oil droplets in water. *Science*, 2006, **313**, 210–213.

S.M. Dammer and D. Lohse, Gas enrichment at liquid-wall interfaces. *Phys. Rev. Lett.*, 2006, **96**, 206101.

B.V. Derjaguin and L.D. Landau, Theory of the stability of strongly charged lyophobic sols and of the adhesion of strongly charged particles in solution of electrolytes. *Acta Physicichim. URSS*, 1941, **14**, 633–662.

I.E. Dzyaloshinskii, E.M. Lifschitz, and L.P. Pitaevskii, The general theory of van der Waals forces. *Adv. Phys.*, 1961, **10**, 165–209.

P.P. Edwards and J.M. Thomas, Gold in a metallic divided state: from Faraday to present-day nanoscience. *Angew. Chem. Int. Ed. Engl.*, 2007, **46**, 5480–5486.

R. Eisenschitz and F. London, Über das Verhältnis der van der Waalsschen Kräfte zu den homöopolaren Bindungskräften. *Z. Phys.*, 1930, **60**, 491–527.

D.H. Everett, *Basic Principles of Colloid Science*. Royal Society of Chemistry, London, 1988.

L. Fisher, *How to Dunk a Doughnut: The Science of Everyday Life*. Arcade Publishing, New York, 2003, p.127.

J. Goodwin, *Colloids and Interfaces with Surfactants and Polymers*, 2nd edition. John Wiley & Sons, Chichester, UK, 2009.

J. Gregory, in *Solid/Liquid Dispersions* (ed T.F. Tadros), Chapter 8. Academic Press, London, 1987.

H.C. Hamaker, The London–van der Waals attraction between spherical particles. *Physica*, 1937, **4**, 1058–1072.

R.G. Horn, D.T. Smith, and W. Haller, Surface forces and viscosity of water measured between silica sheets. *Chem. Phys. Lett.*, 1989, **162**, 404–408.

R.J. Hunter, *Zeta Potential in Colloid Science*. Academic Press, London, 1981.

R.J. Hunter, *Introduction to Modern Colloid Science*. Oxford University Press, Oxford, 1993.

J.N. Israelachvili, The nature of van der Waals forces. *Contemp. Phys.*, 1974, **15**, 159–178.

J.N. Israelachvili, *Intermolecular and Surface Forces*, 3rd edition. Academic Press, 2011.

J.N. Israelachvili and D. Tabor, The measurement of van der Waals dispersion forces in the range 1.5 to 130 nm. *Proc R. Soc. Lond. A* 1972, **331**, 19–38.

J.N. Israelachvili and D. Tabor, Van der Waals forces: theory and experiment. *Prog. Surf. Membr. Sci.*, 1973, **7**, 1–55.

A. Lambrecht, The Casimir effect: a force from nothing. *Phys. World*, 2002, **15**, 29–32.

S.K. Lamoreaux, Casimir forces: Still surprising after 60 years. *Phys. Today*, 2007, **60**, 40–45.

E.M. Lifschitz, The theory of molecular attractive forces between solids. *Soviet Phys. JETP (Engl. Transl.)*, 1956, **2** 73–83.

E.E. Meyer, K.J. Rosenberg, and J. Israelachvili, Recent progress in understanding hydrophobic interactions. *Proc. Natl. Acad. Sci.*, 2006, **103**, 15739–15746.

D.H. Napper, *Polymeric Stabilisation of Colloidal Dispersions*. Academic Press, London, 1984.

B.W. Ninham and V.A. Parsegian, Van der Waals forces across triple-layer films. *J. Chem. Phys.*, 1970, **52**, 4578–4587.

J.L. Parker, P.M. Claesson, and P. Attard, Bubbles, cavities, and the long-ranged attraction between hydrophobic surfaces. *J. Phys. Chem.*, 1994, **98**, 8468–8480.

R.M. Pashley, Hydration forces between mica surfaces in electrolyte solutions. *Adv. Colloid Interface Sci.*, 1982, **16**, 57–62.

A. Poynor, L. Hong, I.K. Robinson, S. Granick, Z. Zhang, and P.A. Fenter, How water meets a hydrophobic surface. *Phys. Rev. Lett.*, 2006, **97**, 266101.

J. Ralston, I. Larson, M.W. Rutland, A.A. Feiler, and M. Kleijn, Atomic force microscopy and direct surface force measurements (IUPAC technical report). *Pure Appl. Chem.*, 2005, **77**, 2149–2170.

A.C. Simonsen, P.L. Hansen, and B. Klösgen, Nanobubbles give evidence of incomplete wetting at a hydrophobic interface. *J. Colloid Interface Sci.*, 2004, **273**, 291–299.

D. Tabor and R.H.S. Winterton, The direct measurement of normal and retarded van der Waals forces. *Proc. R. Soc. Lond. A*, 1969, **312**, 435–450.

D. Thompson, Michael Faraday's recognition of ruby gold: the birth of modern nanotechnology. *Gold Bull.*, 2007, **40**, 267–269.

J.W.G. Tyrell and P. Attard, Images of nanobubbles on hydrophobic surfaces and their interactions. *Phys. Rev. Lett.*, 2001, **87**, 176104.

E.J.W. Verwey and J.T.G. Overbeek, *Theory of the Stability of Lyophobic Colloids*. Elsevier, New York, 1948.

I.U. Vakarelski, R. Manica, X. Tang, S.J. O'Shea, G.W. Stevens, F. Grieser, R.R. Dagastine, and D.Y.C. Chan, Dynamic interactions between microbubbles in water. *Proc Natl. Acad. Sci.*, 2010, **107**, 11177–11182.

B. Vincent, in *Solid/Liquid Dispersions* (ed. T.F. Tadros), Chapter 7. Academic Press, London, 1987.

12

Thin liquid films

12.1 Introduction

Thin liquid films have fascinated people, young and old, for a very long time. Artists have been inspired by bubbles and the painting by Millais of a child playing with bubbles is familiar to many, maybe because it was used in a famous advert for Pears soap. Soap films have also intrigued scientists and mathematicians. C.V. Boyes, a distinguished British physicist, gave public lectures on soap films and wrote a now famous book entitled *Soap Bubbles, Their Colours and the Forces That Mould Them* (Boyes, 1890). The book (see references) remains a classic of popular science and is based on Christmas lectures given (to 'juvenile and popular' audiences) by Boyes at the Royal Institution in London in 1889.

*Joseph Antoine Ferdinand Plateau, 1801–1883. Granger Historical
Picture Archive/Alamy Stock Photo*

The Belgian physicist Joseph Plateau, who studied surface tension and capillary action, carried out extensive experiments with soap films. He lost his sight during the period around 1843–1844, and colleagues and family helped him with his experimentation and his writing. His experiments forming liquid films on wire frames of various shapes, now used as children's toys, are famous (see Fig. 12.1). He became a Fellow of the Royal Society of London in 1870, and his studies led to mathematical work on minimal surfaces.

Soap films tend to contract, the tension of a film being twice (or approximately twice) the surface tension of the soap solution from which it is formed. The colours exhibited by

Surfactants: In Solution, at Interfaces and in Colloidal Dispersions. Bob Aveyard. © Bob Aveyard 2019.
Published in 2019 by Oxford University Press. DOI: 10.1093/oso/9780198828600.001.0001

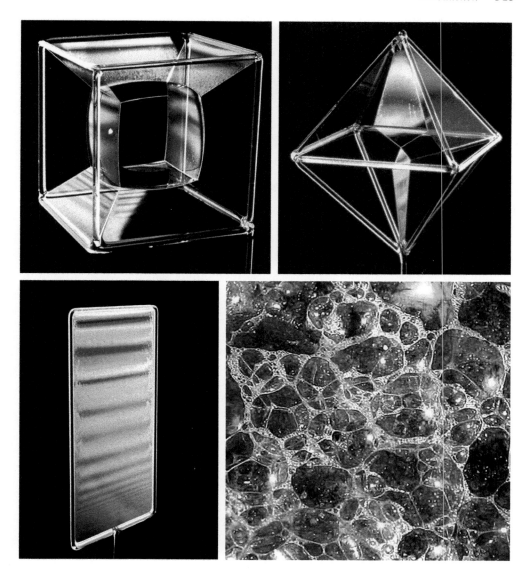

Figure 12.1 *Top left: soap films containing bubble on a cubic glass frame; top right: soap films on a bipyramidal frame; bottom left: vertical soap film; bottom right: foam on beach.*

soap films arise from the interference of (say white) light reflected from front and back surfaces of the film. The two sets of reflected rays have travelled different distances and undergo interference giving rise to the coloured appearance of the film. The interference, hence colour, depends on the thickness of the film. The vertical film shown in Fig. 12.1

Figure 12.2 *Painting of Sir James Dewar lecturing at the Royal Institution, London. Art Collection 3/Alamy Stock Photo.*

(bottom left) is wedge shaped, being thicker at the bottom than the top as a result of liquid drainage under gravity within the film.

Bubbles and thin liquid films also fascinated Sir James Dewar in his later years at the Royal Institution in London (Fig. 12.2). Dewar, famous for his work on the liquefaction of gases and his invention of the thermos (Dewar) flask, was able to keep bubbles 'alive' for many days by blowing them with clean air free of particles (see §14.9.2).

Based on his experimental observations Plateau concluded that soap films (in foams for example) always meet in threes, at an angle of 120°, forming a *Plateau border*. Further, the borders meet in fours at 109.47 degrees (the tetrahedral angle) to form a vertex (Figs. 12.3 and 12.4). For different configurations a system is unstable and will spontaneously rearrange.

A soap film is essentially a thin sandwich of water between two monolayers of adsorbed surfactant in air (i.e. an air/water/air film) as illustrated in Fig. 12.5a. Other thin liquid films include those in which air is replaced by a liquid giving, for example, oil/water/oil or water/oil/water films. These films, like soap films, are symmetrical but unsymmetrical films also exist such as air/oil/water, or solid/liquid/air as illustrated in Fig. 12.5b. Symmetrical thin liquid films also exist between solid particles approaching in a particulate colloidal dispersion and between approaching liquid droplets in an emulsion, as discussed in Chapter 14.

The formation of a thin liquid film on another condensed phase (liquid or solid) is relevant to aspects of wetting. For example (Fig. 12.6), if a drop of liquid (δ) is placed on the surface of another immiscible liquid (β) in air (α) one of several things might happen. One possibility is that it spreads to give a thick, so-called *duplex* film, in which the film surfaces are so far apart that there are no surface forces acting between them (Fig. 12.6a). Alternatively a thin film could be formed in which surface forces do act; this film will be in equilibrium with the remaining δ as a lens (Fig. 12.6b). It is also possible that δ does not spread, all the liquid being present in a lens (Fig. 12.6c), although δ might form an adsorbed layer at the $\alpha\beta$ interface.

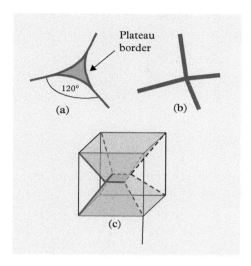

Figure 12.3 *(a) Cross-section of three films meeting at 120° along the Plateau border. (b) Four Plateau borders meeting at an apex at the tetrahedral angle. (c) Soap film on a cubic frame. The film defines the minimum surface area given that it is attached to eight edges of the cube as shown. Four of the Plateau borders have been drawn in full lines to show how they meet.*

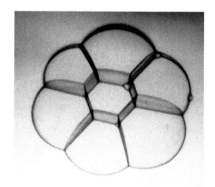

Figure 12.4 *Bubbles on a glass plate showing how three films come together at Plateau borders.*

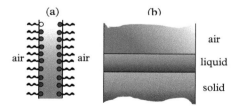

Figure 12.5 *(a) Cross-section of an aqueous soap film. (b) A thin film of liquid between a solid and air.*

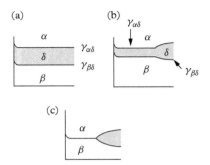

Figure 12.6 *(a) A thick (duplex) film of δ on β in α. (b) A thin film of δ on β in α in equilibrium with a lens of δ. (c) δ does not spread on β in α; it remains as a lens.*

For a duplex film the interfacial tensions $\gamma_{\alpha\delta}$ and $\gamma_{\beta\delta}$ are those between bulk phases. As the film thins however, and the two outer phases begin to interact across the film, the interfacial tensions begin to change; the tensions at the thin film surfaces are denoted $\gamma_{\alpha\delta}^f$ and $\gamma_{\beta\delta}^f$. The *film tension*, γ_f, is the sum of the two interfacial tensions in the film, that is,

$$\gamma_f = \gamma_{\alpha\delta}^f + \gamma_{\beta\delta}^f \tag{12.1.1}$$

The free energy of interaction across unit area of the film, thickness h, is therefore

$$V(h) = \gamma_f(h) - \left(\gamma_{\alpha\delta} + \gamma_{\beta\delta}\right) \tag{12.1.2}$$

When $V(h)$ (which can also be referred to as the *excess film tension*) is negative (the surfaces are mutually attracting), the film tension is less than the sum of the macroscopic tensions, that is, the tensions are reduced as a result of the attractive surface forces.

When, for example, a soap film thins there is the possibility that the surfactant adsorption at the film surfaces changes. In §11.4 it was noted how such changes influence the interaction free energy between the surfaces. A similar conclusion can be arrived at by a slightly different route. The excess Helmholtz free energy of unit area of a surface (A_σ) is related to the surface tension and adsorption at the surface by (see §4.2.2)

$$A_\sigma = \gamma + \sum_i \Gamma_i^\sigma \mu_i \tag{4.2.16}$$

For a soap film $\sum_i \Gamma_i^\sigma \mu_i = \Gamma^\sigma \mu$, where Γ^σ is the relative adsorption of surfactant (see §4.3.1) and μ is its chemical potential in solution. Noting (4.2.16), (12.1.1) can be expressed

$$
\begin{aligned}
V(h) &= 2\left[\left(A_\sigma^f(h) - A_\sigma^b\right) - \left(\Gamma^{\sigma,f}(h) - \Gamma^{\sigma,b}\right)\mu\right] \\
&= 2\left[\Delta_b^f A_\sigma(h) - \Delta_b^f \Gamma^\sigma(h)\mu\right]
\end{aligned} \tag{12.1.3}
$$

Here superscripts f and b refer to quantities for the film surface and for a surface in equilibrium with *bulk* solution, respectively, and Δ_b^f denotes the difference in such quantities. The chemical potential of surfactant in a film in equilibrium with bulk solution does not change with h. If $\Delta_b^f \Gamma^\sigma(h)$ is positive, that is, the adsorption at separation h is greater than the adsorption in a duplex film, this gives a negative (attractive) contribution to $V(h)$. Conversely, if adsorption decreases as the surfaces approach this gives a repulsive contribution. This is consistent with the analysis given in §11.4.

Following these introductory remarks, consideration is now given to various types of thin liquid films in equilibrium with their menisci, and it will be shown how the force–distance curves or equivalently the disjoining pressure isotherms can be determined. Some dynamic aspects of film stability and breakdown will also be introduced.

12.2 Symmetrical thin liquid films

Single liquid films in equilibrium with their liquid menisci, and the forces acting across them, have stimulated much interest. The forces are among those that also determine the stability of colloidal dispersions. Various properties of films have been investigated including equilibrium film thickness, disjoining pressure isotherms (where film thickness is varied), film drainage rates, film elasticity and the way in which films rupture. The films are relatively easy to prepare and their properties have a direct relevance to the films existing between approaching emulsion drops (in the case of say water/oil/water films) and films in foams in the case of air/water/air (soap) films.

12.2.1 Thin film contact angle

If surface forces act in, say, a thin soap film, the surface tension of a film surface differs from that of the bulk soap solution. For an equilibrium film in contact with its meniscus, it follows that there must be a finite contact angle, θ, between the meniscus and the film as illustrated in Fig. 12.7. The horizontal force balance gives

$$2\gamma_{\alpha\delta} \cos \theta = 2\gamma_{\alpha\delta}^f = \gamma_f \tag{12.2.1}$$

and combination of (12.1.2) with (12.2.1) shows that the film interaction energy (per unit area) at the equilibrium thickness is[1]

$$V(h) = 2\gamma_{\alpha\delta} (\cos \theta - 1) \tag{12.2.2}$$

If the contact angle can be measured the film interaction energy can be calculated assuming $\gamma_{\alpha\delta}$ is also known. Measurements can be made for example using large vertical

---

[1] The film is not at equilibrium in the sense that the free energy of the system would be lower if the surfaces did not exist, that is, if the system formed for example a small liquid droplet. The equilibrium of the film is a *metastable* equilibrium.

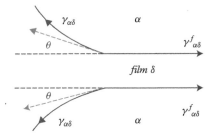

Figure 12.7 *Tensions and contact angle in a system of a thin liquid film in contact with its liquid meniscus.*

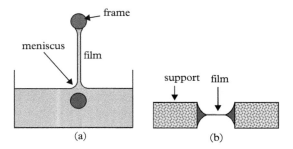

Figure 12.8 *(a) Cross-section of a large vertical soap film drawn from solution on a frame. The film has a meniscus with the surface of the bulk solution. (b) A small horizontal film with its meniscus adjacent to the support.*

soap films drawn on a suitable frame, as illustrated in Fig. 12.8a. The meniscus involved is that between the film and the surface of the bulk solution. Alternatively, much smaller (circular) films can be prepared with a radius of say 0.1–1 mm (Fig. 12.8b) on a suitable glass support (Exerova and Kruglyakov, 1998). Contact angles, as well as film thickness, can be determined interferometrically. Measurement of contact angle in the case of the setup in Fig. 12.8b would involve the determination of the radii of the Newton interference rings caused by the meniscus, observed in reflected monochromatic light. Before giving an idea of the magnitudes of film tensions and contact angles it is useful to consider the shape of a *disjoining pressure isotherm* for a soap film.

12.2.2 Disjoining pressure isotherms

In Chapter 11 interactions between surfaces were discussed mainly in terms of free energy-distance curves, but in the study of thin liquid films it is usual to consider the surface forces in terms of the concomitant pressure between the surfaces (see also §11.6). Since a repulsive pressure tends to 'disjoin' the surfaces the pressure, acting normal to the surfaces, is called the *disjoining pressure* (Derjaguin and Churaev, 1976) which is denoted Π_D. Like interaction energy, the disjoining pressure also depends on the separation of the surfaces; the relationship between the free energy of interaction per unit area, $V(h)$, and $\Pi_D(h)$ is

$$\Pi_D(h) = -dV(h)/dh \qquad (12.2.3)$$

If it is supposed that the main surface forces in the film are van der Waals attraction and electrical repulsion (such as might be the case for a film formed from the solution of an ionic surfactant below the *cmc*) then the shape of the energy curve is that shown in Fig. 12.9a. The assumed values of quantities used to give the shape are given in the legend. For very small separations the energy must rise rapidly, and the nearly vertical line at low h has been added with arbitrary slope. It is seen that there are two minima in the curve and one maximum. The corresponding disjoining pressure isotherm is sketched below in Fig. 12.9b. From (12.2.3) the maximum and the two minima in the interaction energy curve correspond to zero disjoining pressure, as indicated by the dotted vertical lines. The shape of the interaction energy and disjoining pressure curves are similar.

With reference to Fig. 12.10, the pressure, p_m, in the meniscus attached to a planar film is, according to the Laplace equation (3.4.3), less than that, p_2, outside the film. The capillary pressure p_c is the difference between the pressures, that is, $p_2 - p_m$. In the absence of a positive disjoining pressure, and noting that the pressure drop across a plane surface is zero, the pressure in the plane film is equal to p_2. Since $p_2 > p_m$ liquid is sucked *laterally* out of the film into the meniscus. This causes film thinning and when

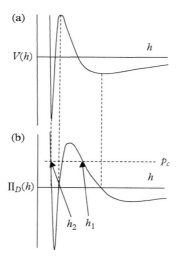

Figure 12.9 *(a) Interaction energy curve. The shape is obtained using (11.2.32) for the electrical repulsion energy in 25 mM 1:1 electrolyte and $\psi(0) = 30$ mV, and (11.2.3) for attraction with $A_H = 3.7 \times 10^{-20}$ J. The nearly vertical line at small h has been added with arbitrary slope. (b) The disjoining pressure isotherm. $\Pi_D(h)$ is zero for the maximum and minima in the energy curve. The horizontal dashed line represents the capillary pressure p_c (see text).*

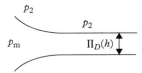

Figure 12.10 *Pressures acting in a system with a meniscus and an adjoining thin film.* $\Pi_D(h)$ *acts normal to the film surfaces and capillary suction acts laterally.*

the film is sufficiently thin, interactions between surfaces begin to occur and a disjoining pressure is generated *normal* to the surfaces. The film reaches an equilibrium thickness (h) when the effects of (positive) disjoining pressure, tending to cause film thickening, and capillary suction tending to thin the film are equal, that is, when $\Pi_D(h) = p_c$. Note that an equilibrium film can only exist on a limb of the disjoining pressure isotherm with a negative slope. If the slope is positive the film spontaneously thins since in thinning $\Pi_D(h)$ would fall. This means that only those parts of a disjoining pressure isotherm with negative slope (i.e. $\Pi_D(h)$ rising with decreasing h) are accessible experimentally.

It follows from Fig. 12.9b that there are two equilibrium thicknesses (where $\Pi_D(h) = p_c$) for a soap film. The film with the larger thickness, h_1 is referred to as a *common black film* (CBF), and the thinner film of thickness h_2 as a *Newton black film* (NBF), since both types of film have a black appearance. Thicker (non-equilibrium) films exhibit the familiar colours. The CBF contains liquid but the NBF is essentially two (hydrated) monolayers of surfactant, head group to head group. The occurrence of a CBF requires the repulsive interactions to be larger than attractive forces at relatively large separations so that, in the case of ionic surfactant systems, the electrolyte concentration must be quite low. Thicknesses of CBFs are of the order of 30 nm, whereas a NBF has a thickness of roughly twice the length of the surfactant molecule (i.e. a few nm). When the capillary pressure reaches the maximum in the disjoining pressure isotherm, the CBF spontaneously thins to the NBF.

A number of studies of contact angles of thin soap films with their menisci have been reported. For example, Kolarov et al. (1968) showed that for CBFs formed from aqueous sodium dodecylsulfate (SDS) solutions in the presence of low concentrations of NaCl, contact angles were low ($<1°$). For the thinner NBFs, θ rose from about 1° at NaCl concentrations <0.334 mol dm^{-3} to about 8° at 0.334 mol dm^{-3}. For even higher NaCl concentration (0.5 mol dm^{-3}), θ rose to nearly 11°. At this salt concentration it was calculated that the excess film tension (i.e. twice the difference between $\gamma^f_{\alpha\delta}$ and $\gamma_{\alpha\delta}$) was 1.2 mN m^{-1}, corresponding (through (12.1.2) and (12.2.2)) to a solution surface tension ($\gamma_{\alpha\delta}$) of 32.6 mN m^{-1}.

A contact angle allows the calculation of a single interaction energy for the equilibrium film thickness that is given under the prevailing experimental conditions, that is, it gives one point on a disjoining pressure isotherm (at thickness h_1 or h_2 in Fig. 12.9b). However if the capillary pressure (hence the disjoining pressure) can be varied, equilibrium film thickness can also be adjusted. This forms the basis of a method for measuring a disjoining pressure isotherm, using an experimental setup such as that shown schematically in

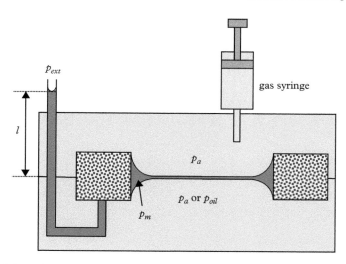

Figure 12.11 *Schematic representation of an experimental set-up for the measurement of a disjoining pressure isotherm of a planar thin aqueous film in air or an asymmetric aqueous film with air above and oil below. The air pressure p_a is adjusted using the gas syringe.*

Fig. 12.11. It is suitable for investigation of plane symmetrical thin liquid films in air and also for asymmetrical films (see §12.3) with air on one side and oil on the other (Bergeron and Radke, 1992). The capillary pressure is equal to the pressure in the air, p_a, less that p_m in the film meniscus. The air pressure is adjusted using the gas syringe and measured using the capillary tube (internal radius r), containing the surfactant solution from which the film is made, attached to the fritted glass film cell. The height of liquid above that of the film is l, and the difference in density between the surfactant solution and air is $\Delta\rho$. The capillary pressure (and therefore $\Pi_D(h)$) is

$$p_c = \Pi_D(h) = p_a - p_m = p_a - (p_{ext} - 2\gamma/r + \Delta\rho g l) \qquad (12.2.4)$$

where γ is the surface tension of the bulk surfactant solution and p_{ext} is the external pressure outside the thermostatted container. Film thicknesses can be measured interferometrically (Scheludko and Platikanov, 1961).[2]

Disjoining pressure isotherms related to formation of CBF and NBF from surfactant solutions at concentrations below the *cmc* (such as that depicted in Fig. 12.9b) have been measured, although as mentioned only the two limbs with negative slope are accessible. Isotherms for films formed from solutions well above the *cmc* are interesting because the presence of micelles in a film can give rise to oscillatory structural forces between surfaces of the type described in §11.3.4. An example isotherm, adapted from Basheva

---

[2] Determination of disjoining pressures and surface forces in fluid/fluid/fluid films with much smaller surface areas is covered in §11.6.

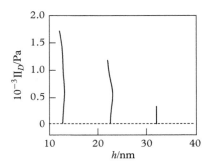

Figure 12.12 *Measured disjoining pressure isotherm for a 100 mmol dm^{-3} aqueous film of the nonionic surfactant Brij 35. Adapted from Basheva et al. (2007). The many data points have been replaced by lines.*

et al. (2007) is shown in Fig. 12.12 for films formed from 100 mmol dm^{-3} aqueous Brij (a commercial nonionic polyoxyethylene lauryl ether). The many data points reported have been represented by the three lines. Three stepwise transitions in film thickness are shown, at thicknesses around $h = 32, 22$, and 12 nm which correspond to metastable thin films containing, respectively, two, one, and zero stratified layers of micelles. Layers of spherical micelles and surfactant bilayers would have very similar thicknesses of course, but Basheva et al. (2007) refer to work which indicates that micelles are likely to be responsible. Thus, measurement of such oscillatory disjoining pressure isotherms can give information on micelle sizes.

12.3   Asymmetrical thin liquid films

12.3.1   Pseudo-emulsion films

It has been seen how a disjoining pressure isotherm for an asymmetric aqueous film, with air on one side and oil on the other (a *pseudo-emulsion* film), can be measured (Fig. 12.11).

Bergeron et al. (1993) studied the metastability of pseudo-emulsion films in connection with the use of foams injected into oil wells as mobility control agents in oil production (see §14.10.2). In the porous rock, gas cells exist in contact with oil wetted by thin aqueous surfactant films, as illustrated in Fig. 12.13. Since nonpolar oils are used as foam breaking agents (see §14.9.3) there is the possibility that the foam films will be ruptured by the crude oil. It is therefore desirable to understand the metastability of such films.

The metastability of a pseudo-emulsion film is related to whether or not a drop of oil placed under the surface of an aqueous surfactant solution in contact with air (or oil vapour in air) will enter the surface. Some possibilities are illustrated in Fig. 12.14. The oil drop (which is less dense than the aqueous surfactant solution) comes to the surface

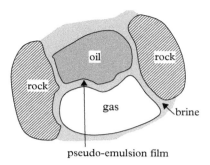

Figure 12.13 *Occurrence of pseudo-emulsion films in porous rock containing foam, oil and, brine.*

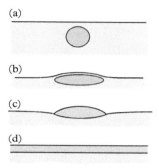

Figure 12.14 *(a) An oil drop under water. (b) Oil drop with aqueous film separating it from air. (c) Oil lens resting at the air/water surface. (d) A film of oil on water.*

and a film of aqueous phase separates it from air (b). If the film is unstable, the drop will enter the surface and possibly form a lens (c) or spread along the surface to give either a duplex film or possibly a thin film in equilibrium with a bulk lens (d), (as illustrated in Fig. 12.6b). Classically, the thermodynamic feasibility of entry and spreading, which is related to the various interfacial tensions γ_{aw}, γ_{ow}, and γ_{oa} (where subscripts o, w, and a refer to oil, water, and air, respectively), has been discussed without reference to thin film forces or capillary pressure in the system. Some results on drop entry are presented in §14.9.3 in connection with foams.

12.3.2   Entry coefficients and the stability of thin pseudo-emulsion films

Consider a hypothetical system, depicted in Fig. 12.15, consisting of a thick layer of oil (shaded), with unit area normal to the paper, immersed in water (a). Suppose now that the oil layer is brought to the surface so that it exhibits unit area of air/oil interface (c). The process is referred to as entry of oil into the air/water interface.

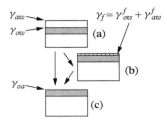

Figure 12.15 *An oil layer (shaded) with unit cross-section immersed in water (a), also with unit cross-section, is brought to the water surface (c) to form unit area of oil/air interface. Unit area of oil/water interface is lost. The process can proceed directly (a) to (c) or via (b) involving the formation a thin water film between air and oil.*

Classically, the thermodynamic feasibility of the process is considered in terms of the *entry coefficient*, $E_{o,aw}$, defined by

$$E_{o,aw} = \gamma_{aw} + \gamma_{ow} - \gamma_{oa} \tag{12.3.1}$$

Since the tensions are the free energies of formation of unit area of the various interfaces, oil entry is feasible if the tension of the oil/air interface formed on entry is less than the sum of the tensions of the air/water and oil/water interfaces removed on entry. Thus, entry is thermodynamically feasible if $E_{o,aw} > 0$. If all phases (oil, water and air) are mutually saturated however, a thermodynamic relationship exists between the three tensions (Box 12.1):

$$\gamma_{oa} \leq \gamma_{ow} + \gamma_{aw} \tag{12.3.2}$$

From this it follows that in equilibrated systems $E_{o,aw} \geq 0$; a zero entry coefficient means entry is not feasible.

Box 12.1 For three fluid phases in mutual equilibrium the three tensions are simply related

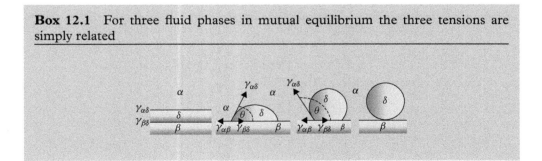

Box 12.1 *Continued*

Imagine a drop of liquid δ is added to the surface of liquid β resting in fluid α. For simplicity suppose the surface of β remains undeformed. This is reasonable if the drop is sufficiently small (see §16.2.3). If δ completely wets β in α then a duplex film of δ is formed (left hand figure). Otherwise the drop exhibits a finite contact angle θ, which can be less than or greater than 90° as illustrated. In the extreme, $\theta = 180°$ (δ completely de-wets β in α). For complete wetting $\theta = 0$. Young's equation for these systems is (§3.3)

$$\gamma_{\alpha\beta} = \gamma_{\beta\delta} + \gamma_{\alpha\delta} \cos \theta \qquad (1)$$

For complete wetting ($\theta = 0$), $\cos \theta = 1$ and so

$$\gamma_{\alpha\beta} = \gamma_{\beta\delta} + \gamma_{\alpha\delta} \qquad (2)$$

For this equality (*Antonoff's rule*) to hold, and for the configuration shown in the left hand figure to be possible at equilibrium, $\gamma_{\alpha\beta}$ must be the largest of the three tensions. For all the other cases illustrated $1 > \cos \theta \geq -1$ so that

$$\gamma_{\alpha\beta} < \gamma_{\beta\delta} + \gamma_{\alpha\delta} \qquad (3)$$

and hence overall

$$\gamma_{\alpha\beta} \leq \gamma_{\beta\delta} + \gamma_{\alpha\delta} \qquad (4)$$

For the inequality any of the three tensions can appear on the left hand side. The result (4) remains valid for systems in which the surface of phase β is deformed by the presence of a drop of δ, and a more rigorous derivation of (4) is given by Rowlinson and Widom (1982)

The equality in (12.3.2) (which as written requires that γ_{oa} is the largest of the three interfacial tensions) is an expression of *Antonoff's rule*. It is interesting that there have been a number of studies reported in the literature of systems with oil and aqueous surfactant solutions where it has been assumed equilibrium exists even though (12.3.2) (or its equivalent) is violated.

The above 'classical' approach takes no account of the intermediate formation of a pseudo-emulsion film, such as that illustrated in Fig. 12.15b, or its role in the entry process. It may be feasible for entry to occur (which would result in the formation of the system represented in Fig. 12.15c but for a thin pseudo-emulsion film to prevent entry. Bergeron et al. (1993) gave an approach to entry which takes explicit account of the disjoining pressure in a thin metastable liquid film, and it is useful to consider the entry process in terms of disjoining pressure isotherms rather than interfacial tensions. The difference between the systems represented in Fig. 12.15b and c is that the oil/air interface in (c), with tension γ_{oa}, is replaced by the thin pseudo-emulsion film with tension γ_f.

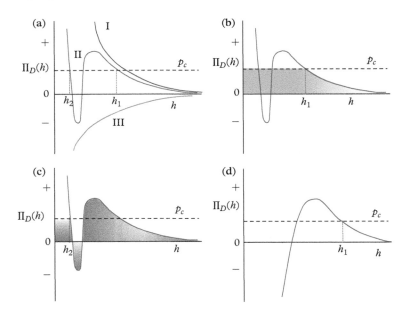

Figure 12.16 *(a) Disjoining pressure isotherms for the formation of: (I) a stable thick film; (II) metastable film of thickness either h_1 or h_2; (III) an unstable film. In (b) and (c) the shaded areas represent the negative of the generalized entry coefficient for pseudo-emulsion films with (b) thickness h_1 and (c) thickness h_2. (d) Isotherm for metastable thin film thickness h_1 but where the classical entry coefficient is large positive. Redrawn from Bergeron et al. (1993).*

When γ_{oa} in the 'classical' entry coefficient defined in (12.3.1) is replaced by γ_f, the 'generalized' entry coefficient, $E_{o,aw}^{g}$, is obtained

$$E_{o,aw}^{g} = \gamma_{aw} + \gamma_{ow} - \gamma_f \tag{12.3.3}$$

so that the generalized entry coefficient is minus the excess film tension (see (12.1.2)).

Some schematic disjoining pressure isotherms are shown in Fig. 12.16a for various kinds of system; the capillary pressure p_c in the systems is indicated by the horizontal dashed line (see §12.2.2). When the disjoining pressure only rises with decreasing film thickness (curve I) a stable film is formed, which means that the aqueous phase spreads on the oil. Curve II corresponds to the formation of a pseudo-emulsion aqueous film with thickness of either h_1 or h_2; such films correspond to the CBF and the NBF given by symmetrical aqueous soap films (§12.2.2) although the pseudo-emulsion films have a white rather than a black appearance. In curve III the disjoining pressure falls continuously as the film thins which gives rise to an unstable film, that is, the aqueous phase does not wet the oil.

The generalized entry coefficient can be related to the disjoining pressure isotherms as follows. For say a thick aqueous film between oil and air, where the tensions of the two surfaces do not depend on film thickness, the Gibbs equation at constant T may be written as

$$d\gamma_f = d\gamma_{ow} + d\gamma_{aw} = -\sum_i \left(\Gamma_{i,ow} + \Gamma_{i,aw}\right) d\mu_i \qquad (12.3.4)$$

where Γ_i are surface excesses of component i at the oil/water and air/water interfaces. When the surfaces interact $d\gamma_f$ has a contribution, $h d\Pi_D$,[3] arising from the change in Π_D with h and the Gibbs equation then becomes

$$d\gamma_f = h d\Pi_D - \sum_i \left(\Gamma_{i,ow} + \Gamma_{i,aw}\right) d\mu_i \qquad (12.3.5)$$

If the Γ_i are supposed to be the same at bulk and film surfaces then[4]

$$\left(d\gamma_f\right)_{\text{nint}} - \left(d\gamma_f\right)_{\text{int}} = h d\Pi_D \qquad (12.3.6)$$

where subscripts $_{\text{nint}}$ and $_{\text{int}}$ indicate films without and with surface interactions respectively. Integrating (12.3.6) gives

$$\gamma_{aw} + \gamma_{ow} - \gamma_{aw}^f - \gamma_{ow}^f = -\int_{\Pi_D(h_\infty)=0}^{\Pi_D(h)} h d\Pi_D \qquad (12.3.7)$$

in which $\Pi_D(h)$ is the disjoining pressure at an equilibrium film thickness h (which is equal to the capillary pressure for planar films). Noting (12.3.3) it is seen that the generalized entry coefficient is given by

$$E_{o,aw}^g = -\int_{\Pi_D(h_\infty)=0}^{\Pi_D(h)} h d\Pi_D \qquad (12.3.8)$$

Therefore, the negative of the generalized entry coefficient is given by the shaded areas shown in Fig. 12.16b, for a film with thickness h_1, and in Fig. 12.16c for a film of thickness h_2. The isotherm shown in Fig. 12.16d results in the formation of a metastable thin film although the classical entry coefficient is large and positive. Values of the generalized coefficients depend on the prevailing capillary pressure, and for films with two equilibrium thicknesses, there are two entry coefficients. When there is a net positive shaded area, $E_{o,aw}^g$ is negative and the oil is non-entering. Note that the equilibrium classical entry coefficient cannot take on negative values. Films which exhibit disjoining pressure isotherms with a strong repulsive branch at large film thicknesses (h_1) give stable films.

---

[3] This contribution is analogous to that arising from changing hydrostatic pressure p on a surface with thickness τ, that is, τdp, as in (4.3.3) in Chapter 4.

[4] It is possible however that adsorption changes with film thickness, as described in §12.1.

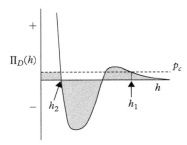

Figure 12.17 *Disjoining pressure isotherm which results in a positive generalized entry coefficient.*

To illustrate this further, positive values of $E_{o,aw}^g$ can result when an isotherm exhibits a large negative well such as that depicted in Fig. 12.17. For a film with thickness h_1, $E_{o,aw}^g < 0$ (non-entering and the excess film tension is positive as a result of repulsive surface forces between film surfaces) but for the film with thickness h_2, $E_{o,aw}^g$ is positive. The latter can be described as an entering system, but there is a very thin film separating oil and gas phases. This film will exhibit a finite contact angle with say a macroscopic water lens on the oil.

In summary, classical entry coefficients are defined in terms of interfacial tensions in macroscopic systems whereas the generalized coefficients are defined in terms of thin film forces. Classical entry coefficients indicate only feasibility for entry or otherwise. It is quite possible, in a system with a positive classical entry coefficient, $E_{o,aw}$, for a metastable thin liquid film to prevent entry. Unlike $E_{o,aw}$, a generalized entry coefficient has two values in systems which exhibit common and Newton thin films. Furthermore, unlike equilibrium classical entry coefficients, $E_{o,aw}^g$ can be negative, and the sign of $E_{o,aw}^g$ for common and Newton films can be opposite. As seen, the value of a generalized entry coefficient depends on the capillary pressure in the system. Classical entry coefficients can have values of several mN m^{-1}, whereas the generalized coefficients tend to be an order of magnitude or more smaller.

12.3.3   Wetting films on solids

Much of the interest in thin liquid films has been focused on the formation of aqueous films on solids. The impetus for the study of wetting (or otherwise) of solids by liquids is its relevance to various important practical processes such as ore flotation, movement of liquids in porous media (e.g. in crude oil recovery and soil science), as well as a range of other chemical technologies (Churaev, 2003).

Imagine bringing a flat smooth solid plate, immersed in liquid, towards the liquid surface, as illustrated in Fig. 12.18. One of three things can happen, which are analogous to what happens when an immiscible liquid is placed on a liquid substrate (see Fig. 12.6). System (a) represents a thick (duplex) wetting film on the solid. The disjoining pressure in such a film only rises if the film is forced to thin. Another possibility is illustrated in (b); here a thin wetting film is formed which can remain in equilibrium with a macroscopic

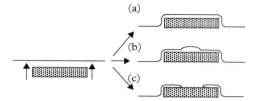

Figure 12.18 *When a flat smooth solid plate immersed in liquid is moved towards the liquid surface one of three situations can arise. (a) A thick wetting film is formed. (b) A thin wetting film is formed possibly in equilibrium with a macroscopic drop; the film surface makes a finite contact angle with the drop. (c) The liquid does not wet the solid and retracts forming a dry patch. After Padday (1970).*

drop, with which it makes a finite contact angle. The disjoining pressure in the film at equilibrium is zero and the interaction energy is minimum (attractive) (see Fig. 12.9). If the solid is not wetted by the liquid then at some stage when the film is quite thick, the liquid retracts to give a dry patch on the solid surface (case (c)).

The forces acting over thin wetting films are among the normal colloid forces discussed in Chapter 11. Electrostatic forces between charged surfaces in asymmetric films can be attractive or repulsive since the sign of charge on the two surfaces can be opposite. Equally, dispersion forces can be attractive or repulsive, depending on the Hamaker constants of the three phases, as discussed in §11.2.1 (see (11.2.8) and Table 11.2).

An apparatus suitable for the investigation of thin wetting films on solids is shown schematically in Fig. 12.19. The thin film is formed when an advancing liquid meniscus in a capillary tube approaches the plane, smooth (transparent) solid surface such as that of glass, mica, quartz or sapphire. The method is similar in principle to that used for the investigation of soap films and pseudo-emulsion films (see Fig. 12.11). The internal diameter, $2r$, of the capillary tube is around 1 mm. The air in the capillary (pressure p_a) is caused to move towards the solid surface either by increasing p_a, or by decreasing the pressure in the liquid using the adjustable side-arm. It was seen that, for an equilibrium film thickness h, the capillary pressure p_c is equal to the disjoining pressure $\Pi_D(h)$ (§12.2.2). The pressure in the meniscus, p_m, differs from p_a by the capillary pressure. If the air pressure in the capillary tube is equal to that above the manometer then

$$\Pi_D(h) = p_c = p_a - p_m = p_a - (p_a - \Delta\rho g l) = \Delta\rho g l \qquad (12.3.9)$$

where $\Delta\rho$ is the difference in density between the solution and air, g is the acceleration due to gravity, and l is the height indicated in Fig. 12.19. The thickness of the films can be determined optically from the interference of monochromatic light reflected from the two sides of the film (Exerova and Kruglyakov, 1998).

An example of a disjoining pressure isotherm for a thin aqueous wetting film on a silicon wafer (taken from Klitzing, 2005) is shown in Fig. 12.20. The isotherm for the

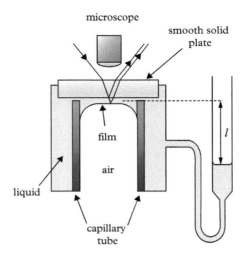

Figure 12.19 *Captive bubble apparatus for investigation of thin wetting films between air and solid surface. After Churaev (2003).*

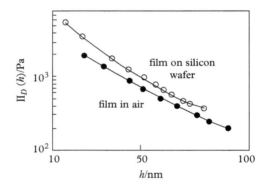

Figure 12.20 *Comparison of disjoining pressure isotherms for a free aqueous film and a wetting film on silicon. The films are stabilized by the nonionic sugar surfactant $C_{12}G_2$. Redrawn from Klitzing (2005).*

corresponding free aqueous (foam) film is also shown for comparison. The films were stabilized by the nonionic sugar surfactant dodecyl β-D-maltoside ($C_{12}G_2$). For a given disjoining pressure the wetting film is thicker than the foam film, and this is attributed to the greater (negative) surface potential (between -70 and -80 mV) at the silicon/water interface than at the air/water surface (-30 to -50 mV).

12.4   Drainage and rupture of thin liquid films

After a liquid film (e.g. a single film on a frame or a film within a foam) is formed it will begin to drain and thin. This results from gravity and through capillary suction into adjoining Plateau borders, driven by the Laplace pressure. In the case of the drainage of a vertical film of a pure liquid there is no velocity gradient normal to the film surfaces and so-called plug flow occurs, as illustrated in Fig. 12.21a, and drainage is rapid. Since there is no adsorbed surfactant present there is minimal repulsion between the surfaces and so the film is unstable and ruptures. It is well-known that stable films cannot be formed from pure liquids.

When adsorbed surfactant layers are present however, (Fig. 12.21b) the surfaces become rigid and behave somewhat like solid walls. There is a parabolic velocity profile normal to the film surfaces and the drainage rate is reduced. Furthermore, movement of liquid close to the surfaces creates a gradient in the surface concentration of surfactant, tending to cause a flow of adsorbed surfactant upwards. Interface movement is arrested (at the no-slip boundary condition) if this concentration gradient is sufficiently large. In addition to slowing down the film drainage, the adsorbed surfactant produces interfacial repulsion when the films are thin enough, leading to film stability.

Surfactant can also stabilize films by another mechanism. Suppose a film has drained and attained an 'equilibrium' thickness. If a patch of the film is subjected to a rapid expansion, such as might occur as a result of mechanical shock, the film will thin locally and potentially rupture. The expansion, however, leads to a decrease in the surface concentration of surfactant and surfactant will flow along the surface from high to low surface concentration. The flow drags along the underlying liquid which in turn causes the film to thicken again. This Marangoni effect (see §6.2.2) is an important mechanism by which potential weak spots in liquid films are healed. The Maragoni flow is opposed by adsorption of surfactant from the interior of the film, since this can also reduce surface concentration gradients. For very thin films, however, the supply of surfactant from the inside of the film may be so low as to be ineffective and the film in the limit becomes purely elastic (Box 12.2).

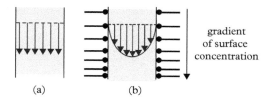

(a)   (b)

Figure 12.21 *Liquid flow velocities across vertical liquid films. (a) Pure liquid. (b) Soap film where the flow is inhibited at the surfaces giving rise to a velocity gradient across the film. Note that the surface concentration of surfactant varies down the film.*

Box 12.2 Gibbs elasticity

It is worth noting that the Gibbs elasticity ε for a thin liquid film of area $\mathcal{A}$ and film tension γ_f, that is,

$$\varepsilon = d\gamma_f/d\ln \mathcal{A} = -d\gamma_f/d\ln \Gamma$$

is defined for the *equilibrium* stretching of an *isolated* film. But even in an isolated film (such as that forming a bubble say) there will usually be surfactant inside the film and the equilibrium stretching will involve surfactant adsorption. The 'elasticity' will then contain a viscous as well as an elastic component. Furthermore ε will depend on film thickness because the supply of surfactant for adsorption will change with thickness.

Up to now thin films have been supposed to be structures with smooth interfaces, but thermal effects give rise to surface roughness in the form of capillary waves (see §8.1). The dynamics of capillary waves are determined mainly by surface tension (as opposed to gravity) which acts as a restoring force. Imagine a thin liquid film with symmetric capillary waves as illustrated in Fig. 12.22. The pressure at A, under the convex liquid surface, exceeds that at B under the concave surface. Liquid therefore tends to flow from A to B which acts to dampen the wave. If only attractive surface forces exist, the thinner parts of the film become thinner under these forces, which increases the chances of forming a hole in the film, leading to rupture. The capillary pressure dampening the capillary waves depends on the amplitude of the waves and the surface tension, and Vrij and Overbeek (1968) showed that there is a critical wavelength, λ_c, given by

$$\lambda_c = \left(\frac{2\pi^2\gamma}{d\Pi/dh} \right)^{1/2} \tag{12.4.1}$$

Above this wavelength the disjoining forces of attraction dominate over the capillary pressure and perturbations will grow. It is supposed that the film will rupture when the amplitude of perturbations is equal to the critical thickness, h_c, of the film, say around a few tens of nm. It is worth noting however that in the study of bubble interactions in water, referred to in §11.6.2 (Vakarelski et al., 2010), it was concluded that surface fluctuations appear not to be implicated in the bubble coalescence process. A review of

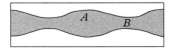

Figure 12.22 *Symmetric wave in a thin liquid film. The pressure under the convex surface near A exceeds that at B under the concave surface.*

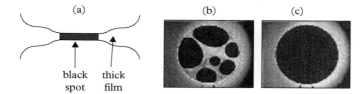

Figure 12.23 *(a) Representation of a black spot coexisting with a thick film. (b, c) Interference images of soap films formed from 2×10^{-4} M aqueous sodium dodecylsulfate solution. (b) Shows black spots from which a black film (c) is formed. Film diameters are 220 μm. Reproduced from K.P. Velikov, O.D. Velev, K.G. Marinova, and G.N. Constantinides, J. Chem. Soc., Faraday Trans., 1997, 93, 2069, with permission from the Royal Society of Chemistry.*

the drainage and rupture of free-standing thin liquid films has been given by Coons et al. (2003).

In soap films, the surfactant layers usually lead to repulsive forces (in addition to the attractive forces) between the surfaces. If the surfactant surface concentration (hence repulsion) is sufficiently high, areas of thin film ('black spots') form rather than the film rupturing (Fig. 12.23). The black spots can subsequently grow to cover the whole film.

REFERENCES

E.S. Basheva, P.A. Kralchevsky, K. Danov, K.P. Ananthapadmanabhan, and A. Lips, The colloid structural forces as a tool for particle characterization and control of dispersion stability. *Phys. Chem. Chem. Phys.*, 2007, **9**, 5183–5198.

V. Bergeron and C.J. Radke, Equilibrium measurements of oscillatory disjoining pressures in aqueous foam films. *Langmuir*, 1992, **8**, 3020–3026.

V. Bergeron, M.E. Fagan, and C.J. Radke, Generalized entering coefficients: a criterion for foam stability against oil in porous media. *Langmuir*, 1993, **9**, 1704–1713.

C.V. Boyes, *Soap Bubbles: Their Colours and the Forces That Mould Them.* The first edition appeared in 1890, and the second in 1911. The latter is available from Dover Publications, and various other sources on the internet.

N.V. Churaev, Surface forces in wetting films. *Adv. Colloid Interface Sci.*, 2003, **103**, 197.

J.E. Coons, P.J. Halley, S.A. McGlashan, and T. Tran-Cong, A review of drainage and spontaneous rupture in free standing thin films with tangentially immobile interfaces. *Adv. Colloid Interface Sci.*, 2003, **105**, 3–62.

B.V. Derjaguin abd N.V. Churaev, On the definition of the concept of wedging pressure and its role in the equilibrium and flow of thin films. *Kolloid. Zh.*, 1976, **38**, 438–448.

D. Exerova and P. M. Kruglyakov, *Foam and Foam Films: Theory, Experiment and Application.* Elsevier, Amsterdam, 1998.

R.V. Klitzing, Effect of interface modification on forces in foam films and wetting films. *Adv. Colloid Interface Sci.*, 2005, **114–115**, 253–266.

T. Kolarov, A. Scheludko, and D. Exerova, Contact angle between black film and bulk liquid. *Trans. Faraday Soc.*, 1968, **64**, 2864–2873.

J.F. Padday, Cohesive properties of thin films of liquids adhering to a solid surface. *Spec. Discuss. Faraday Soc.*, 1970, **1**, 64–74.

J.S. Rowlinson and B. Widom, *Molecular Theory of Capillarity*. Clarendon Press, Oxford, 1982, p.209 *et seq.*

A. Scheludko and D. Platikanov, Untersuchung Dunner Flussiger Schichten Auf Quecksilber. *Kolloid. Z. Z. Polym.*, 1961, **175**, 150.

I.U. Vakarelski, R. Manica, X. Tang, S.J. O'Shea, G.W. Stevens, F. Grieser, R.R. Dagastine, and D.Y.C. Chan, Dynamic interactions between microbubbles in water. *Proc. Natl. Acad. Sci.*, 2010, **107**, 11177–11182.

K.P. Velikov, O.D. Velev, K.G. Marinova, and G.N. Constantinides, Effect of the surfactant concentration on the kinetic stability of thin foam and emulsion films. *J. Chem. Soc., Faraday Trans.*, 1997, **93**, 2069–2075.

A. Vrij and J.T.G. Overbeek, Rupture of thin liquid films due to spontaneous fluctuations in thickness. *J. Am. Chem. Soc.* 1968, **90**, 3074–3078.

Section V

Dispersions stabilized by surfactants

The stabilization of solid dispersions, emulsions, and foams by surfactants, as well as some basic rheology of dispersions and surfactant solutions, is described in this section. Dispersions of solid particles in liquids, and the part played by the various surface forces described in Chapter 11 that are involved in their stabilization, are treated in Chapter 13. The preparation, stability, and breakdown of emulsions and foams are the subjects making up Chapter 14; the way in which unwanted foams and emulsions can be 'broken' are also discussed here. The rheology of dispersions containing surfactants, described in Chapter 15, is important because rheological properties can often be influenced so as to optimize the required behaviour of a wide spectrum of industrial and commercial formulations.

13

Dispersions of solids in liquids

13.1   How do interparticle forces determine the stability and physical properties of dispersions of solids in liquid?

Many systems of practical importance consist of particles dispersed or suspended in a liquid medium and this has stimulated much interest and research. Systems as diverse as paints and printing inks to cosmetics, pharmaceutical and agrochemical preparations, as well as various processed foods are particulate dispersions. Sometimes 'stable' systems, in which the particles do not mutually adhere and form aggregates or flocs, are required, whereas in other cases flocculation is necessary, as, for example, in the clarification of potable liquids. Aggregation and formation of particle networks can often impart desirable rheological properties on a system, as for example in paints.

In a *colloidal dispersion* the particles are small enough not to settle out under gravity. They undergo Brownian motion and when particles come close enough together they experience interactions of the kind described in Chapter 11. The magnitude and sign of the net interaction energy, which depends on the separation of the particles, determines whether or not the particles mutually adhere and hence whether the dispersion is stable or not. If the particles do adhere this ultimately leads to flocculation or coagulation. The size range of colloidal particles is about 1 nm to 1 μm; particles larger than this form *suspensions* in which particles tend to sediment (or cream) due to gravitational forces.

As can be expected, the properties of dispersions depend upon particle concentration. In a dilute dispersion the Brownian motion of the randomly dispersed particles is relatively uninhibited by the pairwise interaction of particles. At the other end of the concentration range, each particle interacts with a number of other neighbours to form a highly ordered 'solid' structure in which the translational movement of the particles is curtailed. In *concentrated dispersions*, which lie between the two extremes, many-body interactions between particles occur and the translational movement of the particles is restricted to a greater or lesser degree. Many industrially important systems are concentrated dispersions.

Van der Waals forces between like particles in a medium are universal and attractive so that if the forces of repulsion between particles are not strong enough, particle aggregation will inevitably occur. In the case of colloidal dispersions, although the

Surfactants: In Solution, at Interfaces and in Colloidal Dispersions. Bob Aveyard. © Bob Aveyard 2019.
Published in 2019 by Oxford University Press. DOI: 10.1093/oso/9780198828600.001.0001

primary particles do not settle under gravity, aggregates of particles can do so if sufficiently large.

Repulsive forces can arise from a variety of sources as discussed in Chapter 11. If the interacting surfaces carry like charge, repulsion results; the DLVO theory of colloid stability treats the balance of electrical repulsion and the attraction due to van der Waals forces in relation to colloid stability and the way in which addition of electrolyte influences the stability. Repulsion between surfaces can also arise from the presence of layers of polymeric/oligomeric moieties chemically attached or physically adsorbed to the surfaces, and such steric repulsion, unlike electrical repulsion, is insensitive to the presence of low concentrations of electrolyte. Of course both electrical and steric repulsion can occur simultaneously. If dissolved polymers or bolaform surfactants (see §1.3.6) are present within a particulate dispersion and are capable of adsorbing onto more than one particle simultaneously, particle bridging can result, leading to instability as the bridged structures grow in size. Polymers in solution can also give rise to attractive depletion forces between particles, which result from the exclusion of polymer molecules from the region between particles as they approach. Other solvent mediated surface forces can also exist, such as hydrophobic (attractive) and hydration (repulsive) forces. The existence of this variety of both attractive and repulsive interactions makes it feasible to fine tune the properties of particulate dispersions and suspensions and hence produce systems with a range of rheological characteristics. Although a full discussion of the rheology of dispersions is beyond the scope of this presentation, a simple account is given in Chapter 15; there are good and more detailed accounts available in the literature (e.g. Goodwin, 1987, 2009).

13.2   Systems stabilized by electrical repulsion

In Chapter 11 the interaction between unit areas of parallel flat plates carrying a surface charge was considered. However, dispersions of interest might more realistically be treated as spheres of equal size, and the interaction between particles is expected to be a function of the particle size. There are also further important considerations as a dispersion of charged particles becomes more concentrated. The particles are associated with their counterions in the double layers surrounding the particles, and as the particle concentration increases the concentration of the counterions increases to a much greater extent. This has a similar effect to adding electrolyte, and the double layers around the particles are compressed, tending to reduce the interparticle repulsion.

In the considerations of electrical interactions in Chapter 11 it was supposed that as two surfaces came together, the surface potential $\psi(0)$ and surface charge density $\sigma(0)$ remain unchanged. For small overlap of potentials the two quantities are related through (11.2.24). However as was mentioned, at close separation of surfaces the surface charge density may change (fall). Thus there are two possible limiting scenarios: the approaching surfaces remain at constant potential or at constant charge density. In practice it is not always obvious which condition is most nearly met, but for small potential overlap the distinction disappears.

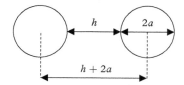

Figure 13.1 *Two identical spheres, radius a, with closest separation between surfaces =: h. The separation between centres is (h + 2a).*

In order to give some numerical examples for interactions in particle dispersions it is convenient to use some simple expressions for the van der Waals attraction and electrical repulsion between like spheres with radius a, as illustrated in Fig. 13.1 where various other distances are indicated. The electrical potential at the innermost part of the diffuse double layer around a particle (e.g. at the Stern plane in the case illustrated in Fig. 11.9) is denoted ψ_d.

For systems with particle radius much larger that the double layer thickness $(a >> \kappa^{-1})$, and where the closest separation of surfaces $h > 2\kappa^{-1}$ (weak overlap of the double layers), the electrical repulsion between the particles is given approximately (for $\psi_d < 25$ mV) by[1]

$$V_R(h) = 2\pi \varepsilon_o \varepsilon_r a \psi_d^2 \ln\left[1 + \exp\left(-\kappa h\right)\right] \tag{13.2.1}$$

In the same system the energy of attraction $V_A(h)$ arising from van der Waals forces can be written

$$V_A(h) = -\frac{A}{12}\left[\frac{1}{x(x+2)} + \frac{1}{(1+x)^2} + 2\ln\left(\frac{x(2+x)}{(1+x)^2}\right)\right] \tag{13.2.2}$$

where A is the Hamaker constant for the interaction of the spheres in the dispersion medium and $x = h/2a$. The variation with separation h of the total interaction energy $V_T(h) = V_R(h) + V_A(h)$, can show similar behaviour to that discussed in Chapter 11 (Fig. 11.12) for the approach of charged flat plates. Some calculated curves obtained using (13.2.1) and (13.2.2) are shown in Fig. 13.2. It has been supposed that the spheres, in aqueous 1:1 electrolyte, have a radius 70 nm and that $\psi_d = 20$ mV. The Hamaker constant for the interaction of the particles across the aqueous phases is taken to be 5×10^{-20} J, similar to that for alumina in water. The repulsion can be modulated by the addition of salt and curves (a), (b), and (c) are for 1:1 electrolyte concentrations of, respectively, $4 \times 10^{-4}, 4.5 \times 10^{-3}$, and 1×10^{-2} mol dm^{-3}. The salt screens the double layer repulsion so the curves become more negative (attractive) as the salt concentration increases. Curve (a), with an energy barrier of about $10kT$ represents the situation in a moderately stable colloidal dispersion, and curve (c) corresponds to an unstable

---

[1] The symbols all have the same significance as in §11.2.

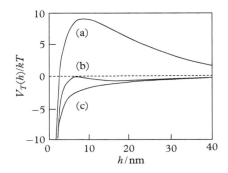

Figure 13.2 *Total energy of interaction between two like spheres, radius 70 nm, in aqueous 1:1 electrolyte. The potential $\psi_d = 20\ mV$ and the Hamaker constant $A = 5 \times 10^{-20}$ J. Curves (a), (b), and (c) are for $c = 4 \times 10^{-4}, 4.5 \times 10^{-3}$, and 1×10^{-2} mol dm^{-3}, respectively. Calculations were made using (13.2.1) and (13.2.2).*

dispersion in which particles tend to adhere on contact. They 'sit' in a deep energy minimum (not shown) and cannot be separated by shearing the system. The particles are said to *coagulate*. The intermediate curve (b) exhibits a very weak secondary minimum of about $1kT$, insufficient in this case to cause *flocculation*. Flocs are loose aggregates that can be readily re-dispersed on application of shear and reform on removal of shear.

It is obvious from Fig. 13.2 that the stability and structure of dilute electrically stabilized systems can be adjusted by addition or removal of electrolyte. If electrolyte is removed (by ion exchange or dialysis) the repulsion between particles increases and a regular ordered lattice can result. This is observed for suspensions of charged latex spheres. If the particle separation is of the order of the wavelength of light the dispersions exhibit iridescent colours. Such an effect can be given at low volume fractions of particles (say of the order 1×10^{-4}) and the repulsion can act over distances of many times the particle diameters. Addition of electrolyte to an initially random arrangement of particles can lead to coagulation or flocculation as described above.

Adsorption of ionic surfactant onto a charged solid surface alters (increases or decreases) the (net) surface charge density, as discussed in §7.5, and hence modifies the stability of a dispersion. This in turn affects the amount of electrolyte (measured as the *critical coagulation concentration*, c_K) that must be added to a charge-stabilized dispersion in order to just cause instability. This can be illustrated by the coagulation of sodium montmorillonite dispersions (Lagaly and Ziesmer, 2003). Montmorillonite is contained in bentonite which is used for example in drilling muds in crude oil production. As discussed in §7.2 clay particles have a *face* charge, negative for sodium montmorillonite, which does not depend on solution conditions, and an *edge* charge which is pH-dependent. The edges in montmorillonite have aluminol and silanol groups which are positively charged at low pH and negatively charged at high pH.

Small additions of the anionic surfactant sodium dodecylsulfate (SDS) at pH $= 6.5$ raise c_K for sodium chloride with 0.025 wt % sodium montmorillonite, as shown in

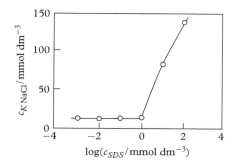

Figure 13.3 *Effect of adsorption of sodium dodecylsulfate (SDS) on the critical coagulation concentration, c_K, for NaCl with 0.025 wt % dispersions of sodium montmorillonite in water at pH = 6.5. Plotted using data of Permien and Lagaly (1995).*

Fig. 13.3. At this pH the edges of the clay plates are on average negatively charged. Small surfactant additions (1 mmol dm^{-3} and less) have little effect on c_K but above this concentration c_K rises substantially (but note the logarithmic scale of the abscissa). It has been argued (Ruprecht and Gu, 1991) that SDS anions can adsorb at sporadically occurring positive sites on the edges of the clay plates and that these adsorbed anions nucleate further adsorption, driven presumably by chain-chain attractions in the adsorbed layers. In this way the charge density is increased which leads to higher values of c_K.

In systems of practical and commercial importance, the stabilization/destabilization of dispersions of solids in liquids is most often controlled by polymeric surfactants, and the remaining sections of this chapter cover the effects of polymers, both adsorbed and dispersed, on dispersion stability.

13.3 Systems stabilized by steric interactions

The origins of steric interactions caused by the presence of adsorbed or grafted polymeric/oligomeric moieties on solid surfaces have been covered briefly in §11.3.1. Various kinds of macromolecules can be effective stabilizers of solid particulate dispersions including block and graft copolymers with anchoring groups and tails protruding into the (good) solvent (see §7.6). It has been seen that steric interactions (energy V_S) can arise in two ways. Mixing of chains attached to the two approaching surfaces (see Fig. 13.4) gives rise to an *osmotic* effect (repulsive energy V_{Sm}) and *volume restriction*, stemming from loss of configurational entropy as the polymer layers overlap, giving an energy of repulsion V_{Se}.

As mentioned in Chapter 11, dispersions can often be stabilized by very little overlap of the chains and if so the osmotic term predominates. The expression for V_{Sm} for the interaction between like spheres with radius a and adsorbed layer thickness δ,

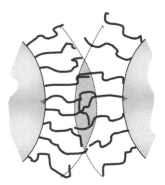

Figure 13.4 *Two spherical particles, radius a, coated with polymer layers, thickness δ. The layers mutually overlap as indicated by the central shaded area.*

$$\frac{V_{Sm}}{kT} = \frac{2\varphi_{2s}^2}{v_1}(0.5 - \chi)X \tag{11.3.2}$$

was given in §11.3.1, in which φ_{2s} is the volume fraction of polymer in the layers, v_1 is the molecular volume of the solvent, and χ is the Flory–Huggins parameter. The overlap volume X for two spheres is related to δ, h, and a by

$$X_{spheres} = (2\pi/3)(\delta - h/2)^2(3a + 2\delta + h/2) \tag{11.3.3}$$

where h is the shortest separation between the surfaces. Plots of V_{Sm}/kT versus h according to (11.3.2) and (11.3.3) for two spheres are shown in Fig. 13.5. The energy V_{Sm} is proportional to $(0.5 - \chi)$ so for a poor solved ($\chi > 0.5$) the steric interaction is attractive (negative). In good solvents however ($\chi < 0.5$), V_{Sm} is repulsive and greater the smaller χ (see curves (c) and (d) in Fig. 13.5). As pointed out in §11.3.1 it is frequently observed, as expected, that particle flocculation is sensitive to the proximity of θ conditions with respect to both solvent and temperature.

Steric repulsion, according to (11.3.2), depends on $(\varphi_{2s})^2$ (see curves (a) and (b) in Fig. 13.5) and for good stabilization polymer adsorption on the particle surfaces needs to be close to saturation. Temperature has a direct effect on steric stabilization, as seen from (11.3.2), and also because χ changes with T.

The way in which polymers can stabilize dispersions and under what conditions is well illustrated by a study reported by Whitby et al. (2003). The stabilizing polymer must have chains which extend into the medium in which the particles are dispersed and at the same time have the ability to adsorb onto the particle surfaces. For this reason, graft (comb) copolymers can be used (see Fig. 7.16). In this study, in which aqueous dispersions of silica were investigated, the stabilizer was a copolymer of poly(acrylic acid) (PAA) constituting the backbone, with grafted polyethylene oxide (PEO)/polypropylene oxide (PPO) moieties providing the chains. The surface charge on silica is pH-dependent

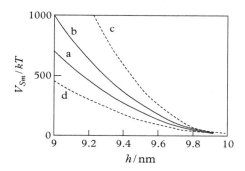

Figure 13.5 *Variation of V_{Sm} with separation h for two spheres, radius a = 100 nm, coated with polymer layers thickness $\delta = 5$ nm. Full lines are for $\chi = 0.4$ and (a) $\varphi_{2s} = 0.8$ and (b) 0.95; dashed lines are for $\varphi_{2s} = 0.9$ and (c) $\chi = 0.3$ and (d) 0.45.*

(§7.2), negative at high pH and positive at pH values less than about 3 (the *pzc*). The polymer adsorption is low at high pH where both the silica surface and polymer backbone are negatively charged. Adsorption is much stronger at low pH however indicating that it is the PAA backbone that adsorbs on the silica surface. The stability/instability of the silica dispersions was followed using rheological measurements (stress viscometry), and forces between silica particles and a flat silica surface were also directly measured concomitantly using atomic force microscopy (AFM) (§11.5.2).

In the presence of low concentrations of the copolymer, where significant steric forces are absent, the silica dispersions are elecrostatically stabilized. As the polymer concentration is increased, and under neutral and basic conditions (where polymer adsorption is low), the dispersions become weakly flocculated. Under these conditions the AFM measurements indicated that there are net attractive forces between the silica surfaces. For low pH a similar flocculation was observed for intermediate surface coverage by the polymer but the systems become stable at higher polymer concentrations/surface coverage as a result of steric stabilization. It was suggested that the instabilities for partial surface coverages are a result either of polymer molecules forming bridges between particles (§13.5) or possibly depletion flocculation (§13.4).

In summary, the foregoing rudimentary discussion of steric forces illustrates the need for certain conditions for effective steric stabilization. The polymer adsorption needs to be high such that the plateau in the adsorption isotherm has been reached (see §7.6). It has been seen how increasing φ increases the repulsion between particles (Fig. 13.5) but in addition, if bare surface exists on particles there is the possibility that a polymer molecule can adsorb simultaneously on two particles forming a bridge (see §13.5). Clearly the polymer must be strongly adsorbed such that it does not tend to desorb on approach of two particles. Further, the thickness δ of the layers is important. Steric forces operate in systems where other forces also exist. For example van der Waals forces between like particles are ubiquitous and attractive. If such forces are sufficiently strong at separations greater than that for the onset of steric repulsion (i.e. for separations greater than 2δ), it is possible that particles may be trapped in an energy minimum at $h \sim 2\delta$. Finally, according to (11.3.2) V_{Sm} is greater the smaller χ, that is, the better the solvent for the stabilizer chains.

13.4   Depletion forces between particles

In §11.3.3 the way in which free polymer molecules in solution can give rise to depletion forces of attraction between surfaces was discussed. The effect was first treated by Asakura and Oosawa (1954, 1958). The exclusion of the centres of polymer molecules from a thin layer next to a solid surface means that when two particles approach in a polymer solution, polymer is excluded from the region between the particles. This causes an osmotic flow of solvent from between the particles, resulting in an attractive depletion force. The force is directly related both to temperature and the polymer volume fraction in solution, φ_2.

The way in which r_g, the polymer radius of gyration, affects V_{dep} as a function of separation h for given φ_2, was illustrated in Fig. 11.19 for the interaction of two flat plates each 250 nm × 250 nm. The results are appropriate for say a cubic particle (side 250 nm) in the colloidal size range, and the attraction is of the order of several 10s of kT which can be sufficient to cause weak flocculation. Of course, aggregation of particles into a floc is associated with a negative entropy change which gives a positive contribution to the free energy change. When $h = 2r_g$, there is no depletion attraction between the plates. Then, attraction increases as h falls, more rapidly the smaller r_g because for a given φ_2 there are more polymer molecules present the smaller r_g, and osmotic pressure is a colligative property.

From what has been said it is clear that to induce depletion forces free polymer must be present in the suspension medium. This will be the case in systems with non-adsorbing polymer present, but if the polymer does adsorb the surfaces should be saturated with polymer and excess polymer must be present in solution, which means that the plateau in the polymer adsorption isotherm must have been attained.

As with the earlier discussion of steric interactions, only a simple description of the depletion force has been given in order to illustrate its origins. However, various issues arise as a result of the rather complex nature of polymer solutions (Goodwin, 2009). In practice, polymers can be expected to be polydisperse and the number of molecules in a given mass will depend on this polydispersity. The depletion force as seen depends on the osmotic pressure of the solution, which is a colligative property as mentioned. The polydispersity also means that there is a range of values of r_g. Further the radius of gyration will be influenced by the solvent quality, hence by the value of χ. In good solvents polymer molecules do not behave as hard spheres, and the segment density changes along a diameter as, therefore, does the solvent content. Other components present in the medium can also affect χ and r_g. Inorganic electrolyte for example can reduce the solubility of polymer moieties (i.e. cause 'salting-out', §5.4.2) and hence increase the value of χ towards 0.5. The radius of gyration of a polymer molecule is also dependent on the presence of other polymer molecules, that is, on polymer concentration.

Depletion interactions between particles are capitalized on in the control and adjustment of the flow (rheological) characteristics of suspensions, and this has no doubt stimulated interest in the study of depletion interactions. Rheological characteristics in concentrated suspensions are determined by the particle concentration, the range, size and magnitude of forces that act between particles, and the structures that are formed if particles mutually adhere (Tadros, 1986, Goodwin, 1987). It can be appreciated that polymer-mediated attraction between colloidal particles can result in phase separation of the mixture, and the concomitant phase behaviour has received much attention (see e.g. Tuinier et al, 2003). It may well be also that depletion forces are implicated in the organization of a wide range of cellular structures (Marenduzzo et al., 2006).

13.5 Bridging of surfaces

It has been seen how polymers/oligomers adsorbed or grafted onto solid surfaces can give rise to steric interactions (repulsive or attractive) between the surfaces and how free polymers in solution can lead to depletion attraction between surfaces. In solid dispersions containing low concentrations of polymer or other molecules that can adsorb simultaneously at available sites on two or more particles, bridging of particles can occur which can lead to flocculation as already indicated. Such instability caused by polymers has been referred to as *sensitization* in the early colloid literature (e.g. Freundlich, 1926). The (steric) stabilization of colloids by polymers was referred to as *protection* by Faraday, who was able to stabilize gold colloids by addition of gelatine (see §11.1).

Flocculation in suspensions by interparticle bridging is of great practical importance. Examples include the clarification of potable liquids, the recovery of solids from liquids (e.g. in mineral processing), and dewatering of suspensions such as sewage (Gregory, 1987). Linear high molecular weight polymers are often used as commercial flocculants, although particle bridging can be effected using low molar mass bolaform materials such as $HOOC(CH_2)_nCOOH$ (Leong, 1997). A prerequisite for bridging is that a flocculant molecule can adsorb on more than one particle simultaneously. Electrostatic forces between a charged polymer (polyelectrolyte) molecule and oppositely charged sites on particle surfaces can give rise to very strong adsorption. Ion binding can facilitate the adsorption of (say) anionic polymers onto negatively charged surfaces in the presence of divalent metal cations such as Ca^{2+}, which are present in hard water. Hydrogen bonding, for example, between OH groups on oxide surfaces and amide groups on polyacrylamide molecules, can also give effective adsorption. Hydrophobic interaction between say alkyl chains in a polymer and hydrophobic surfaces present in water is another possible reason for adsorption (see §7.3).

Since polymeric flocculants are often of high molecular weight, and therefore of similar size to the particles they bind, they can bridge several particles simultaneously (Fig. 13.6), ultimately leading to the formation of large flocs. If the particles carry a charge (in water) they mutually repel, and so a flocculant molecule must be able to span a distance between particles similar to that over which electrical repulsion operates (which can be as much as 100 nm). In practice, aggregates often have to undergo shear and if so

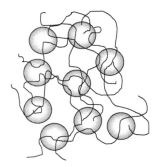

Figure 13.6 *Formation of a floc by bridging of particles by large polymer molecules.*

the particles must be strongly bound to each other if the flocs are to remain intact. This requires that a given particle must be involved in several polymer bridges. For charged particles initially stabilized by electrical repulsion, the repulsion can of course be reduced by addition of electrolyte until aggregation occurs (as discussed in §11.2.4) but the van der Waals attractive forces holding the resulting flocs together are very much weaker than those provided by bridging.

It is found that formation of stable flocs requires an optimum concentration of flocculant to be present. If insufficient adsorption occurs there are too few bridges formed to give stable aggregates whereas if adsorption is too great particles are (re)stabilized, possibly by steric stabilization. However, in aqueous systems with negatively charged particles together with cationic polymers, the optimum dosage may arise for different reasons. In these systems it can be expected that the adsorbed polymer molecules, rather than bridging particles, take up configurations which are relatively flat along the surface. The charge neutralization could therefore be responsible for destabilizing the suspensions. In support of this notion, it is found for a number of systems that the optimum amount of cationic polymer required to destabilize the negatively charged suspensions corresponds to the condition where the solid surface has zero net charge, that is, the particles have zero ζ-potential and electrophoretic mobility. At higher polyelectrolyte concentrations, if excess polymer is adsorbed the net surface charge becomes finite again and the suspensions are re-stabilized.

....

REFERENCES

S. Asakura and F. Oosawa, On interaction between two bodies immersed in a solution of macromolecules. *J. Chem. Phys.*, 1954, **22**, 1255–1256.
S. Asakura and F. Oosawa, Interaction between particles suspended in solutions of macromolecules. *J. Polym. Sci.*, 1958, **33**, 183–192.
H. Freundlich, *Colloid and Capillary Chemistry*. Methuen, London, 1926.
J.W. Goodwin, in *Solid/Liquid Dispersions* (ed. T.F. Tadros), Chapter 10. Academic Press, London, 1987.

J.W. Goodwin, *Colloids and Interfaces with Surfactants and Polymers*, 2nd edition. John Wiley & Sons, Chichester, UK, 2009.

J. Gregory, in *Solid/Liquid Dispersions* (ed. T.F. Tadros), Chapter 8. Academic Press, London, 1987.

G. Lagaly and S. Ziesmer, Colloid chemistry of clay minerals: the coagulation of montmorillonite dispersions. *Adv. Colloid Interface Sci.*, 2003, **100–102**, 105–128.

Y.-K. Leong, Inter-particle forces arising from adsorbed bolaform surfactants in colloidal suspensions. *J. Chem. Soc., Faraday Trans.*, 1997, **93**, 105–109.

D. Marenduzzo, K. Finan, and P.R. Cook, The depletion attraction: an underappreciated force driving cellular organization. *J. Cell Biol.*, 2006, **175**, 681–686.

T. Permien and G. Lagaly, The rheological and colloidal properties of bentonite dispersions in the presence of organic compounds V. Bentonite and sodium montmorillonite and surfactants. *Clays Clay Minerals*, 1995, **43**, 229–236.

H. Ruprecht and T. Gu, Structure of adsorption layers of ionic surfactants at the solid/liquid interface. *Colloid Polym. Sci.*, 1991, **269**, 506–522.

T.F. Tadros, Control of the properties of suspensions. *Colloids Surf.*, 1986, **18**, 137–173.

R. Tuinier, J. Rieger, and C.G. de Kruifa, Depletion-induced phase separation in colloid–polymer mixtures. *Adv. Colloid Interface Sci.*, 2003, **103**, 1–31.

C.P. Whitby, P.J. Scales, F. Grieser, T.W. Healy, G. Kirby, J.A. Lewis, and C.F. Zukoski, PAA/PEO comb polymer effects on rheological properties and interparticle forces in aqueous silica suspensions. *J. Colloid Interface Sci.*, 2003, **262**, 274–281.

14

Emulsions and foams

14.1 Preliminary remarks about emulsions and foams

Emulsions are dispersions of drops of one liquid in a second liquid (the dispersion medium) in which they are insoluble or only sparingly soluble. They typically contain an 'oil' and an aqueous phase, together with a stabilizer, which can be a low molar mass surfactant, a polymer (e.g. a protein or an ABA polymer), or, in the case of Pickering emulsions, solid particles (Chapter 18). Emulsions are often referred to as colloids although droplet sizes are frequently well outside the colloidal size range, which is 1nm - 1μm. Unlike colloidal particles, emulsion droplets (with diameters commonly between say 1 and 50 μm) sediment or cream as a result of their size and the density difference between droplet phase and medium. Thin liquid films exist between droplets in close proximity however, and the thickness of these 'emulsion' films is in the colloidal range (Chapter 12), and both their static and dynamic properties are directly linked to overall emulsion stability. The various forces that act across emulsion films are discussed in Chapters 11 and 12, and arise largely as a result of the presence of stabilizer adsorbed at film interfaces.

Broadly, emulsions can consist of either oil drops dispersed in water (O/W emulsion) or conversely water drops in oil (W/O emulsion). Which type is given in a particular system depends in part on the relative volumes of oil and water present (see also Box 14.1). The phase of larger volume fraction tends to be the continuous phase, and if phase volumes are progressively changed, emulsions can *invert* from one form (say W/O) to the other (O/W)—in this case as the volume fraction of (initially dispersed) water is increased. This type of inversion has been referred to as catastrophic inversion; it occurs abruptly as the volume fraction is changed. The ratio of the viscosities of the phases can also have an effect on emulsion type, with the phase of higher viscosity tending to be the dispersion medium. It is also found that for O/W emulsions the efficiency of homogenization during preparation is significantly dependent on the viscosity of the oil phase.

One of the most important factors that determine emulsion type, however, is the nature of the surface active material used as stabilizer, and the conditions under which it is used, for example, temperature, aqueous phase salt concentration, chemical structure

Surfactants: In Solution, at Interfaces and in Colloidal Dispersions. Bob Aveyard. © Bob Aveyard 2019.
Published in 2019 by Oxford University Press. DOI: 10.1093/oso/9780198828600.001.0001

Box 14.1 What if both oil and water droplets are present? ... multiple emulsions

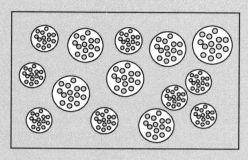

An oil (shaded) in water in oil (O/W/O) multiple emulsion.

So-called *multiple* emulsions also exist and are very useful practically. In a given multiple emulsion both water drops and oil drops are present. In an oil/water/oil (O/W/O) emulsion, for example, the water drops, dispersed in oil, also contain smaller dispersed oil droplets. If such emulsions are stabilized by surfactants, two different surfactants must be used, one favouring O/W emulsions and one W/O emulsions (in the presence of the particular oil used) (§14.2.5). Solid particles (of differing wettability) are particularly useful in the stabilization of multiple emulsions since they are irreversibly adsorbed (Chapter 17) and so cannot readily transfer between water droplet and oil droplet surfaces, which would render them ineffective as stabilizers (§18.2.3).

of the oil phase, and, when surfactant mixtures are used, the ratio of the amounts of the two surfactants present. These factors taken together determine the so-called *hydrophile–lipophile balance* (HLB) of a system, and are the same factors discussed in connection with the behaviour of *micro*emulsions in Winsor systems (§10.1.2). The HLB is discussed in more detail in §14.3.

As seen in Chapter 10, microemulsions are thermodynamically stable whereas emulsions (which can be called *macro*emulsions in the context) are not. Imagine the dispersal of a given volume of oil as microemulsion droplets in a fixed volume of water. The free energy change for the process arises from the creation of droplet interfaces, which have effectively zero interfacial tension (see Box 10.4, Chapter 10), and the positive entropy of dispersion of the drops in water. The net free energy change is negative and microemulsions form spontaneously. If the same amount of oil were to be dispersed in a similar volume of water to form (a much smaller number of) macroemulsion droplets, the work of interface formation would be large and positive since, although the surface area formed is relatively small, the tension at an emulsion drop interface is very much larger (by a factor say of 10^4) than that at a microemulsion drop interface. Further, the positive entropy of mixing the much smaller number of macroemulsion drops with the water would be much less positive, and as a consequence, the net free energy change accompanying macroemulsion formation is positive; the emulsions are not thermodynamically stable. They can however be extremely stable in a kinetic sense. This is because the various processes leading to emulsion breakdown, discussed below, can have substantial activation energies leading to very slow droplet coalescence and phase resolution.

For non-deformable monodisperse spherical drops, the maximum volume fraction of the droplet phase in an emulsion is 0.74. For polydisperse and deformable (Box 14.2) droplets, however, much higher volume fractions can be attained. At sufficiently high volume fractions the continuous phase exists in the form of a network of thin liquid films giving rise to *biliquid foams*. These are foam-like structures in which the gas phase of a foam is replaced by liquid. Systems with volume fractions of dispersed phase in excess of 0.74 are referred to as high internal phase emulsions (HIPEs).

Box 14.2   Are emulsion drops deformable?

There is an excess (Laplace) pressure, Δp, inside a liquid droplet which is directly proportional to the drop interfacial tension, γ, and inversely related to drop radius, r, according to the Laplace equation (see §3.4.1):

$$\Delta p = 2\gamma/r$$

As a result the smaller the drop and the higher the tension, the less deformable the drop becomes. For drops with diameter 1 μm and a fairly typical interfacial tension of 10 mN m^{-1} the excess pressure within the droplets is 40 kPa (about 0.4 atm) and so under many relevant conditions the drops can be considered non-deformable (Walstra, 1996).

Emulsions are ubiquitous in everyday life and are present in many industrial processes and commercial products. Foodstuffs are often O/W emulsions (e.g. milk, mayonnaise) and some are W/O emulsions, such as margarine and low fat spreads (Dickinson, 1992). In many agrochemical products the active materials are present in either O/W or W/O emulsions, and personal care products and cosmetics are often formulated as emulsions. Frequently, unwanted emulsions are encountered, such as dispersed drops of brine in crude oil during processing, and in such cases methods are required to render the emulsions kinetically unstable.

The type of an emulsion once formed is readily determined using a variety of simple methods, one of which is to add a sample of the emulsion to water. An O/W emulsion will mix freely with water whereas a W/O emulsion will not. Alternatively a water-soluble dye can be stirred into the emulsion, and if the emulsion is water-continuous it will become coloured throughout; if not the dye remains un-dissolved. Since an aqueous phase containing even a very low amount of electrolyte is highly conducting (relative to oil) the conductivity of an emulsion indicates the emulsion type. This approach is often used to study the inversion of emulsions from one type to the other (see §14.4 and §18.2.2).

Liquid foams, which are dispersions of gas (usually air) in liquid in the presence of surface active stabilizer, are as familiar in everyday life as are emulsions. Although in many cases foams are desirable or even essential (e.g. in many foodstuffs, and toiletries) there are frequently instances where foams are highly undesirable and must be broken (e.g. in paper manufacture, crude oil treatment and in paint application). The gas regions can be either spherical bubbles (in *kugelschaum* or *wet foam*) or polyhedral structures *(polyederschaum)* consisting of a network of essentially planar thin liquid films joined through Plateau borders (see §12.1). Aqueous foams are perhaps the most frequently encountered type in everyday life, but non-aqueous foams are also common. As in the case of emulsions, foam stability is determined in part by the factors affecting the stability of the individual thin liquid films. In addition bubble disproportionation, which arises from the polydisperse three-dimensional structure of the foam, contributes to foam breakdown. This process is called Ostwald ripening, and also occurs in polydisperse emulsions in which large drops become larger and small drops smaller with time as a result of the differences of Laplace pressures in the dispersed entities. Solid foams (not to be confused with dry foams already referred to) also exist in which the films between gas cavities are solid.

In what follows, emulsions are considered first. A great deal is known and has been written about these systems and the treatment here, given the theme of the book, will emphasize the role played by the surface active stabilizers in the formation, stability and breakdown of emulsions. A discussion of foams follows, and self-evidently the behaviour of thin liquid films is central to foam science.

14.2   How are emulsions made?

Emulsions are usually formed using one of two general methods: *comminution* (dispersion) or *condensation*, although several other approaches are also used depending on whether the process is being carried out industrially or in the laboratory.

14.2.1   Comminution

Comminution involves mixing the components (say oil, water, and stabilizer) and shearing them either by simple shaking or by applying more controlled and intense agitation, which tears one or both fluids into droplets. Hand shaking can produce polydisperse droplets with the smallest drops having diameters of say 20 μm, whereas some laboratory mixers can achieve mean diameters as low as 5–10 μm. Industrial machines can produce emulsions with smaller droplet diameters around 1 μm (Dickinson, 1992). As will be seen, processes by which emulsions break down depend on droplet size, smaller drops being the more stable. Since droplets are stabilized against coalescence by adsorbed surfactant, it is necessary for adsorption onto the newly formed, originally bare droplets to be more rapid than initial mutual contact between drops, which occurs in the order of milliseconds.

Formation of emulsion drops involves the creation of oil/water interface (area $\mathcal{A}$ and interfacial tension γ), and the thermodynamic free energy of formation is therefore $\gamma\mathcal{A}$ plus the relatively small free energy arising from the entropy of dispersion of the droplets in the medium. It might be thought that this equates to the input energy required to form the emulsion. Walstra (1996) presents calculations relating to the formation of 1 m^3 of an O/W emulsion in which the oil volume fraction φ is 0.1, the droplet diameter $d = 1\,\mu$m and the oil/water interfacial tension $\gamma = 0.01$ N m^{-1} (typical for an oil/water interface with surfactant present at its *cmc*). The free energy of forming the droplet interfaces is 6 kJ, whereas the mechanical energy required to form such an emulsion is of the order of 5×10^3 kJ, three orders of magnitude greater. The energy input is required to distort oil/water interfaces and is associated with an increase in local Laplace pressure difference Δp (§3.4.1) over the curved interfaces. In laminar flow the energy for this distortion is provided by viscous forces arising from the flow of the continuous phase. Most of the mechanical energy input ends up as heat, and there is no simple correlation between the interfacial free energy and emulsion stability.

14.2.2   Condensation

In the condensation method the droplet phase (say oil) can be dissolved in a liquid that is water-soluble; this solution is then mixed with the aqueous phase, giving a supersaturated solution of the oil in water. Alternatively, a saturated solution (of say oil in water) can be subjected to a temperature change so that the solution becomes supersaturated and oil comes out of solution as fine droplets. The treatment of nucleation of vapours to give liquid droplets given in §3.4.2 is applicable to the condensation of liquid droplets from supersaturated solution, with vapour pressures replaced by corresponding solution concentrations. Equation (3.4.11) for the free energy of formation of a droplet, radius r and interfacial tension γ may be written in the present context as

$$\Delta_d\mu = \frac{4}{3}\pi r^3 \Delta g + 4\pi r^2 \gamma \tag{14.2.1}$$

in which

$$\Delta g = -\frac{RT}{v}\ln\left(\frac{c}{c_s}\right) \tag{14.2.2}$$

and v is the molar volume of the condensing liquid. The quantity Δg is the free energy of condensation per unit volume of droplet phase (excluding the work of interface formation), and c and c_s are, respectively, the supersaturated concentration of the

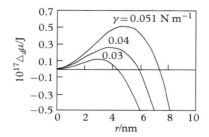

Figure 14.1 $\Delta_d\mu$ *as a function of droplet radius. The supersaturation ratio* c/c_s *is taken as 3, and interfacial tensions assumed are indicated on the curves. Values of parameters used are appropriate for the formation of hexane droplets in water.*

dissolved droplet phase and the saturation concentration under the prevailing conditions. The free energy of droplet formation passes through a maximum, $\Delta_d\mu^*$, with respect to droplet radius at r^*, the radius of the critical nucleus, as shown in Fig. 14.1. By setting $d\Delta_d\mu/dr = 0$ it is readily shown that

$$\Delta_d\mu^* = \frac{16\pi\gamma^3}{3\Delta g^2}; \quad r* = -\frac{2\gamma}{\Delta g} \tag{14.2.3}$$

A droplet formed with $r > r^*$ grows spontaneously, and when $r < r^*$ it re-dissolves. It is seen that the lower the interfacial tension, the lower $\Delta_d\mu^*$ and the smaller r^*. If it is supposed that the above treatment is essentially valid for systems containing surfactant, and that surfactant merely causes a reduction in γ, then surfactant is expected to cause a reduction in both $\Delta_d\mu^*$ and r^*, and hence facilitate droplet formation.

Once a droplet is formed, its growth is expected to be inhibited to a greater or lesser extent by the presence of the adsorbed surfactant monolayer at its surface, which exhibits surface dilational elasticity (see §6.2.2. and Box 12.2 in Chapter 12). At very low surface concentrations, Γ, of surfactant the modulus $\varepsilon = -d\gamma/d\ln\Gamma$ is low. It is also low at high surfactant concentrations because as the interface expands, concomitant adsorption can take place so that Γ remains constant and ε is zero. At intermediate surfactant concentrations, however, the (visco)elasticity is finite and expected to retard droplet growth. Adsorbed surfactant also stabilizes the drops kinetically against coalescence as discussed below.

14.2.3 'Spontaneous' emulsification

In a sense, emulsions formed by condensation are formed spontaneously, but this does not imply the emulsions are thermodynamically stable. In some cases, spontaneous emulsion formation arises from surface turbulence. A well-known example of this involves (ethanol + toluene) mixtures when added to water. In addition to the condensation process discussed above, interfacial turbulence caused as the components cross the various interfaces further contributes to spontaneous emulsion formation.

In Chapter 10 (Box 10.4) it was seen that the interfacial tension at microemulsion drop interfaces is very low, approaching zero, and that microemulsions, which are thermodynamically stable, form spontaneously upon mixing the various components. By the same token, it is also possible for macroemulsions to form spontaneously if *transient* interfacial tensions can be reduced to very low values. The tensions at the drop surfaces after formation however do not remain ultralow, and emulsions so-formed are not thermodynamically stable. An example of such emulsion formation was reported by Gerbacia and Rosano (1973). Pentanol was injected into hexadecane in contact with aqueous sodium dodecylsulfate (SDS). As the alkanol crossed the interface between hexadecane and aqueous SDS, the oil/water tension (which was measured) became very low and remained so for the order of a minute, during which time spontaneous emulsification was observed. This kind of process can occur more generally when an (oil + water) system contains surfactant mixtures or a surfactant and cosurfactant (such as an alkanol). So-called emulsifiable concentrates are systems of this type. A water-insoluble active material (say an agrichemical) is dissolved in an oil which contains surfactant together with cosurfactant that has an appreciable solubility in water. On addition of oil to water the cosurfactant distributes across the oil/water interface into the aqueous phase producing, transiently, very low interfacial tensions resulting in spontaneous emulsification.

14.2.4 'PIT method' and formation of nanoemulsions

The phase behaviour with respect to changes in temperature in systems with oil, water and say a nonionic surfactant has been shown schematically in Fig. 10.22. For a system with roughly equal amounts of water and oil at lower temperatures the aqueous phase is an O/W microemulsion which is in equilibrium with the excess of the oil containing nonionic surfactant monomers at the critical aggregation concentration (*cac*). At high temperatures the systems consists of a W/O microemulsion in equilibrium with excess aqueous phase. At low and high temperatures respectively the systems are Winsor I (WI) and Winsor II (WII) which, when agitated, produce O/W and W/O macroemulsions respectively, as discussed in more detail in §14.3.

As explained in §10.1.3, as the temperature is raised through the WI to the WII region the oil/water interfacial tension passes through a very low minimum in the three-phase WIII regime. The temperature corresponding to the minimum in tension, at which the type of macroemulsion formed inverts from O/W to W/O on heating, is termed the *phase inversion temperature* (PIT). It is found that the emulsion droplet size (Box 14.3) is a minimum around the PIT (Shinoda and Friberg, 1986), but that emulsion stability against coalescence is very low. However, by rapidly cooling an emulsion formed close to the PIT, the small droplets are stabilized against coalescence.

Box 14.3 What's in a name?

The nomenclature of emulsions of various kinds according to drop size can be confusing and appear illogical. The term *micro* in a general sense is taken simply to mean small (from Greek *mikros* meaning small), as in say *micro*scopic. In science micro is a prefix meaning 10^{-6}. Drop diameters of emulsions (also referred to as *macro*emulsion) are in the micron range with diameters from say 1 to 50 μm. The term *micro*emulsion, however, was coined by J.H. Schulman and T.P. Hoar in 1943, implying drop sizes much smaller than those of macroemulsions; microemulsion drop diameters are a few 10s of *nano*metres (as seen in §10.1.2). To confuse matters further there is a now a class of emulsions termed *nanoemulsions* (also sometimes called *mini-emulsions*). These are kinetically (rather than thermodynamically) stable emulsions with droplet diameters in the range of say 50–200 nm.

The PIT method is a convenient way to prepare *nanoemulsions*, sometimes also called *mini-emulsions* (Tadros et al., 2004). These are emulsions with drop diameters often in the range 50–200 nm. Because the drops are so small the emulsions are, like microemulsions, transparent or translucent but, unlike microemulsions they are not thermodynamically stable. Nonetheless, as a result of the small droplet size they do not cream or sediment and are remarkably kinetically stable. Since the droplets are effectively non-deformable (see footnote 2) flocculation tends not to occur and coalescence is prevented. The most common form of instability in nanoemulsions arises from Ostwald ripening in polydisperse systems. Although the PIT method can be used to prepare nanoemulsions they can also be formed by high pressure homogenization or by ultrasonication (see next section).

In emulsions that contain three phases and where the third phase is liquid crystalline, liquid crystals can coat the droplets conferring considerable *mechanical* stability on the emulsion. The liquid crystalline layers on the drops are birefringent and can be observed microscopically using a polarizing microscope.

14.2.5 Some other methods and systems

When a liquid is irradiated ('sonicated') with high intensity ultrasound (frequencies > 20 kHz), the sound waves in the liquid give rise to alternating high and low pressure waves. The high pressures cause compression of the liquid and the low pressures dilation, the latter causing small bubbles to form (cavitation). These cavities collapse when subjected to high pressure giving rise to a microscopic shock wave. If the collapse occurs close to a large drop, the shock wave causes fragmentation into smaller drops. Ultrasound can be used in the laboratory, and it is also used commercially in the production of a large range of emulsified systems and products. Droplet diameters of 1 μm or less can be achieved by this method, depending on input energy.

Comminution methods often involve high shear stresses, which can result in coalescence of drops already formed. A method of emulsification which avoids this problem involves forcing one of the liquids (the droplet phase) through a porous membrane or

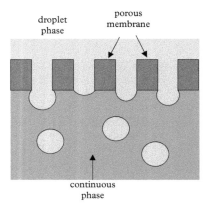

Figure 14.2 *Porous membrane method of emulsification.*

an array of micro-channels into the second, continuous phase (Sotoyama et al., 1999; Kobayashi et al., 2009) (Fig. 14.2). If the pore diameters are uniform then the size distribution of emulsion drops can be fairly sharp. In order to give controlled droplet size the channel surfaces should be non-wetted by the droplet phase, so that for the formation of say W/O emulsions the membrane surface should be hydrophobic.

Micro-sieves suitable for production of monodisperse emulsions in the micron size range are available commercially (see e.g. Aquamarijn, 2016). The silicon membranes can be produced reproducibly by photolithography with uniform pore sizes and shapes.

Multiple (double) emulsions (sometimes called *liquid capsules*—Tadros, 1984) contain drops within drops as mentioned earlier in the chapter. They can be prepared using two surfactant types, one (high HLB) favouring the formation of oil drops in water and the other (low HLB) which gives W/O emulsions (see §14.3). The emulsions are prepared in two stages, the first stage of which is to make the primary emulsion. For example, in the preparation of a W/O/W emulsion, a W/O emulsion is first made and this is then emulsified in water. The emulsions tend not to be very stable however, often coalescing relatively quickly to give a simple emulsion. More stable double emulsions have been obtained by using hydrophobic and hydrophilic particles as stabilizing agents, as described in §18.2.3.

14.3 What type of emulsion is formed? The hydrophile–lipophile balance

Early views about how emulsion type is related to the 'solubility' of surfactants in oil and in water were encapsulated in the well-known *Bancroft's rule* to the effect that the favoured type of emulsion is that in which the surfactant is more soluble in the continuous phase. This understanding, however, needs to be qualified. The way in which surfactants distribute between oil and water in mutual contact has already been discussed in §10.1.1.

It is important to note that surfactant distribution is not simply related to the solubility of a surfactant in each of the two phases separately, that is, not in contact. For example, Aerosol OT (AOT), a twin-tailed anionic surfactant, is over 10 times more soluble in heptane than in water, and yet surfactant *monomers* distribute almost entirely into water when heptane and water are in contact. For low or zero concentrations of NaCl in the aqueous phase ($m_s < 0.05$ mol dm^{-3}) surfactant *aggregates* (above the *cac*) are also formed in the aqueous phase. At higher salt concentrations ($m_s > 0.05$ mol dm^{-3}), however, although surfactant monomers remain almost exclusively in the aqueous phase, aggregates now form in the heptane phase. The locus of surfactant as aggregates (micelles or microemulsion droplets) is determined by the curvature properties of close-packed monolayers of surfactant at the oil/water interface (as explained in §10.2.1) and is therefore unrelated to the monomer distribution.

For systems containing roughly equal volumes of oil and water together with surfactant above its *cac*, a revised statement of the Bancroft rule would be that the type of emulsion formed is that in which the *continuous phase contains the surfactant aggregates*. In some systems this will mean that close to the *cac* the greater part of the surfactant is present as monomers in the droplet phase, as for example in the system $C_{12}E_5$ + water + heptane at 293 K. The ratio of surfactant monomer concentration in oil to that in water is 250:1 and yet surfactant aggregation occurs in the aqueous phase to give O/W microemulsion droplets. When this system is homogenized an O/W emulsion is formed, the continuous phase of which is an O/W microemulsion. The droplet phase contains the surfactant monomer which, close to but above the *cac*, is the major part of the surfactant present.

In order to relate the surfactant molecular structure *empirically* to the type of emulsion that it stabilizes, a so-called HLB number has been assigned to surfactants (Griffin, 1954; Shinoda and Friberg, 1986). The hydrophilic groups and hydrophobic groups each have group numbers, H and L, respectively, which give a measure of the hydrophilicity or lipophilicity of a group; values of H are high and of L are low. An HLB number can then be computed using, for example,

$$HLB\ number = xH - yL + 7 \qquad (14.3.1)$$

where x and y are the numbers of hydrophilic and hydrophobic groups, respectively, in the surfactant molecule. Shinoda and Friberg list the HLB numbers for a large number of commercial surfactants. Broadly, surfactants with a high HLB number (say 12–16) tend to stabilize O/W emulsions and those with low HLB numbers (say 4–8) favour W/O emulsions.

Although the HLB concept as outlined has been widely used in the past, it is important to stress that emulsion type depends not only on the surfactant molecular structure but also on a number of other factors. Since emulsion type depends on the Winsor behaviour of a system and since the latter depends on the variables given in §10.1.2 (Table 10.1), emulsion type also depends on these same variables. For example systems with roughly equal volumes of nonpolar oil, water, and a given nonionic surfactant exhibit WI behaviour at low temperatures, and produce O/W emulsions on agitation. At higher

Table 14.1 *Effect of HLB variables on emulsion type*

Surfactant type	Increase in variable	O/W↔W/O
ionic and nonionic	aqueous salt concentration	→
ionic	temperature	←
nonionic	temperature	→
ionic and nonionic	alkane chain length	←
ionic and nonionic	cosurfactant concentration	→

temperatures systems become WII type that produce W/O emulsions. The ordinate (temperature) on phase diagrams such as that shown in Fig. 10.22 can be replaced by other, what might be termed 'HLB variables' such as aqueous phase salt concentration, chain length of alkane (or average chain length in alkane mixtures), ratio of amounts of hydrophilic and lipophilic surfactants in binary mixtures, or amount of cosurfactant added to the system. The emulsion type formed in relation to these variables is given in Table 14.1, which is the equivalent of Table 10.1 but given in terms of emulsion type rather than Winsor behaviour.

In summary, and importantly, the 'HLB behaviour' relates to the *whole system* and not to surfactant structure alone, and equations such as (14.3.1) are of very limited use and can be misleading.

14.4   What are the origins of emulsion type?

In Chapter 10, the formation of microemulsion droplets was discussed in terms of the curvature properties of close-packed surfactant monolayers at the oil/water interface. As seen above it is observed empirically that, for systems with roughly equal volumes of oil and water, the emulsion type formed is the same as the type (O/W or W/O) of the surfactant aggregates, which are present in the continuous phase.[1] However, the aggregates are much smaller than emulsion drops and their curvature very much greater. So why should the type of emulsion drop, with *low* curvature, depend on the curvature properties of surfactant monolayers?

Kabalnov and Wennerström (1996) give an explanation of the connection between microemulsion and macroemulsion type in terms of the formation and growth of the bridge that must exist between colliding emulsion droplets before coalescence can occur. As shown in Fig. 14.3, prior to coalescence a thin liquid (emulsion) film forms between

---

[1] Although emulsion type has been discussed above in relation mainly to the type of *microemulsion* present, the surfactant aggregates may contain only little oil or water and simply be *swollen micelles* or reverse micelles. For example, the anionic surfactant SDS can stabilize alkane-in-water emulsions but the surfactant aggregates present in the aqueous phase are normal micelles with possibly some solubilized alkane present.

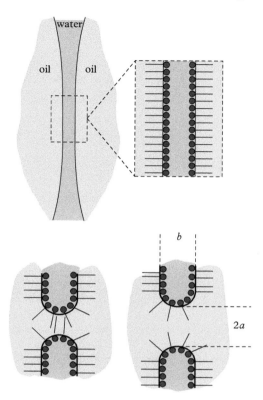

Figure 14.3 *Top: Approach of two oil drops in water with adsorbed surfactant, and formation of a thin aqueous film in oil. Bottom: Formation and growth of an oil neck between the oil droplets. The radius of curvature of the neck parallel to the film is a, and the film thickness is b. After Evans and Wennerström (1999).*

the droplets. In the example shown, the drops are oil drops in water and so an O/W/O film is formed. After initial rupture of the film an oil neck is created (bottom left in the figure), which then grows. This growth entails changes in the area of both the plain film and the curved neck, as well as changes in the mean curvature of the growing neck.

The neck has a radius of curvature a in the direction parallel to the film. If the film thickness b remains constant, the change in curvature of the neck as it grows arises solely from changes in a. Suppose Fig. 14.3 relates to an O/W emulsion stabilized by nonionic surfactant at a temperature below the PIT. For such a system the spontaneous surfactant monolayer curvature is towards the oil phase, since below the PIT O/W microemulsions form, and the stable emulsion type is O/W. The (assumed) semicircular rim around the neck has a curvature (at right angles to the film) towards the water and so is energetically unfavourable. The curvature of the neck parallel to the thin film however is of opposite sign and it falls as the film grows (see Evans and Wennerström, 1999). When these

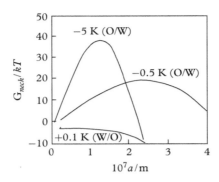

Figure 14.4 *The energy of formation of a neck as a function of the radius a in systems with $C_{12}E_5$. The temperatures are expressed relative to the PIT and the stable form of emulsion is indicated. Redrawn from Evans and Wennerström (1999).*

area and curvature effects are accounted for, the free energy change associated with the growth of the neck, G_{neck}, passes through a maximum with respect to the neck radius of curvature a. If this maximum is sufficiently high (say $30kT$) the neck growth is inhibited and coalescence of the droplets does not occur. The size of the maximum changes with temperature and at about 0.1 K above the PIT a maximum is no longer observed (see Fig. 14.4). However, at just 0.5 K below the PIT the maximum is of the order of $20kT$.

Another often quoted mechanism by which the preferred emulsion type is stabilized relates to interfacial tension gradients within thinning emulsion films. As two drops approach, a thin liquid film is formed, as described, and liquid flows out as the film thins. The flow causes interfacial tension gradients in the film surfaces which arrest interface movement and thereby inhibit film thinning (see §12.4). Adsorption at the film surfaces however will remove the gradients and facilitate film thinning and rupture. Surfactant adsorption to a film surface can occur more readily if the surfactant is present in the bulk of the drops, since adsorption from the continuous phase involves surfactant diffusion into the thin liquid film. This would mean that, in conformity with Bancroft's rule, the stable form of the emulsion is that in which surfactant is present in the continuous phase. Although such a mechanism might contribute to determining the preferred emulsion type, it cannot provide a general explanation. As explained earlier the preferred emulsion type has the *aggregated* surfactant in the continuous phase, but in some stable emulsions the bulk of surfactant can be present as monomers in the droplet phase (see §14.3).

14.5 Phase inversion of emulsions

The discussion of preferred emulsion type above relates to systems with roughly equal volumes of oil and water. For such systems phase inversion (conversion of an O/W to a W/O emulsion or vice versa) can be effected by manipulation of an HLB variable such as temperature, composition of mixed surfactant, or aqueous phase salt concentration. Emulsions can also be inverted however by changing the volume fraction of the dispersed phase, with other parameters left unchanged. If droplets in an emulsion were monodisperse, rigid spheres, then from geometry the maximum volume fraction φ_c of the dispersed phase attainable would be 0.74 and the lowest volume fraction of the medium, $1 - \varphi_c$, would be 0.26. For volume fractions between these extremes either type of emulsion could exist but outside only one type is possible in the simple geometrical approach. As a result of droplet polydispersity and deformability (of larger drops) however, emulsions with droplet volume fractions greater than 0.74 are often found. Inversion caused by increasing φ is accompanied by a sudden change in viscosity which depends on φ, as described in §15.3.

In addition to geometrical effects of changes in volume fraction, the system HLB can also influence emulsion type as φ is varied. Emulsifiers in systems of practical importance are usually mixtures, and components of a mixture can be expected to distribute between oil and water phases differently. As volume fractions change so do concentrations of the various components in each of the phases. This means that adsorption at the droplet interfaces and hence the ratio of surface concentrations also vary, which is expected to influence the preferred curvatures of drop interfaces and emulsion type.

Emulsion phase inversion arising from changes in volume fraction has been referred to as *catastrophic inversion*, because the inversion occurs suddenly as droplet phase is gradually added during homogenization.[2] The process has been successfully described in terms of catastrophe theory (Dickinson, 1981, 1992; Vaessen and Stein, 1995). In this approach emulsion states are represented by points on a folded free energy surface such as that illustrated in Fig. 14.5. The vertical s-axis is an indicator of the structural form of the emulsion, and the two horizontal axes, a and b, are emulsifier concentration and

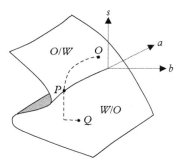

Figure 14.5 *Folded free energy surface to illustrate catastrophic emulsion inversion. The axes represent: s the emulsion structure; a the stabilizer concentration; b ratio of volumes of oil and water. On the upper surface O/W emulsions exist, and on the lower surface W/O emulsions. Redrawn from Dickinson (1992).*

---

[2] Inversion resulting from changes in the HLB of a system has sometimes been referred to as *transitional inversion*.

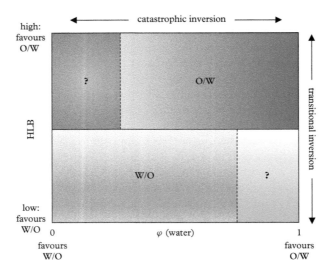

Figure 14.6 *Illustration of favoured emulsion types and the competing effects of (water) volume fraction and system HLB. In the large areas the two effects favour the same emulsion type, but in the two smaller areas the two effects are in opposition.*

oil/water volume ratio respectively. In the figure the upper surface is taken to represent O/W emulsions (with varying emulsifier concentrations and volume fractions) and the lower surface W/O emulsions. Points along, say, the dashed line between O and P represent O/W emulsions which can be changed continuously by changing a and b. If however at point P the oil/water ratio b is increased the system changes discontinuously to a W/O emulsion, which lies on the lower surface, since the underside of the curved surface is inaccessible.

Since there are two broadly independent factors involved in the determination of emulsion type (system HLB and volume fractions of oil and water), there are some circumstances where both factors favour the same emulsion type and others where they do not.

This is illustrated schematically in Fig. 14.6 where the ordinate is the system HLB and the abscissa the volume fraction of water, φ (water). In the large areas the influence of system HLB and water volume fraction both favour (or are consistent with) the formation of the same emulsion type. For example, systems with high HLB favour the existence of O/W emulsions, which are also allowed for φ (water) > 0.26 on the simple geometrical approach. The small areas show where the two factors favour opposite emulsion types. The bottom right area for example has HLB values that favour W/O emulsions but the low volume fraction of oil would favour water being the medium and oil being the droplets, that is, O/W emulsions. In connection with catastrophic inversion, Dickinson (1992) points out that from the model outlined above, it is to be expected that hysteresis[3] can occur and that for a given system emulsions formed in slightly different ways can be of different type. More subtleties and details of representations such as that in Fig. 14.6 are discussed by Abbott (2015).

It has been seen that for monodisperse rigid droplets the maximum droplet volume fraction φ in an emulsion is 0.74. However, the larger the drops and/or the lower the oil/water interfacial tension the lower the Laplace pressure within the droplets, and so it might be expected that in some situations droplets can become deformed and the volume fraction can rise significantly above 0.74. Indeed systems in which φ is as high as 0.99 are possible. These are variously termed HIPEs, gel emulsions, biliquid foams, or aphrons. They are often optically clear and viscous (hence the description gel emulsions) and have a similar structure to foams (see §14.8) but with the internal phase being liquid rather than gas (hence biliquid foams). The continuous phase is a network of thin liquid films in the form of polyhedra. Such emulsions (O/W or W/O) can be prepared using either nonionic or ionic surfactant and have applications in the cosmetics and explosive industries (Binks, 1996).

14.6 Emulsion stability: how does emulsion structure evolve with time?

Although emulsions are not thermodynamically stable, a newly formed emulsion can be kinetically stable for a long period of time, in the sense that droplet coalescence and bulk phase separation do not occur. However other changes, which can occur simultaneously, are usually apparent (Fig. 14.7). Normally, the densities of the droplet and continuous phases will be different and so the droplets will either sediment (fall under gravity) or cream (rise) at a rate dependent on drop size. It is also possible that drops will mutually adhere during encounters, forming aggregates that will sediment or cream more rapidly than individual drops. The types of forces involved in aggregation are the same as those acting between solid particles discussed in Chapter 13, but of course droplets can be deformable unlike solid particles. The solubility of the droplet phase in the medium depends on droplet size according to the Kelvin equation (3.4.9), with solubilities replacing vapour pressures (see also §14.2.2). In polydisperse emulsions therefore, smaller drops effectively distil over to larger drops. This process is termed *Ostwald ripening*, and is an important mechanism for the degradation of emulsions over time. Droplet coalescence, for which the free energy change is negative, is prevented by the presence of energy barriers produced by the interplay of various colloid forces acting across thin liquid emulsion films. Below, the processes of sedimentation/creaming, Ostwald ripening, aggregation and coalescence are considered separately in a little more detail.

---

[3] Hysteresis occurs when, for a given system, the volume fraction at which inversion occurs is different when oil is added to an O/W emulsion and when water is added to W/O emulsion.

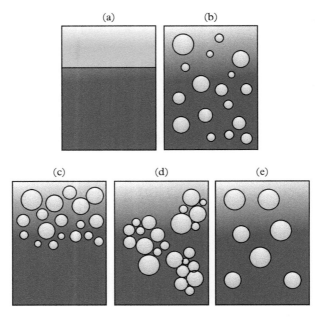

Figure 14.7 *States of an emulsion. (a) Prior to emulsification; (b) initial polydisperse emulsion; (c) creamed emulsion; (d) emulsion with aggregates; (e) system after Ostwald ripening, where the smaller drops have disappeared and the larger drops have grown in size. Aggregation, creaming, and Ostwald ripening can all occur simultaneously.*

14.6.1   Sedimentation

The buoyancy force F_b acting on an undeformed[4] droplet, radius r and density ρ_d in a medium with density ρ_m is given by

$$F_b = \frac{4\pi r^3 \left(\rho_d - \rho_m\right) g}{3} \tag{14.6.1}$$

in which g is the acceleration due to gravity. According to Stoke's law the drag force, F_d, on the drop moving at a velocity v through the medium, with viscosity η_c, is

$$F_d = 6\pi r \eta_c v \tag{14.6.2}$$

When the velocity of the drop is constant (i.e. the acceleration of the drop is zero) the net force on the drop is zero and the two forces, F_b and F_d are equal and opposite, so from (14.6.1) and (14.6.2)

---

[4] The shearing stress caused by the motion of the drop is very small compared to the Laplace pressure within the drop so the spherical drop remains un-deformed by the motion (Walstra, 1996).

$$v = \frac{2r^2 (\rho_d - \rho_m)g\big)}{9\eta_c} \tag{14.6.3}$$

From (14.6.3) it can be shown that for a typical alkane in water emulsion a drop with radius 1 μm, v is approximately 2 cm/day.

In an emulsion, which can be a polydisperse assembly of interacting droplets, even for low droplet volume fractions φ, the treatment of sedimentation/creaming becomes much more complicated (Walstra, 1996). In many emulsions aggregation occurs and the aggregates, made up of drops of varying sizes, sediment/cream *en masse* much more rapidly than individual drops.

14.6.2 Aggregation of droplets

For aggregation to occur, obviously droplets must come into mutual contact, and the rate of aggregation is therefore influenced by the rate of droplet encounters. When droplet collisions are caused by Brownian motion, aggregation is termed *perikinetic*. *Orthokinetic* aggregation occurs when collisions are brought about by shear flow of some kind. In the perikinetic encounter of two droplets the kinetic energy involved is of the order of kT, so if there is an energy barrier on the approach which is greater than say 15–$20kT$ the chance of the drops adhering is very small. In orthokinetic encounters however the kinetic energy can be much greater than kT and fast aggregation can occur.

What happens after a droplet encounter is determined in part by the forces acting between droplets, which in turn depend on the nature of the surfactant stabilizing the emulsion. The droplets may simply move away as a result of net repulsive interactions, or they can rest in a secondary energy minimum, or surmount an energy barrier and sit closer together in a primary minimum. If the thin emulsion film separating drops in an energy minimum ruptures, coalescence can occur. In what follows in this section, the focus is mainly on the interactions between droplets in relation to the formation of aggregates. The various surface forces have been discussed in some detail in Chapter 11, and some properties of symmetric thin liquid films and their drainage and rupture are covered in §12.2 and §12.4. Clearly much of what has been said concerning the stability of solid dispersions in Chapter 13 applies to emulsion stability. However, unlike solid particles, emulsion drops can under some circumstances be deformable, droplet interfaces are fluid and, crucially, droplet coalescence can occur.

For non-deformable oil droplets in an aqueous phase where the predominant repulsive forces arise from surface charge, the DLVO theory can give a reasonable description of droplet interactions. An example of how the interaction potential varies with the closest distance h between droplet interfaces is given in Fig. 14.8. The curves shown are generated using (13.2.1) and (13.2.2) for the electrical double layer forces and van der Waals forces, respectively. The Hamaker constant assumed, $A_H = 5 \times 10^{-21}$ J, is appropriate for say hexadecane droplets in water (Table 11.2). A surface potential ψ_d of 20 mV has been assumed and the curves are for interactions in the presence of three different concentrations of 1:1 inorganic electrolyte in water, as indicated in the figure

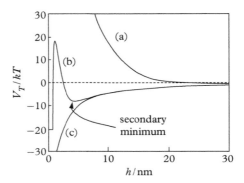

Figure 14.8 *Total energy of interaction between two like oil droplets, radius 0.5 μm, in aqueous 1:1 electrolyte. The potential $\psi_d = 20$ mV and the Hamaker constant $A_H = 5 \times 10^{-21}$ J. Curves (a), (b), and (c) are for $c = 5 \times 10^{-3}, 1 \times 10^{-1},$ and 5×10^{-1} mol dm^{-3}, respectively. Calculations were made using (13.2.1) and (13.2.2).*

legend. For low salt concentration (5×10^{-3} mol dm^{-3}) the repulsion between droplets only rises with droplet approach and droplet aggregation is not expected to occur. For the intermediate concentration, 1×10^{-1} mol dm^{-3}, however, a secondary minimum of about $-8.5\ kT$ at $h \approx 4$ nm is present and at smaller separations an energy maximum of around $16kT$ at $h \approx 0.9$ nm must be surmounted for the droplets to reach the primary minimum (not shown). This suggests that weak (reversible) flocculation can occur (in the secondary minimum). At the highest salt concentration represented, 5×10^{-1} mol dm^{-3}, droplets always experience mutual attraction as they approach and aggregation in a primary minimum occurs.

The DLVO theory cannot be used to describe the stability of W/O emulsions because ions are not soluble in oils with low dielectric constant. In such media the electrical potential at drop surfaces decays only very slowly, and so other repulsive forces must be relied upon to confer stability.

Interactions between droplets can arise from van der Waals forces, electrical repulsion, and solvation forces. In systems with (free) polymers, (attractive) depletion forces can be present and when droplet surfaces have on them adsorbed polymers or oligomers, steric interactions (repulsive or attractive) can occur. The operation of such forces in particulate systems was described in some detail in Chapter 13, and is not covered further here.

The discussion above relates to non-deformable droplets. The flocculation of deformable droplets however involves the formation of plane (or dimpled) circular emulsion films between contacting drops. When many such deformable drops are aggregated (as in a creamed emulsion or HIPE), a polyhedral structure results. The outline of a microscopic image of two mutually adhering dodecane droplets in an aqueous solution of the nonionic sugar surfactant decyl-β-d-glucoside is shown in Fig. 14.9. The droplets are deformable because they are fairly large (diameters around 12 μm) and the interfacial tension is fairly low, about 1 mN m^{-1} (see §14.1). As with

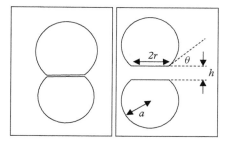

Figure 14.9 *(a) Profile of a microscopic image of two adhering droplets of dodecane (slightly different sizes) in aqueous decyl-β-D-glucoside solution. Droplet diameters are about 12 μm, and the interfacial tension is 1 mN m⁻¹. The contact angle (see text) is about 30°. From original image in Aveyard et al. (1999a). (b) Schematic representation of two adhering droplets of equal size, showing the various dimensions referred to in the text. The film thickness has been exaggerated for clarity.*

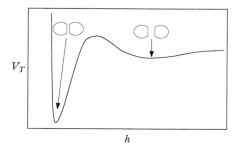

Figure 14.10 *Interaction energy across a thin emulsion film as a function of the film thickness. There are two possible metastable thin films that are analogous to the common black soap film (at the secondary minimum) and Newton black soap film (at the primary minimum).*

soap films for example, there are two possible metastable emulsion films corresponding to the secondary and primary minima in the interaction energy versus film thickness curve, as illustrated schematically in Fig. 14.10.[5]

If it is assumed that all the interactions between droplets occur over the plane emulsion film, that is, interactions that operate between the adjacent curved regions of the drops are ignored, the van der Waals attractive free energy, $V_A(h,r)$, and the electrical free energy of repulsion, $V_R(h,r)$, between two deformed droplets can be written respectively as (Petsev, 1998)

---

[5] The thickness of a metastable film does not correspond exactly to the value of h at a minimum (h_{min}) in V_T since the capillary pressure tends to suck out liquid from the film and so the equilibrium thickness is less than h_{min} (see §12.2.2.)

$$\frac{V_A\,(h,r)}{kT} = -\frac{A_H}{12kT}\left[\frac{3}{4} + \frac{a}{h} + 2\ln\left(\frac{h}{a}\right) + \frac{r^2}{h^2} - \frac{2r^2}{ah}\right] \tag{14.6.4}$$

$$\frac{V_R\,(h,r)}{kT} = \frac{64\pi c}{\kappa}\tanh^2\left(\frac{ze\psi_d}{4kT}\right)\exp\left(-\kappa h\right)\left[r^2 + \frac{a}{\kappa}\right] \tag{14.6.5}$$

The parameters a, h, θ, and r are illustrated on the right in Fig. 14.9 and c in (14.6.5) is the concentration of 1:1 electrolyte in appropriate units (number per unit volume); κ is the inverse of the double layer thickness as elsewhere. The contact angle θ between the thin film and the curved part of the drop is related to a and r (the radius of the circular thin emulsion film) by

$$\theta = \sin^{-1}\,(r/a) \tag{14.6.6}$$

Some illustrative curves of net interaction energies $V_T\,(h,r) = V_A(h,r) + V_R(h,r)$ between two equal deformed dodecane drops in aqueous nonionic surfactant as a function of film thickness h obtained using (14.6.4) and (14.6.5) are shown in Fig. 14.11. The curves are generated using $\psi_d = 5$ mV, $a = 5$ μm, $r = 2.5$ μm (so that $\theta = 30°$) for various salt concentrations ranging from (left to right) 0.2 down to 0.006 mol dm^{-3} as indicated in the figure legend. All the curves exhibit a secondary minimum which becomes deeper (and the corresponding $h = h_{min}$ becomes smaller) the greater the salt concentration. For example for $c = 6 \times 10^{-3}$ mol dm^{-3} the depth of the secondary minimum, which occurs at $h_{min} = 45$ nm, is only about $4kT$; for $c = 0.2$ mol dm^{-3}, however, a minimum with depth of around $29kT$ occurs at $h_{min} = 8$ nm. As expected, the depth of the secondary minimum and the associated value of h_{min} also depend on the value of ψ_d; the higher ψ_d the smaller the depth of the secondary minimum. The surface potential can be adjusted by inclusion of small amounts of ionic surfactant with

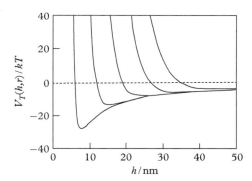

Figure 14.11 *Total free energy of interaction arising from van der Waals and electrical double layer forces for two equal deformed droplets of dodecane in aqueous nonionic surfactant solution. Values of parameters used are: $a = 5$ μm, $r = 2.5$ μm ($\theta = 30°$), $\psi_d = 5$ mV. From left to right the curves are for the salt concentrations, c, 0.2, 0.05, 0.02, 0.01, and 0.006 mol dm^{-3}.*

the nonionic stabilizer, and this means that flocculation can be controlled by adjusting the salt concentration and/or the amount of ionic surfactant present in the emulsion (Aveyard et al., 1999a).

14.6.3 Droplet coalescence

Coalescence in an emulsion does not normally occur during random encounters of droplets undergoing Brownian motion or as a result of the shear induced by creaming. First the emulsion films must form during the aggregation process and then, after a greater or lesser period of time, the films rupture leading to coalescence. The drainage and rupture of soap films is discussed in §12.4; similar considerations apply to the thinning and breaking of emulsion films.

As an emulsion film is being formed, the drainage of liquid from the film is retarded by effects resulting from the presence of adsorbed surfactant stabilizer as described for soap films in §12.4. The liquid flow in the film close to the surface generates a gradient in the surface concentration of surfactant which tends to cause a flow of adsorbed surfactant back into the film. If this gradient is sufficiently large, interface movement is arrested and the surface is effectively rigid, slowing down the egress of liquid from the film. In addition, the surfactant introduces net repulsion between the film surfaces as described above, which tends to stabilize the film. When an emulsion film has attained a metastable state (thickness) surfactant also acts, by the Marangoni effect (§6.2.2), to heal the film if it is stretched and thinned locally.

As remarked in §12.4, the surfaces of thin films are not smooth. Capillary waves create a surface roughness which tends to be dampened by the interfacial tension and interfacial viscoelasticity of the film surfaces. In a theory of Vrij (1966) the criterion for film stability is given as

$$\frac{d^2 V_T(h)}{dh^2} > -\frac{2\pi^2 \gamma}{r^2} \qquad (14.6.7)$$

Here $V_T(h)$ is the net energy of interaction between the film surfaces per unit area (rather than between a pair of droplets). Equation (14.6.7) indicates that, on the basis of the theory, film stability against rupture is expected to be greater the smaller the film (i.e. film radius r), the greater the interfacial tension γ and the greater the net repulsion between surfaces. The higher the tension the more effectively the capillary waves are dampened. It was also seen in §14.2.4 that emulsions made close to the PIT in systems stabilized by nonionic surfactant give the smallest drops which are however very unstable towards coalescence. The oil/water interfacial tension is minimum (and very low) at the PIT and rises when the system is cooled, whereupon the system becomes much more stable with respect to droplet coalescence.

Once a hole (i.e. a neck) has been formed between two droplets, for coalescence to occur the neck must grow, as discussed in §14.4. If the droplets are in an 'unfavoured' emulsion there is no energetic barrier to the growth of the neck and ready coalescence can occur. If on the other hand the droplets are in an emulsion of the 'favoured' type the growth of the neck is energetically inhibited and so coalescence in an emulsion will be a much slower process.

14.6.4   Ostwald ripening

Even if a polydisperse emulsion is stable towards coalescence, the distribution of droplet sizes can nevertheless change over time. This is because the solubility of the droplet phase (component 1) in the continuous medium (component 2) depends on droplet radius (r), as described by the Kelvin equation (see §14.2.2):

$$\ln\left(\frac{x_1}{x_1^o}\right) = \frac{2\gamma v_{m,1}}{rkT} \tag{14.6.8}$$

in which x_1 is the (mole fraction) solubility of the droplet phase in the medium, and x_1^o is the normal solubility of component 1 in component 2; $v_{m,1}$ is the molecular volume of droplet phase. From (14.6.8) it is seen that the solubility x_1 of the droplet phase in the medium is greater the smaller the droplet size (r). Therefore in a polydisperse emulsion the droplet phase (assuming it is not completely insoluble) 'distils' from the smaller drops to the larger drops (by molecular diffusion through the medium),[6] and the small droplets become smaller and large drops larger with time.

Ostwald ripening can be effectively retarded by adding a component to the droplet phase which has very sparing or zero solubility in the continuous phase. Thus for example a simple inorganic salt can be added to the droplet phase of a water-in-alkane emulsion, or hexadecane to oil droplets in say a hexane-in-water emulsion. This reduces the activity of the major component of the droplets, and hence the driving force for dissolution of the droplets. It turns out that only very little additive is needed to suppress Ostwald ripening. Imagine a water droplet, containing uni-univalent salt (molarity m) in a W/O emulsion, were to shrink by outflow of water from the drop. The Laplace pressure would increase, but there is also an osmotic pressure effect due to the presence of the salt, which would give rise to a backflow of water into the drop. When the two effects are equal the drop size remains constant and it can be shown that under these circumstances

$$2\gamma = 3rmRT \tag{14.6.9}$$

Walstra calculates on this basis that, for a system with $\gamma = 0.01$ N m^{-1} and drop radius $r = 1$ μm, a NaCl concentration of less than 0.01 wt % would be needed to halt Ostwald ripening.

The theory of the kinetics of Ostwald ripening has been well studied and is described in an approach due originally to Lifshitz and Slezov, and Wagner (LSW theory). If it

---

[6] It is also possible that, if surfactant micelles are present in the medium, they could solubilize and transport some material between droplets. For systems with *ionic* surfactant however this is less likely to happen since the charged micelles will be repelled by the emulsion droplets carrying the same charge. Under these circumstances oil molecules in the droplets are unlikely to be directly transferred to micelles (Kabalnov, 1994).

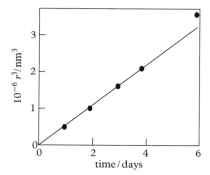

Figure 14.12 *The cube of the number average droplet radius as a function of time for the Ostwald ripening of dodecane-in-water emulsions stabilized by 0.21 wt % SDS and 0.15 wt % pentanol. The volume fraction of dodecane in the emulsions is 0.004. Redrawn from Taylor and Ottewill (1994).*

is assumed that the separation of droplets is large compared to droplet size and that transport between droplets is limited by diffusion in the medium, the LSW theory gives for the rate of ripening

$$\frac{\mathrm{d}r_c^3}{\mathrm{d}t} = \frac{8c_1^o \gamma v_1 D_1}{9RT} \qquad (14.6.10)$$

Here c_1^o is the normal solubility of the droplet phase (component 1) in the continuous phase (2), v_1 is the molar volume of bulk liquid 1, and D_1 is the diffusion coefficient of 1 in the continuous phase 2. The quantity r_c is the critical droplet radius (see Fig. 3.10) in the system at a given time and is approximately equal to the number average drop radius. It is seen from (14.6.10) that a feature of Ostwald ripening is the linear dependence of r_c^3 on time t, as illustrated in Fig. 14.12 for the Ostwald ripening of dodecane-in-water emulsions stabilized by SDS and pentanol (Taylor, 1998).

14.7   Emulsion breaking

Unwanted emulsions are frequently encountered and effective methods are required to destabilize ('break') them, by causing droplet coalescence. An obvious strategy is to deactivate the processes, or remove the conditions, that are responsible for the initial stability of the emulsion. A number of procedures have been used that can loosely be divided into mechanical, thermal and chemical methods, although as will be seen this categorization is not altogether satisfactory.

If an emulsion is sensitive to shear it can be agitated in some way, say in a blender, which will bring droplets rapidly into energetic contact and enhance droplet coalescence. Centrifugation can also be used to force droplets together and facilitate coalescence. Another mechanical method involves the introduction of a solid into the emulsion which

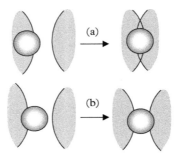

Figure 14.13 *Approach of two droplets, one with an adsorbed particle. (a) Particle is preferentially wetted by droplet phase so on contact the droplets coalesce. (b) Particle preferentially wetted by continuous phase and the particle bridges the two droplets preventing coalescence.*

is preferentially wetted by the droplet phase. For example, a filter bed with a hydrophilic surface tends to cause the breakdown of W/O emulsions. In this case a thin asymmetric oil film between a water droplet and the solid surface must be ruptured whereupon the water droplet spreads and wets the solid surface. In this way the solid removes droplets from the oil phase. Alternatively, the solid can be in the form of particles which are much smaller than the droplets.

Imagine a drop with a spherical particle adsorbed that is more wetted by the drop than the continuous phase (Fig. 14.13a). When a second drop approaches and contacts the particle, the two droplet interfaces will tend to cross as illustrated, causing coalescence of the drops. However, if the particle preferentially wets the continuous phase it is able to bridge droplets (Fig. 14.13b) preventing coalescence, and form either flocs or extended networks of droplets. The former would result in an enhanced rate of sedimentation or creaming whereas the latter would inhibit sedimentation or creaming. An example of a chemical demulsification is the addition of polyvalent ions (e.g. Ca^{2+}) to an emulsion stabilized by an alkali metal carboxylate. If the calcium salt of the carboxylic acid is insoluble in water it comes out of solution, which destabilizes the emulsion. A similar example is the addition of a cationic surfactant to an emulsion stabilized by an anionic surfactant, where an insoluble complex is formed between the two surfactants. In cases where an O/W emulsion is stabilized by ionic surfactant, the addition of electrolyte may break the emulsion by screening the surface charge and hence reducing the repulsion between droplets.

The origins of the favoured type of emulsion were discussed above in §14.3 and §14.4 in terms of the curvature properties of the adsorbed monolayers of surfactant stabilizer, and the concomitant system HLB. If by changing the conditions (by adding electrolyte or a cosurfactant or second surfactant, or by changing the temperature) of an originally stable O/W emulsion such that the favoured emulsion type now becomes W/O, it can be expected that the emulsion will become unstable and break. This approach to demulsification is well-illustrated by the breaking of brine-in-crude oil emulsions by addition of a suitable commercial surfactant (demulsifier).

The brine content of crude oil can be quite high and when an oil + brine mixture passes through well head chokes and valves, the shear can produce brine-in-crude oil emulsions. Such emulsions must be resolved to remove both the water and salt before transportation and processing of the oil. The emulsions are stabilized by, *inter alia*, polyaromatic materials ('asphaltenes') occurring in the crude oil. The interfacial tension of an aged crude oil/brine interface is high, around 25–30 mN m^{-1}. The stability appears to arise mainly from the mechanical strength of the layers coating the droplets. Since the tension is relatively high it is possible to displace the coating by adsorption of suitable surfactants, when the tension can fall to as low as say 10^{-1} mN m^{-1}.

Some commercial chemical demulsifiers appear to act like common low molar mass surfactants in the way they effect emulsion breakdown, and crude oil behaves in this respect rather like a pure alkane. Further, it is observed that under some circumstances 'overdosing' can occur in which, above a certain concentration the demulsifier becomes ineffective. The demulsification behaviour and overdosing can be reproduced using for example the simple pure anionic surfactant (AOT) as demulsifier for brine in crude oil emulsions. As a measure of emulsion breakdown, the initial rate of aqueous phase resolution is used and can be determined as a function of surfactant concentration and concentration of NaCl in the aqueous droplets. Some illustrative results, taken from Aveyard et al. (1990) are shown in Fig. 14.14. Fig. 14.14a is for high NaCl concentration in the droplets (0.5 mol dm^{-3}) and Fig. 14.14b is for [NaCl] = 0.1 mol dm^{-3}. The arrows indicate the *cac* for the AOT in the emulsion system. For high salt concentrations it is seen that the resolution rate rises with surfactant concentration up to the *cac*. In this regime the surfactant is progressively adsorbed at droplet interfaces, displacing the indigenous stabilizing layer. At the *cac* and beyond, the adsorbed layer is effectively saturated and the emulsion becomes progressively more stable. This is the phenomenon of overdosing and occurs because the emulsion (brine-in-oil) is the favoured one at high salt concentration (see Table 14.1). At the lower salt concentration however

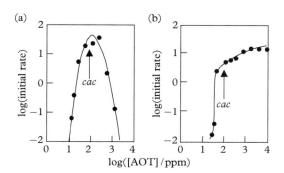

Figure 14.14 *Resolution of brine-in-crude oil emulsions by AOT: log of initial rate of resolution against log AOT concentration. (a) [NaCl] = 0.5 mol dm^{-3}; (b) [NaCl] = 0.1 mol dm^{-3}. Redrawn from Aveyard et al. (1990).*

(Fig. 14.14b) the favoured emulsion type is oil-in-water and so when the adsorbed AOT monolayer around the droplets becomes close-packed the W/O emulsion becomes unstable.

14.8   Foams: what are they and how do they resemble emulsions?

Foams are dispersions of gases (e.g. air, CO_2) in liquids; their structures parallel those of emulsions, and the processes that control the two types of dispersion have much in common. Aqueous foams are the most frequently observed (e.g. in foodstuffs and beer) but non-aqueous foams are common. Biofoams, stabilized by surfactant proteins, also occur in nature (Cooper and Kennedy, 2010). Like emulsions, foams can be stabilized by simple low molar mass surfactants, and also by polymeric surfactants (§1.3.4). Emulsions have inverted forms (O/W and W/O) which can be interconverted as seen. The inverted form of a foam is a liquid in gas dispersion, that is, an aerosol. Obviously foams and aerosols cannot be interconverted.

Foams have varied practical uses, in part because they have interesting rheological properties. Foam viscosity is high relative to that of its continuous (film) phase and in addition foams exhibit a yield stress and can be shear thinning. The use of foams in crude oil displacement capitalizes on the high viscosity of foams, and is referred to further below. A major application of foams is in mineral processing in which hydrophobized ore particles become attached to air bubbles (Chapter 17) rising through a slurry containing frothing agents (e.g. ethoxylated alcohols) to stabilize the gas bubbles (see later). Solid foams (where the liquid films have solidified or become gels after formation) have also found many uses, for example in the form of cellular plastics.

In the laboratory foams can be formed readily by passing gas (e.g. air, nitrogen) through a sinter into surfactant solution in a column as illustrated in Fig. 14.15. At the bottom of the column the foam consists of roughly spherical gas bubbles dispersed in the solution (*wet* foam or *kugelschaum*). As the bubbles rise under the influence of gravity, liquid drains from between them and some coalescence occurs giving rise, at the top of the tube, to a polyhedral network of thin liquid lamellae (*polyederschaum* or *dry* foam) in which the volume fraction of liquid is very low. The persistent foam formed for example during the washing of dishes by hand is an example of a polyhedral foam. Wet foams and dry foams are the analogues of, respectively, droplet emulsions and HIPEs. Foams are considered to be colloids because one of the dimensions, that is, the thickness of the liquid films, can be in the colloidal size range.

Although the structure of foams and emulsions are similar in some respects, there are also significant differences. The cell or bubble size in foams (typically 50 μm to several mm) is much larger than in emulsions. Further, the interfacial tensions in emulsions and foams are very different, often being an order of magnitude higher in foams. The excess pressure Δp inside spherical bubbles or droplets, radius r and interfacial tension γ, is given by the Laplace equation (see §3.4.1):

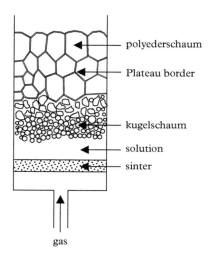

Figure 14.15 *Formation of foam by blowing gas through a sinter into surfactant solution. At the bottom of the column the foam is 'wet' and referred to as* kugelschaum. *Towards the top, the liquid has drained from between the bubbles and the foam consists of a network of thin liquid films joined at Plateau borders* (polyederschaum).

$$\Delta p = \frac{2\gamma}{r} \qquad (3.4.2)$$

Since γ in foams exceeds that in emulsions by only a factor of about 10 whereas radii differ by a much greater factor, this means emulsion droplets are usually much less deformable than are gas bubbles in a foam. Thus, polyhedral structures are much easier to form in foams than in emulsions, where droplets often remain spherical even in a creamed or sedimented layer. It also follows that it is much more difficult to form small gas bubbles in foams than small liquid droplets in emulsions.

14.9   Methods of foam formation

Above, a simple laboratory method for production of foams using a foaming column was described. More broadly foams can be produced in three general ways: (i) by agitation of an 'unlimited' amount of gas with liquid, (ii) dispersion of controlled volume of gas in liquid, and (iii) by in situ bubble generation (Dickinson, 1992; Prins, 1988).

14.9.1   Agitation of unlimited amount of gas with liquid

This method is often used in the preparation of foodstuffs in the home, obvious examples being whipping of cream or egg-white in an open bowl. Air is initially incorporated into the liquid in the form of large bubbles, which are subsequently reduced in size by shear forces generated by further agitation. Although the air supply is unlimited, once the whipping rod is covered with foam, it cannot incorporate more air. Egg-white foam is stabilized by adsorbed proteins and glycoproteins. Ovalbumin is denatured when the foam is baked (together with sugar) to give meringue, which is an example of a solid foam.

14.9.2 Dispersion of controlled volume of gas in liquid

In the preparation of foams for experimental purposes, using a foaming column such as that illustrated in Fig. 14.15, as well as in the production of foams industrially (where there are quality control requirements), the amount of gas mixed with the liquid is controlled. Bubbles can be generated by passage of gas at a controlled flow rate through a sintered disc and into the foaming liquid. The range of bubble sizes formed initially depends on various factors, including the range of pore sizes in the disc, the gas/liquid interfacial tension and the gas flow rate. When bubbles emerge from orifices in the disc, the air/solution interface is not at equilibrium and so the surface tension is higher than the equilibrium tension; the interface possesses a dilational viscosity (see §6.2.2). Further, at low flow rates (pressures) the gas may not pass through the smaller pores as a result of the larger Laplace pressure across the more highly curved liquid/gas interfaces within the pores. Bubble sizes initially produced can be reduced by subsequent mixing/agitation. In the manufacture of ice cream, the ingredients (water, fat, sugar, and emulsifiers) are emulsified and then foamed and frozen simultaneously giving a product with about twice the volume of the original emulsion. In this case the gas is introduced into the emulsion in a cooled tube containing a rotating blade which whips the mixture into a foam. The stability of the foam results from the high viscosity of the aqueous phase as well as from the presence of ice crystals. The foamed formulation has a softer and lighter texture than could be obtained by simply freezing the initial emulsion.

14.9.3 In situ bubble generation

Some gases have appreciable solubilities in liquids, the solubility being approximately proportional to the gas pressure. When the pressure over the liquid is reduced gas nucleates heterogeneously to form bubbles. In some systems bubbles can be formed from gas evolved during a chemical reaction.

In carbonated soft drinks and champagne, bubbles form when the bottle is opened. However in these cases the liquids do not contain surface active materials able to stabilize the bubbles against coalescence and so no stable foam is formed (or required). Beer on the other hand contains surface active proteins and polypeptides which derive from the malt used in brewing, and these materials stabilize any foam formed, giving the beer a desired 'head'. In bottled and canned beer the bubbles come from dissolved CO_2. In draught beer air is also introduced by turbulent flow in the small holes in the tap when the beer is dispensed from the pump.

An important application of foaming is in the commercial production of cellular plastics and metals using so-called *blowing agents*, which can be classed as either physical or chemical. In some instances the two types are used in a combination. Liquid CO_2 and various hydrocarbons (e.g. pentane and cyclopentane) are examples of physical blowing agents. To form a foam the liquid blowing agent must be volatilized; the heat for this can come from say an exothermic chemical reaction such as cross-linking in polymer formation. In contrast, chemical blowing agents undergo chemical reaction to give a gaseous product during foam formation. For example titanium hydride powder is used in the formation of metal foams; at sufficiently high temperatures it decomposes to form hydrogen gas and titanium.

Polyurethane foams are produced in very large quantities using blowing agents, largely as flexible products for cushioning materials and rigid materials for insulation and construction uses. Annual production worldwide was about 18 million tonnes in 2015. Polyurethanes are synthesized by the reaction between alcohols containing at least 2 reactive OH groups in the molecule and isocyanates with more than one reactive (–NCO) groups per molecule, for example,

$$OCN\text{-}R_1\text{-}NCO + HO\text{-}R_2\text{-}OH \rightarrow [\text{-}CO\text{-}NH\text{-}R_1\text{-}NH\text{-}CO\text{-}O\text{-}R_2\text{-}O\text{-}]_n$$

A case where a mixture of physical and chemical blowing agents is used is in the production of very low density flexible polyurethane foams for mattresses. Isocyanate and water (which react to form carbon dioxide) are used in combination with liquid carbon dioxide which boils producing vapour in the reacting mixture. The use of the two blowing agents allows the control of temperature since the chemical reaction is exothermic and the heat produced is used to vaporize the liquid CO_2 (which is endothermic).

Cellular plastics like polyurethane can be either closed-celled structures in which the bubbles are isolated, or open-celled in which the cells are interconnected by holes in the lamellae between bubbles. Closed structures make good thermal insulating materials whereas open-celled materials are more compressible and are used for cushioning materials.

14.10   Foam stability and breakdown

14.10.1   Measurement of the stability of a foam

Two quantities of interest in the experimental investigation of foams are *foamability* and *foam stability*. The former is measured by the volume of foam formed from a solution just after generation has ceased. This is often determined by a simple 'shake test' in which a given volume of solution is agitated in a given way for a given time. On the other hand, *foam stability* relates to persistence of a foam once formed, and can be investigated using a graduated foaming column similar to that shown in Fig. 14.15. Gas is passed through the solution at a known flow rate and the changes in the foam volume with time are observed after the flow of gas has ceased. As liquid drains through the foam, the volume of the liquid layer at the bottom of the column can be measured, as can the volume of

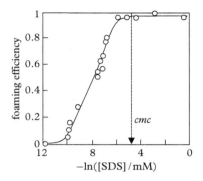

Figure 14.16 *Foaming efficiency of aqueous solutions of SDS as a function of bulk concentration. The foaming efficiency approaches unity as the adsorbed monolayers become close-packed, which is well below the* cmc. *Redrawn from Aveyard et al. (1993).*

foam above this layer. The latter is measured to the top of the polyhedral foam and does not give direct information about the internal foam structure and breakdown. Bubble or cell volumes can be determined microscopically as a function of time and of foam height, and the area of lamellar surfaces estimated.

A third useful quantity related to stability is the *foaming efficiency*, defined as the ratio of the volume of foam formed per unit time to the gas volume flow rate through the column. The foaming efficiency depends on the surfactant concentration in a foaming solution because the surface concentration, Γ, of surfactant is related to its bulk concentration. As shown in Chapter 5, Γ becomes effectively constant, and the monolayer effectively close-packed, at bulk concentrations that are significantly below the critical micelle concentration (see Fig. 5.2b). As an example, the foaming efficiency of SDS solutions is shown as a function of the solution concentration of SDS in Fig. 14.16. The foaming efficiency reaches a value close to unity at a surfactant concentration which is about 35 per cent of the *cmc*, which for SDS is 8 mM.

Since foams are made up of individual lamellae, information on aspects of foam stability and breakdown can be obtained from the study of single films formed say on a glass frame. The films can be in contact with the solution from which they are drawn (as in Fig. 12.5a), or they can be isolated, that is, in contact only with the menisci around the edge of the frame.

14.10.2   Bubble disproportionation

Just as polydisperse emulsions undergo Ostwald ripening, so polydisperse foams exhibit the analogous *bubble disproportionation* in which larger bubbles grow at the expense of smaller bubbles. For a given value of γ, Δp and hence the solubility of the gas in the medium are greater the smaller r. As a consequence, if the tension remains constant, gas diffuses through the medium from small bubbles to larger bubbles, that is, the smaller bubbles shrink and the larger bubbles grow. Since the solubilities of dispersed gases in (say aqueous) foams are greater than solubilities encountered in emulsions (e.g. of hydrocarbons in water) bubble disproportionation is usually a much more rapid process than Ostwald ripening.

For gas to flow through the medium from one bubble to another it must first permeate through the adsorbed surfactant monolayers around the bubbles. For simple low molar mass surfactants, the resistance afforded by the monolayers is small compared to the resistance offered by the bulk liquid between droplets. Nonetheless, for more effectively close-packed monolayers the surface resistance can become important. This has been illustrated by observing the rate of decrease of the radius of a single soap bubble resting on the surface of an aqueous solution of SDS. The rate was reduced by a factor of three when dodecanol was present together with the surfactant. The monolayer permeability depends on the surface pressure (hence surface tension), the temperature and the size (Box 14.4) of the bubble (see Princen and Mason, 1965).

Box 14.4 **What if the surface tension of a bubble changes with its size?**

The implicit assumption of constant surface tension as bubble size changes during disproportionation may not always be appropriate. Imagine two interconnected gas bubbles in liquid, as illustrated in Fig. 14.17. If the surface tensions of the two bubbles are equal, the bubbles are at unstable equilibrium when the sizes are equal (shaded bubbles). If, however, a fluctuation in pressure were to cause gas flow say from bubble A to bubble B, if the surface tensions remained unchanged, the smaller bubble would continue to shrink and the larger bubble would continue to expand. However, if growth of a bubble (increase in its surface area) were to be accompanied by an increase in surface tension as a result of its surface viscoelasticity, then depending on the size of the increase, it is possible that Δp for the bubble would increase to such an extent as to force gas back into the smaller bubble.

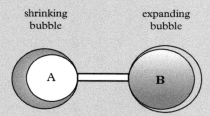

Figure 14.17 *Two interconnecting bubbles (i) initially at unstable equilibrium (shaded) and (ii) after flow of gas from A to B.*

The tension, and hence Δp for the smaller bubble would be falling due to surface elasticity. It is seen therefore that surface viscoelasticity can inhibit the rate of bubble disproportionation. If the bubble surfaces were perfectly elastic, the disproportionation could be completely halted (Prins, 1988).

14.10.3   Film drainage and rupture

The stability of foams and emulsions is determined by similar factors, including the operation of surface forces (Chapter 11) and the elasticity and drainage of the thin liquid films (§12.4). However, the stability of emulsions formed with a given (say) aqueous surfactant solution is usually much greater than that of foams generated using the same surfactant solution. In part this arises because encounters between un-flocculated small emulsion drops that occur during Brownian motion, result in the formation of only transient emulsion films of small area. In flocculated systems however, films exist for a relatively long period of time, and the likelihood of film rupture is greater the more extensive the film (see §14.5.3). Since areas of films in polyhedral foams are much larger than those in flocs of small emulsion drops, it is to be expected, on this count, that foams will be less stable than the corresponding emulsions.

In polyhedral foams (Box 14.5), as in HIPEs, according to Plateau's laws (§12.1) the liquid lamellae meet along Plateau borders, at an angle to each other (in the most stable form) of 120° and the borders themselves meet in fours at a vertex at the tetrahedral angle of 109.47°, as described in §12.1. In addition to gravity, liquid drainage from the lamellae of a polyhedral foam results from capillary suction (see §12.2.2). The cross-section of a Plateau border is represented in Fig. 14.18. The border has a radius of curvature of r in the plane of the paper; the other radius of curvature, normal to this, is infinite (zero curvature), that is, each surface in the border is cylindrically curved. The pressure in the gaseous regions is denote p_2, and the pressure inside the border (shaded) is p_1. Since the lamellae are (assumed to be) planar the pressure inside a film is also p_2. From the Laplace equation p_2 exceeds p_1 by an amount γ/r and so liquid is sucked from films into the Plateau borders.

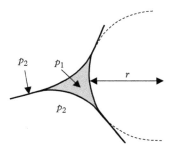

Figure 14.18 *Pressures within a polyhedral foam. Within the shaded Plateau border the pressure is p_1 and outside the border it is p_2. The radius of curvature of the surface of the border in the plane of the paper is r; normal to this the radius of curvature is infinite.*

Box 14.5   Lowest surface area of monodisperse dry foams: from Kelvin to Weaire and Phelan

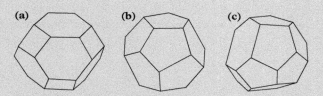

Foam cell shapes. (a) A tetradecahedron with eight hexagonal faces and six square faces, as in the Kelvin structure. (b, c) The two cells in the WP structure; (b) is an irregular pentagonal dodecahedron and (c) is a tetradecahedron with 12 pentagonal faces and two hexagonal faces.

A real polyhedral foam network is made up of a collection of different types and size of polyhedra; it is not geometrically possible to fill the space with an assembly of any single type and size of regular polyhedron. Kelvin posed the question in 1887 of how the space within a dry foam can be filled by cells of *equal volume* with the *lowest area of surface* between the cells, the system being consistent with Plateau's laws (§12.1). He proposed that a space filling array of equal-sized tetradecahedra[†] with six square faces and eight hexagonal faces fulfilled these requirements (see (a) in the figure). In such a structure it turns out that the hexagonal faces would have to be slightly curved. This proposition held sway for over 100 years until it was discovered by Weaire and Phelan (1994) that a structure of lower area exists. The Weaire–Phelan (WP) structure, with surface area 0.3 per cent less than that of the Kelvin structure, has two different kinds of cell of equal volume. One is an irregular pentagonal dodecahedron (12 pentagonal faces, (b) in the figure) and the other a tetradecahedron with two hexagonal faces and 12 pentagonal faces ((c) in the figure). The Kelvin structure has been realized experimentally (Weaire, 2009), and more recently Gabbrielli et al. (2012) have experimentally created the WP structure. This was achieved by making a mould with faceted walls that conformed to the foam geometry, which induced the perfect formation of the WP structure. The foams contained about 1500 bubbles. The WP structure inspired the structural support system for The Beijing National Aquatics Centre at the 2008 Olympics. The structure is lightweight and strong, with all the joints being close to tetrahedral angles.

[†]A tetradecahedron (also called a tetrakaidecahedron) is in general a polyhedron with 14 faces, of which there are a number.

As the films thin as a result of drainage, at some (non-equilibrium) stage they appear coloured in white light as a result of interference of light reflected from front and back surfaces of the lamellae. If the films remain intact long enough, they ultimately appear black. At this stage surface forces act across the films and two (meta)stable thicknesses are possible, the thicker films being common black films and the thinner (ca. 2 nm) Newton black films. In both cases the disjoining pressure in the films is equal to the capillary pressure (§12.2.2).

For a foam to be persistent, the rate of drainage of liquid must be slow, and it has already been explained in §12.4 how adsorbed surfactant at film surfaces inhibits the drainage by rendering the surfaces immobile. Drainage under *gravity* depends on the density ρ of the liquid as well as on its viscosity η, as can be expected, and the velocity of flow v_f depends on the film thickness d according to the expression (Mysels et al., 1959)

$$v_f = \frac{\rho g d^2}{12\eta} \tag{14.10.1}$$

It is seen that the velocity is proportional to the square of the film thickness. For water at 298 K v_f is approximately 80 μm s^{-1} for $d = 10$ μm, and for $d = 1\mu$m it is about 0.8 μm s^{-1}. It can be shown from (14.10.1) that the time t for a film initially with thickness d_o to reach a thickness d where $d << d_o$ is given approximately by (Prins, 1988)

$$t = \frac{6\eta H}{\rho g d^2} \tag{14.10.2}$$

where H is the height of the film. From this equation it can be calculated that for a film with $H = 1$ mm it takes about 10 min to reach a thickness of 1 μm and about 70 days to thin to 10 nm.

Liquid drains from the thin films into the Plateau borders and must therefore ultimately leave the foam through the network of Plateau borders. Drainage takes place until the capillary suction at height h (in say a column of foam) becomes equal to the hydrostatic pressure of the liquid at that height. This occurs when

$$\frac{\gamma}{r} = \rho g h \tag{14.10.3}$$

The rate of drainage in such a complex polyhedral structure is however complicated to treat theoretically. An empirical relationship that has been used to describe the drainage is

$$v = v_o \left(1 - e^{-Kt}\right) \tag{14.10.4}$$

Here v is the volume of liquid in the foam at time t, v_o is the volume of the un-foamed initial solution, and K is a constant.

Film rupture becomes more likely as the films drain and become thinner. Rupture can result from mechanical shock, and the breaking of one film in a foam can trigger rupture of neighbouring films. In the case of very thin films, the presence of capillary waves at the surfaces can lead to rupture as discussed in §12.4.

14.11 Foam and film breaking by solid particles and liquid droplets

There are many instances where stable foams are required, as outlined above, but in some technologies foams are unwanted and must be destabilized, as for example in pulp and paper production, waste water treatment and in some processes in the oil industry. A number of products such as paints and washing powders contain additives to control unwanted foam formation (Garrett, 1993; Denkov, 2004). *Antifoams* are formulations that are pre-dispersed within a foaming solution to control foam formation, whereas *defoamers* are added to break foams that have already been formed. In both cases, for aqueous foams, typical formulations contain an oil together with dispersed hydrophobic solid particles with dimensions comparable to lamellar thickness (say 1–10 μm). Oils and solid particles can each cause foam breakdown but act synergistically when mixed. Examples of solids include silica, alumina, ethylene bis-stearamide, aluminium stearate, and polypropylene. Many oils have been employed including mineral oils, silicone oils, and aliphatic alkanols, acids, and amides. The details of how antifoams and defoamers work can differ since, in the case of antifoams, the liquid/solid mixture must break through into the liquid/gas interface of lamellae from the liquid interior, whereas defoamers enter the surfaces of lamellae from gas (air). However, defoamers can become dispersed in the foaming solution over time and then behave effectively as an antifoam. Possible mechanisms for the rupture of liquid films and foam breaking by solid particles and/or oils are now described.

14.11.1 Film and foam breaking by solids

When solids (particles or thin rods) are used to break aqueous films and foams the most important property of the solid is its wettability by the film-forming solution, which is expressed in terms of the contact angle θ that the solid makes with the air/solution interface (§3.3). A simple demonstration of this is the way in which a single soap film responds when a thin rod is passed at right angles through the film (Fig. 14.19). It is found that hydrophilic rods ($\theta < 90°$) are readily incorporated without the film being immediately ruptured, whereas hydrophobic rods ($\theta > 90°$) tend to cause film rupture on contact. In the case of a hydrophilic rod (Fig. 14.19a) the Laplace pressure causes liquid in the film to flow into the meniscus formed at the contact with the rod, as indicated by the arrow. Conversely, at the meniscus where the hydrophobic rod meets the film, the Laplace pressure repels liquid away from the rod and (c) causes film rupture.

Some data for the inclusion of thin cylindrical glass rods into single films formed from 0.2 mM aqueous hexadecyltrimethylammonium bromide (CTAB) solution are shown in Fig. 14.20. The rods were treated ('silanized'; Box 14.6) with octadecyltrichlorosilane (OTS) to give a range of contact angles θ with the surfactant solution. The graph shows the percentage of thin films which ruptured within 30 s of inclusion of the rod, as a function of the contact angle θ. The rods for which θ is close to 90° and above are effective in causing film rupture whereas for lower contact angles the rods have little effect on film lifetime.

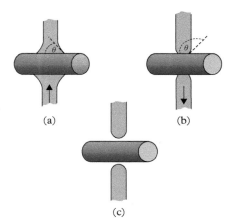

(a)   (b)

(c)

Figure 14.19 *A cylindrical rod in a vertical soap film. (a) The rod is hydrophilic ($\theta < 90°$) and the Laplace pressure draws liquid towards itself as indicated by the arrow. (b) The rod is hydrophobic ($\theta > 90°$) and liquid is expelled from the region of the rod. (c) The rod is completely de-wetted and the film is ruptured.*

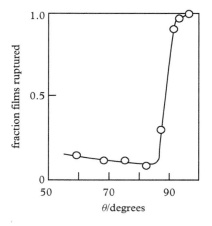

Figure 14.20 *Fraction of single thin films formed from 0.2 mM aqueous CTAB solution ruptured during the first 30 s after incorporation of OTS-coated cylindrical glass rods, diameter 0.5 mm. Redrawn from Aveyard et al. (1994a).*

The situation in which solid particles are present in foams is a little more complicated. Before film rupture can occur a particle (e.g. a spherical glass bead) must first adsorb at one surface of a lamella, after which as the film thins, the particle engages with the second surface and bridges the film. Obviously the particle diameter must be similar to or larger than the thickness of the liquid film it ruptures. Further, as the foam drains,

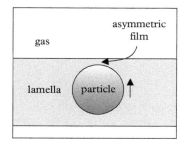

Figure 14.21 *A spherical particle in a lamella approaching the lamella surface. A thin asymmetric solid/liquid/gas film is formed across which surface forces act.*

Box 14.6 How to render glass hydrophobic—silanization

'Silanization' of surfaces which contain hydroxyl groups (e.g. glass, mica) involves reaction with silane derivatives, for example, dichlorodimethylsilane or hexamethyldisilazane. The silane derivative forms a covalent –Si–O–Si– bond at the surface and the hydrophobic groups (e.g. methyl groups) are exposed at the surface rendering it hydrophobic. The degree of coverage by the hydrophobic groups determines the contact angle θ. The glass rods referred to in Fig. 14.20 were treated with solutions of OTS in cyclohexane, of differing concentrations and over different periods of time, to give a range of contact angles with the surfactant solutions.

---

solid particles can under some circumstances be swept through the foam and collect in the Plateau borders and slow the drainage down.

Whether particles can adsorb at the surfaces of lamellae from the inside of the films depends on the nature of the surface forces that act between the particle/liquid interface and the lamella liquid/gas interface, as illustrated in Fig. 14.21. Assuming adsorption does occur, what happens then depends on the particle wettability by the surfactant solution. Suppose a wetting spherical particle ($\theta < 90°$) bridges a liquid film and that the liquid/gas interfaces are planar up to contact with the particle. As the film drains further the interfaces become curved as shown in Fig. 14.22a, and the Laplace pressure generated tends to pull the liquid back towards the particle as indicated by the arrow.[7] Since $\theta < 90°$ there is one lamella surface on each side of the particle equator and the particle does not rupture the film. If however the particle is non-wetting (Fig. 14.22b) the lower surface moves around the particle as the film thins, and it will ultimately contact the upper lamella surface (which is on the same side of the particle equator) and the

---

[7] In addition to the meniscus curvature in the plane of the paper (as drawn), there is also a curvature at right angles to this and of opposite sign. It is implicit in the arguments proposed that the curvature in the plane of the paper is the greater of the two curvatures.

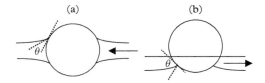

Figure 14.22 *(a) Wetting particle bridging a liquid lamella. Liquid is drawn toward the particle by the Laplace pressure resulting from the curved meniscus around the particle. (b) Non-wetting particle bridging a liquid film. Liquid is driven from the region of the particle by the Laplace pressure. The lower surface of the lamella ultimately contacts the upper surface and the film ruptures.*

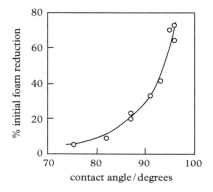

Figure 14.23 *Foam reduction in aqueous 0.2 mM CTAB solutions caused by Ballotini beads (diameters between 15 and 79 μm) with varying contact angles with the surfactant solution. Redrawn from Aveyard et al. (1994b).*

film will rupture. In this case the Laplace pressure in the meniscus expels liquid from the vicinity of the particle, aiding film thinning close to the particle.

The presence of particles in foams can affect both the *foamability* (§14.10.1) and the foam stability. The per cent reduction in the initial foam volume of 0.2 mM aqueous solutions of CTAB caused by the presence of spherical silanized glass beads (diameters between 15 and 79 μm) with a range of contact angles through 90°, is shown in Fig. 14.23. For contact angles less than about 70° the foamability is not much affected by the particles, but above this angle the reduction in foamability rises increasingly steeply with θ.

The effect of the particles on *foam stability* (in similar systems but covering a range of surfactant concentrations between 0.1 and 1.6 mM) is expressed in terms of a half-life ratio τ in Fig. 14.24. A foam half-life, $t_{1/2}$, is the time it takes for an initial foam volume to fall to half its value, and a quantity τ is defined as

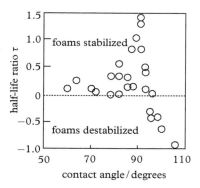

Figure 14.24 *Half-life ratios τ as a function of contact angle of Ballotini beads with solutions of CTAB with concentrations ranging from 0.1 to 1.6 mM. Redrawn from Aveyard et al. (1994b).*

$$\tau = \left[\left(t_{1/2} \right)_{particles} - \left(t_{1/2} \right)_{noparticles} \right] / \left(t_{1/2} \right)_{particles} \qquad (14.11.1)$$

Positive values of τ indicate increased stability and negative values decreased stability. The value of τ rises from around zero (no effect on stability) at $\theta = 80°$ to a maximum of about 1.5 at 90° and then falls rapidly to around -1 at 105°. The maximum suggests two distinct processes are involved, one enhancing stability and the other destabilizing the foam, with both effects being dependent on θ. In regions where the foam stability is significantly enhanced (τ positive) it is observed (from the turbidity) that there are only few particles in the drained liquid beneath the foam, and particles are seen to collect in the Plateau borders. This reduces the rate of liquid drainage through the foam. For $\theta > 90°$ foam films are rapidly broken by the particles. It is noted that the strength of particle attachment to liquid (lamella) surfaces is very dependent on θ, and is maximum for $\theta = 90°$ (see §17.1). For $\theta < 75°$ many particles collect in the subnatant liquid which becomes turbid; more than half of an adsorbed particle resides in the lamella interior and is therefore readily swept through the liquid interior of the lamellae to the subnatant liquid.

14.11.2 Effects of insoluble liquids on film and foam stability

Small liquid droplets of oil (e.g. alkane) dispersed in aqueous surfactant solution can act in a variety of ways to destabilize thin films and foams. To rupture a film a droplet inside the film must first enter a surface. If, after entry, the oil *spreads* (§16.2.1) on the surface it could rupture the film by dragging along underlying liquid hence causing Marangoni thinning (§6.2.2) and potentially film rupture. A further possibility is that molecules of the oil are solubilized in the surfactant monolayers (§5.6) at surfaces of the lamellae and destabilize them by reducing or removing the surface elasticity which is necessary for stability. If a droplet does not spread after entry, it can ultimately bridge the film,

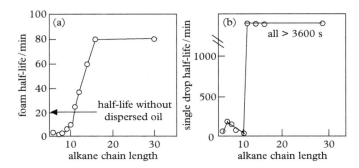

Figure 14.25 *(a) Foam half-life in systems with 3.8 mM AOT in 0.03 M aqueous NaCl containing dispersed alkane droplets. Chain length 30 represents squalane. (b) Half-life of single alkane drops placed under the interface between 3.8 mM AOT in 0.03M NaCl and air saturated with alkane vapour. Redrawn from Aveyard et al. (1999b).*

somewhat like a solid particle although obviously a liquid drop is deformable. A bridge once formed can be either mechanically stable or unstable.

An example of the effects of oil droplets on foam stability is illustrated in Fig. 14.25a. The half-life of foams formed from 3.8 mM AOT in 0.03 M NaCl containing dispersed alkane droplets is seen to depend markedly on the alkane chain length N_c. Alkanes up to decane ($N_c = 10$) lower the half-life (i.e. reduce foam stability) whereas undecane ($N_c = 11$) and upwards progressively stabilize the foams. It is found that the half-life of single drops of an alkane placed in the aqueous AOT solutions, just beneath the interface with air saturated with the alkane vapour, is low for $N_c \leq 10$, that is, the drops enter the interface relatively rapidly. On the other hand for drops for which $N_c \geq 11$, the drop half-life is >3600 s in all cases. It is clear then that the 'entering' (Box 14.7) alkanes destabilize the foams, whereas 'non-entering' droplets enhance the stability, possibly by collecting in the Plateau borders and inhibiting drainage.

Box 14.7   To enter or not to enter . . .

As explained in §12.3.2 the *feasibility* of drop entry into an interface can be discussed in terms of a classical entry coefficients, $E_{0,aw}$, defined in terms of interfacial tensions in the system. For mutually saturated phases at equilibrium $E_{0,aw} \geq 0$; positive values of the coefficient indicate drop entry is feasible. When the coefficient is zero, entry is not feasible in an equilibrated system. Feasibility however does not mean that entry will necessarily occur, since it can be prevented by forces acting over a thin oil/water/air pseudo-emulsion film as discussed in §12.3.2. If the phases are not equilibrated the constraints on the values of entry coefficients (as well as spreading coefficients) no longer hold.

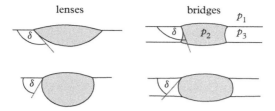

Figure 14.26 *Schematic representation of the shape of oil lenses on one surface of a thinning lamella and bridges across a lamella. Left: Non-spreading oil lenses. Right: The corresponding oil bridges over thin liquid films of the aqueous phase. The angle δ determines whether or not the bridge causes rupture of the film (see text).*

The shape of an oil lens resting on an aqueous phase is related to the values of the oil/water tension, γ_{ow}, air/solution tension γ_{aw}, and oil/air tension γ_{oa} (at equilibrium or otherwise) as outlined in §16.2.3. For very small lenses the oil/air and oil/water interfaces are spherically curved. With reference to Fig. 14.26, where the slight curvature of the air/water surface in the vicinity of the lens is ignored, lenses for which $\delta > 90°$ give rise to bridges in which the curvature (in the plane of the paper) of oil in contact with the aqueous phase is concave towards the lamella. That this is a mechanically unstable configuration can be seen by considering the pressures in the system. The pressure within the bridge, p_2, is greater than that, p_1, in the air surrounding the lamella; the pressure inside the lamella, p_3 is greater than that in the bridge. In an equilibrium system the pressure difference over a plane surface is zero. It follows that a system in which p_3 exceeds p_1 is unstable, and the effects of Laplace pressure force liquid in the lamella away from the oil bridge, causing film thinning and rupture. When δ is less than 90° no such effect occurs and the bridge is mechanically stable.

It is instructive to consider what systems could give rise to lenses for which δ is either above or below 90°, and if such systems are likely to exist at equilibrium. First, assume a set of tensions similar to those for an aqueous surfactant solution at the *cmc* in contact with an alkane at equilibrium, for example, $\gamma_{ow} = 5$ mN m^{-1}, $\gamma_{aw} = 27$ mN m^{-1}, and $\gamma_{oa} = 24$ mN m^{-1}. For these tensions the calculated value of $\delta = 131.2°$ (see §16.2.3). The entry coefficient is $+8$ mN m^{-1}, which means entry is feasible, and the spreading coefficient (§16.2.1) is -2.0 mN m^{-1}, so the oil is non-spreading. Thus an equilibrium system with these tensions could give a lens similar to that depicted in Fig. 14.26, top left. The use of a different set of assumed tensions, for example, $\gamma_{ow} = 1$ mN m^{-1}, $\gamma_{aw} = 27$ mN m^{-1}, and $\gamma_{oa} = 27.5$ mN m^{-1}, gives a value of $\delta = 60.9°$. The lens shape would be similar to that shown in Fig. 14.26 bottom left; the entry coefficient is positive ($+0.5$ mN m^{-1}) meaning entry of the drop into the air/water surface is feasible, and the spreading coefficient is negative (-1.5 mN m^{-1}) so the drop does not spread. As with the previous example, the interfacial tensions assumed could be encountered in hydrocarbon oil + aqueous surfactant systems.

14.11.3 Synergism between oil drops and particles in foam breaking

It is known that oil and particles act synergistically in preventing foam formation or in breaking existing foams, and commercial antifoams are often dispersions of small hydrophobic solid particles in a mineral oil. Oil drops are formed when antifoam is agitated in a foaming aqueous solution, and particles adsorb at drop surfaces. A possible origin of the synergism is that the particles, under some circumstances (*vide infra*), facilitate the entry of oil drops into the surfaces of lamellae; it is supposed that the oil drops then rupture the films as described above.

A drop of oil, with adsorbed particles, within an aqueous film is shown schematically in Fig. 14.27 (left). As depicted, a particle on the drop has entered the lamella surface and bridges a thin aqueous film between the lamella surface and the oil drop. An enlarged view a bridging particle is shown on the right of the figure. The particle has a contact angle θ_{aw} with the air/water interface of the lamella, and θ_{ow} with the oil/water interface. Assuming for the moment that dynamic effects are unimportant, the oil is expected to enter the lamella surface if $\theta_{aw} + \theta_{ow} \geq 180°$, that is, if the oil/water and air/water surfaces coincide.

In Fig. 14.28 the half-lives of heptane drops, coated with hydrophobic silica particles, placed under the interface between air and aqueous solutions of AOT (of various concentrations in 0.03 mol dm^{-3} NaCl) are shown as a function of $\theta_{aw} + \theta_{ow}$. There is clearly a dependence of drop half-life on the sum of the angles. For a sum less than 50°, drop entry is retarded; the thin aqueous film between oil drop and air that is bridged by the particle (Fig. 14.27 right), is relatively thick and stable. This would mean that oil and particles act antagonistically in film rupture. As $\theta_{aw} + \theta_{ow}$ is increased above 50°, however, and the aqueous film thickness is progressively decreased, drop entry is facilitated (low half-lives). The sum of angles of 50° is considerably less than that of 180° expected for synergy in the absence of dynamic effects. The large discrepancy could be due to dynamic wetting effects occurring when particles on a drop meet and enter the air/water surface of a lamella.

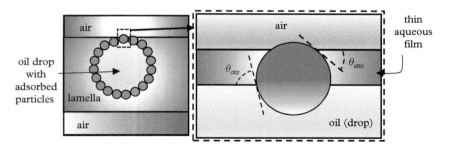

Figure 14.27 *Left: An oil droplet, with a layer of adsorbed particles, approaching the surface from within a lamella. Right: A single particle bridging the thin aqueous film between oil and air. The contact angles between the particle and the oil/water and air/water interfaces are shown.*

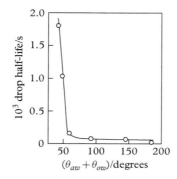

Figure 14.28 *Half-life of heptane drops, coated with hydrophobic silica particles, placed under the surface of AOT solutions of various concentrations in aqueous 0.03 mol dm^{-3} NaCl. Redrawn from Aveyard and Clint (1999).*

14.12   Some uses of foams

The occurrence and some uses of foams have already been briefly mentioned earlier. Here, two of the major applications are elaborated upon a little further: mineral processing and crude oil recovery.

14.12.1   Mineral processing

Froth flotation is used in the processing of minerals, and exploits the different chemistry of mineral surfaces to effect the separation of different minerals within an ore. In summary, ore is ground to give small particles which ultimately consist of a single mineral type. The powder is added to water and the slurry so formed is treated (with surfactants called *collectors*) in order to render the surfaces of certain particles hydrophobic. The slurry is then added to flotation tanks containing *frothing agents* through which air is bubbled. Particles with the required hydrophobicity become attached to the air bubbles and rise to the surface as a froth, which is skimmed off and added to separating tanks.

Collectors exhibit strong specific adsorption (§7.3.2) onto the mineral surfaces via their hydrophilic groups, the tails exposed to water rendering the mineral surfaces hydrophobic. Examples of collectors are xanthates: $ROCS_2^-Na^+$ or K^+ (designated X^-) (that are used with sulfide minerals such as PbS, CuS, and SbS). The alkyl group R ranges from ethyl to hexyl depending on the mineral being processed, and the adsorption takes place by ion exchange:

$$S_2^-\ (\text{surface}) + 2X^-\ (\text{aq}) \leftrightarrow 2X^-\ (\text{surface}) + S_2^-\ (\text{aq})$$

Many collectors are available for specific minerals, including anionic, cationic, and nonionic surfactants.

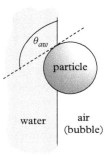

Figure 14.29 *Particle in a solution/air interface. For $\theta_{aw} > 90°$ more than half the particle resides in air (bubble).*

The function of the frothing agents is to stabilize the air bubbles rising through the flotation tanks so that mineral particles can become attached and rise with the bubbles to form a froth, which should collapse readily after it has been skimmed off the surface and added to separating tanks. In order for a particle to adsorb at a bubble surface, the thin aqueous film between particle and bubble surface must first thin sufficiently rapidly and rupture and the kinetics of adsorption are very important. Particles need to be hydrophobic in order to become strongly attached and remain at the bubble surface. The adsorption of solid particles to fluid/fluid interfaces is described in §17.1 where it is shown that the free energy of adsorption, $\Delta_a^{w\to aw}G$, of a spherical particle, radius r, from water to the (planar) air/water interface is given by

$$\Delta_a^{w\to aw}G = -\pi r^2 \gamma_{aw}(1 - \cos\theta_{aw})^2 \tag{14.12.1}$$

As θ_{aw} increases from 0 to 180° $\Delta_a^{w\to aw}G$ becomes increasingly more negative, that is, the more hydrophobic the particle the more strongly it is adsorbed from water. Further as θ_{aw} increases, a progressively smaller part of the particle resides in the aqueous phase (see Fig. 14.29), which means that the force on the particle from the aqueous phase as the particle plus bubble rise (which tends to dislodge the particle from the interface) is less the greater θ_{aw}.

The choice of suitable frothers can be difficult and depends on a number of factors; terpene alcohols (e.g. pine oil) and ethoxylated alcohols have commonly been used.

14.12.2   Crude oil recovery and some simple elements of foam rheology

A schematic cross-section of a dome-shaped oil well is shown in Fig. 14.30. Crude oil is contained in porous rock (e.g. sandstone or limestone) which is permeable to both oil and water. The oil-bearing layer is sandwiched between an aquifer and a gas region, all of which is bounded by a cap of impermeable rock. In an oilfield, several wells are drilled, some of which are for water injection (not shown), and the others for oil production. The injected water sweeps oil through the porous rock towards the production wells.

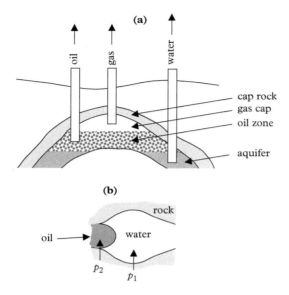

Figure 14.30 *(a) Schematic cross-section through an oil well showing a water layer (aquifer), oil zone and gas cap, and the production of gas, oil and water. (b) Curved oil/water interface in a rock pore.*

The pore sizes in the rock are small; cavities with diameters of about 10 μm are interconnected by pores with diameters of the order of 1 μm. Oil droplets can become trapped in the rock by capillary forces, since the crude oil/water interfaces in the pores (Fig. 14.30b) are highly curved. If the interface (with tension γ) is spherically curved with radius r the pressure difference across the oil/water interface, $p_2 - p_1$, is given by the Laplace equation as $2\gamma/r$ (§4.3.1). This pressure difference, Δp, has to be overcome by the pressure of the injected water. Clearly, Δp is directly proportional to the interfacial tension, and this is why in the past there has been considerable interest in the attainment of ultralow oil/water interfacial tensions in oil recovery systems (see §10.1.3).

The viscosity of water is much less than that of a typical crude oil. This can lead to *viscous fingering,* in which the water preferentially flows through the porous rock, ahead of the main oil/water front, causing a drastic fall in the rate of oil production. An obvious approach to preventing this is to increase the viscosity of the aqueous phase (*mobility control*). One method is to add water-soluble polymers (e.g. polyacrylamides) to the injection water. Another possibility is to inject aqueous foam, because interestingly foam viscosity is very much greater than that of the continuous aqueous phase which forms the foam lamellae.

The subject of foam rheology is broad and complex (Höhler and Cohen-Addad, 2005), and beyond the scope of this presentation.[8] Briefly, foams can exhibit either solid-like or liquid-like mechanical characteristics, depending on the applied stress and the gas

---

[8] Rheology of colloids is treated generally in Chapter 15.

volume fraction φ in the foam. For zero stress the transition from liquid-like to solid-like behaviour occurs at a gas volume fraction close to 0.64, which is that corresponding to the random close-packing of *polydisperse* hard spheres (Farr and Groot, 2009). Below $\varphi \sim 0.64$ the systems lose their rigidity and can be described as bubbly liquids (wet foam). At high volume fractions the foams become dry (polyhedral) (§14.6.1) and exhibit a yield stress, as illustrated in Fig. 14.31. In the wet foam region, the bubbles can be regarded as interacting spheres with elastic repulsion and viscous friction between first neighbours (Durian, 1997). When the stress on a dry foam is increased, irreversible bubble rearrangements occur at the yield stress and thereafter the foam flows, behaving as a non-Newtonian viscous fluid. Below the yield stress the system is (visco)elastic. Three dimensional polyhedral foams are disordered, but under 'equilibrium' conditions Plateau's rules hold (§12.1). If not, a foam will dissociate into structures in which the rules do hold. During normal bubble disproportionation and foam coarsening, local strains are produced which change the length of bubble edges. When an edge length goes to zero the configuration becomes unstable and switching of bubble neighbours occurs. Such

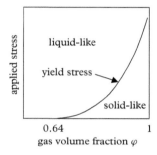

Figure 14.31 *Schematic diagram of the mechanical behaviour of foams as a function of applied stress and gas volume fraction.*

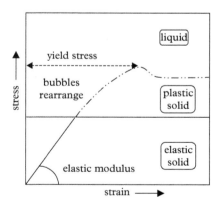

Figure 14.32 *Schematic representation of the stress–strain relationship for a polyhedral foam. Redrawn from Drenckhan et al. (2005).*

topological changes are called T1 processes (see Box 14.8) and can occur individually or as avalanches. These same changes can also occur when a macroscopic strain is applied.

A summary of the stress–strain relationship for a polyhedral foam with a fixed volume fraction φ is illustrated schematically in Fig. 14.32 (after Drenckhan et al., 2005).

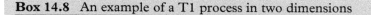

Box 14.8 An example of a T1 process in two dimensions

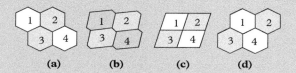

(a) (b) (c) (d)

Figure 14.33 *Two-dimensional foam structures formed by application of an increasing shear strain. The T1 rearrangement occurs when bubbles denoted 2 and 3 are separated and the topology changes. Redrawn from Höhler and Cohen-Addad (2005).*

For simplicity, imagine a two-dimensional foam of initially regular hexagons (Fig. 14.33a). The total edge length in a unit foam cell increases with the applied strain (b) and is associated with a stored elastic energy. Initially, if the strain is removed the foam returns elastically to its original configuration. When the strain reaches its yield value the length of the edge separating bubbles 2 and 3 becomes zero and structural rearrangement occurs (T1 process). The stress needed to induce the T1 process is the yield stress.

GENERAL READING

Encyclopedia of Emulsion Technology (ed. P. Becher). Marcel Dekker Inc., New York: Vol. 1 Basic Theory, 1983; Vol. 2 Applications, 1985; Vol. 3 Theory, Measurement and Applications, 1988; Vol. 4 Emulsification, Stability and Rheology, 1996.
Surfactants (ed. T.F. Tadros). Academic Press, London, 1984. A useful collection of articles on surfactant behaviour at a relatively simple level.

REFERENCES

S. Abbott, *Surfactant Science: Principles and Practice*. DEStec Publications, 2018, also available as a free eBook.
Aquamarijn, High-porosity microsieves, 2016. Available at: https://www.aquamarijn.nl/shop/high-porosity-microsieves/.
R. Aveyard and J.H. Clint, Liquid droplets and solid particles at surfactant solution interfaces. *J. Chem. Soc., Faraday Trans.*, 1995, **91**, 2681–2697.

R. Aveyard, B.P. Binks, J. Esquena, P.D.I. Fletcher, R. Buscall, and S. Davies, Flocculation of weakly charged oil–water emulsions. *Langmuir*, 1999a, **15**, 970–980.

R. Aveyard, B.P. Binks, P.D.I. Fletcher, and J.R. Lu, The resolution of water-in-crude oil emulsions by the addition of low molar mass demulsifiers. *J. Colloid Interface Sci.*, 1990, **139**, 128–138.

R. Aveyard, B.P. Binks, P.D.I. Fletcher, T.G. Peck, and C.E. Rutherford, *Adv. Colloid Interface Sci.*, 1994a, **48**, 93.

R. Aveyard, B.P. Binks, P.D.I. Fletcher, and C.E. Rutherford, Contact angles in relation to the effects of solids on film and foam stability. *J. Dispersion Sci. Technol.*, 1994b, **15**, 251–271.

R. Aveyard, B.P. Binks, and P.D.I. Fletcher, in *Foams and Emulsions* (ed. J.F. Sadoc and N. Rivier), p.21. Kluwer Academic Publishers, 1999b.

R. Aveyard, P. Cooper, P.D.I. Fletcher, and C.E. Rutherford, Foam breakdown by hydrophobic particles and nonpolar oil. *Langmuir*, 1993, **9**, 604–613.

B.P. Binks, Emulsions. *Ann. Rep. Progr. Chem., Sect. C.*, 1996, **92**, 97–134.

A. Cooper and M.W. Kennedy, Biofoams and natural protein surfactants. *Biophys. Chem.*, 2010, **151**, 96–104.

N.D. Denkov, Mechanisms of foam destruction by oil-based antifoams. *Langmuir*, 2004, **20**, 9463–9505.

E. Dickinson, Interpretation of emulsion phase inversion as a cusp catastrophe. *J. Colloid Interface Sci.*, 1981, **84**, 284–287.

E. Dickinson, *An Introduction to Food Colloids*. Oxford University Press, Oxford, 1992.

W. Drenckhan, S. Hutzler, and D. Weaire, Foam physics: the simplest example of soft condensed matter. *AIP Conf. Proc.*, 2005, **748**, 22–28.

D.J. Durian, Bubble-scale model of foam mechanics: melting, nonlinear behavior, and avalanches. *Phys. Rev. E*, 1997, **55**, 1739.

D.F. Evans and H. Wennerström, *The Colloidal Domain Where Physics, Chemistry and Biology Meet*. Wiley-VCH, Weinheim, 1999, Section 11.3.6.

R.S. Farr and R.D. Groot, Close packing density of polydisperse hard spheres. *J. Chem. Phys.*, 2009, **131**, 244104.

R. Gabbrielli, A.J. Meaghera, D. Weaire, K.A. Brakke, and S. Hutzler, An experimental realization of the Weaire–Phelan structure in monodisperse liquid foam. *Phil. Mag. Lett.*, 2012, **92**, 1–6.

P.R. Garrett, *Defoaming: Theory and Industrial Applications*. Marcel Dekker, New York, 1993.

W.W. Gerbacia and H.L. Rosano, Microemulsions: formation and stabilization. *J. Colloid Interface Sci.*, 1973, **44**, 242–248.

W.C. Griffin, Calculation of HLB values of nonionic surfactants, *J. Soc. Cosmet. Chem.*, 1954, **5**, 249–256.

R. Höhler and S. Cohen-Addad, Rheology of liquid foam. *J. Phys.: Condens. Matter*, 2005, **17**, R1041.

A. Kabalnov, Can micelles mediate a mass transfer between oil droplets? *Langmuir*, 1994, **10**, 680–684.

A. Kabalnov and H. Wennerström, Macroemulsion stability: the oriented wedge theory revisited. *Langmuir*, 1996, **12**, 276–292.

I. Kobayashi, Y. Murayama, T. Kuroiwa, K. Uemura, and M. Nakajima, Production of monodisperse water-in-oil emulsions consisting of highly uniform droplets using asymmetric straight-through microchannel arrays. *Microfluid Nanofluid*, 2009, **7**, 107.

K.J. Mysels, K. Shinoda, and S. Frankel, *Soap Films*. Pergamon Press, New York, 1959.

D.N. Petsev, in *Modern Aspects of Emulsion Science* (ed. B.P. Binks), Chapter 10. Royal Society of Chemistry, London, 1998.

H.M. Princen and S.G. Mason, Shape of a fluid drop at a fluid–liquid interface. I. Extension and test of two-phase theory. *J. Colloid Sci.*, 1965, **20**, 156–172.

A. Prins, in *Advances in Food Emulsions and Foams* (ed. E. Dickinson and G. Stainsby), Chapter 3. Elsevier Applied Science, London, 1988.

J.H. Schulman and T.P. Hoar, Transparent water-in-oil dispersions: the oleopathic hydro-micelle. *Nature*, 1943, **152**, 102.

K. Shinoda and S. Friberg, *Emulsions and solubilization*. Wiley-Interscience, New York, 1986.

K. Sotoyama, Y. Asano, K. Ihara, K. Takahashi, and K. Doi, Water/oil emulsions prepared by the membrane emulsification method and their stability. *J. Food Sci.*, 1999, **64**, 211–215.

T.F. Tadros, in *Surfactants* (ed. T.F. Tadros), Chapter 13. Academic Press, London, 1984.

T.F. Tadros, P. Izquierdo, J. Esquena, and C. Solans, Formation and stability of nano-emulsions. *Adv. Colloid Interface Sci.*, 2004, **108–109**, 303–318.

P. Taylor, Ostwald ripening in emulsions. *Adv. Colloid Interface Sci.*, 1998, **75**, 107–116.

P. Taylor and R.H. Ottewill, The formation and ageing rates of oil-in-water miniemulsions. *Colloids Surf. A*, 1994, **88**, 303–316.

G.E. Vaessen and H.N. Stein, The applicability of catastrophe theory to emulsion phase inversion. *J. Colloid Interface Sci.*, 1995, **176**, 378–387.

A. Vrij, Possible mechanism for the spontaneous rupture of thin, free liquid films. *Disc. Faraday Soc.*, 1966, **42**, 23–33.

P. Walstra, in *Encyclopedia of Emulsion Technology*, Volume 4 (ed. P. Becher), Chapter 1. Marcel Dekker, New York, 1996.

D. Weaire, Kelvin's ideal foam structure. *J. Phys.: Conf. Ser.*, 2009, **158**, 012005.

D. Weaire and R. Phelan, The structure of monodisperse foam. *Phil. Mag. Lett.*, 1994, **70**, 345–350.

15

Rheology of colloids

15.1 Some rheological quantities

Rheology is the study of the deformation and flow of materials, and the rheological characteristics (viscosity, elasticity, and viscoelasticity) of colloidal dispersions depend on the properties of the dispersed particles, droplets or surfactant aggregates. For example the shape, size and state of aggregation as well as the concentration of dispersed particles all influence the rheological properties of a dispersion. Structure can form throughout a phase, as in a gel, and this structure can be broken up during flow and possibly reform when static. The rheology of the continuous phase of a dispersion is also of obvious relevance. For example the presence of water soluble polymers in aqueous systems can impart stability to dispersed solid (or liquid) particles (Dickinson, 1992). The rheological behaviour can be tailored to the use and properties of a wide spectrum of industrial and commercial formulations such as toiletries, foodstuffs, pharmaceutical preparations, paints, and cosmetics. Some fundamental definitions of rheological quantities for liquid *surfaces* with adsorbed or spread amphiphilic molecules present were presented in Chapter 6. Here a brief and simple account is given of some relevant aspects of the rheology of *bulk* systems in order to point to its importance in the study and behaviour of colloidal systems (Box 15.1).

In connection with the rheology of surfaces, the shear deformation of a square area of surface (Fig. 6.3) was considered; for bulk systems it is appropriate to imagine the deformation of a rectangular block of material. With reference to Fig. 15.1, suppose a *shear stress* τ is applied tangentially along the top surface, with area $\mathcal{A}$, keeping the bottom face fixed. The stress is $F/\mathcal{A}$, where F is force applied, and the *shear strain* γ that results is[1]

$$\gamma = \Delta l/h = \tan\theta \tag{15.1.1}$$

---

[1] As in Chapter 6, although the symbol γ is used for interfacial tension, the same symbol (but bold and non-italic $\boldsymbol{\gamma}$) is also used for strain in conformity with other authors. The meaning of the symbol will be obvious in the context.

Surfactants: In Solution, at Interfaces and in Colloidal Dispersions. Bob Aveyard. © Bob Aveyard 2019.
Published in 2019 by Oxford University Press. DOI: 10.1093/oso/9780198828600.001.0001

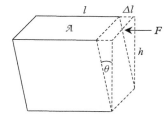

Figure 15.1 *Shear deformation of a rectangular block. The force F acts tangentially to the area $\mathcal{A}$; the bottom face is fixed. The angular deformation is denoted θ; $\tan \theta = \Delta l/h$.*

It is seen that the strain is a dimensionless quantity. For a linear elastic solid, the *shear modulus*, G,[2] is the constant ratio of the shear stress to shear strain, that is,

$$G = \tau/\gamma \tag{15.1.2}$$

Box 15.1   Flow and mistletoe

The term *rheology* stems from the Greek noun *rheos* meaning a stream or anything flowing, which in turn comes from the verb *rhein* to flow. On the other hand the term *viscosity* derives from the Latin, *viscum*, meaning anything sticky such as birdlime, a kind of glue made from mistletoe berries. The species *Viscum album* is the European mistletoe. Viscosity is also referred to as *dynamic* viscosity; the cgs unit is the poise (P) and $1 P = 10^{-1}$ Pa s. The viscosity of water at 20°C is 1.002 mPa s = 1.002 cP. The symbol used here for viscosity is η, but μ is also widely used in the literature to denote viscosity. In this chapter however μ is used for *apparent* viscosity. *Kinematic* viscosity (ν) is viscosity/fluid density and has SI units $m^2 s^{-1}$. One $cm^2 s^{-1}$ is 1 stokes (St), the unit named after Irish mathematician Sir George Gabriel Stokes (1819–1903).

For liquids the stress is related to the *rate* of deformation and stress and strain rate are related by

$$\tau = \eta(\mathrm{d}\,(\Delta l/h)\,/\mathrm{d}t) = \eta\dot{\gamma} \tag{15.1.3}$$

The coefficient η is the viscosity (SI unit = Pa s = kg/(m s) and $\dot{\gamma}$ is the shear strain rate. A Newtonian (i.e. linear) viscous fluid is one in which shear stress and shear strain rate are linearly related, so that the viscosity is independent of shear rate as illustrated in Fig. 15.2a. Water is a Newtonian liquid and some very dilute dispersions in water

---

[2] The word *modulus* has a variety of meanings but here it is a quantity expressing the degree to which a system possesses a specific property (e.g. elastic modulus). It can also mean the absolute value of a complex number (see later).

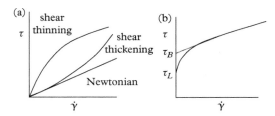

Figure 15.2 *Plots of stress versus strain rates for viscous fluids. (a) The relationship is linear for Newtonian fluids. Curves for shear thinning (pseudoplastic) and shear thickening (dilatant) fluids are also shown. (b) The behaviour of a plastic material; a minimum yield stress τ_L is needed for flow. The Bingham yield stress is the extrapolated value τ_B.*

are also Newtonian (see §15.3). Dispersions of sufficient concentration in which there are significant net interactions between the dispersed entities exhibit non-Newtonian behaviour.

The stress–strain rate behaviour of shear thinning (*pseudoplasic*) and shear thickening (*dilatant*) systems is illustrated in Fig. 15.2a. In shear thinning systems (e.g. many polymer solutions and some modern paints) the viscosity falls with increasing shear rate (possibly due to breakdown of structure under shear) whereas for shear thickening liquids the viscosity rises with shear rate. The latter behaviour can arise in concentrated non-aggregated dispersions as a result of structural changes. In these systems the slope of a plot of stress against strain rate at a particular strain rate gives an *apparent viscosity, μ*.

There are two interesting *time-dependent* (as opposed to shear rate-dependent) rheological responses. Systems in which the viscosity falls with time when a constant shear rate is applied are termed *thixotropic;* the viscosity in such systems falls when the system is shaken or stirred. Examples include gels and other colloids. The opposite of thixotropic is *rheopectic*; the viscosity in rheopectic systems increases the longer the systems are sheared. This behaviour is much less common than thixotropy, but potentially important practically since the systems are shock-absorbing.

The behaviour of a *plastic* material is illustrated in Fig. 15.2b. The material does not flow until the yield stress (τ_L in the figure) is reached. There is then a curved region of the stress–strain rate curve before a linear (Newtonian) regime is attained at high strain rates. The stress extrapolated from the linear region to zero strain rate, τ_B, is called the *Bingham yield stress*. In practice it can be difficult to extract the exact behaviour of a system at low strain rates.

So far elastic and viscous behaviour has been discussed, but in very many colloidal systems the two types of behaviour coexist, that is, the systems exhibit *viscoelasticity*. Suppose a step stress τ_o is applied to a system and the subsequent deformation (shear strain), $\gamma(t)$, is followed as a function of time t (the procedure being referred to as a *creep* experiment). Various possible responses are illustrated in Fig. 15.3, in which (a) shows the sudden application (and removal) of the stress. The abscissa in all diagrams is time t. In (b) to (e) the ordinate is the strain, $\gamma(t)$. The response of a purely *elastic* solid is shown

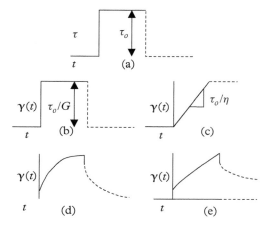

Figure 15.3 *Responses in creep experiments in various types of rheological system. In all cases the abscissa is time t. (a) represents the sudden application of shear stress τ_o and here the ordinate is τ. In (b) to (e) the ordinate is the strain, $\gamma(t)$, The full lines give the creep and broken lines the recovery. (b) shows the response of a (linear) elastic solid, (c) is for a (linear) viscous fluid, and (d) and (e) are for a viscoelastic solid and viscoelastic liquid, respectively.*

in (b); the strain rises instantaneously to a *constant strain* τ_o/G (see (15.1.2)) and remains at this value until the stress is removed, when the strain falls immediately to zero. The behaviour of a Newtonian *viscous* fluid is shown in Fig. 15.3(c), where $\gamma(t)$ is seen to rise linearly with t (with slope $\tau_o/\eta = \dot{\gamma}$ —see (15.1.3)) and then remain constant when the stress is removed. This *constant strain rate* is achieved immediately in a purely viscous fluid.

Curves (d) and (e) refer to the behaviour of *viscoelastic* materials. For a viscoelastic solid (curve (d)) a constant equilibrium strain is achieved (i.e. the curve reaches a plateau), whereas for a viscoelastic liquid a constant rate of strain is achieved (the line becomes rectilinear). For *linear* viscoelastic systems (observed below a critical value of applied stress), γ is linearly related to τ_o:

$$\gamma(\tau_o, t) = \tau_o \mathcal{J}(t) \tag{15.1.4}$$

where $\mathcal{J}(t)$, the shear *creep compliance*, is independent of τ_o. Inspection of (15.1.2) shows that $\mathcal{J}(t)$ is the inverse of a shear modulus. In nonlinear systems $\mathcal{J}(t)$ is dependent on τ_o, usually falling as τ_o increases.

In order to characterize a viscoelastic system it is necessary to extract the elastic (storage) and viscous (loss) components of the behaviour. In rheological experiments, timescales involved are of great significance. If an *oscillating* strain is applied to a fluid, whether or not the fluid flow can 'keep up' depends on the angular frequency ω

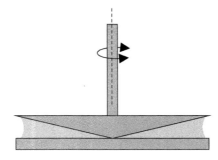

Figure 15.4 *Cone and plate arrangement for rheological measurements. The plate is fixed and the cone is oscillated; the fluid sample under test fills the gap between cone and plate.*

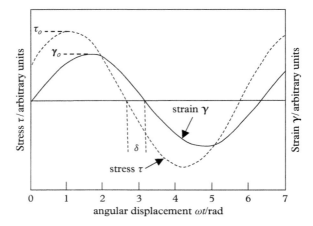

Figure 15.5 *Sinusoidal stress or strain as a function of angular displacement ωt in an oscillatory experiment. The frequency ω is in rad s⁻¹ and t is time. The phase angle δ = 0° if the fluid is purely elastic and 90° if it is purely viscous; for a viscoelastic fluid δ lies between 0° and 90°.*

(rad s^{-1})^3 of the oscillation. Imagine that a sinusoidal strain wave is applied to a purely elastic material, contained say between a flat plate and a cone (Fig. 15.4) or in the gap between concentric cylinders.[4] It has been seen (Fig. 15.3b) that the strain and resulting stress occur simultaneously and so the stress wave is completely in phase with the strain wave i.e. the phase angle δ in Fig. 15.5 is zero. On the other hand for a purely viscous fluid the stress and strain are 90° (π/2 radians) out of phase. The value of δ lies between

---

[3] The angular frequency $\omega = 2\pi f$ where f is the frequency i.e. the rate at which an oscillation or wave repeats. If the period of an oscillation is T then $\omega = 2\pi/T$.

[4] In both the cone and plate and concentric cylinder configurations, the shear strain rate is constant across the whole gap.

0 and 90° for viscoelastic fluids, and the relative extents of elastic and viscous responses depend on the frequency of the applied oscillating strain.

The quantities obtained from an oscillatory experiment at a given frequency are the amplitudes (maximum values) of stress and strain, that is, τ_o and γ_o, (see Fig. 15.5) and the phase angle δ, and all these are determined as a function of frequency.

The *complex modulus*, G^*, is a measure of the total shear stiffness of a material, the stiffness arising from both viscous flow and elastic strain. Its absolute value is given as the ratio (maximum stress/maximum strain) (see (15.1.2)), and is a function of the frequency, that is,

$$|G^*(\omega)| = \tau_o/\gamma_o \tag{15.1.5}$$

The complex modulus has contributions from the storage (elastic) modulus $G'(\omega)$ (the ratio of in-phase stress to strain), and the loss (viscous) modulus $G''(\omega)$ (the ratio of 90° out of phase stress to strain) and is expressed as a complex number (see Appendix for a note on complex numbers) by

$$G^*(\omega) = G'(\omega) + iG''(\omega)$$
$$= |G^*|(\cos\delta + i\sin\delta) \tag{15.1.6}$$

The storage and loss moduli are obtained from τ_o, γ_o, and δ (for a given frequency) since

$$G'(\omega) = (\tau_o/\gamma_o)\cos\delta$$
$$G''(\omega) = (\tau_o/\gamma_o)\sin\delta \tag{15.1.7}$$

It is seen from (15.1.7) that $G''(\omega)/G'(\omega) = \tan\delta$ (the 'loss tangent'), which reflects the ratio between lost and stored energy in a cyclic deformation. It is a measure of the relative importance of viscous and elastic responses in a material (Bohlin et al., 1984). With reference to Fig. 15.6 (see also Appendix A15.1) it can be seen that the absolute magnitude of G^*, $|G^*|$, is related to $G'(\omega)$ and $G''(\omega)$ by

$$|G^*| = \sqrt{G'^2 + G''^2} \tag{15.1.8}$$

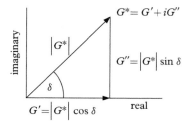

Figure 15.6 *Representation of the complex modulus G^* and the storage (G') and loss (G'') moduli.*

As an alternative to using the complex modulus, $G^*(\omega)$, the phase relationship can be expressed in terms of a *complex viscosity*, $\eta^*(\omega)$ defined as $G^*(\omega)/i\omega$ so, noting that the reciprocal of i is $-i$,

$$\eta^*(\omega) = \eta'(\omega) - i\eta''(\omega) \tag{15.1.9}$$

Alternatively one may write

$$G^*(\omega) = G'(\omega) + i\omega\eta' \tag{15.1.10}$$

The quantity $\eta'(\omega) = G''(\omega)/\omega$ is the dynamic viscosity (real in phase part of the complex viscosity), and $\eta''(\omega) = G'(\omega)/\omega$. The dynamic viscosity can approach η, the ordinary steady-flow viscosity at low frequencies (see footnote 4) (Franklin and Krizec, 1969).

As indicated earlier, a non-oscillatory method for the determination of the viscoelastic characteristics of a material involves the measurement of the shear creep compliance $\mathcal{J}(t)$ (see Fig. 15.3e). The material is subjected to a constant stress τ_o and the resulting shear strain $\gamma(\tau_o, t)$ is measured over time. The stress, strain and creep compliance are related, as seen in (15.1.4), by $\gamma(\tau_o, t) = \tau_o\mathcal{J}(t)$. As mentioned earlier, for a linear viscoelastic system $\mathcal{J}(t)$ is independent of the imposed stress τ_o.

15.2 The Maxwell model: a simple approach to the behaviour of a viscoelastic fluid

All that has been said above about viscoelasticity is of course independent of any assumed model used to describe the behaviour of a system. It is however useful to have mechanical analogies to see what rheological behaviour might be expected based on certain assumptions about the rheological response of a material. Models have been proposed combining springs (for elasticity) and dashpots (for viscosity) in various numbers and configurations. A dashpot is a cylinder containing viscous liquid (viscosity η) in which a piston moves.

For a perfect (Hookean) spring which is subjected to an instantaneous stress the deformation also occurs instantaneously and, as discussed earlier, stress and strain are related to the spring constant G, as in

$$G = \tau/\gamma \tag{15.1.2}$$

Stress and strain for a Newtonian viscous liquid are related through the viscosity by

$$\tau = \eta\dot{\gamma} \tag{15.1.3}$$

In the *Maxwell model* for a viscoelastic fluid a spring is connected in series with a dashpot, as illustrated in Fig. 15.7. In an oscillating experiment the complex stress and strain are

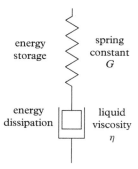

energy storage — spring constant G

energy dissipation — liquid viscosity η

Figure 15.7 *Spring and dashpot (Maxwell model) representation of a viscoelastic fluid.*

denoted by an asterisk and it can be appreciated from (15.1.2) and (15.1.3) therefore that the complex shear strain rate $\dot{\gamma}^*$ is given by[5]

$$\dot{\gamma}^* = \tau^*/\eta + \dot{\tau}^*/G \tag{15.2.1}$$

where the first term on the right is for the dashpot and the second for the spring.

The oscillating strain and resulting stress (see Fig. 15.5) can be expressed in terms of the amplitudes (maximum values during a cycle), γ_o and τ_o, respectively, by

$$\gamma^* = \gamma_o \left(\cos(\omega t) + i \sin(\omega t) \right) = \gamma_o e^{(i\omega t)} \tag{15.2.2}$$

$$\tau^* = \tau_o \left(\cos(\omega t + \delta) + i \sin(\omega t + \delta) \right) = \tau_o e^{i(\omega t + \delta)} \tag{15.2.3}$$

From this, $\dot{\gamma}^*$ and $\dot{\tau}^*$ are

$$\dot{\gamma}^* = i\omega\gamma^*; \dot{\tau}^* = i\omega\tau^* \tag{15.2.4}$$

Noting that $G^*(\omega) = \tau^*/\gamma^*$ and substituting for $\dot{\gamma}^*$, $\dot{\tau}^*$ and τ^* in (15.2.1), using (15.2.2) to (15.2.4) it is shown that

$$\frac{1}{G^*(\omega)} = \frac{1}{G} + \frac{1}{i\omega\eta} \tag{15.2.5}$$

[5] In the context, equations such as (15.1.2), (15.1.3) and (15.2.1) are often referred to as *constitutive* equations, that is, ones that relates two physical quantities for a material, for example stress and strain. The equation describes the response of the material to applied forces.

To proceed, a relaxation time t_r is defined:[6]

$$t_r = \eta/G \qquad (15.2.6)$$

In the Maxwell model it represents the time for the energy stored in the spring to travel to the dashpot and dissipate. If the simple model is exact then t_r is independent of the frequency, otherwise not. From (15.2.4) and (15.2.6) it seen that

$$G^*(\omega) = G\left(\frac{i\omega t_r}{1 + i\omega t_r}\right) \qquad (15.2.7)$$

From (15.2.6) and (15.2.7) and noting that $G^*(\omega) = G'(\omega) + iG''(\omega)$ ((15.1.6)) it follows that

$$G'(\omega) + iG''(\omega) = G\left[\frac{(\omega t_r)^2}{1 + (\omega t_r)^2}\right] + iG\left[\frac{\omega t_r}{1 + (\omega t_r)^2}\right] \qquad (15.2.8)$$

At sufficiently high frequencies the response of the model becomes purely elastic and therefore G (the spring constant) is written $G(\infty)$. In the limit of zero frequency the dynamic viscosity ($\eta'(\omega) = G''(\omega)/\omega$) reaches the value $\eta(o)$ which is the viscosity of the dashpot. Plots of $G'/G(\infty)$, $G''/G(\infty)$ and $\eta'/\eta(o)$ against log of the angular frequency are shown in Fig. 15.8. From the figure, and by inspection of (15.2.7), it is seen that when $\omega = 1/t_r$, $G'(\omega) = G''(\omega) = G(\infty)/2$ and $G''(\omega)$ has its maximum value. Since at this frequency the system exhibits equal amounts of elastic and viscous behaviour the phase angle $\delta = 45°$.

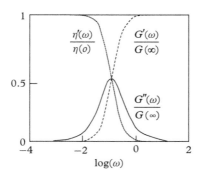

Figure 15.8 *The dependence of $G'(\omega)$, $G''(\omega)$, and $\eta'(\omega)$ on frequency ω according to the Maxwell model. The frequency corresponding to maximum in $G''(\omega)$ is $1/t_r$. For the construction of these curves the following values were assumed: $t_r = 8\ s$ and $G(\infty) = 15\ Pa$, hence $\eta(o) = 120\ Pa\ s$.*

---

[6] Since the units of viscosity are Pa s and of the modulus Pa, it is seen that the unit of t_r is time.

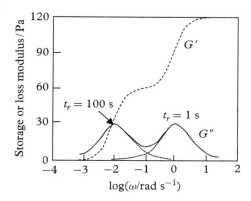

Figure 15.9 *G' and G'' as a function of log ω for a Maxwell system showing two distinct relaxation times t_r of 1 s and 100 s. G is taken as 60 Pa for both processes.*

The Maxwell model as presented has a single relaxation time. It may well be that in systems of practical interest there are several relaxation processes occurring simultaneously, and in which case G' and G'' can be represented as the sum of terms for the various processes. Suppose there are two distinct relaxation processes occurring. The system can be represented by two (spring + dashpot) pairs (see Fig. 15.7) connected in parallel so that $G' = G_1' + G_2'$ and $G'' = G_1'' + G_2''$. A plot of G' and G'' against $\log(\omega/\text{rad s}^{-1})$ is shown in Fig. 15.9 for a system exhibiting relaxation times of 1 s and 100 s as indicated; G in both cases has been taken to be 60 Pa. There is a large spread around each relaxation time and even for a 100-fold difference in times there is significant overlap in the values of G''. As the relaxation times become closer together the distinctness of the peaks becomes increasingly obscured.

It is seen from (15.2.8) that

$$G' = \left(\frac{G(\omega t_r)^2}{1 + (\omega t_r)^2} \right); \quad G'' = \left(\frac{G\omega t_r}{1 + (\omega t_r)^2} \right)$$

which is consistent with

$$\left(G' - \frac{G}{2} \right)^2 + G''^2 = \left(\frac{G}{2} \right)^2 \tag{15.2.9}$$

It follows from (15.2.9) that for a system conforming to the Maxwell model with a single relaxation time the points on a plot of G'' against G' fall on the arc of a circle,[7] and such

---

[7] The equation of a circle, with its centre at $(x,y) = (a,b)$ and of radius r, is $(x - a)^2 + (y - b)^2 = r^2$. Cole–Cole plots have been widely used in the representation of relative permittivities, which are also complex

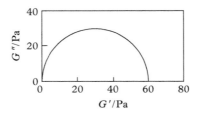

Figure 15.10 *A Cole–Cole plot of G'' versus G' calculated on the basis of the Maxwell model with a single relaxation time. The value of G is 60 Pa.*

plots are referred to as Cole–Cole plots. An example is shown in Fig. 15.10 where G has been taken as 60 Pa.

Creep experiments, which are non-oscillatory (see Fig. 15.1.3), have already been referred to. The shape of a creep compliance curve on the basis of the Maxwell model is now considered; the constitutive equation has already been given as

$$\dot{\gamma} = \tau/\eta + \dot{\tau}/G \tag{15.2.1}$$

The solution of this differential equation is

$$\gamma(t) = \left(\frac{\tau_o}{\eta(o)}\right) t + \frac{\tau_o}{G(\infty)}. \tag{15.2.10}$$

where τ_o is the instantaneous stress applied at zero time, as indicated in Fig. 15.11. The viscosity $\eta(o)$ and spring constant $G(\infty)$ have the same significance as previously. In terms of creep compliance, $\mathcal{J}(t) = \gamma(t)/\tau_o$

$$\mathcal{J}(t) = \left(\frac{1}{\eta(o)}\right) t + \frac{1}{G(\infty)} \tag{15.2.11}$$

The compliance curve according to (15.2.11) is shown in Fig. 15.11. When the stress is applied instantaneously at $t = 0$ there is a purely elastic response and the compliance rises instantaneously to $1/G(\infty)$. After this, during the creep phase, the compliance rises linearly with slope $1/\eta(o)$. At time t_1 the stress is removed and the stored elastic energy is released, and thereafter the compliance remains constant at a value $t_1/\eta(o)$. In real systems the instantaneous responses shown in the figure can be difficult to achieve or observe.

The Maxwell model with a single relaxation time is obviously rather crude. It has however been widely used as a first approximation in the treatment of oscillatory

quantities with real and imaginary parts. Each point on a circular arc represents a given frequency and the intercepts with the abscissa (real axis) give the values of the real part at zero and infinite frequency.

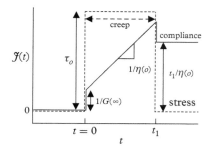

Figure 15.11 *Creep compliance curve according to the Maxwell model.*

experimental data for viscoelastic colloidal systems and is often quite successful in capturing the main features, at least over a limited range of frequencies.

15.3   Dilute particulate and polymer dispersions

In the preceding sections the definitions of some rheological quantities have been introduced, and a useful model for the behaviour of viscoelastic systems discussed. In the next section some experimental results to illustrate the rheological behaviour of various colloidal dispersions are presented. Before this however a very brief description of the treatment of the viscosity of dilute particulate dispersions and dilute polymer solutions is given.

A dispersion of solid spherical particles in a Newtonian liquid is Newtonian at sufficiently low particle volume fractions, φ. In such a state there are effectively no attractive or long-range repulsive forces between the particles, which act independently with respect to the hydrodynamic flow of the medium; particle aggregation is also absent. The viscosity, which is dependent on φ, is however independent of particle size. Einstein famously treated this situation and showed that the *relative viscosity* of the dispersion, η_r, is given by

$$\eta_r = \eta/\eta_o = 1 + 2.5\varphi \qquad (15.3.1)$$

where η is the viscosity of the dispersion and η_o that of the medium. The relative viscosity, which being a ratio of viscosities is dimensionless, is clearly a measure of the effect of the particles on the viscosity of the medium.

As mentioned in §15.1, at higher concentrations particulate dispersions become non-Newtonian, shear-thinning at intermediate concentrations and possibly shear-thickening at higher concentrations. Particle interactions become evident; the effects of these interactions and of the imposed shear flow are opposed by Brownian motion (with translational thermal energy $3kT/2$ per particle) (see e.g. Dickinson, 1992 or Goodwin, 2009). For a given volume fraction φ, small particles have a greater thermal energy, and for low shear rates the structure (spatial arrangement of particles) of a dispersion remains relatively undisturbed by the flow, the behaviour being more or less Newtonian. For larger particles however Brownian motion becomes increasingly unimportant and the shear forces cause disruption of the colloidal structure, leading to non-Newtonian behaviour.

For non-Newtonian systems, where the viscosity is the *apparent* viscosity μ for a given strain rate (as mentioned in §15.1) and the relative viscosity $\mu_r = \mu/\eta_o$, further terms in φ must be added to (15.3.1):

$$\mu_r = 1 + 2.5\varphi + k_2\varphi^2 + \dots. \tag{15.3.2}$$

The constant k_2 takes account of the hydrodynamic interactions between pairs of spherical particles and its value depends on the importance of the Brownian motion of the particles. It turns out that for small particles and low shear rates (when the thermal energy of the particles dominates over effects due to the other forces) the value of $k_2 = 6.2$. For large particles, when Brownian motion is unimportant, k_2 take a value of about 5.22. Equation (15.3.2), with the appropriate value of k_2, describes systems well for $\varphi < 0.15$. For higher particle concentrations various empirical equations for the relative viscosity are available, but a discussion of such equations is beyond the scope of this simple account.

In characterising the viscosity of dilute dispersions of *polymer molecules*, in addition to the relative viscosity, two further measures of solution viscosity are defined. The *specific viscosity* η_{sp} is given by

$$\eta_{sp} = \frac{\eta - \eta_o}{\eta_o} = \eta_r - 1 \tag{15.3.3}$$

and is the increase in viscosity caused by the presence of polymer, normalised to the solvent viscosity; like the relative viscosity the specific viscosity is dimensionless. More generally η, η_{sp}, and η_r can be replaced by the apparent quantities μ, μ_{sp}, and μ_r for non-Newtonian systems. The specific viscosity of course depends on the polymer concentration c (expressed as mass volume^{-1}, e.g. kg dm^{-3}) but for non-ideal solutions η_{sp}/c is not constant, that is, the viscosity does not rise linearly with c. It is therefore useful to define the *intrinsic viscosity* $[\eta]$, which is the limiting value of η_{sp}/c as the concentration tends to zero, that is,

$$[\eta] = \lim_{c \to 0} \left(\eta_{sp}/c \right) \tag{15.3.4}$$

The intrinsic viscosity, which has the units of reciprocal concentration, is of particular interest since it is related to the molecular weight of the dissolved polymer.

The viscosity of a polymer solution is expressed in terms of the polymer concentration c and the intrinsic viscosity of the solution in the Huggins equation:

$$\eta = \eta_o \left[1 + [\eta]\, c + k_H[\eta]^2 c^2 + \dots \right] \tag{15.3.5}$$

Here k_H is the Huggins coefficient, which as might be anticipated is associated with the interaction of pairs of polymer molecules. Ignoring terms beyond that in $[\eta]^2$, (15.3.5) can be expressed

$$\eta_{sp}/c = [\eta] + k_H[\eta]^2 c \qquad (15.3.6)$$

so that a plot of η_{sp}/c versus c, which is linear for a dilute polymer solution, gives $[\eta]$ as the intercept and k_H can be obtained from the slope. The intrinsic viscosity is related to the polymer molecular weight by the empirical expression

$$[\eta] = K M_v^x \qquad (15.3.7)$$

in which subscript v indicates that the molecular weight M is an average value obtained from solution viscosities; it lies between the number average and the weight average molecular weights (Goodwin, 2009). The constants K and x, which have been obtained experimentally, are tabulated in the literature for given polymer–solvent pairs and can be related to the polymer backbone stiffness and the quality of the solvent for the polymer.

15.4 Some illustrative results for the rheological behaviour of colloidal dispersions

As seen, there are various rheological parameters that can be measured, but the most interesting quantities in many practical cases are those that directly relate to a process of interest or condition of a system (Cunningham, 2013). For dispersions in storage or that are slowly creaming or sedimenting, for example, the zero shear viscosity will be of interest. Experiments yielding viscoelastic parameters e.g. storage and loss moduli, as well as creep compliance measurements can also be relevant. On the other hand, systems can be fast-flowing such as paints during brushing or material being pumped through a pipe. In this case the viscosity at a fixed shear rate and that at infinite shear are likely to be informative. In between there is a transition region relevant for example to the sagging of wet paint or a dispersion in a pump at start-up. Here the yield stress and yield strain in the system are expected to be important. Examples of experimentally determined rheological properties of some colloidal systems are now presented briefly.

15.4.1 Dilute dispersions of spherical particles

Dilute dispersions of non-interacting spherical solid particles are the simplest rheological systems to deal with and a good place to start. As discussed earlier, up to volume fractions of about 0.15 these systems tend to obey the Einstein viscosity equation (15.3.1). This can be seen in Fig. 15.12 where the relative viscosity of aqueous particulate dispersions is plotted against φ. The dotted line in the figure is drawn with slope 2.5 and the experimental data fall on the line for φ up to about 0.15. Various kinds of spherical particles are represented on the graph. The open symbols are for yeast particles (roughly

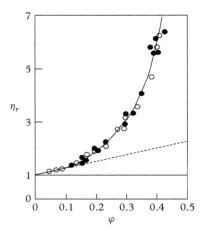

Figure 15.12 *Relative viscosity of aqueous dispersions as a function of volume fraction. Open symbols are for yeast cells and filled symbols are for spherical solid particles, for example, polystyrene and glass, with radii of around 1 μm. The dotted line has slope = 2.5. Redrawn from Toda and Furuse (2006).*

spherical with radius about 5μm) and the filled points for various other solid particles including glass and polystyrene spheres all with diameters of the order of microns. Beyond $\varphi = 0.15$ there are strong positive deviations from the Einstein equation.

Toda and Furuse (2006) extended the Einstein theory for the dependence of relative viscosity on particle volume fraction to give the expression

$$\eta_r = \frac{1 - 0.5\varphi}{(1 - \varphi)^3} \tag{15.4.1}$$

It was found that this equation described the viscosities of aqueous solutions of glucose and sucrose well but modifications were needed to fit data for large particles such as those represented in Fig. 15.12. The full line in the figure is generated by the expression

$$\eta_r = \frac{1 + 0.5\kappa\varphi - \varphi}{(1 - \kappa\varphi)^2\,(1 - \varphi)} \tag{15.4.2}$$

with $\kappa = 1 + 0.6\varphi$. Equation (15.4.2) reverts to (15.4.1) for $\kappa = 1$.

15.4.2 Suspensions of minerals

Aspects of the rheology of suspensions of mineral particles have frequently been studied. Rheology can give information on, for example, the flocculation of particles to give networks, brought about by adjustment of salt concentration or pH, or by addition of polymer. Rheological data are often considered in combination with other findings on say interparticle forces, phase behaviour or ζ-potentials.

As an example the viscous behaviour of aqueous suspensions of goethite (hydrous ferric oxide α-FeOOH, average particle length and diameter 0.43.and 0.16 μm, respectively) is considered (Blakey and James, 2003). At neutral pH the systems, with particle volume fractions from 0.01 to 0.08, were flocculated, and were both viscous and markedly shear thinning (Fig. 15.13). Water at 20°C has a viscosity of 1 mPa s so that at low shear rates even the least concentrated of the dispersions has an apparent viscosity μ over 10 times that of water. For $\varphi = 0.08$ and a shear rate of between 1 and 10 s^{-1} the apparent viscosity is about 10^4 that of water.

The faces of the goethite particles are, like, for example, alumina, covered with hydroxyl groups and the surface charge is dependent on the pH (see §7.2). The *p.z.c.* of the goethite sample used was at pH = 9.5, close to that of corundum (alumina) at pH = 9. It was proposed that the different crystal faces have different charge densities, due to different types of OH group resulting from the modes of coordination to iron atoms. The different faces can even carry the opposite *sign* of charge. This gives rise to the possibility that attractive forces can exist resulting from electrostatic interactions between different faces of neighbouring particles, in addition to van der Waals forces and electrostatic repulsion.

Some apparent viscosities of goethite suspensions are shown in Fig.15.14 as a function of pH for the conditions given in the legend. The results were shown to be essentially independent of the salt concentration up to 0.1 mol dm^{-3} NaCl. The viscosities were correlated with the state of flocculation of the suspensions, observed microscopically. At pH $\approx$ 5.5, at which the ζ-potential is about 75 mV, the goethite particles are well dispersed, as might be expected, but by pH $\approx$ 6.5 flocculation is beginning to be very significant even though the ζ-potential is still high ($\approx$55 mV). The apparent viscosity rises considerably in this pH range as a result of the aggregation, which it is suggested results from of the attraction between positive and negative regions of surface charge, such as occurs between clay particles which have edges and faces of different sign charge (§7.2). Maximum apparent viscosity is given around the *p.z.c.* as can be anticipated and

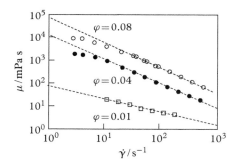

Figure 15.13 *Apparent viscosity as a function of shear rate at 25°C for aqueous suspensions of goethite at neutral pH with the particle volume fractions as indicated. The aqueous phase is 0.1 mol dm^{-3} NaCl. Redrawn from Blakey and James (2003).*

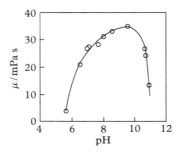

Figure 15.14 *Apparent viscosity (at a shear rate of 100 s^{-1} and 25°C) of goethite suspensions ($\varphi = 0.02$) in 10^{-3} mol dm^{-3} NaCl as a function of pH. Redrawn from Blakey and James (2003).*

as the pH is increased beyond this the extent of flocculation begins to decrease. The shear thinning illustrated in Fig. 15.14 for neutral pH results from the progressive break up of flocs as the shear rate increases.

15.4.3   Solid dispersions in the presence of polymers

Solid particulate dispersions in which polymers are also present are frequently encountered and their rheology studied. The polymer alone can of course have a drastic effect on rheological properties, as for example with hydrophobically modified copolymers used as aqueous borne 'associative' thickeners. Polymers present in particulate dispersions can act in various ways and some examples were given in Chapter 13. The dispersions can be stabilised by steric effects resulting from adsorbed polymer, or destabilised by free polymer in the medium by depletion forces. Polymer molecules can also bridge particles causing flocculation.

Burns et al. (2003) investigated the rheology associated with the depletion flocculation of aqueous concentrated electrostatically-stabilised polystyrene latex particles in the presence of non-adsorbing poly(acrylic acid) (PAA). It is often observed that depletion-flocculated systems show the behaviour illustrated in Fig. 15.2b, that is, Bingham flow. The constant viscosity at higher shear rates is termed the plastic viscosity and denoted η_P and so the stress, τ, in a Bingham fluid may be written as

$$\tau = \tau_B + \eta_P \dot{\gamma} \tag{15.4.3}$$

where τ_B is the Bingham yield stress. A typical plot used to obtain τ_B in systems with a range of polymer concentrations (and molar mass) is shown in Fig. 15.15. The suspensions start to flow at stresses below the Bingham yield stress, but τ_B is regarded as the stress at which all permanent interparticle contacts have been broken by the flow. The variation of τ_B with polymer concentration for a sample with molar mass 2.5 $\times$ 10^5 is shown in Fig. 15.16.

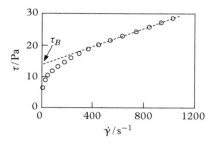

Figure 15.15 *Shear stress versus shear rate for 20% w/w latex dispersion in the presence of 0.2 g dm^{-3} PAA with molar mass 2.5 × 10^5. The extrapolated stress τ_B is the Bingham yield stress. Redrawn from Burns et al. (2003).*

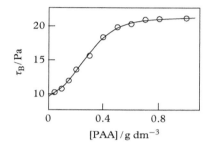

Figure 15.16 *Dependence of Bingham yield stress in 20% w/w latex dispersions as a function of PAA concentration. The polymer molar mass is 2.5 × 10^5. Redrawn from Burns et al. (2003).*

As seen, τ_B initially rises with increase of polymer concentration and eventually reaches a plateau value. Since the yield stress is a measure of the mechanical strength of the flocs it can be related to the energy required to break the flocs up into single particles. The connection between rheology and the surface forces acting between particles is of particular interest in colloid science.

15.4.4 Effects of temperature on zero shear viscosity of wormlike micellar solutions

In some surfactant solutions, micelle growth can lead to the formation of very long threadlike aggregates, as discussed in (§9.5.3). The entanglement of such wormlike micellar chains confers viscoelastic properties on the solutions, and this type of behaviour is discussed in §15.4.5 below. In the present section however it is shown how the simpler zero shear viscosities of wormlike micellar solutions can give insights into effects of temperature on the formation of threadlike micelles.

Although like polymers in some respects, threadlike micelles are able to exchange surfactant molecules with those in the coexisting solution. Consequently, micellar length (aggregation number) can respond to changes in solution conditions such as temperature. The usual situation is that when the micellar solution is heated, micellar length decays exponentially (Kalur et al., 2005). It has been mentioned in §9.5.3 that the free energy, per surfactant molecule, of forming the micelle caps is more positive than that of forming the cylinder, and this difference is implicated in limiting the growth of micelles. It has been suggested that as the temperature rises the energy of forming the hemispherical caps becomes less unfavourable and so the micellar chains become smaller. In this context it is worth remembering however that micellar length can be remarkably high, up to and beyond 1 μm.

Kalur et al. (2005) studied the effects of temperature on the zero shear viscosity, η_o, of aqueous solutions containing the C_{22}-tailed cationic surfactant erucylbis(hydroxyethyl) methylammonium chloride (EHAC),

$$C_8H_{17}CH{=}CHC_{12}H_{24}\overset{+}{N}(C_2H_4OH)_2CH_3Cl^-$$

together with the organic salt sodium hydroxynaphthalenecarboxylate (SHNC). Zero shear viscosities of 40 mM aqueous EHAC in the presence of a range of concentrations of SHNC from 200 to 450 mM SHNC were determined at 298 K, and in this range of SHNC concentrations η_o fell from about 10^3 Pa s (at which the solutions were highly viscoelastic) to about 10^{-3} Pa s (about that of water) for 450 mM SHNC. Some results for the effects of temperature on η_o are shown in Fig. 15.17 for 40 mM EHAC solutions containing 240 and 280 mM SHNC. In the case of the lower concentration of SHNC the viscosity falls monotonously with T as expected, in line with the behaviour of a number of other systems containing threadlike micelles, and the fall is associated with the shortening of the micelle length (Raghavan and Kaler, 2001).

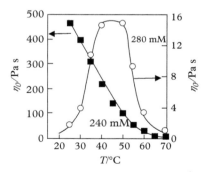

Figure 15.17 *Zero shear viscosities as a function of temperature of 40 mM aqueous solutions of EHAC in presence of SHNC at the concentrations indicated (see text for abbreviations). Results from Kalur et al. (2005).*

An unusual feature of the work reported by Kalur et al. (2005), however, was that in some systems, the viscosity was found to *rise* with increasing T, as seen in Fig.15.17. For systems containing 240 mM SHNC, the viscosity only falls but for [SHNC] = 280 mM, although importantly the viscosity is overall much lower (see the ordinates in Fig.15.17), the viscosity first rises with T and then after attaining a maximum, falls. For 360 mM SNHC (not shown) the viscosity (which is in the 10s of mPa s range) only rises with T over the same 50 degree interval. If sodium salicylate is used rather than SHNC to provide the counterion, the viscosity only falls with T.

An explanation given by the authors for their findings runs as follows. The counterion provided by SHNC, that is, HNC^-, is very hydrophobic as a result of the naphthalene ring, which is able to penetrate into the micelle interior giving a highly (negatively) charged micelle surface. The SHNC concentrations used are very high relative to that of the surfactant EHAC. At lower temperatures, where the viscosity is low, it is likely the micelles are small spherical or slightly elongated structures, which would be favoured by aggregates with highly charged surfaces. As the temperature rises, the solubility of HNC^- in water rises so that some 'bound' entities will leave the micelles and enter the aqueous phase. This would lead to a reduction of surface charge on the micelles which in turn would favour the formation of more elongated or cylindrical micelles. The maximum in viscosity shown in Fig.15.17 is explained by noting the natural effect of increasing T is to create shorter micelles (as explained) so when there are two competing effects, that is, T *versus* micelle charge, a maximum in viscosity might be expected.

15.4.5 Viscoelastic surfactant solutions and the Maxwell model

A final example of the rheology of colloidal systems involves surfactant solutions. The formation of wormlike micelles was discussed in §9.5.3, where it was mentioned that these systems exhibit interesting viscoelastic properties.

In dilute solutions (concentration $c < $ a critical concentration c^*) the micelles are well separated and there is no overlapping. For $c > c^*$ however (i.e. in *semi-dilute* solutions) overlap of micelles becomes pronounced and viscoelastic behaviour is observed. For rapid oscillatory shear where the time period is less than the relaxation time required for the solution to return to its equilibrium state, the sample behaves like an elastic solid. For sufficiently slow deformations however, relaxation becomes possible and the system behaves as a viscous fluid. If the amplitudes of (sinusoidal) stress and strain are small enough, the viscous and elastic responses are linear, i.e. as previously discussed, the fluid behaviour is Newtonian and the elastic behaviour Hookean. A relaxation time t_r has been defined as

$$t_r = \eta/G \qquad (15.2.6)$$

and as seen is determined by the balance of the viscosity and the elasticity of the viscoelastic material. The ratio of t_r to the experimental time t, say of an oscillation, (i.e. t_r/t) is the *Deborah number*, De (Box 15.2).

> **Box 15.2** The mountains flowed...
>
> The Deborah number *De* was introduced by Markus Reiner; it is named after the prophetess Deborah in the Bible (Judges 5:5) who sang 'The mountains flowed before the Lord'. This is taken to mean that even apparently solid materials flow if they are observed over a long enough period. For solid-like behaviour *De* >> 1, and for liquid behaviour *De* << 1; viscoelastic behaviour corresponds to *De* ~ 1. Reiner, along with Eugene C. Bingham, also coined the term *rheology*. Reiner became professor at the Technion in Haifa after the founding of the state of Israel; he died in 1976.

The viscoelasticity of the wormlike micellar solutions can be treated usefully (over a limited frequency range) using the Maxwell model. Some experimental results of oscillatory linear rheological measurements on a 100 mM aqueous solution of hexadecyltrimethylammonium tosylate (CTAT) (in the semi-dilute region) with 10 per cent of the tosylate counterion replaced by hydroxynaphthalenecarboxylate (HNC^-) anion (taken from Hassan et al., 1998) are shown in Fig. 15.18a. In the figure G' and G'' are plotted against frequency ω. As seen from the figure, $G' < G''$ for $\omega < 1/t_r$, and $G' = G''$ for $\omega \sim 0.35$ rad s^{-1}, giving a stress relaxation time $t_r \sim 3$ s. The Maxwell model predicts a monotonous fall in G'' with increasing ω (Fig. 15.8) whereas the experimental data in Fig. 15.18a shows an increase in G'' at high frequencies. The Cole–Cole plot for similar data (but without added HNC^- anions) is shown in Fig.15.18b. The Maxwell model predicts that the data for different frequencies should fall on the arc of a circle, and the deviations from this are clearly seen again for higher frequencies.

The behaviour in the region at higher frequencies before G'' begins to rise with ω have been explained using the theory of Cates (1987) for 'living' polymers. For densely

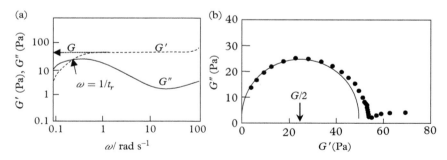

Figure 15.18 *Rheological parameters for a solution of 100 mM CTAT at 30°C. (a) Storage modulus G' and loss modulus G'' as a function of oscillation frequency ω. (b) Plot of loss modulus versus storage modulus (Cole–Cole plot). Points are experimental and the full line is for a Maxwell element with a single relaxation time. Redrawn from Hassan et al. (1998).*

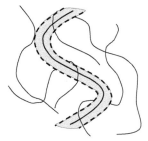

Figure 15.19 *Reptation of a polymer molecule through an imaginary tube (dashed lines) passing through the polymer network.*

entangled long, unbreakable linear polymers the main relaxation method is *reptation* (Fig. 15.19).[8] For chains with length L the relaxation time of the system, t_{rep}, is

$$t_{rep} = L^2/D_c \tag{9.5.10}$$

in which D_c is the diffusion constant of a chain along its own contour. The polymer chain completely diffuses through the hypothetical tube in time t_{rep}. In the case of wormlike micelles (i.e. *living* polymers) however, in addition to reptation, the micelles can break up into shorter lengths and reform (randomly) with other fragments. Thus, an additional characteristic time t_b, the average time for a chain of length L to break into two fragments, becomes important. Given two characteristic times there are various times of interest depending on the value of (t_b/t_{rep}). For the case in which $t_b \ll t_{rep}$, micelle break-up and recombination will occur many times during the period it would take a micelle to pass through the original tube by reptation. The relaxation will be exponential with a new characteristic time t_r given by (Cates, 1987)

$$t_r = \left(t_b t_{rep}\right)^{0.5} \tag{15.4.4}$$

In this regime at low frequencies the behaviour is Maxwellian, giving circular curvature on the Cole–Cole plot as observed experimentally. For slow scission of micelles $(t_b \gg t_{rep})$ however stress relaxation occurs by reptation and the micelles behave like polydisperse unbreakable polymers; now the stress relaxation function is non-exponential, that is,

---

[8] De Gennes (1971) supposed that a polymer molecule entangled with its neighbours is effectively confined to diffuse within a tube (defined by neighbouring chains) rather than move perpendicular to the tube. Such movement along the tube, by analogy with the way a snake moves, is referred to as reptation.

$$\tau \approx \exp\left[-\left(t/t_{rep}\right)^{0.25}\right] \qquad (5.4.5)$$

and deviations from the semicircular line occur. The onset of these deviations is at a frequency of the order of the inverse of the micellar breaking time t_b.

15.4.6 Human blood

Blood is a remarkable and complex fluid, not least in its rheological properties. Human blood consists largely of erythrocytes (red blood cells, RBC) suspended in blood plasma. In a human adult there are about 5 million RBC mm^{-3} blood compared to 4000–11,000 white blood cells (leucocytes) and about 150,000–400,000 platelets. The percentage volume occupied by packed RBC in blood (the *haematocrit*) is about 40 per cent for women to 50 per cent for men. The RBC are deformable biconcave discs with a dumb-bell cross-section (Fig. 15.20a). They are viscoelastic (Puig-de-Morales-Marinkovic et al., 2007) and have the tendency to aggregate reversibly into stacks called rouleaux (Fig. 15.20b). They undergo pulsatile flow, taking about 20 seconds to complete one circuit of the body. The function of RBC is to take up oxygen in the lungs and release it throughout the body. The cells must be capable of passing through microcapillaries with diameter ca. 5 μm, that is, less than that (6–8 μm at rest) of the RBC. The flow velocity is lowest in the microcapillaries because they are so numerous.

Whole blood is viscoelastic and, as mentioned, so are the RBC themselves. Blood viscosity and its shear dependence result from the formation and disintegration of the aggregates of RBC, the deformability of the cells and the haematocrit. Another non-Newtonian effect is the dependence of viscosity on the vessel diameter. For tube diameters < 12–15 μm the viscosity falls markedly (the Fahraeus–Lindqvist effect) which is due in part to the haematocrit being smaller in such small capillaries. When RBC flow in a tube with a diameter that is a low multiple of the cell diameter, the cells flow down the centre of the tube, liquid near the vessel surface being particle-free. Although the rheology of blood *in vivo* is complicated, some insights can be gained from data obtained for human blood *in vitro*.

The results shown in Fig.15.21 were obtained for normal human blood using an oscillatory rheometer over a range of oscillatory shear rates, with a frequency of 10 Hz

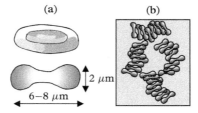

Figure 15.20 *(a) Representation of a red blood cell with approximate dimensions. Lower figure is a cross-section. (b) Stacks of aggregated red blood cells (rouleaux).*

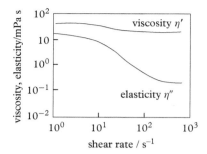

Figure 15.21 *The viscosities and elasticities of 50% haematocrit human blood obtained in oscillatory measurements at a frequency of 10 Hz at 22°C. Results closely mirror those reported by Thurston (1972).*

(Thurston, 1972). Both elasticity η'' and viscosity η' are high for blood at rest or under low shear. Here the RBC form large aggregates (Fig. 15.20b) which begin to disaggregate as the shear increases. At intermediate shear rates (from say 10 to 100 s^{-1}) the elasticity progressively falls as the aggregates continue to break up; the cells tend to align in the direction of flow and become deformed. At the highest shear rates illustrated the cells are stretched and form layers which are separated by layers of plasma. Once all the aggregates have broken the elasticity is low, and the energy coming from the shear is dissipated rather than stored.

Appendix A15: Complex numbers

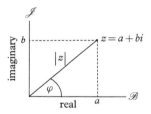

A complex number, z, can be expressed as $z = a + bi$ where a and b are real numbers and $i = \sqrt{-1}$. It can be represented, as shown in the figure, on an *Argand diagram*, which represents the complex plane. The numbers a and b are referred to as the *real* and *imaginary* parts of z respectively. The absolute value (modulus) of z, $|z|$, is given as $|z| = \sqrt{a^2 + b^2}$. The *phase* φ of z is the anticlockwise angle between the positive real axis and the line joining the origin to the point z, so that $a = |z| \cos \varphi$ and $b = |z| \sin \varphi$ and hence

$$z = a + bi = |z| (\cos \varphi + i \sin \varphi)$$

The bracket containing the trigonometric terms is given by Euler's formula (identity) as:

$$e^{i\varphi} = (\cos \varphi + i \sin \varphi)$$

....

REFERENCES

B.C. Blakey and D.F. James, The viscous behaviour and structure of aqueous suspensions of goethite. *Colloids Surf. A*, 2003, **231**, 19–30.

L. Bohlin, P.-O. Hegg, and H. Ljusberg-Wahren, Viscoelastic properties of coagulating milk. *J. Dairy Sci.*, 1984, **67**, 729–734.

J.L. Burns, G.J. Jameson, and S. Biggs, The rheology of concentrated suspensions of depletion-flocculated latex particles. *Colloids Surf. A*, 2003, **214**, 173–180.

M.E. Cates, Reptation of living polymers: dynamics of entangled polymers in the presence of reversible chain-scission reactions. *Macromolecules*, 1987, **20**, 2289–2296.

N. Cunningham, Centre for Industrial Rheology, 2013. Available at: http://www.rheologyschool. com. This gives useful insights into the relevance of rheological data to practical systems.

E. Dickinson, *An Introduction to Food Colloids*. Oxford University Press, Oxford, 1992, Chapter 3.

P.-G. de Gennes, Reptation of a polymer chain in the presence of fixed obstacles. *J. Chem. Phys.*, 1971, **55**, 572–579.

J.W. Goodwin, *Colloids and Interfaces with Surfactants and Polymers*, 2nd edition. John Wiley & Sons, Chichester, UK, 2009.

A.G. Franklin and R.J. Krizec, Complex viscosity of a kaolin clay. *Clays Clay Minerals*, 1969, **17**, 101–110.

P.A. Hassan, S.J. Candau, F. Kern, and C. Manohar, Rheology of wormlike micelles with varying hydrophobicity of the counterion. *Langmuir*, 1998, **14**, 6025–6029.

G.C. Kalur, B.D. Frounfelker, B.H. Cipriano, A.I. Norman, and S.R. Raghavan, Viscosity increase with temperature in cationic surfactant solutions due to the growth of wormlike micelles. *Langmuir*, 2005, **21**, 10998–11004.

M. Puig-de-Morales-Marinkovic, K.T. Turner, J.P. Butler, J.J. Fredberg, and S. Suresh, Viscoelasticity of the human red blood cell. *Am. J. Physiol. Cell Physiol.*, 2007, **293**, C597–C605.

S.R. Raghavan and E.W. Kaler, Highly viscoelastic wormlike micellar solutions formed by cationic surfactants with long unsaturated tails. *Langmuir*, **2001**, **17**, 300–306.

G.B. Thurston, Viscoelasticity of human blood. *Biophys. J.*, 1972, **12**, 1205–1217.

K. Toda and H. Furuse, Extension of Einstein's viscosity equation to that for concentrated dispersions of solutes and particles. *J. Biosci. Bioeng.*, 2006, **102**, 524–528.

Section VI

Wetting of liquids and solids

This section consists solely of Chapter 16, although various aspects of wetting are covered in Chapter 12 devoted to thin liquid films. To get a broader feel for how liquids wet solids and other liquids, the two chapters should be read in conjunction with each other. In addition, some basic aspects of wetting and adhesion are covered in Chapter 3. Much of the interest in wetting phenomena has arisen from the importance of wetting in many industrial processes and in biological phenomena as well as in domestic cleaning of various kinds. Often surfactants are added in order to control (enhance or reduce) wetting. Some solid surfaces however are highly water-repellent, due not to surfactant adsorption but as a result of the surface roughness. Such surfaces are common in nature.

16

Wetting

16.1 Background

The subject of wetting is of importance in large areas of the chemistry, physics, and engineering of surfaces, and deals with the ways in which a liquid behaves when in contact with a second (immiscible) liquid or solid phase. Various aspects of wetting have already been introduced in previous chapters. The idea of the cohesion of a material and the mutual adhesion of different materials was introduced in §3.3, as was the contact angle θ that a liquid makes with a solid or liquid sub-phase. It was seen that the work of adhesion W_A between a solid and a liquid is related to the contact angle through the Young–Dupré equation (3.3.3).

In Chapter 12, the formation of asymmetric thin liquid films on a solid or a liquid phase was described and the concept of the entry coefficient E was developed. As illustrated in Fig. 12.3, a drop of liquid placed on a smooth surface can either completely wet the surface, forming a thick (i.e. duplex) film, spread out to form a thin liquid film (possibly coexisting with a bulk lens or meniscus of the liquid) or remain as a droplet with no spreading over the surface. As will be seen in Chapters 17 and 18, the adsorption of solid particles at fluid interfaces is strongly influenced by the relative wettability of the particle surface by the two fluid phases. Since wettability and contact angle are determined by the various interfacial tensions (as shown in Young's equation (3.3.1) for example), and since adsorption lowers interfacial tension (as described by the Gibbs equation), the presence of surfactant in a system can be expected to influence wetting in some way. The details of this influence are determined by the relative adsorption of surfactant at the various interfaces present in the system.

Much of the surface-chemical interest in wetting relates to the way in which extensive solid surfaces or solid particles, colloidal or otherwise, are wetted by liquid (very often aqueous) phases. In §7.2 it was shown how solid surfaces can be broadly hydrophilic (polar) or hydrophobic (nonpolar), and how this can influence the adsorption of surfactants from aqueous solution (§7.5). Many of the surface-chemical investigations into the wetting of solids and liquids have been driven by a need to understand the behaviour of systems of practical interest such as those encountered in detergency, crude oil recovery and ore recovery by froth flotation. As will be seen, because of the complexity of such systems semi-empirical approaches have been devised to aid understanding and prediction of wetting behaviour.

Surfactants: In Solution, at Interfaces and in Colloidal Dispersions. Bob Aveyard. © Bob Aveyard 2019.
Published in 2019 by Oxford University Press. DOI: 10.1093/oso/9780198828600.001.0001

The areas already covered elsewhere in the book relate to equilibrium, or at least static systems. However, dynamic aspects of wetting can be of equal importance, for example, in the coating of surfaces with liquids, particularly at high speed as for example in the production of photographic film. A very brief account of dynamic wetting will be given later in the chapter.

It is convenient to consider the wetting of liquids and of solids separately, although obviously the two areas have some common themes. A major difference between the two types of system is that the interfacial tensions in liquid/fluid systems can be measured directly, whereas solid/fluid tensions cannot. Indeed solid surfaces are unlikely to be in an equilibrium state. Also, the surfaces of practical interest are often rough and chemically heterogeneous whereas liquid/fluid interfaces are molecularly smooth and homogeneous. The wetting of liquids by liquids is considered first and is followed by the treatment of the wetting of solids.

16.2   Equilibrium wetting of liquids by liquids

16.2.1   The spreading coefficient

Consider three fluid phases, α, β, and δ (such as water, vapour, and oil, respectively) in an enclosed space at mutual thermodynamic equilibrium and suppose a small drop of δ is placed on the $\alpha\beta$ interface (Fig. 16.1). Phase δ will not spread at the $\alpha\beta$ interface (removing the $\alpha\beta$ interface)[1] to give a thick film if the sum of the final interfacial tensions, $\gamma_{\alpha\delta} + \gamma_{\beta\delta}$, is greater than the tension of the $\alpha\beta$ interface, $\gamma_{\alpha\beta}$. A *classical spreading coefficient* for the spreading of δ at the $\alpha\beta$ interface, $S_{\delta,\alpha\beta}$, can usefully be defined as

$$S_{\delta,\alpha\beta} = \gamma_{\alpha\beta} - (\gamma_{\alpha\delta} + \gamma_{\beta\delta}) \tag{16.2.1}$$

so that spreading will not occur if $S_{\delta,\alpha\beta}$ is negative. Suppose that $\gamma_{\alpha\beta}$ is the largest of the three interfacial tensions then, since at thermodynamic equilibrium (see §12.3.2),[2]

$$\gamma_{\alpha\beta} \leq \gamma_{\alpha\delta} + \gamma_{\beta\delta} \tag{16.2.2}$$

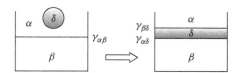

Figure 16.1 *Addition of liquid δ to the $\alpha\beta$ fluid/fluid interface. It is supposed that δ spreads to form a duplex film.*

[1] Of course, if the volume of *d* is sufficiently large it will 'spread' on *a* as a result of gravity.

[2] The inequality in (16.2.2) can be written with any of the three tensions on the left hand side so that $\gamma_{\alpha\delta} < \gamma_{\alpha\beta} + \gamma_{\beta\delta}$ and $\gamma_{\beta\delta} < \gamma_{\alpha\beta} + \gamma_{\alpha\delta}$. Obviously the equality can only be true if the largest of the tensions appears on the left hand side; this is explained further in §12.3.2.

it follows that $S_{\delta,\alpha\beta} \leq 0$. Spreading to give a duplex film of δ corresponds to $S_{\delta,\alpha\beta} = 0$, and positive values of the classical spreading coefficient are not possible in systems at thermodynamic equilibrium. Noting footnote 2 and the definition of the spreading coefficient in (16.2.1) it can also be shown that $S_{\delta,\alpha\beta} \geq -2\gamma_{\alpha\delta}$ (the equality holding if $\gamma_{\alpha\delta}$ is the largest of the three tensions) and therefore the possible values that the equilibrium spreading coefficient can take are expressed by

$$-2\gamma_{\alpha\delta} \leq S_{\delta,\alpha\beta} \leq 0 \tag{16.2.3}$$

If thermodynamic equilibrium does not exist between the α, β, and δ phases, an *initial* spreading coefficient, S_i, can be defined that can be positive, but if so it will fall as a system moves towards equilibrium. It is noted with reference to Fig. 16.1 that, for example, the vapour of phase δ can adsorb at the $\alpha\beta$ interface lowering $\gamma_{\alpha\beta}$ in the equilibrated system, and it is possible for δ to spread initially but then retract as equilibrium is approached.

Spreading to give thick films can also be understood in terms of the more familiar work of adhesion (of δ with β in α), $W^A_{\beta\delta,\alpha}$, and the work of cohesion of δ in α, $W^C_{\delta,\alpha}$. The work of adhesion is defined by (see §3.3)

$$W^A_{\beta\delta,\alpha} = \gamma_{\alpha\delta} + \gamma_{\alpha\beta} - \gamma_{\beta\delta} \tag{16.2.4}$$

and so it is seen from (16.2.1) and (16.2.4) that

$$S_{\delta,\alpha\beta} = W^A_{\beta\delta,\alpha} - 2\gamma_{\alpha\delta} = W^A_{\beta\delta,\alpha} - W^C_{\delta\alpha} \tag{16.2.5}$$

where the work of cohesion of δ in α, $W^C_{\delta\alpha} = 2\gamma_{\alpha\delta}$. From (16.2.5) it follows that the spreading coefficient is negative (i.e. spreading does not occur) when the work of cohesion of δ in α is greater than the work of adhesion between β and δ in α, as expected.

The entry of a liquid immersed in another liquid into a fluid/liquid interface was discussed in §12.3.2 where a *classical entry coefficient*, E was defined. For the entry of say a drop of δ in α into the $\alpha\beta$ interface (Fig. 16.1), the entry coefficient $E_{\delta,\alpha\beta}$ is defined as

$$E_{\delta,\alpha\beta} = \gamma_{\alpha\beta} + \gamma_{\alpha\delta} - \gamma_{\beta\delta} \tag{16.2.6}$$

Clearly, if δ can enter the $\alpha\beta$ interface from α to give a thick film between α and β phases, α does not spread at the $\beta\delta$ interface. Conversely if α does spread on δ, entry of δ cannot occur. The spreading coefficient for α on δ is

$$S_{\alpha,\beta\delta} = \gamma_{\beta\delta} - \left(\gamma_{\alpha\beta} + \gamma_{\alpha\delta}\right) \tag{16.2.7}$$

and it is seen from (16.2.6) and (16.2.7) therefore that $S_{\alpha,\beta\delta} = -E_{\delta,\alpha\beta}$. Since at equilibrium, spreading of α corresponds to $S_{\alpha,\beta\delta} = 0$, a value of $E_{\delta,\alpha\beta}$ of zero indicates that δ will not enter the $\alpha\beta$ interface. Positive values of the entry coefficient indicate entry is feasible to give a thick layer of δ between the α and β phases, although the formation of a metastable thin film of α between δ and β may inhibit entry in practice.

For spreading to give a *thin* film of δ, account has to be taken of the surface forces between the two surfaces ($\beta\delta$ and $\alpha\delta$) of the film since the interfacial tensions are modified by the proximity of the surfaces. Just as a generalized entry coefficient has been defined and discussed for the case of a thin film (see §12.3.2 and (12.3.8)), so a *generalized spreading coefficient* (for say δ at the $\alpha\beta$ interface) can be defined as

$$S^g_{\delta,\alpha\beta} = \int_{\Pi_D(h_\infty)=0}^{\Pi_D(h)} h d\Pi_D = -E^g_{\alpha,\beta\delta} \qquad (16.2.8)$$

where, as in (12.3.8), Π_D is the disjoining pressure in the film and h is the 'equilibrium' film thickness (Bergeron and Langevin, 1996). The sign of S^g depends on the shape of the disjoining pressure isotherm, and as for E^g can be positive or negative and in systems where stable thin films with two different thicknesses can exist there are two different generalized spreading coefficients. It is clear from (16.2.8) that the generalized spreading coefficient is simply the excess film free energy per unit area arising from the surface forces between the two surfaces.

A number of studies of spreading of oil on aqueous surfactant solutions have been made, and it has been seen that oil molecules can be solubilized in surfactant monolayers (§10.2.1). Kellay et al. (1992, 1994) studied the spreading of straight chain liquid alkanes (Box 16.1) on aqueous NaCl solutions containing AOT (sodium di(2-ethylhexyl)sulfosuccinate)) above the critical micelle concentration (*cmc*). In these systems, which can exhibit Winsor behaviour, the oil/water interfacial tension can become ultralow (§10.1.3). It was found that for chain lengths less than 11, alkanes spread to give a thin film which is in equilibrium with bulk alkane, which resides in the meniscus at the edge of the container and with which it presumably has a finite contact angle. The thin films retain their thickness as bulk alkane is progressively added. For a given alkane the mean film thickness was observed to be dependent on the NaCl concentration in the surfactant solution. For decane spread on aqueous solutions over a range of salt concentration, the film thickness was maximum (about 7 nm excluding the surfactant chain region) when the oil/water interfacial tension was minimum (ca. 3×10^{-3} mN m^{-1}). These effects were interpreted as arising from the fluctuations of the oil/water interface which are greater the lower the oil/water tensions (see §8.1 and §12.4). The oil/vapour interface, which has a high tension, remains essentially flat, as illustrated in Fig. 16.2.

Box 16.1 Alkanes and water—the simplest systems?

Arguably, some of the simplest systems studied in wetting experiments have contained liquid n-alkanes and water in contact with an equilibrium vapour phase; the phases are designated δ, β, and α respectively in Fig. 16.1. In the context the systems are considered simple because

Box 16.1 *Continued*

only dispersions forces act between alkanes and water at the oil/water interface. It is known that the vapours of the lower alkanes can *adsorb* at the water/vapour interface (see §2.3) leading to the formation of monolayers and a reduction in γ_{aw} for vapour pressures of alkane, p less than the saturated vapour pressure p^o. Hauxwell and Ottewill (1970) further observed that pentane, hexane, and heptane adsorb from their *saturated* vapour phases ($p = p^o$) onto the surface of water to give films exhibiting interference colours, indicating that these alkanes spread macroscopically on water. Octane on the other hand was found usually to give small lenses on the water surface in equilibrium with an adsorbed monolayer. Liquid alkanes higher than octane do not spread macroscopically on water, nor does water spread macroscopically on any of the liquid alkanes. These systems are discussed further in §16.2.2 in terms of classical spreading coefficients and Hamaker constants, the latter relating to surface forces and the formation of *thin* liquid films.

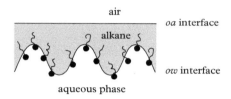

Figure 16.2 *Thin liquid film of alkane on aqueous solution of AOT in NaCl. The oil/air interface has a high tension and is planar whereas the oil/water interface has an ultralow tension and is undulating.*

16.2.2   Wetting and the non-retarded Hamaker constant

Wetting has already been discussed in terms of the spreading coefficient S in §16.2.1 and the disjoining pressure isotherm in §12.3.2; disjoining pressure $\Pi_D(h)$ is related to the Hamaker constant A_H (defined in §11.2.1) for the interaction between thin film surfaces. The spreading coefficient is defined in terms of interfacial tensions of macroscopic phases and indicates whether or not macroscopic spreading of one liquid on (say) another liquid, to give a *thick* (duplex) film, is feasible. On the other hand treatments involving disjoining pressure isotherms and the generalized spreading coefficient (defined in (16.2.8)) include the formation of *thin* liquid films where there is an interaction between the two film surfaces, and where the tensions of the surfaces are modified as a consequence.

Imagine a film of oil (o) between water (w) and a vapour phase (a) (Fig. 16.3) and suppose that only dispersion forces act between the film surfaces. If a 'wets' w in the presence of o then a film of o between w and a will spontaneously thin. This means that the free energy of interaction per unit area of the a/o and o/w interfaces, $V(h)$, is negative (attractive); it is given by (see §11.2.1)

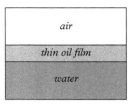

Figure 16.3 *Water (w) interacts with air (a) (through dispersion forces) over a thin film of oil (o).*

$$V(h) = -A_H/12\pi h^2 \qquad (16.2.9)$$

where A_H is the non-retarded Hamaker constant (positive for attraction between surfaces) and h is the thickness of the film of o. The Hamaker constant can be obtained, say, by use of (11.2.10). The disjoining pressure $\Pi_D(h)$ within the film is equal to $-\mathrm{d}(V(h)/\mathrm{d}h$ and is therefore related to A_H by

$$\Pi_D(h) = -A_H/6\pi h^3 \qquad (16.2.10)$$

The disjoining pressure is negative for a spontaneously thinning film and positive for a thickening film. (Disjoining pressure isotherms for the formation of various kinds of liquid films are illustrated in Fig. 12.13 in §12.3.2.) Therefore, wetting or otherwise can be predicted from Hamaker constants, but it has to be born in mind that the constants relate to surface forces acting over only small distances, say a few nanometres.

To illustrate the observation that the spreading coefficient S is relevant to macroscopic spreading whereas the Hamaker constant relates only to the formation of very thin films, the wetting of water (w) by a series of liquid n-alkanes (o) in the presence of vapour (a) is considered. As already mentioned in §16.2.1, it is known from experiment that heptane and lower homologues spread macroscopically on water in equilibrated systems. Octane is a borderline case and higher alkanes do not spread as thick films on water; water does not spread to give a thick film on any of the alkanes. The vapours of the lower alkanes (up to octane) adsorb significantly on the surface of water (as monolayers) lowering its surface tension at equilibrium by an amount π, the surface pressure of the adsorbed alkane.

The spreading coefficient for spreading of o at the aw interface, $S_{o,aw}$, is defined as

$$S_{o,aw} = \gamma_{aw} - (\gamma_{ow} + \gamma_{oa}) \qquad (16.2.11)$$

For macroscopic spreading at equilibrium, $S_{o,aw} = 0$. The surface tension γ_{aw} is that of the clean water surface *less* the surface pressure, π, of adsorbed alkane vapour. Since γ_{ow} and γ_{oa} (assuming no adsorption of water vapour to the alkane surface) are known, and $S_{o,aw} = 0$ for the lower alkanes, π in these cases can be deduced (see Table 16.1). Further, for $S_{o,aw}$ to be zero, γ_{aw} must be the largest of the 3 interfacial tensions (see §12.3.2). The spreading coefficient, $S_{w,oa}$, for water on those alkanes for which $S_{o,aw} = 0$, that is,

Table 16.1 *Non-retarded Hamaker constants, from (11.2.10), and spreading coefficients S in systems with n-alkanes and water. For water on alkanes, $S_{w,oa} \sim -100$ mN m^{-1} for all alkanes.*

| Alkane | Water–oil–air films | | | | | Oil–water–air films |
	T_1 $(10^{-21}$ J)	T_2 $(10^{-21}$ J)	$A_{w\text{-}o\text{-}a}$ $(10^{-21}$ J)	$S_{o,aw}$ (mN m^{-1})	π (mN m^{-1})	$A_{o\text{-}w\text{-}a}$ $(10^{-21}$ J)
pentane	−0.86	1.68	0.82	0	6.56	1.22
hexane	−0.88	3.21	2.33	0	3.46	−1.44
heptane	−0.91	4.60	3.68	0	1.19	−1.30
octane	−0.93	6.05	5.11	−0.8		−2.46
decane	−0.95	7.82	6.87	−3.3		−3.80
dodecane	−0.97	9.14	8.17	−5.4		−4.75
tetradecane	−0.98	9.76	8.78	−7.1		−5.14
hexadecane	−0.99	10.42	9.43	−8.4		−5.59

$$S_{w,oa} = \gamma_{oa} - (\gamma_{ow} + \gamma_{aw}) \tag{16.2.12}$$

must therefore be negative. That is, for those cases where an alkane spreads macroscopically on water, water does not form a duplex film on that alkane.

Non-retarded Hamaker constants for the alkane + water systems calculated using (11.2.10) are shown in Table 16.1. For the interaction of vapour with water over alkane, $A_{w\text{-}o\text{-}a}$ for all systems is positive, suggesting that none of the alkanes, pentane to hexadecane spread as *thin* films on water whereas pentane, hexane and heptane (and possibly octane) are known to spread as *thick* films. Except in the case of pentane, Hamaker constants $A_{o\text{-}w\text{-}a}$ for the interaction of vapour with alkane over a water film are negative (repulsive) suggesting water spreads on the alkanes to give *thin* films (except for pentane). However the spreading coefficients $S_{w,oa}$ are large and negative (about −100 mN m^{-1} in all cases) meaning water does not spread on any of the alkanes to give *thick* films.

The differences between predictions based on non-retarded Hamaker constants and experimental observations for the macroscopic spreading of alkanes on water has been explained by Israelachvili (1991), using the cases of pentane and dodecane on water as examples. Pentane wets water macroscopically and dodecane does not, as seen. The non-retarded Hamaker constants for these systems are both positive (attractive) as shown in Table 16.1. The Hamaker constant obtained from (11.2.10) is made up of two terms, T_1 and T_2, the former containing ε (static relative permittivities) and the latter the refractive indices n (in the visible region). Values of these terms are given in Table 16.1, and are referred to as the 'zero frequency' (T_1) and the 'dispersion' (T_2) contributions. Only the interactions involving dispersion forces (hence T_2) become retarded (see §11.2.1), falling off more rapidly with increasing film thickness. For pentane $A_{w\text{-}o\text{-}a}$ is small

($+8.2 \times 10^{-22}$ J) with $T_1 = -8.6 \times 10^{-22}$ J and $T_2 = +16.8 \times 10^{-22}$ J. At small film thickness, the positive T_2 dominates, but becomes retarded with increasing h so that at some film thickness the Hamaker constant changes sign. For the higher alkanes (including dodecane), although the value of T_1 remains small and negative, T_2 becomes much larger and remains positive. Therefore the Hamaker constants remain positive, consistent with the observation that the higher alkanes do not spread (as thin or thick films) on water. The signs and values of the spreading coefficients $S_{o,aw}$ and $S_{w,ao}$ are consistent with experimental findings on macroscopic spreading. Note however that the *generalized* spreading coefficient, defined in terms of the excess film energy (and the disjoining pressure isotherm), like the Hamaker constant can also change sign with film thickness (§12.3.2).

16.2.3   Lens shape

With reference to Fig. 16.4, the shape of a lens resting at a liquid/fluid interface is dependent on the various interfacial tensions a, b, and c between the phases. For sufficiently small lenses, that is, those with radius r (say 10s of microns) much less than the capillary length l (Box 16.2), the surfaces of the lens are spherically curved, and the surface in which the lens rests remains planar up to the lens. Much larger lenses become distorted by gravity, as does the surrounding liquid/fluid interface.

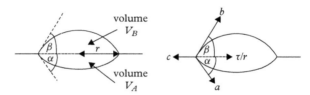

Figure 16.4 *Circular lens, radius r, at a liquid/fluid interface showing lens angles α and β; δ is the sum of these angles. The various interfacial tensions are denoted a, b, and c, and τ is the line tension in the three-phase contact line.*

Box 16.2   Gravity *versus* interfacial tension—the capillary length

The capillary length l is a characteristic length defined as

$$l = (\gamma/\rho g)^{0.5}$$

where ρ is the liquid density and g the acceleration due to gravity. It is a measure of the competing forces of gravity and interfacial tension on, for example, a drop resting on a surface. The drop (lens) shape is mainly determined by g if the drop is large relative to l and by γ if it is small relative to l. For water $l \approx 2.7$ mm and so if the dimensions of a drop are much less than this, the drop or lens surface is effectively spherically curved.

In the consideration of lens shape below, it will be assumed for simplicity that line tension (Box 16.3) is unimportant, so that the shape is determined solely by values of the interfacial tensions. The volumes V_A and V_B of the spherically curved lens caps (see Fig. 16.4) are given by simple geometry as

$$V_A = \frac{\pi r^3}{3 \sin^3 \alpha} \left[\cos^3 \alpha - 3 \cos \alpha + 2 \right]; \qquad (16.2.13)$$

$$V_B = \frac{\pi r^3}{3 \sin^3 \beta} \left[\cos^3 \beta - 3 \cos \beta + 2 \right] \qquad (16.2.14)$$

so that the lens volume $V = V_A + V_B$ can be calculated from values of a, b, c, and r.

Box 16.3   Line tension—does it influence wetting?

The lens angle δ ($= \alpha + \beta$), shown in Fig. 16.4 for small lenses, is dependent on the *line tension* τ acting in the circular three-phase contact line around the lens. The origin of τ is discussed in Chapter 17 in connection with adsorbed solid particles (§17.2); it is an excess free energy per unit length of contact line (i.e. a force) resulting from the action of surface forces in the vicinity of the contact line. It can be positive or negative. Resolving forces acting at the contact line horizontally (Fig. 16.4) gives

$$c = a \cos \alpha + b \cos \beta + (\tau / r) \qquad (16.2.15)$$

and it follows from this that the angles α and β are given by (Pujado and Scriven, 1972):

$$\cos \alpha = \frac{c^2 (1 - \overline{\tau})^2 + a^2 - b^2}{2ac (1 - \overline{\tau})}; \cos \beta = \frac{c^2 (1 - \overline{\tau})^2 + b^2 - a^2}{2bc (1 - \overline{\tau})} \qquad (16.2.16)$$

where $\overline{\tau}$ is the dimensionless reduced line tension τ / cr. For positive τ the contact line around a lens tends to contract and cause an increase in δ; for negative τ the contact line tends to expand and reduce δ. The effects of line tension are only important for very small lenses. When line tension is important however, it should be included in the definition of the spreading coefficient (now denoted S^*) so that say for an oil lens on water

$$S^*_{o,aw} = c (1 - \overline{\tau}) - (a + b) \qquad (16.2.17)$$

The spreading coefficient therefore depends on lens radius, as does δ (as seen from (16.2.16)). It has been shown (Aveyard and Clint, 1997) that, in principle, the operation of negative line tension can cause the spreading of drops with radius less than a critical value, when drops of the same material with radius greater than this value will not spread (a wetting transition). The measurement of the dependence of δ on r has allowed the measurement of line tension acting around lenses of dodecane on water. Lens angles δ (measured by observing the interference rings produced by the lenses) were ca. 2° and lens radii varied between about 10 and 150 μm; the line tension was found to be about 2×10^{-11} N. (Aveyard et al., 1999).

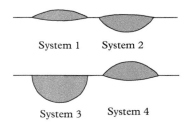

Figure 16.5 *Shapes of oil lenses resting at air/aqueous phase interfaces. The various tensions and lens angles for systems 1–4 are given in Table 16.2; clearly the lens volumes differ.*

Table 16.2 *Quantities used in the calculation of the lens shapes illustrated in Fig. 16.5. Lens radius r was taken as 30 μm in all cases*

Quantity	System 1	System 2	System 3	System 4
a/mN m^{-1}	52.8	6.8	0.1	0.41
b/mN m^{-1}	25.4	31.0	25	0.175
c/mN m^{-1}	72.8	34.0	25	0.49
S/mN m^{-1}	−5.4	−3.8	−0.1	−0.095
α /deg	14.5	58.6	89.9	20.0
β/deg	31.4	10.8	0.23	53.2
V/pL	17.7	30.3	56.5	28.4

Some examples of possible lens shapes are given in Fig. 16.5 for the four systems represented in Table 16.2. In the calculations it has been assumed that the lens radius was the same (30μm) in all cases so the lens volumes (given in the Table) obviously vary. The quantities assumed for System 1 are appropriate for a dodecane lens resting on water in air at 293 K. The lens is rather flat, and the spreading coefficient is −5.4 mN m^{-1}. The values assumed in System 2, where the spreading coefficient for the oil is −3.8 mN m^{-1}, could reflect those for mineral oil (surface tension 31 mN m^{-1}) resting on an aqueous surfactant solution (surface tension 34 mN m^{-1}, and oil/water tension 6.8 mN m^{-1}). The lower cap of the lens is much larger than the upper cap because $a << b$; the lower the tension, the larger the surface area and volume of the lens cap.[3] The difference between a (0.1 mN m^{-1}) and b (25.0 mN m^{-1}) in System 3 is even greater, and the oil/air interface of the lens is virtually flat and all the lens lies below the air/aqueous solution interface. Systems containing oil, water, and surfactant that produce microemulsions often display tensions similar to those for System 3 (see e.g. §10.1).

---

[3] Ignoring effects of gravity, the pressure throughout a lens is constant. From the Laplace equation (3.4.2) it follows that $a/R_A = b/R_B$ where R_A and R_B are the radii of curvature of the lower and upper lens caps respectively. Thus the larger the tension, the larger the radius of curvature of the cap.

Similar kinds of system can also form three phases consisting of a surfactant-rich phase coexisting with excess oil and aqueous phases, as discussed in §10.1.3 and the tensions assumed for System 4 correspond to such a system.

16.3   'Equilibrium' or static wetting of solids by liquids

16.3.1   Introductory remarks

Whether or not, and how liquids spread on solid surfaces is of everyday concern, as for example when writing on paper with a pen, painting the house or using an aerosol spray to apply an agrochemical liquid to leaf surfaces in the garden. Figure 16.6 represents a drop of liquid resting on a smooth, horizontal solid surface. As shown, the liquid *partially* wets the solid, that is, the contact angle θ has a value other than 0 or 180°. If the liquid is volatile, then at equilibrium in a closed system there is likely to be an adsorbed layer present on the solid/air interface. Wetting in this case is referred to as *moist* wetting. If adsorption on the solid surface is negligible then *dry* wetting results (Cazabat, 1987). Wetting takes place in a third phase, which can be a gas (commonly air, as in Fig. 16.6) or another (immiscible) liquid.

Young's equation for the dry wetting system depicted in Fig. 16.6, comprising a drop of liquid on a plane horizontal solid surface (Box 16.4) in air is

$$\gamma_{sa} = \gamma_{sl} + \gamma_{la} \cos \theta \qquad (16.3.1)$$

For moist wetting, the air phase contains the vapour of the liquid phase, which can adsorb on the solid lowering the tension γ_{sa} by an amount π, the surface pressure of the adsorbed film (see §4.4). Young's equation is then

$$\gamma_{sa} - \pi = \gamma_{sl} + \gamma_{la} \cos \theta \qquad (16.3.2)$$

Since the contact angle is an important measured quantity it is useful to relate it to the spreading coefficient. As discussed above, the spreading coefficient (in a dry wetting system) is

$$S_{l,sa} = \gamma_{sa} - (\gamma_{sl} + \gamma_{la}) \qquad (16.3.3)$$

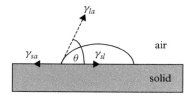

Figure 16.6 *Drop of liquid (l) resting on a solid (s) in air (a). The interfacial tensions, γ, are for the interfaces denoted by the subscripts.*

so that from (16.3.1) and (16.3.3)

$$S_{l,sa} = \gamma_{la} \left(\cos \theta - 1 \right) \tag{16.3.4}$$

For partial wetting (finite θ) it is seen from (16.3.4) that $S_{l,sa} < 0$; for $\theta = 0$ (wetting) $S_{l,sa} = 0$, as for the spreading of liquids on liquid.

Box 16.4   The highs and lows of surface energy

In the discussion of wetting of solids (usually in air) it is common to refer to solid surfaces as being either *high energy* or *low energy* surfaces (Studart et al., 2007). 'Energy' in this context is the surface free energy per unit area of the solid. High energy surfaces include those of many metals and ceramics, in which the bulk phases are strongly held together by covalent, ionic or metallic bonds. Consequently they have surface free energies per unit area as high as several $1000 \ mJ \ m^{-2}$, and are readily (partially) wetted by liquids. Indeed, many metals (with clean polished surfaces) exhibit contact angles with water of less than 10°. Various organic polymers, such as poly(methyl methacrylate), poly(vinyl chloride), and poly(ethylene), tend to have low energy surfaces and contact angles of say between 70 and 110° with water. The bulk phases of low energy solids cohere through physical forces, such as van der Waals forces and hydrogen bonds, and they usually have surface free energies of the less than 100 mJ m^{-2}. Ceramics like alumina, silicon carbide, silica and zirconia, have values of θ with water ranging up to 90°. Many organic liquids have low surface tensions, say between 20 and 30 $mN \ m^{-1}$, whereas the surface tensions of liquid metals are much higher (e.g. 486.5 $mN \ m^{-1}$ for mercury).

Surfaces, especially high energy surfaces, can readily become contaminated through adsorption of impurities resulting in spurious values of θ. Surface roughness and surface heterogeneity can also have a substantial effect on contact angle (see §16.3.5). Further, very often a value of θ depends on whether the liquid has approached its static state by advancing or by receding along the solid surface, so that *advancing* and *receding contact angles* are often reported. The advancing angle is larger than the receding angle, and the difference between the two angles is the *hysteresis* in θ. Contact angle hysteresis can be used to probe surface heterogeneity and roughness.

16.3.2   Measurement of contact angle

Two common approaches to the measurement of contact angles of liquids on plane, smooth, non-porous solid surfaces are (a) *force tensiometry*, in which the force on a vertical plate, or fibre, is measured as it is lowered into and retracted from the liquid, and (b) the optical observation (*goniometry*) of the profile of a sessile drop of a liquid on the horizontal solid surface. Different approaches must be used for porous materials and powders as seen below.

In the tensiometric method a vertical smooth non-porous plate (such as a rectangular glass slide or polished metal plate, with wetting perimeter P) is suspended from a balance. The liquid can be moved up and down relative to the plate, so that the plate can be immersed to an adjustable extent, as illustrated in Fig. 16.7. The weight w of the solid plate is first recorded in air (1 in Fig. 16.8). Then the liquid is raised until the plate just comes into contact with the liquid surface (Fig. 16.7a and the beginning of line 2 in Fig. 16.8); this gives the zero for the depth of immersion. A liquid meniscus is formed and the force F increases, as indicated by line 2 in Fig. 16.8. The liquid is then raised so the plate is immersed, and the force falls (line 3) as a result of the buoyancy of the plate. The plate is then retracted by lowering the liquid (line 4 in Fig. 16.8). Lines 3 and 4 are extrapolated back to zero depth of immersion to give values of forces F_{adv} and F_{rec} associated with the advancing and receding contact angle respectively. Then, for the advancing contact angle, for example, the extrapolated force is given by

$$F_{adv} = w + \gamma_l P \cos \theta_A \tag{16.3.5}$$

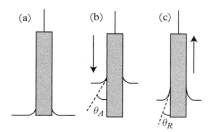

Figure 16.7 *Determination of the advancing and receding contact angles, θ_A and θ_R, tensiometrically by immersion of a vertical smooth non-porous plate in a liquid. The plate is attached to a balance.*

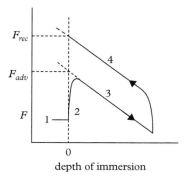

Figure 16.8 *Force F as a function of the depth of immersion of the solid plate as it advances into and then recedes from the liquid.*

Thus, from a knowledge of the wetting perimeter (P) and the surface tension, γ_l of the liquid, θ_A can be calculated. The receding contact angle θ_R can be obtained in a similar fashion from F_{rec}.

Goniometry entails the determination of the profile of an axisymmetric sessile drop resting on the horizontal solid surface under test. The complete drop shape can be obtained optically and the profile fitted in some way. The most theoretically rigorous method is to use the Young–Laplace equation (see §A3.2); the fitting takes account of the effects of gravity as well as of surface tension on the drop shape. Once the equation of the profile is available, the first derivative at the point where the drop surface meets the solid substrate gives the contact angle. Advancing angles can be obtained by increasing the size of the drop, and receding angles by withdrawing liquid from the drop.

Although horizontal and vertical solid surfaces have been referred to above, the surfaces can be tilted. One method for the determination of θ involves tilting an originally vertical plate until the liquid surface on one side meets the solid horizontally, so that the tilt angle of the plate from the horizontal gives θ, as illustrated in Fig. 16. 9a. A solid plate with a sessile drop can also be tilted, in which case both the advancing and receding contact angles are attained just before slippage occurs (Fig. 16. 9b). Possible origins of the difference between advancing and receding contact angles (i.e. contact angle hysteresis) are mentioned in §16.3.5.

An alternative strategy is needed to probe the wettability of solid particles in the form of powders. A common approach is to observe the capillary uptake of liquid into a packed column of the powder. The column, with say a sintered glass bottom to retain the powder, is attached to a balance and the bottom of the tube brought into contact with the liquid (Fig. 16.10). The mass of the tube is measured as a function of time, and the contact angle is related to the liquid uptake by

$$t = \frac{m^2 \eta}{c \rho^2 \gamma_l \cos \theta} \tag{16.3.6}$$

which is a modification of the Washburn equation. Essentially the equation describes the Poiseuille flow of liquid, viscosity η, density ρ and surface tension γ_l, along capillaries under the influence of the Laplace pressure. The parameter t is the time after contact of the liquid with the powder and m is the mass of liquid taken up by the bed of particles

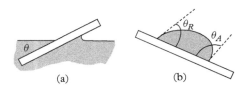

(a)   (b)

Figure 16.9 *(a) Tilting plate partially immersed in liquid. (b) Sessile drop on tilting plate giving both advancing and receding contact angles.*

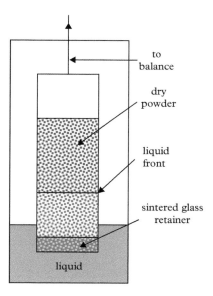

Figure 16.10 *Liquid advancing in a packed column of solid powder in a glass tube attached to a balance.*

at time t. In a packed bed, pores with a range of diameters can exist within the (porous) particles and in addition there are spaces *between* the particles.

Poiseuille's equation is for flow in a single capillary and obviously not directly applicable to a bed of porous particles; the factor c in (16.3.6) is introduced to describe the 'structure' of the bed and is experimentally determined. A plot of t versus m^2 according to (16.3.6) is linear with slope $\eta/c\rho^2\gamma_l\cos\theta$. If an experiment is carried out with a liquid which is known to have zero contact angle with the solid, that is, that completely wets the solid so that $\cos\theta = 1$, then the slope can be used to obtain a value of c for that particular packed column. If a column is packed with the same powder using exactly the same protocol, then knowing (assuming) c for such a column, the slope of an experimental plot using any liquid of interest can be used to give $\cos\theta$.

A number of commercial instruments are available to measure contact angles using the methods described above.

16.3.3   Surface tension of solids and its components

In the discussion of wetting of solids so far, relevant expressions are presented in terms of the measurable contact angle rather than the experimentally inaccessible surface tension of the solid/fluid interface. As seen, the contact angle of a liquid with a solid can depend on the way in which the configuration of a system has been approached. Surface roughness and chemical heterogeneity influence the contact angle, and an observed value may represent a metastable condition, determined in part by the intrinsic angle but also by the local geometry and/or chemical composition of the surface. Nevertheless, useful insights can be obtained by using an approach to wetting that retains interfacial tensions of solid/fluid interfaces.

It was shown in §(4.2.2) that surface tension, γ, and excess surface free energy per unit area ('surface free energy'), A_σ, are not in general the same but are related at equilibrium by

$$A_\sigma = \gamma + \sum_i \Gamma_i^\sigma \mu_i \qquad (4.2.16)$$

where Γ_i^σ are surface excess concentrations. In the unique case of a system with only one component (the solid in contact with a vacuum) and with the Gibbs dividing surface chosen such that the surface excess of the solid is zero, it is seen from (4.2.16) that $A_\sigma = \gamma$. Otherwise surface tension and surface free energy are different, although they are obviously dimensionally equivalent. However, even if adsorption of material at a solid surface is ignored, solid surfaces are not expected in general to be at equilibrium. If the surface tension of a solid is defined as the force per unit length acting within the surface, the work $\mathrm{d}W$ of forming an area $\mathrm{d}A$ of the surface is $\gamma\,\mathrm{d}A = \mathrm{d}(AA_\sigma)$ and so

$$\gamma = A_\sigma + A\,(dA_\sigma/dA) \qquad (16.3.7)$$

So, γ is only equal to A_σ if $\mathrm{d}A_\sigma/\mathrm{d}A = 0$. For a non-equilibrium solid surface, A_σ changes with extension and so $\mathrm{d}A_\sigma/\mathrm{d}A$ is not zero and A_σ and surface tension are not equivalent. Nonetheless, the terms surface tension and surface free energy are often used interchangeably. In what follows reference will be made to the surface tension of solids such as was used (16.3.1), defined as a force per unit length within the solid interface.

Although γ for a solid surface is not readily measurable directly, the *change* in γ resulting from adsorption (i.e. the surface pressure π) can be obtained if the extent of adsorption is known independently. For the adsorption of surfactant from solution at concentration c for example, the Gibbs equation may be written

$$\pi = RT \int_{c=0}^{c} \Gamma\,\mathrm{d}\ln c \qquad (16.3.8)$$

The surface concentration of adsorbed surfactant, Γ, can be measured as a function of c as described in §7.1.

A useful concept in wetting is that of the *critical surface tension* for a solid or a liquid (Shafrin and Zisman, 1967). It is found empirically that, say, for a homologous series of non-spreading alkanes in contact with a smooth solid surface (or with water), $\cos\theta$ is a linear function of the alkane surface tension. This is shown in Fig. 16.11 for alkanes in contact with solid surfaces exhibiting only CF_3 moieties.[4] Extrapolation of the line to

---

[4] Long fluorocarbon chain acids can be adsorbed from solution in decane onto polished platinum foil to give close-packed oriented monolayers which exhibit only the end groups of the chains. Contact angles and wetting are determined in the main by the outermost chemical groups and are independent of the nature of the underlying solid.

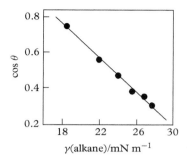

Figure 16.11 *Cos θ against surface tension for a series of homologous alkanes on solid surfaces exhibiting CF$_3$ groups. Redrawn from Ellison et al. (1953).*

$\cos \theta = 1$ ($\theta = 0$) gives the surface tension (termed the critical surface tension, γ_c) of the (hypothetical) alkane that will just spread on the solid, which is 12.9 mN m^{-1} for the case illustrated. The significance of the critical surface tension, which depends on both the solid and the liquids used, can be understood as follows. When spreading occurs (in an equilibrium system) $S_{l,sa} = 0$ and $\gamma_{la} = \gamma_c$ so that from (16.3.3)

$$\gamma_c = \gamma_{sa} - \gamma_{sl} \tag{16.3.9}$$

where γ_{sl} refers to the solid in contact with the liquid with surface tension γ_c. Clearly γ_{sl} and hence γ_c depend on the interactions between solid and liquid which in turn depend on the chemical constitution of *both* phases. If liquid alkanes are used, only dispersion forces act at the solid/liquid interface. If $\gamma_{sl} \ll \gamma_{sa}$ the critical surface tension *approximates* to the surface tension of the solid.

In Chapter 3 (§3.3) it was shown how it can be useful to treat interfacial tensions between liquids as having components that arise from the different types of interaction at the interface (Fowkes, 1963). For example, the surface tension of a liquid can be split into contributions from dispersion forces, γ^d, and from 'polar' forces γ^p. The example of a nonpolar liquid (1) forming an interface with water (2) was given. For this case only dispersion forces act between the phases at the interface and it was shown that the interfacial tension between the liquids can be expressed

$$\gamma_{12} = \gamma_1 + \gamma_2 - 2\sqrt{\gamma_1^d \gamma_2^d} \tag{3.3.5}$$

A similar approach has been used for solids (s) in contact with liquids (l) (see e.g. Clint, 2001). If, in addition to dispersion forces, polar forces also act across a solid/liquid interface, the term $-2\sqrt{\gamma_1^p \gamma_2^p}$ is added to the right hand side of the expression for the interfacial tension, so that

$$\gamma_{sl} = \gamma_l + \gamma_s - 2\sqrt{\gamma_l^d \gamma_s^d} - 2\sqrt{\gamma_l^p \gamma_s^p} \tag{16.3.10}$$

Combination of (16.3.10) with Young's equation,

$$\gamma_s = \gamma_{sl} + \gamma_l \cos\theta \tag{16.3.11}$$

yields

$$(1 + \cos\theta)\,\gamma_l = 2\left(\sqrt{\gamma_s^d \gamma_l^d} + \sqrt{\gamma_s^p \gamma_l^p}\right) \tag{16.3.12}$$

In order to obtain the components of the surface tension of the solid at least two measured values of θ for liquids of different polarity on the solid are required, so that simultaneous versions of (16.3.12) can be solved for γ_s^d and γ_s^p. In practice a number of liquids (with known γ_l^d and γ_l^p) are used and the whole data set analysed to obtain the 'best' values of γ_s^d and γ_s^p. Some results obtained from experimental contact angles for various liquids with a range of polarities on a series of solid polymers are given in Table 16.3. All the polymers represented have small polar components of surface tension so $\gamma_s \approx \gamma_s^d$. Noting that the critical surface tension γ_c, also given in Table 16.3b, is the difference between the solid surface tension and the solid/liquid tension (with a liquid that just spreads) as expressed in (16.3.9), the solid/liquid tensions ($= [\gamma_s - \gamma_c]$) range from approximately zero to 13 mN m^{-1}.

Such low values are to be expected for (mainly) organic liquids interacting with nonpolar organic solids.

The term polar has been used in a rather general way above. In order to relate the treatment of the surface tension of solids more closely to the chemistry of a system,

Table 16.3 (a) *Properties of probe liquids and (b) the calculated surface tensions and their polar and dispersive components for a series of solid organic polymers. Tensions are given in* mN m^{-1}

(a)				(b)				
Probe liquid	γ_l	γ_l^d	γ_l^p	**polymer[†]** γ_s		γ_s^d	γ_s^p	γ_c
water	72.8	22.6	50.2	PE	36.8	36.6	0.2	31
methylene iodide	50.8	49.0	1.8	PTFE	18.8	18.5	0.3	18
formamide	58.2	36.0	22.2	PVF	41.8	39.6	1.9	28
α-bromonaphthalene	44.6	44.6	0	PVC	44.2	43.8	0.4	39
glycerol	63.4	40.6	22.8	PS	43.4	42.9	0.4	33
				PMMA	44.1	43.2	0.9	39

[†]PE = polyethylene; PTFE = poly(tetrafluoroethylene); PVF = poly(vinyl fluoride); PVC = poly(vinyl chloride); PS = polystyrene; PMMA = poly(methyl methacrylate).

van Oss et al. (1988), expressed the polar component of the surface tension in terms of acid–base interactions. This component, γ^{AB}, is made up of two non-additive parts, γ^+, the electron acceptor (acid) contribution, and γ^- the electron donor (base) contribution such that

$$\gamma^{AB} = 2\sqrt{\gamma^+\gamma^-} \tag{16.3.13}$$

Using this approach the analogue of (16.3.12) is

$$(1 + \cos\theta)\,\gamma_l = 2\left(\sqrt{\gamma_s^d\gamma_l^d} + \sqrt{\gamma_s^+\gamma_l^-} + \sqrt{\gamma_s^-\gamma_l^+}\right) \tag{16.3.14}$$

The calculated acid–base components of a test liquid depend on the split assumed for water, often supposed to have equal acid and base components (Della Volpe and Siboni, 1997); the consistency of the approach has been questioned by Kwok (1999).

16.3.4 Wetting and surfactant adsorption

Since wetting and adhesion are determined by various interfacial tensions, and (positive) adsorption always lowers interfacial tension, surfactant in a system can be expected to influence wetting behaviour. Adsorption of surfactant from a solution will often be different at the various interfaces present. For simplicity, adsorption of a single nonionic surfactant from dilute solution is considered here, and it is supposed that its activity in solution is equal to its concentration.[5] The Gibbs adsorption equation for such a system, at constant T and p, is (§4.3.1)

$$d\gamma/d\ln c = -RT\Gamma \tag{16.3.15}$$

Here Γ is the molar surface concentration of surfactant, and the equation applies to each of the interfaces present. Below, aqueous systems without an 'oil' present are considered first.

Young's equation for a drop of aqueous solution (w) resting on a plane surface of solid (s), in air (a), is

$$\gamma_{sa} = \gamma_{sw} + \gamma_{aw}\cos\theta \tag{16.3.16}$$

where θ is the contact angle of the solution with the solid. Combination of (16.3.16) with (16.3.15) gives (Lucassen-Reynders, 1963)

$$d\left(\gamma_{aw}\cos\theta\right)/d\gamma_{aw} = d\gamma_{sa}/d\gamma_{aw} - d\gamma_{sw}/d\gamma_{aw} = \left(\Gamma_{sa} - \Gamma_{sw}\right)/\Gamma_{aw} \tag{16.3.17}$$

---

[5] The same results are obtained for ionic surfactants such as AOT, SDS and CTAB, although the Gibbs equation in these cases has a 2 on the right hand side (see §4.3.3).

Wetting of hydrophobic solids.

Some contact angles of aqueous solutions of cetyltrimethylammonium bromide (CTAB) on PTFE (a hydrophobic solid) in air are shown in Fig. 16.12 (lower curve). The angles vary with surfactant concentration and straddle 90°. A plot according to (16.3.17) of $\gamma_{aw} \cos \theta$ against γ_{aw} is shown in Fig. 16.13 (open circles), and is seen to be reasonably linear. If it is supposed that surfactant is not adsorbed at the solid/air interface (*i.e.* that $\Gamma_{sa} = 0$) in the wetting experiments,[6] then the slope of the plot of $\gamma_{aw} \cos \theta$ against γ_{aw} gives the ratio $-\Gamma_{sw}/\Gamma_{aw}$ designated $-\xi$, which is given in Table 16.4. Thus if the assumption that $\Gamma_{sa} = 0$ is correct it appears that the adsorption of CTAB at the PTFE/water interface is very close to that at the air/water interface. The same is true for other hydrophobic solids, as seen in Table 16.4.

The Young–Dupré equation (3.3.3) can be expressed

$$\gamma_{aw} \cos \theta = -\gamma_{aw} + W_{air}^{A} \tag{16.3.18}$$

where W_{air}^{A} is the work of adhesion between a surfactant solution and solid in air. Also, when the relationship between $\gamma_{aw} \cos \theta$ and γ_{aw} is linear, (16.3.18) can be written (Blake, 1984)

$$\gamma_{aw} \cos \theta = -\xi \gamma_{aw} + I \tag{16.3.19}$$

For $\xi = 1$, comparison of (16.3.18) and (16.3.19) shows that the intercept I is equal to W_{air}^{A}, which on the assumptions made is the same for all surfactant concentrations. The contact angle of pure water ($c = 0$) on PTFE in air is about 112°, as seen in Fig. 16.12, so taking the surface tension of water to be 72.8 mN m^{-1}, the work of adhesion

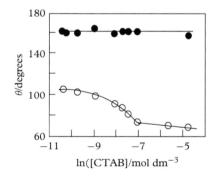

Figure 16.12 *Contact angle θ as a function of ln([CTAB]/mol dm^{-3}) for aqueous CTAB on PTFE in air (o) and under dodecane (•). Redrawn from Aveyard et al. (1993).*

---

[6] Although there may be no obvious means of transport of surfactant from an advancing drop to a dry solid surface close to the TCL, such adsorption has been invoked in the treatment of the spreading kinetics of drops on hydrophobic substrates (e.g. Lee et al., 2008). See §16.4.2.

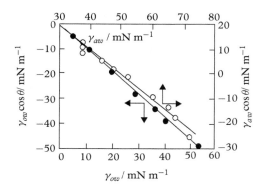

Figure 16.13 $\gamma \cos \theta$ *versus* γ *for aqueous CTAB solutions on PTFE in air (o) and under dodecane ($\bullet$).
Redrawn from Aveyard et al. (1993).*

Table 16.4 *Wetting data for aqueous surfactant solutions on hydrophobic
solids (a) Slopes ξ which are equal to Γ_{sw}/Γ_{aw} or Γ_{sw}/Γ_{ow} if Γ_{sa} or $\Gamma_{so} = 0$
(see text). (b) Work of adhesion of solid with aqueous phase in air. Superscript ^{o}
refers to pure water.*

(a)

Surfactant	Fluid/fluid interface	Wax	EBS	PTFE
CTAB	air/aqueous solution	0.97	1.03	0.99
	dodecane/aqueous solution	0.80	0.80	0.95
AOT	Air/aqueous solution	0.96	1.05	1.03
	dodecane/aqueous solution	0.83	0.82	0.96

(b)

Quantity	Surfactant	Wax	EBS	PTFE
W_{air}^A/ mJ m^{-2}	CTAB	50	49	49
	AOT	50	45	47
θ /deg		110	111	112
$\left(W_{air}^A\right)^o$/ mJ m^2		47	46	45

of water with PTFE is calculated from (16.3.18) to be 45 mJ m^{-2}. This is reasonably
close to the (constant) work of adhesion for the range of aqueous CTAB solutions in
contact with PTFE obtained from the intercept of the plot in Fig. 16.13, which is 49 mJ
m^{-2} (as in Table 16.4b). Inspection of (16.3.19) shows that the work of adhesion can

also be obtained as the value of γ_{aw} for $\gamma_{aw} \cos\theta = 0$, that is, when $\theta = 90°$, which gives $W^A_{air} = 50$ mJ m^{-2}.

Some data for systems with other hydrophobic solids (EBS[7] and paraffin wax) and with the anionic surfactant sodium bis(2-ethylhexyl) sulfosuccinate (AOT) are shown together with data for CTAB solutions on PTFE in Table 16.4. Both surfactants with all three solids give essentially the same behaviour.

The results for systems containing oil (dodecane) are quite different from those with air; the contact angles of aqueous CTAB solutions with PTFE immersed in dodecane are high (160°) and independent of the surfactant concentration, as seen in Fig. 16.12. For paraffin wax and EBS in otherwise similar systems the contact angle is 140°, also independent of surfactant concentration. Clearly, if $\cos\theta$ is constant, a plot of $\gamma_{ow} \cos\theta$ versus γ_{ow} must be linear (with slope $\cos\theta$) and pass through the origin (Fig. 16.13). The work of adhesion under oil, W^A_{oil}, can be calculated using the Young–Dupré equation and some sample results are shown in Fig. 16.14. The range of interfacial tensions represented is appropriate for surfactant concentrations from zero to the *cmc*. The work of adhesion of aqueous surfactant solution with a hydrophobic solid in, for example, an alkane is seen to be dependent on γ_{ow} (hence surfactant concentration) and much lower than it is in air, becoming zero when $\theta = 180°$.

Wetting of hydrophilic solids.

The examples of wetting given so far have involved hydrophobic (nonpolar) solids. For hydrophilic solids (e.g. ceramic materials and ionic solids such as silver iodide) in contact with water there are additional factors present that can affect wetting. The solid/liquid interface will in general carry a charge, the magnitude and sign of which depends on solution conditions, for example, pH, and concentration of potential determining ions,

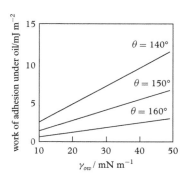

Figure 16.14 *Work of adhesion of aqueous surfactant solution with hydrophobic solid under oil (e.g. dodecane) as a function of the oil/water interfacial tension. The (invariant) contact angle of the aqueous solution with the solid is indicated on the various lines.*

---

[7] Ethylene bis-(stearamide), EBS, is a waxy solid used in commercial antifoam formulations (see §14.7.1).

as discussed in §7.5.2. If ionic surfactant is present, the surfactant ion can carry the same or opposite sign charge to that on the solid/water interface. Suppose the charges are of opposite sign as for example for a sodium alkylsulfonate in aqueous solution in contact with alumina below the *p.z.c.* (at pH = 9.1) when the alumina/water interface carries a positive charge (see Fig. 7.9). At fixed surfactant concentration and ionic strength, the lower the pH the greater the positive charge at the interface and the greater the adsorption of the anionic surfactant due to Coulombic interactions. Of course, surfactant adsorption will increase with increasing surfactant concentration (at fixed pH) up to the *cmc*. In Fig. 16.15, the variation of the air/aqueous solution tension, γ_{aw}, and contact angle, θ, is shown schematically as a function of surfactant concentration c for, say, aqueous solutions of an anionic surfactant resting on a positively charged hydrophilic solid (after Johnson and Dettre, 1969). The contact angle of pure water with the solid is taken to be very low, or zero (as would be the case for water on fused quartz) and as the surfactant concentration increases, the contact angle passes through a maximum (or reaches a plateau) when the surface tension of the surfactant solution is falling quite rapidly.

The significance of the changes in θ with surfactant concentration can be understood as follows. Combination of Young's equation (16.3.16) and Gibbs adsorption equation (16.3.15) yields

$$\gamma_{aw} \sin\theta \frac{d\theta}{d\ln c} = RT \left[\Gamma_{sa} - \Gamma_{sw} - \Gamma_{aw} \cos\theta \right] \tag{16.3.20}$$

The quantity $\gamma_{aw} \sin\theta$ is always positive so that the sign of $d\theta/d\ln c$ is determined by the relative values of the various adsorptions Γ. For positive values of $d\theta/d\ln c$, $\Gamma_{sa} > \Gamma_{sw} + \Gamma_{aw} \cos\theta$. This means that surfactant must adsorb at the solid/air interface if the contact angle is to rise with increasing concentration. Adsorption (deposition) of surfactant at the solid/air interface could occur if the three-phase contact line (TCL) recedes.

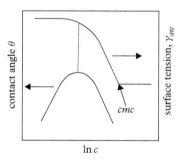

Figure 16.15 *Schematic variation of the surface tension γ_{aw} and contact angle θ with ln(surfactant concentration, c) for a system of aqueous ionic surfactant solution resting on a hydrophilic solid in air. It is supposed the solid/liquid interface has charge of opposite sign to that of the surfactant ion. After Johnson and Dettre (1969).*

Surfactant adsorption at low concentrations onto the solid/solution interface will occur through interactions of the charged head group with the solid/solution interface so that the hydrophobic alkyl chains of the surfactant are exposed to the solution, progressively creating a hydrophobic surface. This is expected to cause an increase in θ, as observed. The maximum in contact angle corresponds to the presence of a close-packed monolayer of surfactant at the solid/solution interface. Further adsorption forms a second layer oriented chain-to-chain with the first layer so that the surfactant heads are outermost, in contact with the solution. The surface becomes progressively more hydrophilic as adsorption proceeds, and the contact angle falls. From (16.3.20), for negative $d\theta$ /dln c, $\Gamma_{sw} > \Gamma_{sa} - \Gamma_{aw} \cos \theta$. It is seen in Fig. 16.15 that when θ is falling with surfactant concentration, the magnitude of $d\gamma_{aw}$/dln c, and so of Γ_{aw} is large.

16.3.5   Contact angles of liquids on solids with surface texture: superhydrophobic surfaces

The solids considered in §16.3.4 have smooth and chemically homogeneous surfaces, and a given liquid on such a surface has a unique contact angle at equilibrium. The angle is determined by the chemical constitution of the liquid and the solid surface, the interaction between the phases at the interface and the prevailing conditions, and it is this angle which appears for example in Young's equation. If however the solid surface is rough and/or chemically heterogeneous contact angle hysteresis occurs. When liquid is progressively removed from a drop resting on a horizontal heterogeneous surface, the contact area of the drop with the solid surface initially remains constant as both the volume of the drop and the contact angle fall. Ultimately the drop begins to recede with a constant contact angle, which is the receding angle θ_R. If rather than withdrawing liquid, liquid is added to the drop, the contact angle initially increases whilst the contact area remains constant. At some stage the droplet begins to advance along the surface with a constant contact angle, which is the advancing angle, θ_A (see §16.3.1 and §16.3.2).

Hydrocarbon and fluorocarbon moieties on a smooth, chemically homogeneous solid surface produce some of the strongest water repellency observed, resulting in contact angles between about 115 and 120°. Nevertheless contact angles considerably in excess of this are frequently encountered. In Fig. 16.16a, an image is shown of a water drop (diameter about 2 mm) resting on a smooth film of alkylketene dimer (AKD, a waxy material) coated on glass. The contact angle is 109°. Image (b) is of a drop of water on the same chemical material but now with a rough surface, which is formed when a liquid AKD film solidifies. SEM images of the rough surface are shown taken from above and in cross-section in images (c) and (d), respectively. The contact angle is dramatically increased to 174° by the surface roughness. Such surfaces have been described as 'superhydrophobic', and are known to exist in nature.

Structuring of surfaces can also lead to greater (super) *wettability* (Extrand et al., 2007). Liquids are observed to wick and spread over large areas of the surface of graphite with regular arrays of square pillars on the surface. Large differences in the inherent wettability of the graphite had little effect on the spreading, which was however

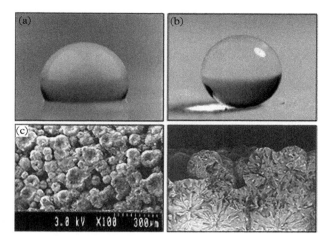

Figure 16.16 *(a) Water drop on smooth AKD surface, θ = 109°. (b) Water drop on fractal AKD surface θ = 174°. (c, d) SEM images of fractal AKD surface from top and cross-section, respectively. Adapted with permission from T. Onda, S. Shibuichi, N. Satoh, and K. Tsujii, Langmuir, 1996, 12, 2125. Copyright 1996 American Chemical Society.*

influenced by the depth and width of the channels. Shallow and narrow channels were the most effective in promoting wicking.

When a liquid drop is added to, say, a crenelated surface, as shown diagrammatically in Fig. 16.17, if the water enters the recesses under the drop (Fig. 16.17a, the area covered by the water is greater than the area that the drop projects onto the surface. (Representations of some model hydrophobic surfaces with designed patterns, for use in wetting studies, are shown diagrammatically in Fig. 16.18.) The ratio of the area of solid wetted by water to the planar area projected by the drop on the surface is designated r, the *roughness factor*, and is obviously greater than unity. According to Wenzel (1936, 1949)

$$\cos \theta_W = r \cos \theta \qquad (16.3.21)$$

where θ_W is the observed contact angle and θ is the equilibrium contact angle. The latter is the angle that would be observed on a perfectly smooth surface of the same chemical composition. The Wenzel equation (16.3.21) predicts that for systems with $\theta < 90°$, θ_W is smaller than θ, whereas for $\theta > 90°$, θ_W exceeds θ. The effects can occur for only slight surface roughness.

If a water drop straddles protrusions, leaving pockets of air between drop and solid, as illustrated in Fig. 16.17b, the drop is resting on an effectively heterogeneous surface consisting of patches of solid (smooth or rough) and of air. The system is similar to one in

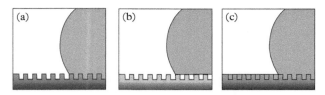

Figure 16.17 *Water drop on crenellated hydrophobic surface. (a) Water fills recesses, covering all the surface under the drop (Wenzel state). (b) Recesses under drop are not filled (Cassie state). (c) Recesses over the whole surface are filled with water (prior to placing the drop).*

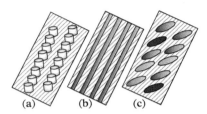

Figure 16.18 *Representation of surface structures with controlled design, such as those used by Bico et al. (1999). (a) Cylindrical columns, (b) linear stripes, and (c) shallow 'holes', all crenelated on the micron-scale.*

which the solid surface exhibits patches of two different wettabilities with contact angles θ_1 and θ_2, for which Cassie and Baxter (1944) proposed the equation

$$\cos \theta_{CB} = \varphi_1 \cos \theta_1 + (1 - \varphi_1) \cos \theta_2 \qquad (16.3.22)$$

where φ_1 is the fraction of solid surface area with contact angle θ_1. If the solid surface is rough then $\varphi_1 \cos \theta_1$ in (16.3.22) becomes $r\varphi_1 \cos \theta_1$. The Cassie–Baxter equation has been applied to the situation where one of the surfaces, say 2, is air rather than solid, in which case $\theta_2 = 180°$ and $\cos \theta_2 = -1$ and (16.3.22) becomes

$$\cos \theta_{CB} = \varphi_s \cos \theta - (1 - \varphi_s) = -1 + \varphi_s (1 + \cos \theta) \qquad (16.3.23)$$

Here φ_s is the fraction of the surface area which is solid and θ is the contact angle of the liquid with a smooth surface of the same solid. Introduction of areas of air always increases θ_{CB} over θ.

If, as depicted in Fig. 16.17c, liquid fills the recesses over the whole surface, not only under the liquid drop, then $\theta_2 = 0$ and $\cos \theta_2 = 1$ and the appropriate form of (16.3.22) is

$$\cos \theta_{CB} = \varphi_s \cos \theta + (1 - \varphi_s) \qquad (16.3.24)$$

where, as before, θ is the equilibrium contact angle of water with the smooth solid surface. It is clear from (16.3.24) that in this case θ_{CB} is always smaller than θ.

For water-repellent surfaces ($\theta > 90°$), both the Wenzel equation (16.3.21) and Cassie–Baxter equation (16.3.23) predict an increase in contact angle over θ for a smooth surface of the same material, as seen. However the equations represent different physical states; droplets in the 'Wenzel state' have a large contact area with the rough solid surface and so adhere strongly to the solid. For systems in the 'Cassie state' (i.e. described by (16.3.23)) there are trapped air pockets between drop and solid surface so the drops exhibit strong anti-adhesive properties, and can readily roll off a surface, say under the effect of wind or gravity (Yang et al., 2006) (Fig. 16.19). For micro- and nanostructured surfaces (see also Box 16.5), droplets often remain in the Cassie state but can be switched into the Wenzel state by forcing a drop against the solid substrate (Carbone and Mangialardi, 2005); superhydrophobic properties are lost in the Wenzel state (Li et al., 2017).

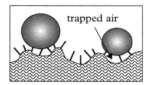

Figure 16.19 *Schematic representation of droplets resting on a solid surface with micron-scale roughness and nanoscale asperities. Droplets sit on the small asperities and have little contact or adhesion with the solid surface.*

Box 16.5   Lotus leaves, cicada wings, and jumping drops

The leaves of the lotus plant have the remarkable ability to self-clean (the *lotus effect*) as a result of their superhydrophobic surfaces. Water droplets (rain, dew) are able to roll off the leaves, adsorbing or ingesting particulate contaminants and pathogens (e.g. spores) along the way. The lotus effect is widespread in nature (Barthlott and Ehler, 1977) and exhibited by the surfaces of many plants (e.g. raspberry and rice leaves) as well as by insects. The feet of water striders are superhydrophobic as are the wing surfaces of cicadas and of some butterflies and various body surfaces of geckos. Many of the surfaces possess hierarchical structures; there is roughness on the micron scale and these rough surfaces are in turn covered with asperities on the nanometre scale (e.g. waxy tubules in the case of the lotus leaf). Water droplets sit on the asperities, trapping air between drop and solid surface (Fig. 16.19). There is little contact between water and solid surface and the droplet adhesion to the surface is low and so small tilt angles can cause droplets to roll off under gravity. Intriguingly in some cases, external forces appear to be unnecessary in the removal process, for example, from the surfaces of cicada wings (Wisdoma et al., 2013). The surface energy released by the coalescence of water droplets can result in self-propelled jumping of the fused droplets away from the surface. For two coalescing like water droplets, radius 50 μm and surface tension 72.8 mJ m^{-2}, the

---

Box 16.5 *Continued*

---

energy released is about 3.5×10^{-9} J, equal to the work of raising a coalesced drop vertically against gravity by about 30 cm. Ejected droplets can carry adsorbed solid particles with them. Initially, the particles adhere to the solid surface by (at least) van der Waals dispersion forces, and must be detached before removal can occur. Imagine a spherical particle, radius 15 μm resting on, and interacting with, a plane smooth solid surface (rather than sitting on asperities); suppose the effective Hamaker constant for the interaction is 10^{-19} J (a typical value). The (adhesion) energy required to move a particle from a closest separation from the solid surface of 1 nm effectively to infinity is around 2.5×10^{-16} J, corresponding to a pull-off force of 250 nN, similar to that (150–200 nN) reported for the pull-off force for a spherical silica particle, radius 15 μm, from the skin of the dorsal region of a gecko (Watson et al., 2015). The adsorption free energy of such a particle (say with a contact angle θ with water of 90°) into a droplet surface (see Chapter 17) is about -5×10^{-11} J, several orders of magnitude larger that the likely energy of adhesion of the particle with the solid, so adsorption of the particle at the droplet surface can easily remove it from contact with the solid surface.

The Cassie–Baxter equation has been widely used in the consideration of superhydrophobic surfaces, but it predicts only a single 'equilibrium' contact angle. It gives an area average of $\cos\theta$ and does not predict, or furnish an explanation of, contact angle hysteresis. As Gao and McCarthy (2006) argue, when a drop of liquid, originally at rest, moves a small distance (relative to the drop size) across a solid surface to another position at rest, as illustrated in Fig. 16.20, many of the liquid molecules (black filled circles) in contact with the solid surface, remain in place. The open circles represent molecules of the liquid that either leave or enter the solid/liquid interface. These molecules are all in the vicinity of the TCL. It might be expected therefore that the contact angle hysteresis (seen in the transitional distorted drop) is related in some way to the behaviour close to the TCL rather than over the macroscopic areas of contact between liquid and solid, such as represented in the Cassie–Baxter equation. Although the TCL at the advancing or receding edge of a drop is distorted (e.g. on surfaces such as those shown in Fig. 16.18), Choi et al. (2009) maintain that it is not this distortion per se that is responsible for hysteresis, but rather the *differential* area fraction of solid substrate encountered by

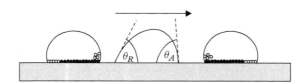

Figure 16.20 *Movement of a liquid drop on a solid surface through a small displacement from left to right via a transitional distorted drop. The black circles represent molecules of the liquid which remain in contact with the solid throughout the process. The open circles represent molecules that leave or enter the solid/liquid interface.*

the contact line as it traverses the surface. The differential parameters can be calculated from the surface structure if the *direction* of movement of the contact line is known. For example the wetting behaviour on surfaces such as that illustrated in Fig. 16.18b, depends on the direction of liquid movement, and hysteresis disappears when movement is parallel to the stipes (see also Box 16.6).

Box 16.6 Sliding drops

Small liquid droplets, for example, raindrops, can adhere to surfaces that are not horizontal, such as window panes. When a horizontal plate with a small sessile drop resting upon it is progressively tilted by an angle α (Fig. 16.21), the drop becomes distorted as does the originally circular three-phase contact line around the drop. The leading angle, θ_2, increases and the rear angle θ_1 falls until the drop just begins to slide, when $\theta_2 = \theta_A$ and $\theta_1 = \theta_R$, the advancing and receding contact angles, respectively. A simple treatment of drop slippage is given by Quéré et al. (1998). In this the relationship between the contact angle hysteresis, $\Delta\theta \ (= \theta_A - \theta_R)$, and the angle α at which a drop (of specified volume) just begins to slip is derived; α is volume dependent. At the point of slippage, the upward capillary forces on the drop due to surface tension acting around the contact line just balance the downward forces due to gravity acting on the drop.

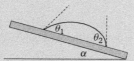

Figure 16.21 *Sessile drop on a tilting plate.*

16.4 Some dynamic aspects of wetting

16.4.1 How do drops spread? The precursor film

The discussion so far has, in the main, addressed equilibrium aspects of wetting rather than the way in which a liquid spreads over a surface before reaching an equilibrium or steady state. In equilibrium systems, spreading is feasible if the equilibrium spreading coefficient, S, is zero; S is minus the free energy change that accompanies the covering of a unit area of (dry) solid surface with a unit area of duplex liquid film.

Suppose a drop, which is small relative to the capillary length of the liquid (see §16.2.3) is placed upon a smooth horizontal solid surface. Initially the system is not expected to be at equilibrium, and irrespective of the value of the equilibrium spreading coefficient, the *initial* spreading coefficient is likely to be positive so it is thermodynamically feasible for the drop to spread. By assuming Poiseuille flow of liquid driven by the excess (Laplace) pressure in the spherically curved sessile drop, the evolution of the drop (contact line) radius, maximum drop height and apparent contact angle with time can be correctly described (Bonn et al., 2009). Interestingly however, this approach does not give an explanation of why the spreading kinetics do not depend on the magnitude of S_i, which might be expected to give a measure of the energy driving the spreading. To understand this it is necessary to consider details of the drop shape at the spreading front.

Many years ago Hardy (1919) observed that a dynamical 'precursor' film precedes a spreading droplet. The film is of submicron thickness (say 10–100 nm) and moves ahead of the macroscopically observed contact line during spreading (Fig. 16. 22). The films are formed even when the liquids are non-volatile, and so do not result from vapour condensation (as Hardy originally supposed). Rather, they are formed by liquid flow from the bulk drop. Suppose a non-volatile, perfectly wetting liquid interacts with a solid through van der Waals forces alone, with a positive Hamaker constant. The flow from the drop into the film is caused by a negative disjoining pressure (see (16.2.10)) in the film, resulting from the attractive van der Waals forces, and the capillary pressure associated with the concave cusp-shaped liquid surface (Fig. 16.22). Note that the curvature of the bulk drop, conveniently assumed to have a wedge shape close to the contact line, is much less than that of the cusp shaped surface. All the interfacial free energy gained in spreading on the (dry) solid surface is taken up by friction in the precursor film, which explains the independence of the macroscopic spreading kinetics on the spreading coefficient. The bulk drop effectively spreads on the precursor film. The thickness of the precursor film can be related to the surface energy of the solid, lower energy surfaces giving rise to thicker precursor films (Tiberg and Cazabat, 1994).

16.4.2   Spreading of surfactant solutions

As might be expected, surfactants can have a considerable effect on the dynamics of spreading of (say) water on solids or on other liquid surfaces, and they are used to control wetting in many technological processes. Spreading rate is influenced by the kinetics of surfactant adsorption to relevant interfaces as well as to the orientation which surfactant molecules adopt at surfaces.

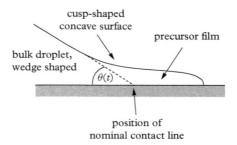

Figure 16.22 *Precursor film ahead of a spreading drop on a solid surface; θ (t) is the observed contact angle which falls with time as the drop flattens.*

Spreading on hydrophobic surfaces.

An example of spreading on a *hydrophobic* solid surface is shown in Fig. 16.23. The results, taken from Starov et al. (2000) are for the spreading of a drop of 0.05 wt % aqueous solution of sodium dodecylsulfate (SDS) on PTFE. The surfactant concentration is about 20 per cent of the *cmc*. There were no instabilities present (e.g. dendrite formation), the surface of the small drop was spherically curved, and the base was circular with radius R. In the case illustrated, the spreading is slow. When the drop is first placed on the smooth dry surface the contact angle is similar to that for pure water, after which an 'initial' value, θ_o, is established. This angle remains constant for 1–15 s, depending upon surfactant concentration. Spreading then starts, resulting in increasing base radius R and decreasing contact angle until the completion of spreading. In the explanation of these results it was supposed that surfactant adsorbed at the *dry* solid surface in the vicinity of the contact line (illustrated in Fig. 16.24) which causes the solid in this region to become hydrophilic, which aids spreading. This adsorption process is supposed to be slow and rate determining. On the basis of the model, Starov et al. (see also Lee et al., 2008) obtained the expression for the contact angle θ_t at any time t:

$$\cos \theta_t = \cos \theta_o + (\cos \theta_\infty - \cos \theta_o)\,(1 - \exp\,(-t/\tau)) \qquad (16.4.1)$$

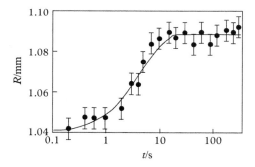

Figure 16.23 *The base radius of a spreading drop (2.5 µl) of 0.05 wt % aqueous SDS on a hydrophobic PTFE wafer as a function of time. Redrawn from Starov et al. (2000).*

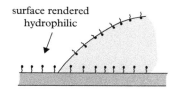

Figure 16.24 *Schematic representation of proposed surfactant adsorption for a drop of aqueous surfactant solution on a hydrophobic solid surface.*

in which θ_∞ is the final contact angle after spreading and τ is a time for the transfer of surfactant from the drop surface to the bare hydrophobic surface. Equation (16.4.1) together with the condition of constant volume of the drop allowed the derivation of an expression for the change in R with t which successfully reproduces the shape of the curve in Fig. 16.23.

If by some means a non-wetting liquid film is formed on a solid, it can of course *dewet* by retraction at the edge of the film. If the film is sufficiently thin (a few nm) however surface forces can be important and the growth of thermal fluctuations can lead to spontaneous breaking and the creation of holes in the film (see §12.4) and hence dry patches on the solid.

Superspreading.

A particularly interesting phenomenon associated with hydrophobic surfaces (which are low energy surfaces) is the so-called *superspreading* of surfactant solutions. Ananthapadmanabhan et al. (1990) found that aqueous solutions of 'hammer-shaped' trisiloxane polyethylene oxide surfactants spread very rapidly and extensively on Parafilm, which has a paraffin wax coating with which water exhibits a contact angle of 99°. No correlation was found between spreading and the surface tension of the solutions, and the rapid spreading only occurred in air which was humid, and not in dry air. Subsequently, a broader picture has emerged (Hill, 1998). Parameters of interest are the extent of spreading, that is, area $(\mathcal{A})$ covered by surfactant solution at a specified time, and the rate of spreading $(\mathrm{d}\mathcal{A}/\mathrm{d}t)$. For superspreading (also called surfactant-enhanced spreading) the spreading area increases linearly with time and the rate is found to depend on both surfactant concentration and on the nature (surface free energy) of the solid substrate. Further, silicon containing surfactants are now known not to be the only ones giving rise to rapid spreading (Stoebe et al., 1997); for example, hydrocarbon chain nonionic surfactants of the type C_nE_m can also spread rapidly, as can the di-chain surfactants AOT and didodecyldimethylammonium bromide. Aqueous solutions of the single chain ionic surfactants SDS and dodecyltrimethylammonium bromide, however, do not superspread on low energy surfaces.

To give an idea of the rapidity of spreading and its dependence on surfactant concentration and the surface free energy of the solid surface (represented by $\cos\theta$, where θ is the contact angle of water with the solid) results for the spreading of $C_{10}E_3$ are illustrated in Fig. 16.25. It is seen that for this surfactant, spreading rates as high as 60 mm^2 s^{-1} are reached, the maximum being attained for the more hydrophilic of the surfaces represented ($\theta \approx 53°$ for a surfactant concentration ≈ 0.5 wt %). The complexity of the results is clear from the figure, which is representative of the behaviour in other superspreading systems. Although a number of attempts have been made over the last two decades to understand the origins of the behaviour, there appears as yet to be no complete picture of the mechanism(s) of superspreading (Venzmer, 2011), although the Marangoni effect may well be implicated in some way.

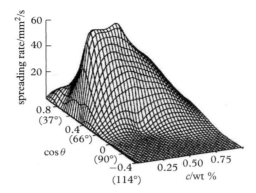

Figure 16.25 *Variation of spreading rate of aqueous solutions of* $C_{10}E_3$ *with surface energy of solid substrate (represented by cos θ) and surfactant concentration. Values in brackets below cos θ are values of θ. Redrawn from Hill (1998).*

Spreading on hydrophilic surfaces.

Although the emphasis above has been on more or less hydrophobic surfaces, there has been much research on the dynamic aspects of wetting of hydrophilic surfaces by surfactant solutions. Of particular interest has been the formation of dendrites and of the unsteady stick-slip motion at the spreading edge of drops (Marmur and Lelah, 1981; Frank and Garoff, 1995). Pure water spreads completely on clean surfaces of, for example, sapphire and silicon wafers (with a native oxide layer). The interfaces between water (at neutral pH) and solid carry a surface charge (§7.2), which is positive in the case of sapphire and negative for silicon oxide. The dynamic wetting behaviour depends of several factors, including the interactions between surfactant molecules and the solid/water interfaces, and the 'autophobing'[8] of the surfaces in the vicinity of a spreading drop.

Stick-jump spreading is characterized by an advancing contact line moving by first pinning to the surface, and then de-pinning and jumping forward. It can be readily observed when a vertical substrate is pushed down into a suitable surfactant solution. Frank and Garoff show that stick-jump spreading is observed in systems where there is a strong attraction between surfactant head groups and the solid surface, as with a cationic surfactant (such as CTAB) and a negatively charged silicon (oxide) substrate, or with an anionic surfactant (e.g. SDS) and the positively charged interface between water and sapphire. When a drop of CTAB solution (at say a concentration of 0.1 *cmc*) is placed on

---

[8] When surfactant molecules adsorb onto a solid surface with their head groups interacting with the solid (e.g. through Coulombic forces), the hydrophobic surfactant tails are oriented outwards rendering the surface hydrophobic, so the solution retracts and attains a non-zero contact angle with the solid. This process is termed *autophobing*.

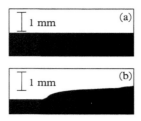

Figure 16.26 *An advancing edge of a drop of solution of CTAB (black area from above), concentration 0.1 cmc, on silicon oxide. (a) Contact line just before jump and (b) wave propagating along contact line from right to left during a jump. Adapted with permission from B. Frank and S. Garoff, Langmuir, 1995, 11, 87. Copyright 1995 American Chemical Society.*

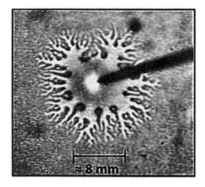

Figure 16.27 *A 2 μl drop of 1mM aqueous AOT on a pre-moistened glass slide 0.5 s after deposition. The radius of the inner ring of the drop is about 4 mm and thickness of the preformed water layer is approximately 100 nm. Reprinted with permission from S.M. Troian, X.L. Wu, and S.A. Safran, Phys. Rev. Lett., 1989, 62, 1496. Copyright 1989 American Physical Society.*

a silicon oxide surface there is an increase in contact angle of 5–10° over 30 s and a final contact angle of 25° is attained as a result of autophobing. An advancing contact line of a drop can pin at a hydrophobic region and the meniscus distorts, until the contact line de-pins and then moves on. This is illustrated in Fig. 16.26 where the edge of a drop of aqueous CTAB (black area) on silicon oxide is seen from above. In (a) the contact line is shown just before a jump and in (b) a wave is seen propagating from right to left during a jump.

When the interaction between surface and surfactant is repulsive, as between sapphire (positively charged) and a cationic surfactant, or a negatively charged glass surface and an anionic surfactant, surfactant does not adsorb significantly at the solid/liquid interface, and the dynamic wetting behaviour is quite different. A photograph (taken from above) of a 2 μl drop of a 1 mM aqueous solution of the anionic surfactant AOT spreading on glass is shown in Fig. 16.27 (from Troian et al., 1989). The glass was moist, with a continuous wetting film of water 100 nm thick, before addition of the drop. The photograph was taken 0.5 s after drop addition, and shows very clearly the fingering of the drop at the contact line (i.e. *dendritic spreading*). Frank and Garoff (1995) note that even if a surface is not pre-wetted, a thin aqueous layer (precursor film) spreads rapidly across the surface ahead of the dendrites, and is necessary for their formation. Surfactant concentration gradients at the liquid/vapour surface in the vicinity of the spreading front possibly drive the formation of the dendrites by the Marangoni effect (§6.2.2). Adsorbed surfactant flows into the liquid surface of the dendrites dragging along underlying liquid. In the case of systems where the solid surface and surfactant have opposite charge, strong surfactant adsorption causes autophobing of the solid, which prevents the spreading of the thin precursor film necessary for dendrite formation.

· ·

REFERENCES

K.P. Ananthapadmanabhan, E.D. Goddard, and P. Chandar, A study of the solution, interfacial and wetting properties of silicone surfactants. *Colloids Surf. A*, 1990, **44**, 281–297.

R. Aveyard and J.H. Clint, Liquid lenses at fluid/fluid interfaces. *J. Chem. Soc., Faraday Trans.*, 1997, **93**, 1397–1403.

R. Aveyard, J.H. Clint, D. Nees, and V. Paunov, Size-dependent lens angles for small oil lenses on water. *Colloids Surf. A*, 1999, **146**, 95–111.

R. Aveyard, P. Cooper, P.D.I. Fletcher, and C.E. Rutherford, Foam breakdown by hydrophobic particles and nonpolar oil. *Langmuir*, 1993, **9**, 604–613.

W. Barthlott and N. Ehler, Raster-Elektronenmikroskopie der Epidermis-Oberflächen von Spermatophyten. *Akad. Wiss. Lit. Mainz*, 1977, **19**, 110.

V. Bergeron and D. Langevin, Monolayer spreading of polydimethylsiloxane oil on surfactant solutions. *Phys Rev. Lett.*, 1996, **76**, 3152.

J. Bico, C. Marzolin, and D. Quéré, Pearl drops. *Europhys. Lett.*, 1999, **47**, 220.

T.D. Blake, Wetting, in *Surfactants* (ed. T.F. Tadros). Academic Press, London, 1984.

D. Bonn, J. Eggers, J. Indeku, J. Meunier, and E. Rolley, Wetting and spreading. *Rev. Mod. Phys.*, 2009, **81**, 739.

G. Carbone and L. Mangialardi, Hydrophobic properties of a wavy rough substrate. *Eur. Phys. J. E*, 2005, **16**, 67–76.

A.B.D. Cassie and S. Baxter, Wettability of porous surfaces. *Trans. Faraday Soc.*, 1944, **40**, 546–551.

A.-M. Cazabat, How does a droplet spread? *Contemp. Phys.*, 1987, **28**, 347–364.

W. Choi, A. Tuteja, J.M. Mabry, R.E. Cohen, and G.H. McKinley, A modified Cassie–Baxter relationship to explain contact angle hysteresis and anisotropy on non-wetting textured surfaces. *J. Colloid Interface Sci.*, 2009, **339**, 208–216.

J.H. Clint, Adhesion and components of solid surface energies. *Curr. Opin. Colloid Interface Sci.*, 2001, **6**, 28–33.

C. Della Volpe and S. Siboni, Some reflections on acid–base solid surface free energy theories. *J. Colloid Interface Sci.*, 1997, **195**, 121–136.

A.H. Ellison, H.W. Fox, and W.A. Zisman, Wetting of fluorinated solids by hydrogen-bonding liquids. *J. Phys. Chem.*, 1953, **57**, 622–627.

C.W. Extrand, S.I. Moon, P. Hall. and D. Schmidt, Superwetting of structured surfaces. *Langmuir*, 2007, **23**, 8882–8890.

F.M. Fowkes, Additivity of intermolecular forces at interfaces. I. Determination of the contribution to surface and interfacial tensions of dispersion forces in various liquids. *J. Phys. Chem.*, 1963, **67**, 2538–2541.

B. Frank and S. Garoff, Origins of the complex motion of advancing surfactant solutions. *Langmuir*, 1995, **11**, 87–93.

L. Gao and T.J. McCarthy, Contact angle hysteresis explained. *Langmuir*, 2006, **22**, 6234–6237.

W.P. Hardy, The spreading of fluids on glass. *Phil. Mag.* 1919, **38**, 49–55.

F. Hauxwell and R.H. Ottewill, A study of the surface of water by hydrocarbon adsorption. *J. Colloid Interface Sci.*, 1970, **34**, 473–479.

R.M. Hill, Superspreading. *Curr. Opin. Colloid Interface Sci.*, 1998, **3**, 245–254.

J.N. Israelachvili, *Intermolecular and Surface Forces*, 2nd edition. Academic Press, 1991.

R.E. Johnson and R.H. Dettre, in *Surface and Colloid Science* (ed E. Matijevic), Vol. 2, pp.85–153. Interscience, New York, 1969.

H. Kellay, J. Meunier, and B.P. Binks, Wetting properties of *n*-alkanes on AOT monolayers at the brine–air interface. *Phys. Rev. Lett.*, 1992, **69**, 1220.

H. Kellay, B.P. Binks, Y. Hendrik, L.T. Lee, and J. Meunier, Properties of surfactant monolayers in relation to microemulsion phase behaviour. *Adv. Colloid Interface Sci.*, 1994, **49**, 85–112.

D.Y. Kwok, The usefulness of the Lifshitz–van der Waals/acid–base approach for surface tension components and interfacial tensions. *Colloids Surf. A*, 1999, **156**, 191–200.

K.S. Lee, N. Ivanova, V.M. Starov, N. Hilal, and V. Dutschk, Kinetics of wetting and spreading by aqueous surfactant solutions. *Adv. Colloid Interface Sci.*, 2008, **144**, 54–65.

Y. Li, D. Quéré, C. Lv, and Q. Zheng, Monostable superrepellent materials. *PNAS*, 2017, **114**, 3387–3392.

E.H. Lucassen-Reynders, Contact angles and adsorption on solids. *J. Phys. Chem.*, 1963, **67**, 969–972.

A. Marmur and M.D. Lelah, The spreading of aqueous surfactant solutions on glass. *Chem. Eng. Commun.*, 1981, **13**, 133–143.

T. Onda, S. Shibuichi, N. Satoh, and K. Tsujii, Super-water-repellent fractal surfaces. *Langmuir*, 1996, **12**, 2125–2127.

P.R. Pujado and L.E. Scriven, Sessile lenticular configurations: translationally and rotationally symmetric lenses. *J. Colloid Interface Sci.*, 1972, **40**, 82–98.

D. Quéré, M.-J. Azzopardi, and L. Delattre, Drops at rest on a tilted plane. *Langmuir*, 1998, **14**, 2213–2216.

E.G. Shafrin and W.A. Zisman, Critical surface tension for spreading on a liquid substrate. *J. Phys. Chem.*, 1967, **71**, 1309–1316.

V.M. Starov, S.R. Kosvintsev, and M.G. Velarde, Spreading of surfactant solutions over hydrophobic substrates. *J. Colloid Interface Sci.*, 2000, **227**, 185–190.

T. Stoebe, Z. Lin, R.M. Hill, M.D. Ward, and H.T. Davis, Enhanced spreading of aqueous films containing ethoxylated alcohol surfactants on solid substrates. *Langmuir*, 1997, **13**, 7270–7275.

A.R. Studart, U.T. Gonzenbach, I. Akartuna, E. Tervoort, and L.J. Gauckler, Materials from foams and emulsions stabilized by colloidal particles. *J. Mater. Chem.*, 2007, **17**, 3283–3289.

F. Tiberg and A.M. Cazabat, Spreading of thin films of ordered nonionic surfactants. Origin of the stepped shape of the spreading precursor. *Langmuir*, 1994, **10**, 2301–2306.

S.M. Troian, X.L. Wu, and S.A. Safran, Fingering instability in thin wetting films. *Phys. Rev. Lett.*, 1989, **62**, 1496.

C.J. van Oss, R.J. Good, and M.K. Chaudhury, Additive and nonadditive surface tension components and the interpretation of contact angles. *Langmuir*, 1988, **4**, 884–891.

J. Venzmer, Superspreading—20 years of physicochemical research. *Curr. Opin. Colloid Interface Sci.*, 2011, **16**, 335–343.

G.S. Watson, B.W. Cribb, L. Schwarzkopf, and J.A. Watson, Contaminant adhesion (aerial/ground biofouling) on the skin of a gecko. *J. R. Soc. Interface*, 2015, **12**, 20150318.

R.N. Wenzel, Resistance of solid surfaces to wetting by water. *Ind. Eng. Chem.*, 1936, **28**, 988–994.

R.N. Wenzel, Surface roughness and contact angle. *J. Phys. Chem.*, 1949, **53**, 1466–1467.

K.M. Wisdoma, J.A. Watson, X. Qua, F. Liua, G.S. Watson, and C.-H. Chena, Self-cleaning of superhydrophobic surfaces by self-propelled jumping condensate. *PNAS*, 2013, **110**, 7992–7997.

C. Yang, U. Tartaglino, and B.N.J. Persson, Influence of surface roughness on superhydrophobicity. *Phys. Rev. Lett.*, 2006, **97**, 116103.

Section VII

Systems with particles at liquid interfaces

Particles act in some respects like molecular surfactants. They adsorb at fluid interfaces and can form monolayers, the behaviour of which can be described in terms of surface equations of state (see Chapter 5). If the particles have diameters in the micron range the monolayers can be viewed using an optical microscope and their structure observed directly. Like surfactants, particles can also stabilize thin liquid films, emulsions and foams. Particles in monolayers and thin films are described in Chapter 17, and in Chapter 18 the role of particles in stabilizing emulsions is considered.

17

Particles in monolayers and thin liquid films

The interest in solid particles in previous chapters has centred on their dispersion and stability in bulk liquid media. It has long been known, however, that solid particles readily become attached (adsorbed) to liquid interfaces. In the absence of surfactant, they can stabilize emulsions (so-called *Pickering emulsions*) and influence the preferred emulsion type, that is, determine if water-in-oil (W/O) or oil-in-water (O/W) emulsions are formed. Particles are also used commercially, as dispersions in oil, to destabilize aqueous foams (§14.9.4). There are many systems of practical importance in which solid particles are known to be present at liquid surfaces, including, for example, fat crystals in food emulsions, asphaltenes in crude oil emulsions, and particles attached to the surfaces of air bubbles in water in the process of ore flotation.

Clearly then, adsorbed particles can act in some respects like adsorbed surfactant or polymer molecules. Before attempting to understand the role of particles in influencing stability of thin liquid films and emulsions, the properties of adsorbed particle layers at planar liquid interfaces are first explored. Interest is centred on why solid particles should adsorb at liquid surfaces, what determines the strength of that adsorption and what the position of particles in the interface is. Just as with surfactants, the nature of an adsorbed layer of particles is influenced by the lateral interactions acting between the particles. The origins of attractive and repulsive forces between particles are considered. Particle monolayers can be compressed on a Langmuir trough in much the same way as insoluble surface active molecules (see §4.4).

17.1 Why do solid particles adsorb at liquid interfaces? The adsorption free energy

Dense hydrophobic solid objects, such as a wax-coated pin, will often 'float' on the less dense water, just as hydrophobized solid particles can be lifted through water in mineral flotation processes. It is obvious, therefore, that solid particles are strongly attached to liquid/fluid interfaces. For simplicity spherical particles are considered here. Although in various contexts particles are not expected to be spherical, a number of fundamental

Surfactants: In Solution, at Interfaces and in Colloidal Dispersions. Bob Aveyard. © Bob Aveyard 2019.
Published in 2019 by Oxford University Press. DOI: 10.1093/oso/9780198828600.001.0001

laboratory experiments have been performed using such particles, which are readily prepared or purchased. Initially it will be supposed that the surface of the particles is homogeneous, that is, all of the same wettability, and unlike surfactant molecules such particles are not amphiphilic. However, *Janus* particles,[1] in which there are well-defined parts (e.g. two hemispherical areas) of different wettability are often prepared and are well-studied. The adsorption of Janus particles will be covered later in the chapter (see also Box 17.1).

Box 17.1 'Plenty of room at the bottom'—nanoscience and technology

---

Richard Feynman, the American physicist, in a lecture entitled 'There's Plenty of Room at the Bottom' to the American Physical Society in 1959, observed: 'A biological system can be exceedingly small . . . the cells are very tiny . . . they walk around; they wiggle; and they do all kinds of marvellous things. Consider the possibility that we too can make a thing very small which does what we want – that we can manufacture an object that manoeuvres at that level'. Here Feynman is imagining the birth of what is now called nanoscience, nanotechnology, and nanoengineering. Of course, colloid science as described elsewhere in this book (see e.g. §11.1) is part of nanoscience, and nano-sized particles were made by Faraday long ago. Nanoparticles have at least one dimension in the size range of about 1–100 nm (compared to the usual definition of the colloidal size range of 1–1000 nm). (The particle diameters in Faraday's ruby coloured gold sol were about 6 nm). Like colloid science, the development of modern nanoscience had to await the development of appropriate 'measuring' instruments, notably in this case scanning probe microscopy (STM and AFM—see §8.2). The basic processes in the biological world work largely in the nanosize range. There have been notable milestones in nanoscience in recent years, including the discovery of buckminsterfullerenes (buckyballs) in 1985 and carbon nanotubes in 1991. Nanoparticles are of course the basic building blocks of nanomaterials. In Chapters 17 and 18, the focus is on spherical particles, some falling outside the nano-range, in order to address some basic ideas concerned with adsorption and monolayer structures at fluid interfaces. However there has been amazing skill and ingenuity in producing 'a whole zoo of micro- and nanoparticles' (Dasgupta et al., 2017) with a range of geometries and functionalities. There are also many 'particles' occurring in nature such as the egg-shaped malaria parasite (micron-sized), casein micelles that stabilize fat globules and various viruses (dimensions below 100 nm). Reference is made in this chapter to spherical Janus particles, but again other shapes have been made; for example, Ruhland et al. (2013) have made not only spherical Janus particles but corresponding Janus cylinders and Janus discs and studied their adsorption kinetics at liquid/liquid interfaces. Nanotechnology already impacts significantly on our lives in a host of diverse areas through products such as sun creams, plastics resistant to gas permeation, water filters, pharmaceutical products, and sports equipment. Nanotechnology is also leading to new medical diagnostic equipment, chemical sensors and chemotherapy treatments.

---

[1] They are referred to as *Janus* particles after the double-faced Roman god. The month of January is named after Janus since this month looks backwards to the old year with one face and forward to the new with the other face.

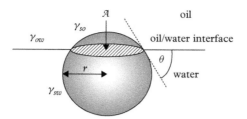

Figure 17.1 *A homogeneous spherical particle, radius r, resting at an oil/water interface. Interfacial tensions, γ, for the various interfaces are denoted by the subscripts. The particle is sufficiently small that the oil/water interface remains planar up to the particle surface.*

Imagine a spherical particle, radius r, in a system with oil, water and an oil/water interface between the liquids. Suppose the surface of the particle is polar and preferentially wetted by water, that is, it is hydrophilic. When the particle is adsorbed from water to the interface, as illustrated in Fig. 17.1,[2] part of its surface is removed from water and immersed in oil, which in itself is energetically unfavourable. So why does the particle adsorb?

The relative wettability of the particle by the oil and water is characterized by the contact angle θ (conventionally measured through water) that the particle makes with the oil/water interface. For $\theta < 90°$ (as shown in Fig. 17.1) the solid surface is more wetted by water than by oil and more than half of the particle is immersed in water. It is unusual for θ to be zero (particle completely in water) or 180° (particle completely in oil), that is, for a particle not to adsorb. The area of the spherical cap immersed in water is $2\pi r^2(1 + \cos\theta)$, and that immersed in oil is $2\pi r^2(1 - \cos\theta)$. The circular area $\mathcal{A}$ of plane oil/water interface, with tension γ_{ow}, removed by the presence of an adsorbed particle is $\pi r^2(1 - \cos^2\theta)$. The free energy change, $\Delta_{ads}^{w \to ow} G$, accompanying adsorption of a particle from water to an equilibrium position at the interface is therefore

$$\Delta_{ads}^{w \to ow} G = 2\pi r^2 \left(1 - \cos\theta\right)\left(\gamma_{so} - \gamma_{sw}\right) - \pi r^2 \left(1 - \cos^2\theta\right)\gamma_{ow} \qquad (17.1.1)$$

where γ_{so} and γ_{sw} are the interfacial tensions of the solid/oil and solid/water interfaces, respectively. It can be seen from (17.1.1) that it is the negative free energy change accompanying the removal of oil/water interface, $-\pi r^2(1 - \cos^2\theta)\gamma_{ow}$, that provides the driving force for adsorption.

The tensions for the various interfaces (shown in Fig. 17.1) are related through Young's equation (§3.3):

$$\gamma_{so} - \gamma_{sw} = \gamma_{ow} \cos\theta \qquad (17.1.2)$$

---

[2] It is supposed that the particle is sufficiently small that the interface remains planar up to contact with the particle as shown in Fig. 17.1. A sufficiently large and dense particle will however cause a local deformation of the interface through gravitational effects.

From (17.1.1) and (17.1.2) therefore

$$\Delta_{ads}^{w \to ow} G = -\pi r^2 \gamma_{ow} (1 - \cos\theta)^2 \qquad (17.1.3)$$

For adsorption from the oil phase the sign in brackets becomes positive so that in general

$$\Delta_{ads} G = -\pi r^2 \gamma_{ow} (1 \pm \cos\theta)^2 \qquad (17.1.4)$$

Some values of adsorption free energies for a spherical particle ($r = 100$ nm), calculated using (17.1.4), are shown in Fig. 17.2 as a function of θ. The interfacial tension is taken to be 0.05 N m^{-1}, typical of that for an alkane/water interface. The left limb is for adsorption from water (for $0 < \theta < 90°$) and the right limb for adsorption from oil ($90° < \theta < 180°$). The energies are expressed in units of $10^5 kT$ and for a particle with radius of 100 nm the magnitude of the adsorption free energy (at its maximum, for $\theta = 90°$) is of the order of $10^5 kT$. In effect such a particle is irreversibly adsorbed, unlike an adsorbed low molar mass surfactant molecule which can exchange with molecules in bulk. Nonetheless, the adsorption free energy falls off rapidly with contact angle either side of 90°. Equation (17.1.4) indicates that the strength of particle attachment to an interface depends linearly on the liquid interfacial tension, and varies with the square of the particle radius. The variation of the free energy of adsorption with particle radius is illustrated in Fig. 17.3. As anticipated, the order of magnitude of the adsorption free energy approaches that expected for a single low molar mass surfactant molecule when r is similar to a molecular size (Box 17.2). In the example shown in Fig. 17.3 the adsorption free energy is about $-17 kT$ for $r = 1$ nm.

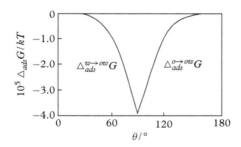

Figure 17.2 *Free energies (in units of kT) of adsorbing a spherical particle, radius 100 nm, from water (left limb) or oil (right limb) to the oil/water interface (γ ow = 0.05 N m$^-$ 1) as a function of contact angle θ measured through the aqueous phase.*

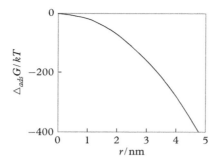

Figure 17.3 *Variation, with radius r, of the free energy of adsorption of a spherical particle from water to a liquid interface with tension 0.05 N m^{-1}. The contact angle has been taken as 70°.*

Box 17.2 How can $\Delta_{ads}G$ for a particle be compared to that for a single surfactant molecule?

Is the 'free energy of adsorption' of a single particle comparable to that of a surfactant molecule of similar size? The *values* of the standard molar free energies of adsorption of surfactant molecules referred to in Chapter 5, $\Delta_{ads}\mu^\circ$, depend on the chosen standard states for molecules in bulk solution and at the surface. That is, the standard free energy of adsorption relates to a hypothetical process of taking surfactant molecules from their chosen standard state in solution (e.g. 1 mol dm^{-3}) to a chosen standard state at the interface (e.g. surface pressure $\pi = 1$ mN m^{-1}). It follows that a standard free energy of adsorption of a molecular group within a surfactant molecule cannot be obtained by simply chopping $\Delta_{ads}\mu^\circ$ into bits! As explained in Chapter 5 however if $\Delta_{ads}\mu^\circ$ is determined for an homologous series of alkyl chain surfactants, using the *same* standard states in all cases, the methylene group increment is independent of the standard states chosen. For a methylene group $\Delta_{ads}^{w \rightarrow ow}G^\circ \sim -3$ kJ mol^{-1} which is equivalent to about $-1.25kT$ per CH$_2$ group. Therefore for a dodecyl chain for example (as in SDS) the adsorption free energy is $-15kT$.

To calculate the adsorption free energy of a particle using (17.1.4) a value of the contact angle is required. The experimental determination of θ however can be somewhat problematic, depending on particle size and shape. For large particles, θ can be determined photographically, as illustrated in Fig. 17.4 for a glass sphere (radius $r = 53\mu$m) made hydrophobic by coating with CF$_3$ groups. The particle is resting in the surface of a much larger pendent drop of water in air. The contact angle is related to the dimensions shown in Fig. 17.4 by

$$\theta = \arcsin\left(\frac{2xy}{x^2 + y^2}\right) - \arcsin\left(\frac{2xY}{X^2 + Y^2}\right) \tag{17.1.5}$$

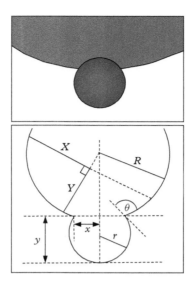

Figure 17.4 *Above is a reconstruction of a photographic profile of a spherical glass bead, radius 53 μm coated with CF$_3$ groups, at the surface of a drop of water in air; θ (through water) is clearly larger than 90°. Below, the relevant dimensions for the calculation of the contact angle are indicated. Redrawn from Aveyard et al. (1996).*

Particles of interest, such as those used to stabilize emulsions (see Chapter 18) are often in the colloidal size range however and simple optical methods cannot be used. Discussions of various possible procedures for the determination of contact angles of colloidal particles with liquid interfaces can be found in a dedicated collection of articles on colloidal particles at fluid interfaces (Binks, 2007).

17.2   What is line tension and can it affect particle adsorption significantly?

Just as there is an excess free energy associated with the formation of an interface between two phases, so there is a free energy change arising from the formation of a *contact line* where three phases meet. This free energy, per unit length of the three-phase contact line, is termed the *line tension* (symbol τ) and it has the units of force. It is the one-dimensional analogue of (the two-dimensional) surface (interfacial) tension. Unlike surface tension, however, line tension can be either positive or negative depending on whether work is required to form the line (positive τ) or energy is released when a contact line is created (negative τ). Obviously, when a particle is adsorbed a solid/fluid/fluid contact line is created. Although this was not accounted for in the preceding analysis leading to the expression for the free energy of adsorption, (17.1.4), as particles become smaller (and small particles are often the ones of practical interest) the effects of line tension, relative to those involving interfacial tensions, become increasingly important. A simple explanation of the origins of line tension is given below, and consideration is given to how important it might be in influencing the contact angle θ, that is, the wettability and position of a particle adsorbed at a fluid interface. Effects of line tension on the shape of oil lenses on water and on their wetting behaviour is discussed in §16.2.3.

17.2.1 Origins of line tension

For simplicity consider for the moment a drop of liquid resting on a plane smooth solid surface in air, as illustrated in Fig. 17.5. The dashed line represents the profile the drop would have in the absence of surface forces. However, in the vicinity of the meniscus surface forces act between the liquid/air and the solid/liquid interfaces and so the meniscus shape (represented by the full line in the figure) deviates from the extrapolated profile. The meniscus is in effect an asymmetric (vapour/liquid/solid) thin liquid film of continuously varying thickness around the perimeter of the drop. The line tension arises from the action of the surface forces, for example, van der Waals, electrostatic, and structural forces (Chapter 11). It can be appreciated that the magnitude and sign of the line tension is related in some way to the shape of the disjoining pressure isotherm (the relationship between disjoining pressure $\Pi_D(h)$ and meniscus thickness h normal to the solid surface—see §12.2.2). To illustrate the point, in Fig. 17.6 two *calculated* disjoining pressure isotherms are shown, one corresponding to a positive and one to a negative line tension, obtained for the contact line around a liquid oil lens resting on water. The height of the maximum in an isotherm is related to the magnitude (and sign) of τ, the higher the maximum the more positive τ.

17.2.2 How does line tension affect the adsorption of spherical particles?

The contact line formed around an adsorbed spherical particle at equilibrium in, say, an oil/water interface is now considered (Fig. 17.7). A knowledge is sought of the ways in which τ and particle size affect θ. As mentioned, particles can be very effective in the stabilization of emulsions and very often the particle size is very small, in the colloidal size range.

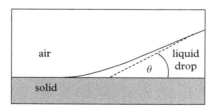

Figure 17.5 *Liquid meniscus close to a plane solid surface (shaded area). The dashed curve is the profile the liquid drop would have in the absence of surface forces. The actual profile (full line) differs from this as a result of the action of surface forces.*

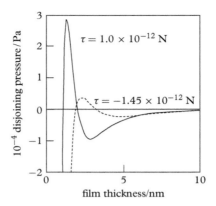

Figure 17.6 *Examples of shapes of disjoining pressure isotherms that correspond to a negative and a positive line tension as indicated. The curves relate to the meniscus around a liquid oil lens resting on water. Redrawn from Aveyard et al. (1999).*

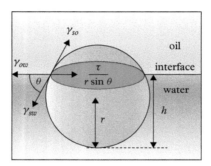

Figure 17.7 *Forces acting at the contact line formed by an adsorbed spherical particle radius r adsorbed at an oil/water interface. The particle is immersed in the water phase to a depth h and the contact angle the particle makes with the oil/water interface is θ (through the aqueous phase).*

The height h to which a spherical particle is immersed in water, and correspondingly the value of θ, are influenced by the sign and magnitude of the line tension τ. For $\theta < 90°$, as shown in Fig. 17.7, if the line tends to contract (positive τ) the particle will tend to be pushed lower into the water, reducing the length of the contact line and the value of θ. For the case where $\theta > 90°$, the particle tends to be forced more into the oil and θ (through water) is increased. Negative line tension will have the opposite effects and hence force the contact angle closer to 90°, tending to maximize the length of the contact line. It can be imagined that for sufficiently large positive line tension a particle will be excluded from the interface, i.e. that it will undergo a *wetting/drying transition*. Since τ affects θ, in what follows the contact angle in the *absence* of effects of line tension is denoted θ_0.

With reference to Fig. 17.7, the balance of forces along the tangent to the solid surface at the contact line (i.e. along the direction the contact line would move if the system were not at equilibrium) gives

$$\gamma_{so} + \frac{\tau \cos \theta}{r \sin \theta} - \gamma_{sw} - \gamma_{ow} \cos \theta = 0 \tag{17.2.1}$$

which, noting Young's equation (17.1.2) (in which the contact angle is θ_o), leads to

$$\frac{\tau}{\gamma_{ow} r} = \bar{\tau} = \sin \theta \left(1 - \frac{\cos \theta_o}{\cos \theta} \right) \tag{17.2.2}$$

The quantity $\bar{\tau}$ is termed the *reduced* line tension and it varies between zero and ± 1, depending on the sign of τ. Equation (17.2.2) shows how for a given γ_{ow} and r, θ depends on τ, and how for a given τ and γ_{ow}, θ depends on r. Curves of θ versus τ obtained by the use of (17.2.2) are shown in Fig. 17.8, for positive τ and for $\theta_o = 70°$ and $110°$ with γ_{ow} taken as 0.050 N m^{-1}.

In the case where $\theta_o < 90°$, as τ *increases* the contact angle initially falls (solid line), but after reaching a maximum value of τ, τ_m (at $\theta = \theta_m$) θ apparently falls with *decreasing* τ (dashed line). It is clear from what has been said that the dashed part of the curve represents unstable configurations of a particle in the interface. For a given τ, the stable (or metastable) configuration is that with the larger θ. Further, for line tensions greater than τ_m the particle has no stable position in the interface and so cannot be adsorbed. It might be thought therefore that the wetting/drying transition occurs at $\tau = \tau_m$. However,

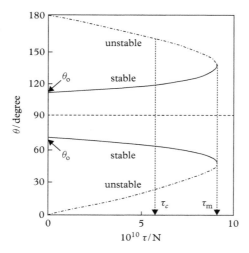

Figure 17.8 *Variation of contact angle with line tension according to (17.2.2). The upper curve is for $\theta_o = 110°$ and the lower curve for $\theta_o = 70°$. The interfacial tension γ_{ow} has been taken as 0.050 N m^{-1}, and r has been set at 50 nm.*

a consideration of the way in which the free energy of adsorption varies with line tension shows that the transition occurs at a critical line tension, τ_c, less than τ_m. The free energy of a particle in the interface, G^s, which now includes a contribution from the line tension, is

$$G^s = 2\pi r^2 \left[\gamma_{sw} (1 + \cos\theta) + \gamma_{so} (1 - \cos\theta) - 0.5\gamma_{ow} \sin^2\theta + \tau \sin\theta / r \right] \qquad (17.2.3)$$

The free energy of the particle completely immersed in water, G^w, is $4\pi r^2 \gamma_{sw}$, and subtracting this from G^s gives the free energy of adsorption of the particle from water. Then, noting (17.2.2) it can be shown that, for $\theta < 90°$

$$\Delta_{ads}^{w \to ow} G = \frac{2\pi r\tau}{\sin\theta} (1 - \cos\theta) - \pi r^2 \gamma_{ow}(1 - \cos\theta)^2 \qquad (17.2.4)$$

For $\theta > 90°$ the sign in the brackets on the right hand side of (17.2.4) becomes positive.

The wetting/drying transition for the particle occurs when the free energy in bulk (water for $\theta < 90°$) is equal to that in the surface i.e. when the free energy of adsorption is zero. To obtain adsorption free energies using (17.2.4) θ is calculated for assumed values of τ and θ_o using (17.2.2) and then substituted in (17.2.4). A plot of adsorption free energy against line tension obtained in this way is shown in Fig. 17.9. The system represented is the same as that in Fig. 17.8. As seen, the adsorption free energy is negative for the lower values of τ and becomes zero at $\tau = \tau_c$. For $\tau_c < \tau < \tau_m$ the free energy is positive, and as seen from Fig. 17.8 the contact angle still falls with increasing positive line tension. In this region of *metastability*, a local balance of forces on the particle in the interface exists but nonetheless the Gibbs energy of the system with the particle at the surface is more positive than that with the particle submerged in one (the more wetting) of the bulk phases. The upper dashed limb in Fig. 17.9 represents unstable configurations and corresponds to the dashed part of the curve in the lower graph in Fig. 17.8.

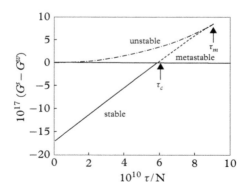

Figure 17.9 *Variation of adsorption free energy with line tension according to (17.2.4) for a particle with $\theta_o = 70°$. The interfacial tension γ_{ow} has been taken as 0.050 N m^{-1}, and r has been set at 50 nm.*

Expressions for τ_m and θ_m in terms of θ_o are readily obtained by noting that at the maximum in the curves shown in Fig. 17.8, $d\tau/d\theta = 0$. Differentiation of τ with respect to θ in (17.2.2), and setting the result to zero gives

$$\frac{\tau_m}{\gamma_{ow} r} = \bar{\tau}_m = \left[1 - (\cos\theta_o)^{2/3}\right]^{3/2} = \sin^3\theta_m; \cos\theta_m = (\cos\theta_o)^{1/3} \qquad (17.2.5)$$

An expression for the critical line tension is arrived at by setting the free energy of adsorption in (17.2.4) equal to zero whence (for $\theta_o < 90°$)

$$\frac{\tau_c}{\gamma_{ow} r} = \bar{\tau}_c = 0.5 \sin\theta_c (1 - \cos\theta_c) \qquad (17.2.6)$$

For $\theta_o > 90°$ the sign in brackets is reversed. From (17.2.2) and (17.2.4) it can be shown that θ_c and θ_o are related by

$$\cos\theta_o = 0.5 \cos\theta_c (1 \pm \cos\theta_c) \qquad (17.2.7)$$

in which the positive sign now refers to systems where $\theta_o < 90°$.

It is useful to know in what circumstances positive line tension could feasibly inhibit particle adsorption at fluid interfaces. Just as interfacial tensions between oil and water (in the presence of surfactant) can vary by several orders of magnitude (§10.1.3), reported experimental values of line tension (in the absence of surfactant) vary enormously. For solid + liquid + vapour systems, reported values are usually positive and for systems with particles, τ has ranged from 10^{-9} to 10^{-6} N (Amirfazlia and Neumann, 2004).

An early attempt to estimate the value of line tension in contact lines around glass microspheres in a liquid/vapour interface was made by Mingins and Scheludko (1979). The particles were suspended in pendent drops of 0.01 M aqueous octyltrimethylammonium bromide, and their penetration (or otherwise) into the air/solution interface under gravity was observed. By noting the size of the smallest particles which were able to enter the drop surface and form a three-phase contact line, an estimation was made of the magnitude of the (positive) line tension, which was given as 10^{-10} N.[3] To give an idea of the magnitude of line tension which can inhibit the adsorption of a spherical particle into a fluid interface, some critical and maximum line tensions, calculated using (17.2.5), (17.2.6), and (17.2.7), are shown in Fig. 17.10 as a function of particle radius for an oil/water interface with tension 0.050 N m^{-1} and $\theta_o = 70°$. The critical and maximum line tensions are linear functions of particle radius and, for a sphere with diameter 1 μm, τ_m is about 10^8 N. For a line tension of 10^{-9} N particles with diameters less than about 100 nm are expected to be excluded from the interface.

---

[3] Surfaces of solids are often chemically heterogeneous and physically rough so that there can be difficulty in obtaining a meaningful value of θ_o. Further, it is possible that particle entry into an interface can be inhibited by the formation of a (meta)stable thin liquid film between the particle surface and fluid interface. A number of experimental values of line tension quoted in the literature are perhaps best regarded as broad estimates.

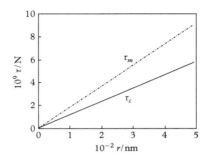

Figure 17.10 *Critical and maximum line tensions as a function of spherical particle radius for systems in which* $\gamma_{ow} = 0.050$ N m^{-1} *and* $\theta_o = 70°$.

17.3 Adsorption of Janus particles

17.3.1 Introductory remarks

If, say, a hydrophobic cap is put on an otherwise hydrophilic spherical particle, a Janus particle results. As seen, a homogeneously wetted particle is already surface active (unless unusually $\theta = 0°$ or $180°$); addition of the hydrophobic cap renders the particle amphiphilic in addition and this enhances the surface activity of the particle. Janus particles in some respects resemble surfactant molecules, with the hydrophobic cap acting rather like the hydrophobic chain of a surfactant.

The hydrophilic surface of a particle is referred to here as *polar*, and the hydrophobic part as *apolar*. The extent of an apolar cap is defined by the angle α, as illustrated in Fig. 17.11(a) and the contact angle of the polar surface with the oil/water interface is denoted θ_p. The wettability of the apolar surface is expressed by its contact angle with the oil/water interface, θ_a. This is the angle (through the aqueous phase) that for example a purely apolar particle would have at the liquid interface.

For polar particles with apolar caps the angle α is less than $90°$. When $\alpha > 90°$ however the apolar surface area of the particle exceeds that of the polar part. Below, systems with $\alpha < 90°$ are referred to as examples, but it is simple to extend the treatment to cases where α exceeds $90°$. For $\alpha < 90°$, it can be appreciated from Fig. 17.11a that for a particle to have an equilibrium contact angle with the oil/water interface θ_p must exceed α. Further, as the hydrophobicity of the particle is increased by increasing θ_p, at some stage the particle will become more wetted by the oil phase than by water, and particles will transfer to the oil. This will occur when $\theta_p = (\theta_p)_t$ and so in order to consider adsorption of a particle from water (as the more wetting or 'preferred' phase) to the oil/water interface, contact angles in the range $\alpha < \theta_p < (\theta_p)_t$ are considered. If $\alpha > \theta_p$, the polar part of the particle cannot access an equilibrium contact angle with the oil/water interface and in this case it is supposed that a Janus particle is pinned at the oil/water interface at the circular junction between polar and apolar regions of the particle surface, as illustrated in Fig. 17.11b (Ondarçuhu et al., 1990).

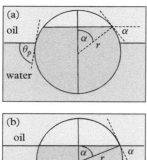

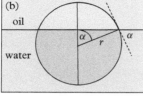

Figure 17.11 *(a) Janus particle resting at the oil/water interface. The angle α expresses the size of the apolar cap (yellow shaded) and θp is the contact angle of the polar part of the sphere with the oil/water interface. (b) No equilibrium contact angle is accessible and the particle is pinned at the oil/water interface around the circular junction of polar and apolar regions of the particle surface.*

The treatment below is a macroscopic one, and may not apply to systems with nanoparticles, say with radius of the order of 10 nm. This is because very small particles are expected to have considerable rotational freedom at an interface (Cheung and Bon, 2009). For larger Janus particles however, with radius ≥100 nm say, it is reasonable to suppose that particles will be oriented at oil/water interfaces, with each of the two surfaces on a particle being (mainly) in contact with the more wetting of the two liquid phases, as illustrated in Fig. 17.11.

Expressions for the free energies of adsorption and of transfer of Janus particles can be obtained from a knowledge of the areas of solid surfaces transferred between phases (during particle adsorption or transfer), and the difference between the final and initial solid/liquid interfacial tensions, which are related to contact angles and the oil/water interfacial tension via Young's equation. The approach is similar to that used earlier to obtain expressions for the adsorption free energies for homogeneously wetted particles. The cases where, for an adsorbed Janus particle, (i) an equilibrium contact angle is accessible and (ii) where it is not and the particle is pinned at the polar/apolar boundary, are now treated separately.

17.3.2 Systems in which an equilibrium contact angle is attained

The free energy of adsorption of a Janus particle from water to the oil/water interface, as illustrated in Fig. 17.11a, can be shown to be given by

$$\Delta_{ads}^{w \to ow} G = 2\pi r^2 \gamma_{ow} \left\{ (1 - \cos\alpha)(\cos\theta_a - \cos\theta_p) - 0.5(1 - \cos\theta_p)^2 \right\} \quad (17.3.1)$$

If θ_p is increased, with θ_a and α held constant, the particles will become preferentially wetted by oil when $\theta_p = (\theta_p)_t$, at which point the free energy of transfer of a particle from water to oil, $\Delta_t^{w \to o} G$, is zero. The transfer free energy is given by

$$\Delta_t^{w \to o} G = 2\pi r^2 \gamma_{ow} \left\{ (1 - \cos\alpha)\cos\theta_a + (1 + \cos\alpha)\cos\theta_p \right\} \tag{17.3.2}$$

and setting $\Delta_t^{w \to o} G = 0$ shows that

$$\cos\left(\theta_p\right)_t = \cos\theta_a \left(\frac{\cos\alpha - 1}{\cos\alpha + 1} \right) \tag{17.3.3}$$

For $\theta_p > (\theta_p)_t$ adsorption of particles from the oil phase is considered. For adsorption from oil the apolar region remains in the oil and the expression for the adsorption free energy $\Delta_{ads}^{o \to ow} G$ is readily shown to be

$$\Delta_{ads}^{o \to ow} G = -\pi r^2 \gamma_{ow} \left(1 + \cos\theta_p\right)^2 \tag{17.3.4}$$

which, as expected, is identical to (17.1.4) for a homogeneous particle taking the positive sign in brackets.

When the free energy of transfer between oil and water is zero and $\theta_p = (\theta_p)_t$, $\Delta_{ads} G$ is obviously the same for adsorption of a particle from water and from oil. The quantity $\Delta_{ads}^{o \to ow} G / \Delta_{ads}^{w \to ow} G$, termed the *Janus balance*, $\mathcal{J}$, has been defined by Jiang and Granick (2007), and is equal to unity for $\theta_p = (\theta_p)_t$. As will be seen (Fig. 17.12), the magnitude of the free energy of adsorption is maximum when $\theta_p = (\theta_p)_t$ ($\mathcal{J} = 1$).

The variation of the free energies of adsorption with θ_p (for $\alpha < 90°$) is illustrated in Fig. 17.12. The particle radius is taken to be 100 nm and the oil/water interfacial tension

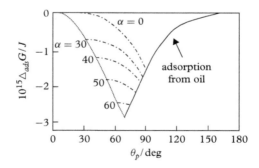

Figure 17.12 *Free energies of adsorption from water as a function of θ_p for spherical Janus particles, radius 100 nm, at the oil/water interface with $\gamma_{ow} = 0.05 \ N \ m^{-1}$. The contact angle of the apolar cap, θ_a, is taken as 150° and its size is defined by α, as indicated on the curves. The curves are truncated at α at low angles (dotted line). The full curve with positive slope is for the adsorption of the particles from oil for $\theta_p > (\theta_p)_t$ (see text). Redrawn from Aveyard (2012).*

as 0.050 N m^{-1}, appropriate for an alkane/water interface. The curves (calculated using (17.3.1)) are truncated at lower angles at α, and at higher angles at $(\theta_p)_t$. The full curve on the right (obtained using (17.3.4)) represents the adsorption of the particles from the oil phase after transfer from water. It is seen then that the introduction of an apolar (polar) cap onto a polar (apolar) particle increases the magnitude of the (negative) adsorption free energy. The minimum of a curve of free energy of adsorption versus θ_p for a given α and θ_a (as in Fig. 17.12) corresponds as seen to the condition where the distribution of particles is equal between oil and water phases and $\theta_p = (\theta_p)_t$. As α is increased, α and $(\theta_p)_t$ mutually approach and the minimum free energy of adsorption, $\Delta_{ads}^{w \rightarrow ow} G_m$, that can be attained for a given value of θ_a, is obtained when $\alpha = (\theta_p)_t$; these quantities, which correspond to the minimum of the minima in adsorption free energy, are denoted α_m and $(\theta_p)_{tm}$. The relationship between α_m ($= (\theta_p)_{tm}$) and θ_a can, by reference to (17.3.3), be seen to be

$$\cos\theta_a = \cos\alpha_m \left(\frac{\cos\alpha_m + 1}{\cos\alpha_m - 1} \right) \tag{17.3.5}$$

and noting (17.3.1), $\Delta_{ads}^{w \rightarrow ow} G_m$ is given by

$$\Delta_{ads}^{w \rightarrow ow} G_m = 2\pi r^2 \gamma_{ow} \left\{ (1 - \cos\alpha_m)(\cos\theta_a - \cos\alpha_m) - 0.5(1 - \cos\alpha_m)^2 \right\} \tag{17.3.6}$$

To obtain α_m as a function of θ_a, the required root of (17.3.5) is

$$\cos\alpha_m = 0.5 \left\{ (\cos\theta_a - 1) + \left[(1 - \cos\theta_a)^2 - 4\cos\theta_a \right]^{0.5} \right\} \tag{17.3.7}$$

In Fig. 17.13, minimum adsorption free energies are presented as a function of α for various θ_a (in the range 90–180°); the energies are given relative to the minimum free

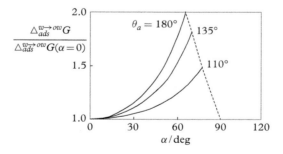

Figure 17.13 *Minimum free energies of adsorption, relative to that at $\alpha = 0°$, as a function of α. Each curve is for a given value of θ_a as indicated and is truncated at $\alpha = \alpha_m = (\theta_p)_{t\,m}$, where the free energy is $\Delta_{ads}^{w \rightarrow ow} G_m$ (shown by the dashed line on the right). Assumed values are $r = 100$ nm and $\gamma_{ow} = 50$ mN m^{-1}. Redrawn from Aveyard (2012).*

energy for $\alpha = 0°$, that is, for a homogeneous polar particle. Each of the curves is truncated at α_m corresponding to a given θ_a, and the dashed line in the Figure passes through the lowest minimum adsorption free energies, $\Delta_a^{w\to o} G_m$ for the various θ_a. For $\theta_a = 180°$, α_m is $65.53°$, and interestingly the free energy of adsorption for $\alpha = \alpha_m$ is twice that for $\alpha = 0°$. That is, for a system with given particle size and oil/water interfacial tension, the largest possible magnitude of the adsorption free energy attainable is that corresponding to $\theta_a = 180°$ and $\theta_p = (\theta_p)_{t,m} = \alpha_m$, given by (17.3.5) and (17.3.7). The magnitude of this free energy is twice that which can be obtained for a purely polar particle ($\alpha = 0°$).

17.3.3   Systems without an equilibrium contact angle

If an equilibrium contact angle cannot be attained it is supposed here that a Janus particle rests in the oil/water interface with the boundary between its polar and apolar regions coplanar with the interface, as illustrated in Fig. 17.11b. In this case, the free energies of adsorption of a particle from water and from oil respectively, are given by

$$\Delta_{ads}^{w\to ow} G = -\pi r^2 \gamma_{ow} \left[2\left(\cos\alpha - 1\right)\cos\theta_a + \left(1 - \cos^2\alpha\right) \right] \tag{17.3.8}$$

$$\Delta_{ads}^{o\to ow} G = -\pi r^2 \gamma_{ow} \left[2\left(1 + \cos\alpha\right)\cos\theta_p + \left(1 - \cos^2\alpha\right) \right] \tag{17.3.9}$$

As an example, adsorption of a particle from water is considered using (17.3.8) in the range $\alpha < \theta_a < (\theta_a)_t$ where $(\theta_a)_t$ is given by (cf. (17.3.3))

$$\cos(\theta_a)_t = \cos\theta_p \left(\frac{\cos\alpha + 1}{\cos\alpha - 1} \right) \tag{17.3.10}$$

Some illustrations are given in Fig. 17.14 of the way in which the adsorption free energy varies with θ_a. The free energies are expressed relative to that for a homogeneous particle of the same size with a contact angle of $90°$ ($-\pi r^2 \gamma_{ow}$), which is the maximum energy for a homogeneous particle.

The two dashed curves in Fig. 17.14 are for particles with $\theta_p = 30°$; the lower of the curves is for $\alpha = 80°$, and the particles do not transfer to oil below $\theta_a = 180°$. The upper dashed curve is for particles with $\alpha = 100°$, for which oil becomes the more wetting phase when $\theta_a = 127.5°$. The full curve is given for $\alpha = 90°$ and $\theta_p = 0$, and as seen the magnitude of the adsorption free energy reaches its maximum of three times that for adsorption of a homogeneous particle with $\theta = 90°$. This is also readily appreciated by inspection of (17.3.8), for which the term in brackets becomes equal to 3.

It is worth noting that a Janus particle pinned at the oil/water interface, although unable to access an equilibrium contact angle, can nonetheless be trapped in an energy minimum. Suppose for example that a vertical upward force F is applied to an adsorbed Janus particle (with say $\alpha = 90°$). If the contact line slips around the particle, polar surface represented by the hatched area in Fig. 17.15a is transferred from water to oil

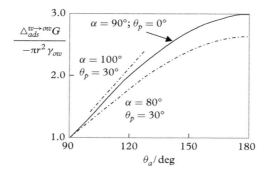

Figure 17.14 *Free energies of adsorption of Janus particles from water relative to the free energy of adsorption of a homogeneous particles with $\theta = 90^\circ$. The curves are for values of α and θ_p indicated. Redrawn from Aveyard (2012).*

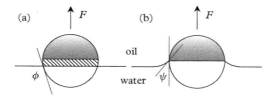

Figure 17.15 *An upward force F is applied to a Janus particle (with $\alpha = 90^\circ$) at an oil/water interface. (a) It is supposed the contact line slips around the particle and the hatched area of polar surface is transferred from water to oil, giving an increase in ϕ above 90°. (b) The contact line is pinned to the oil/water interface and the meniscus around the particle is formed as illustrated. After Aveyard (2012).*

and the angle ϕ increases above 90°. The free energy change $\Delta \uparrow G$ for such an upward displacement can be shown to be given by

$$\Delta \uparrow G = -\pi r^2 \gamma_{ow} \cos\phi \left[2\cos\theta_p - \cos\phi\right] \tag{17.3.11}$$

For a particle with $r = 100$ nm and $\theta_p = 30^\circ$ at an interface with $\gamma_{ow} = 0.050$ N m^{-1} the free energy change for an increase in ϕ from 90 to 95° is about $6 \times 10^4 kT$. If the particle is pushed vertically down, apolar surface is transferred from oil to water, again at an energy cost. For displacement up or down, the area of oil/water interface increases at a further energy cost.

If the contact line around an adsorbed particle, rather than slipping remains pinned to the oil/water interface a vertical force acting on a particle will create a meniscus around the particle, as illustrated in Fig. 17.15b. Neglecting buoyancy effects for such small particles, the (capillary) force F needed to create the meniscus, is

$$F = 2\pi r \gamma_{ow} \cos \psi \qquad (17.3.12)$$

In summary, the magnitude of the free energies of adsorption of Janus particles can be considerably larger than for homogeneous particles of the same size and at the same interface. The greatest magnitude of the free energy of adsorption of a homogeneous particle (which is for $\theta = 90°$) is $-\pi r^2 \gamma_{ow}$. On the other hand the maximum free energy of adsorption of a Janus particle which exhibits an equilibrium contact angle with the oil/water interface is $-2\pi r^2 \gamma_{ow}$. For Janus particles that consist of two hemispheres of different wettability, the maximum possible free energy of adsorption is $-3\pi r^2 \gamma_{ow}$.

In Chapter 18, the stabilization of emulsions, both by homogeneous and by Janus particles, is considered.

17.4 Particle monolayers at fluid interfaces

17.4.1 Experimental study of monolayers using a Langmuir trough

As already seen, in the study of the behaviour of adsorbed surfactant monolayers the surface concentration of (soluble) surfactant, Γ, can be varied simply, by changing the bulk concentration of surfactant in solution. From a knowledge of the variation of interfacial tension with the activity of surfactant in solution, surface concentrations can be obtained by use of the Gibbs adsorption equation (§5.3). Adsorbed surfactant exchanges readily with surfactant in solution and adsorption equilibrium is usually rapidly established. For particles, monolayers must be formed in other ways. One way is to add (spread) a known quantity of particles, dispersed in a volatile spreading solvent, directly onto a known area of interface. Surface concentration of particles can be varied by the number of particles added to the interface. Monolayers of say water-insoluble amphiphiles (e.g. long chain alkanols or alkanoic acids) on a water surface can be similarly prepared. Such *insoluble* or *spread* monolayers, as well as monolayers of particles, can be contained by suitable barriers in the fluid interface, and the area between the barriers varied at will to give monolayers of required surface concentrations. The interfacial tension in such systems is usually discussed in terms of a surface pressure, π, which is the lowering of the interfacial tension γ caused by the presence of the surface film (§4.4):

$$\pi = \gamma_0 - \gamma \qquad (4.4.1)$$

where γ_0 is the tension in the absence of the monolayer.

The *Langmuir trough* or *film balance* is an apparatus which allows the variation of surface concentration of particles and the measurement of the concomitant changes in surface pressure. A trough which is suitable for the study of particle monolayers spread at the oil/water interface is shown schematically in Fig. 17.16. A monolayer is spread and confined at an oil/water interface between moveable barriers, and the whole trough

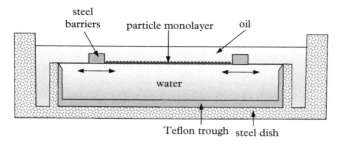

Figure 17.16 *Schematic representation of the Langmuir trough used for the study of particle monolayers spread at the oil/water interface. The trough can be placed on the stage of an optical microscope so photographic images can be obtained of the monolayers at various stages of compression. After Aveyard et al. (2000b).*

can be placed on the stage of an optical microscope so that photographic images of the monolayer in various stages of compression can be obtained. Individual particles can be seen if diameters are of the order of a micron and upwards. The surface pressure of the monolayer can be measured using a Wilhelmy plate placed vertically through the oil/water interface (see §A3.1.1).

In passing, it is noted that in the Langmuir trough experiment particle monolayers are compressed mechanically, and are confined by the barriers. It is possible however to observe particle monolayers under compression, for example, at an oil/water interface, in systems where the monolayer is not confined by barriers, say in a tube. When droplets of (an unstable) Pickering emulsion (see Chapter 18) contained in a vial coalesce the monolayers at the oil/water interface become more concentrated and develop a significant surface pressure since the particles are irreversibly adsorbed. This is observed to induce the spontaneous ascent of a thin film up the wall of the container. The container wall must be pre-wetted for the phenomenon to occur, and the film comprises an oil and a water layer with a particle monolayer adsorbed between them (Fig. 17.17). It appears that even small surface pressures (dilute films) are sufficient, and that particle size is not important. The effect is observed both with nanoparticles (e.g. gold colloids, see Binks et al., 2006) and with micron-sized particles (Cheng and Velankar, 2008).

17.4.2 What do particle monolayers look like? Structure of dilute monolayers

Interest in particle monolayers at fluid interfaces was stimulated by the work of Pieranski (1980), where it was shown that small latex particles at water/air interfaces form ordered monolayers. A great deal of work has been done since then probing the melting of two-dimensional crystals as well as investigating the mechanisms of particle aggregation at air/water interfaces. More recently there has been increased scientific interest in particle monolayers at *oil*/water interfaces. As with surfactants, adsorbed particle monolayers can be responsible for stabilization (or destabilization) of thin liquid films and foams and of

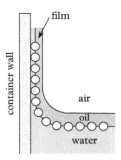

Figure 17.17 *Thin vertical film spontaneously formed at container wall. It consists of a layer of water and of oil with an adsorbed particle monolayer between. The water phase is an unstable O/W Pickering emulsion.*

emulsions and emulsion films. Various kinds of particles have been studied including polystyrene latex spheres, with different chemical groups at the surface, and spherical silica particles, which can be prepared with a range of hydrophobicities.

As seen, the behaviour of monolayers of surfactants adsorbed at fluid interfaces can be explored indirectly by observing conformity or otherwise to adsorption isotherms or surface equations of state based on assumed models (§5.3.2). Lateral intermolecular interactions between adsorbed surfactant molecules affect behaviour in various ways. Ionic surfactants experience mutual repulsion through Coulombic interactions that can be modulated by electrolyte present in the aqueous phase, which screens the repulsion. A given surfactant behaves differently at air/water and oil/water interfaces. For example, attractive interactions between alkyl chains which act at the air/water interface are damped out when the chains are immersed in oil at an oil/water interface. For particle monolayers in which the particles are large enough to be observed directly using an optical microscope, the effects of lateral interactions on monolayer structure can be simply observed. Examples of the structures of monolayers of monodisperse polystyrene latex particles (diameter 2.6 μm) present at air/water and octane/water interfaces (with and without NaCl present in the aqueous phase) are illustrated in Fig. 17.18.[4] The surfaces of the particles carry sulfate groups with a surface charge density (in water) of 8 μC cm^{-2}, corresponding to an area per sulfate group of about 2 nm^2. Images (a) to (c) are for particles at air/aqueous NaCl solution interfaces. For low concentration of NaCl (10 mmol dm^{-3}) particles are well dispersed in the monolayer with little aggregation occurring (image (a)), showing the existence of a net lateral repulsive force between the particles. Addition of further NaCl to the aqueous phase begins to cause aggregation indicating that the repulsion between particles is Coulombic in origin; the salt damps

---

[4] The estimated contact angle for the particles resting in the octane/water interface is between 70 and 80° (measured through the aqueous phase) so that a substantial part of a particle is immersed in the oil phase. The contact angle of a particle at the air/aqueous solution interface is however much smaller and most of the particle rests in the aqueous phase.

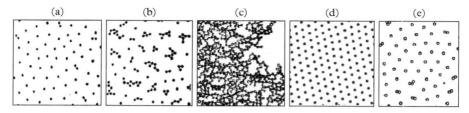

Figure 17.18 *Images of monolayers of 2.6 μm diameter spherical polystyrene particles at fluid interfaces. Images (a), (b), and (c) are of particles spread at the air/aqueous NaCl interface. Salt concentrations for (a), (b), and (c) are respectively 10, 100, and 1000 mmol dm⁻³. Images (d) and (e) are of particles at octane/aqueous solution interfaces; in (d) the aqueous phase is water and for (e) it is 1 mol dm⁻³ aqueous NaCl. Reprinted with permission from R. Aveyard, J.H. Clint, D. Nees, and V.N. Paunov,* Langmuir, *2000,* **16,** *1969. Copyright 2000 American Chemical Society.*

out the electrical repulsion through the aqueous phase. For high concentrations of NaCl (e.g. 1 mol dm⁻³ corresponding to image (c)) the particles are completely aggregated in an interconnecting network.

The situation for particles at the oil/water interfaces is, perhaps surprisingly, significantly different. At the octane/water interface (image (d)) the particles form a highly ordered hexagonal structure (a *colloidal crystal*), indicating strong long-range mutual repulsion between particles. The fractional surface coverage for the monolayer shown is only 0.03 and the surface pressure is almost zero; the distance between centres of neighbouring particles is 14μm. High concentration of NaCl in the aqueous phase (e.g. 1 mol dm⁻³) although reducing the repulsion, does not remove it sufficiently for aggregation to occur and the particles remain largely discreet (image (e)). Since the salt cannot enter the nonpolar oil phase these findings show that strong electrical (Coulombic) repulsion between particles at the oil/water interface takes place *through the oil phase*; the particle/oil interface carries a residual charge (see e.g. Boneva et al., 2009).

The amount of charge on a particle at the oil/particle interface depends on the contact angle. For $\theta < 90°$ more than half of a spherical particle resides in the aqueous phase. As the contact angle rises the proportion of the particle surface in oil increases, and so therefore does the lateral electrical repulsion between particles and concomitant order within the film. The effect of particle wettability (i.e. of θ) on monolayer structure is illustrated in Fig. 17.19 for dilute monolayers of silica spheres (diameter 1μm) that have been rendered hydrophobic to different extents by silanization (see §14.7.2).

17.4.3   A closer look at the lateral force between charged particles at a liquid interface

Above, reference was made to the lateral Coulombic interaction, in oil, between two like charged particles. However, a charge q on one particle in oil (taken as a point charge at a vertical distance ζ from the oil/water interface—see Fig. 17.20) gives rise to an *image*

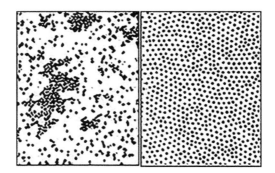

Figure 17.19 *Monolayers of silica particles, diameter 1μm, at the octane/water interface. Contact angles of the particles with the interface are 115° (left image) and 129°. Reprinted with permission from T.S. Horozov, R. Aveyard, J.H. Clint, and B.P. Binks,* Langmuir, *2003,* **19**, *2822. Copyright 2003 American Chemical Society.*

charge q_{im} in the aqueous phase also a distance ζ from the interface. The charge on a particle in oil can therefore also interact with the image charge, in water, of the other particle. The image charge q_{im} is related to q by

$$q_{im} = -q \left(\frac{\varepsilon_{wat} - \varepsilon_{oil}}{\varepsilon_{wat} + \varepsilon_{oil}} \right) \qquad (17.4.1)$$

where ε_{wat} and ε_{oil} are the relative permittivities of the aqueous and oil phases respectively. For particles at an alkane/water interface $\varepsilon_{wat} \approx 80$ and $\varepsilon_{oil} \approx 2$, so that $q_{im} \approx -q$. The force f_{inter} of interaction between the two spherical particles, lateral separation L between centres, can be expressed (Aveyard et al., 2000a)

$$f_{inter} \approx \frac{q_{oil}^2}{4\pi \varepsilon_{oil} \varepsilon_o} \left[\frac{1}{L^2} - \frac{L}{\left(4\varsigma + L^2 \right)^{3/2}} \right] \qquad (17.4.2)$$

where ε_o is the permittivity of free space. The charge q_{oil} is the charge on the particle/oil interface taken to be a point charge (distance ζ from the oil/water interface; Fig. 17.20). With reference to (17.4.2) it is seen that the force consists of two terms, one in L^{-2} corresponding to the interaction of the two charges in oil, the other to the interaction between the charge on particle 2 with the image charge of particle 1.

Long-range repulsive forces between two charged polystyrene latex particles (2.7 μm diameter) at the alkane/water interface have been measured experimentally as a function of separation L using a laser tweezer method (Aveyard et al., 2002). The results, shown in Fig. 17.21, were found to be consistent with the repulsion arising from a very small net electric charge at the particle/oil interface, corresponding to a fractional dissociation of the total ionizable surface (sulfate) groups of about 4×10^{-4}. The full line in the figure is generated by (17.4.2). The quantity ζ is related to the particle radius R and contact angle θ by

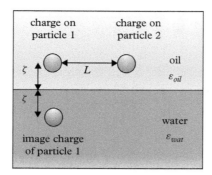

Figure 17.20 *The residual charge at the particle/oil interface, taken to be a point charge distance ζ from the oil/water interface, has an image charge in the aqueous phase, a distance ζ from the interface. The charge on particle 2 is horizontal distance L from that on particle 1.*

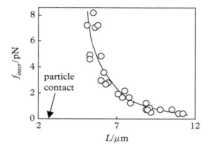

Figure 17.21 *Interparticle force between two charged polystyrene latex spheres (diameter 2.7 mm) at the alkane/water interface. The full line is generated by (17.4.2) and the points are experimentally determined using the laser tweezer technique. Redrawn from Aveyard et al. (2002).*

$$\zeta = 0.5R\,(3 + \cos\theta) \tag{17.4.3}$$

From a knowledge of the interaction force between particles it is possible to obtain a theoretical expression for the π–A isotherm for a particle monolayer. The surface pressure π is the force exerted per unit length on, say, the barrier of a Langmuir trough.

On the basis of the interaction force between two particles being given by (17.4.2), the following isotherm equation can be obtained for a monolayer of spherical particles, radius R, in which the separation between centres of neighbouring particles is D (Aveyard et al., 2000a):

$$\pi = \frac{q^2}{2\sqrt{3}\,\varepsilon_{oil}R^3 x^{3/2}} \left[1 - \frac{1}{(1 + 4\beta/x)^{1/2}} + \ln\left\{ \frac{1 + (1 + 4\beta/x)^{1/2}}{2} \right\} \right] \tag{17.4.4}$$

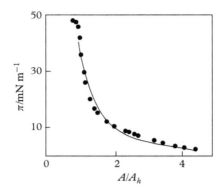

Figure 17.22 *Surface pressure vs reduced trough area for a monolayer of polystyrene latex spheres, diameter 2.6 μm, spread at the octane/10 mmol dm^{-3} aqueous NaCl interface. Filled circles represent experimental data and the full line is the best fit generated using (17.4.4) assuming 1.03% of the sulfate groups at the oil/particle interface are dissociated. Redrawn from Aveyard et al. (2000b).*

In (17.4.4),

$$x = A/A_h = (D/2R)^2; \beta = (\zeta/R)^2 \qquad (17.4.5)$$

where A is the trough area $(N\sqrt{3}D^2/2)$ and A_h is the trough area at close-packing of the monolayer $(N2\sqrt{3}R^2)$; N is the number of particles in the monolayer. The fit of an experimental isotherm, using (17.4.4), for a monolayer of polystyrene particles at the octane/10 mmol dm^{-3} aqueous NaCl interface, is shown in Fig. 17.22. The fit, obtained by assuming the percentage ionization of sulfate groups at the particle/oil interface to be about 1.03 per cent, is seen to be reasonable.

It is interesting that the presence of the surface charge can also give rise to an *attractive* force (Nikolaides et al., 2002) referred to as an *electrodipping* force (Boneva et al., 2009). Particles with diameter of the order of a few microns are too small to deform an oil/water or an air/water interface as a result of gravitational effects alone. It turns out however that the presence of electrical charge at the particle surface gives rise to a force which tends to push the particle towards the aqueous phase (which has the higher relative permittivity). This causes an electrically induced curvature of the interface around the particle, which in turn gives rise to an attractive *capillary force* between neighbouring like particles (see Fig. 17. 23a, (1) and (2)).

Capillary forces between *dissimilar* particles (e.g. a hydrophobic and a hydrophilic particle) that give (by whatever means) menisci such as those illustrated in Fig. 17.23a, (3) result in repulsion. Capillary forces are often discussed in terms of the meniscus slope angle ψ, defined in Fig. 17.23b; the choice of signs for the angles is arbitrary (except one is positive, the other negative). Capillary attraction between like particles arises because the liquid meniscus deforms in such a way that the gravitational potential energy of the two particles decreases when they approach (Kralchevsky and Nagayama,

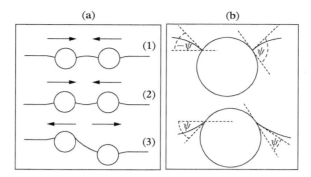

Figure 17.23 *(a) In (1) and (2), two like particles at deformed liquid interfaces attract through capillary forces. In (3), unlike particles repel. (b) The definition of meniscus slope angle, ψ. The choice of signs for the angles is arbitrary. Note that the slope angles are different from the contact angles θ.*

1994). In terms of the meniscus slope angles, ψ_1 and ψ_2 for particles 1 and 2, the capillary interactions are attractive when the slope angles have the same sign, and repulsive when the signs are opposite.

17.4.4 What happens to monolayers when they are compressed?

The particle monolayers represented in Fig. 17.18 are dilute but as explained the monolayers can be compressed, for example on a Langmuir trough. Questions of particular interest relate to how the structure and surface pressure vary with surface coverage of the particles and how, under sufficient applied lateral pressure, the monolayers ultimately 'collapse'.

The π–A isotherm shown in Fig. 17.24 for spherical polystyrene particles (diameter 2.6 μm) at the octane/water interface can be divided broadly into three regions, denoted *A*, *B* and *C*. At high areas (region *A*) the isotherm rises only slowly, as a result of long-range Coulombic repulsion between particles. In this region the structure of the monolayer is hexagonal (Fig. 17.18d). In region *B* the surface pressure rises more steeply with reduction in surface area, and the hexagonal structure becomes distorted. In region *C* the rate of pressure increase begins to fall, and the locus of the rapid change in slope corresponds to monolayer collapse, at a surface pressure π_c.

Interestingly for this system, collapse does not consist of particle ejection from the monolayer, but rather the monolayer folds progressively as the surface area decreases, as seen in Fig. 17.25. The monolayer appears to collapse when π_c approaches the interfacial tension γ_o of the particle-free oil/water interface. The tension of the particle-free interface can be varied by addition of surfactant to the aqueous phase, and a plot of π_c versus γ_o for particle monolayers on aqueous surfactant solutions is shown in Fig. 17. 26. Since $\pi_c = \gamma_o - \gamma_c$ the effective interfacial tension γ_c at monolayer collapse is close to zero (see also §18.3.2).

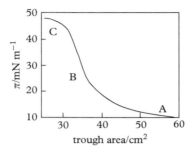

Figure 17.24 *Surface pressure–area isotherm for a monolayer of monodisperse polystyrene latex spheres, diameter 2.6 μm, spread at the octane/water interface. Redrawn from Aveyard et al. (2000a).*

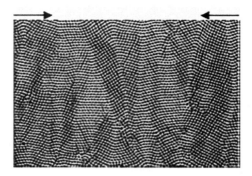

Figure 17.25 *Collapsed monolayer of 2.6 μm diameter polystyrene spheres at the octane/water interface. Direction of compression is indicated by the arrows. Reprinted with permission from R. Aveyard, J.H. Clint, D. Nees, and V.N. Paunov, Langmuir, 2000, 16, 1969. Copyright 2000 American Chemical Society.*

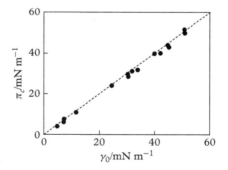

Figure 17.26 *Variation of collapse pressure with oil/water interfacial tension γ_0 for monolayers of polystyrene latex spheres, diameter 2.6 μm, formed at the interface between aqueous surfactant solutions and octane. The dashed line has unit slope. γ_0 is varied using various concentrations of four different surfactants: decyl-β-glucoside, hexadecyltrimethylammonium bromide, sodium dodecylsulfate, and hexadecylpyridinium chloride. Redrawn from Aveyard et al. (2000b).*

17.5   Thin liquid films stabilized by particles

17.5.1   Introductory remarks

As already mentioned small particles, like surfactants, can act as emulsion stabilizers. It is also known that the relative wettability of the particles by oil and water influences the type of stable emulsion that is formed. When two particle-covered drops approach in an emulsion, a thin liquid (emulsion) film is formed between them (Fig. 17.27). If this film is unstable the drops can coalesce and the emulsion is unstable. If on the other hand the film is stable at some particular thickness or in some configuration, the drops may adhere rather than coalesce, leading to flocculation in the emulsion; of course in a stable emulsion the droplets may part again after an encounter. Films of interest can be water films between oil phases (O/W/O), which are relevant to oil-in-water (O/W) emulsions, or oil films between aqueous phases (W/O/W), occurring in water-in-oil (W/O) emulsions.

It is to be expected that the properties of the isolated monolayers will influence the stability of thin liquid films formed with adsorbed particles. The contact angle of the particles at the oil/water interface is central to stability. The particle surface concentration in the monolayers bounding a film is also likely to be important. As seen above, repulsive electrical forces acting laterally between particles can determine the monolayer structure; these same forces can also act between particles carrying a charge on *opposite faces* of a film.

17.5.2   Experimental methods for the study of films

If the particles have diameters of around a micron or above the liquid films, like the particle monolayers, can be studied microscopically. In addition the films can be observed in a vertical as well as a horizontal configuration, so that effects due to particle sedimentation in a film can be observed. A setup for the study of horizontal films used by Stancik et al. (2004) is illustrated in Fig. 17.28. In the configuration shown, an O/W/O film is formed by bringing a particle-coated oil drop in water up to a plane particle monolayer at an oil/water interface. W/O/W films can be formed similarly by bringing

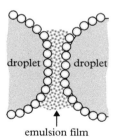

Figure 17.27 *Two particle-coated droplets in an emulsion with an intervening emulsion film, stabilized (or otherwise) by particle monolayers.*

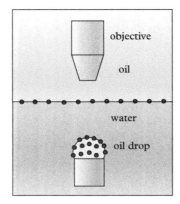

Figure 17.28 *Setup for the microscopic study of horizontal thin liquid films stabilized by particles. As shown, the films studied are O/W/O, but a water drop can be formed in oil and W/O/W films investigated. Redrawn from Stancik et al. (2004).*

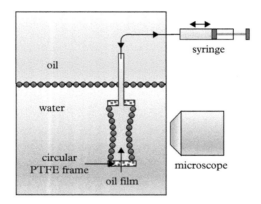

Figure 17.29 *Schematic diagram of apparatus for the formation and study of vertical thin particle-stabilized oil films in water. The film is formed on a circular PTFE frame (seen here from the side) by passage through the monolayer at the plane oil/water interface. Oil can be added or withdrawn from the film by use of the syringe, and the film is viewed microscopically. Redrawn from Horozov et al. (2005).*

together a particle-coated water drop in oil up to the planar interface. These authors observed the behaviour of thin liquid films formed in systems containing decane and aqueous NaCl, stabilized by polystyrene particles (3.1 μm diameter).

Horozov et al. (2005) studied vertical films in systems with octane and water stabilized by silica particles (diameter 3 μm) with a range of wettabilities, achieved by silanization of the particles, and like Stancik et al. investigated both O/W/O and W/O/W films. A schematic representation of the apparatus used is given in Fig. 17.29. The setup shown is used for the study of W/O/W films which are formed by passing a circular PTFE frame (diameter about 6 mm), originally in the oil, through a particle-laden oil/water interface. A simple variant of the apparatus allows the investigation of O/W/O films by pulling, in this case a *glass* frame from water through the plane oil/water interface into the oil phase. The thickness of a film can be controlled, for example by pumping oil into or out of a W/O/W film using the syringe (Fig. 17.29). Some of the results reported by Horozov et al. are described below simply as examples of the types of behaviour that can be exhibited by particle-stabilized thin films.

17.5.3 Behaviour and properties of particle-stabilized thin films

It will be useful, before looking at experimental findings, to consider what the effects of particles on the stability of thin liquid films might be. In Fig. 17.30a, a hydrophilic particle is represented bridging an O/W/O film. For a hydrophilic particle the contact angle (through water) at an oil/water interface is less than 90°, as shown. This angle can be achieved for the particle with each surface of the film and so the particle tends to keep the two oil/water interfaces apart, hence stabilizing the film. The situation for a hydrophilic particle in a W/O/W film however is quite different. In this case, $\theta > 90°$ and in order to attain such an angle for each oil/water interface, the interfaces would tend to cross, obviously leading to instability. So, hydrophilic particles are expected to rupture oil films in water. By similar arguments hydrophobic particles tend to stabilize W/O/W films (Fig. 17.30b), but rupture O/W/O films.

Films without particles.

Before looking at the effect of particles on film stability, the behaviour of the films in the absence of particles is considered. Once formed, films are made to thin (opened up) by sucking liquid out. Octane films formed in water are very unstable and rupture at large thicknesses soon after formation. Water films in octane on the other hand survive several minutes, and it is noteworthy that a dimple is initially formed in the film, as shown in image (a) in Fig. 17.31, obtained in reflected light. The dimple is formed because the exit of the water from the centre of the film is impeded as a result of the relatively high viscosity of water. In the latter stages of thinning however, the water films become plane parallel with thicknesses of about 100 nm, which is considerably less than a particle diameter (3 μm for the systems to be described below). If particles were present it would be expected that they would bridge and stabilize a water film if $\theta < 90°$. The film stability (in the absence of particles) probably arises from the negative charge at each of the oil/water interfaces, caused by adsorption of hydroxyl ions.

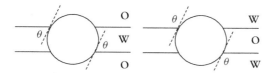

Figure 17.30 *(a) A hydrophilic particle ($\theta < 90°$) resting in an O/W/O film. The two O/W surfaces attain their required contact angle with the particle when the surfaces are a finite distance apart. (b) For a W/O/W film, θ must be $>90°$ for the surfaces to be held apart.*

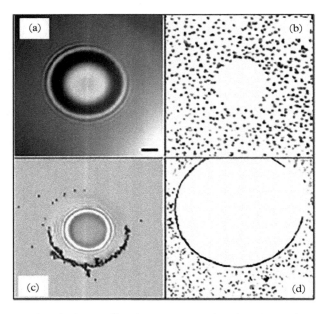

Figure 17.31 *Images of vertical water films in octane taken in reflected (a and c) and transmitted (b and d) light. (a) In the absence of particles and (b to d) in the presence of silanized silica particles (diameter 3μm). Contact angles are in (b) 99°, (c) 85°, and (d) 65°. In all cases, a dimple is formed which is clearly seen in reflected light. The scale bar in (a) is equal to 50 μm and the scale is the same for all images. Reprinted with permission from T.S. Horozov, R. Aveyard, and J.H. Clint,* Langmuir, *2005, 21, 2330. Copyright 2005 American Chemical Society.*

Vertical water films in octane with dilute silica particle monolayers.

The stability of the water films in the presence of dilute particle monolayers (with of the order of 10^5 particles cm^{-2}) depends on the particle wettability. As expected for $\theta = 152°$ the films are unstable. Image (b) in Fig. 17.31 (obtained using transmitted light) is of a water film in which the particles have $\theta = 99°$, and as seen particles have been expelled from the dimple. For particles with lower contact angles (85° and 65°), when the films are opened up *rapidly*, a ring of bridging particles is trapped at the periphery of the dimple as seen in images (c) and (d) of Fig. 17. 31 and as illustrated in Fig. 17.32b. If the films are opened up *slowly* however the particles at the periphery of the dimple do not bridge the film. This indicates that there is some repulsion between the particle surfaces in the water film (carrying a negative charge) and the (opposite) surface of the film which also carries a negative charge as already seen. Opening the film slowly allows the particles to flow out of the dimple without penetrating the oil/water interfaces and hence without bridging the film.

After formation, stable films which have a bridging ring of particles can be closed down by pumping water back into the film. Images of the vertical water films (stabilized by particles with $\theta = 65°$) that result as water is progressively added to the film, are shown in Fig. 17.33. The image in Fig. 17.33a is for the same film as that shown in Fig. 17.31d.

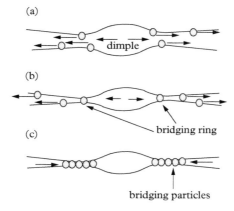

Figure 17.32 *Water films in oil. (a) Opening up the film with dimple. (b) Rapid opening up, with bridging particles at periphery of the dimple. (c) Closing the film with build-up of particles around dimple periphery. Horizontal arrows indicate direction of liquid and particle flow.*

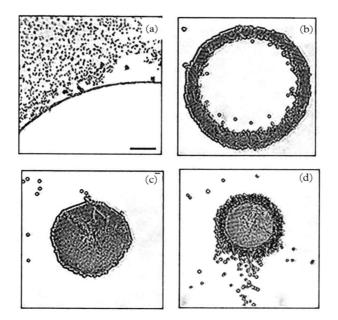

Figure 17.33 *Progressive closing (from (a) to (d)) of vertical water films in octane stabilized by silica particles with contact angle 65°. Reprinted with permission from T.S. Horozov, R. Aveyard, and J.H. Clint, Langmuir, 2005, 21, 2330. Copyright 2005 American Chemical Society.*

The large ring of particles in Fig. 17.33a shrinks and the 'wall' thickness increases (Fig. 17.33b and Fig. 17. 32c) until ultimately a two-dimensional crystalline disc of bridging particles is formed (Fig. 17.33c). The particles at the edge of this disc then begin to break away from one of the oil/water interfaces of the film, remaining in the other, and sediment under gravity, as seen in Fig. 17.33d.

Octane films in water stabilized by hydrophobic silica particles.

In contrast to water films in octane, which are stabilized by hydrophilic particles, octane films in water require hydrophobic particles for their stabilization. The behaviour described below is for octane films with silanized silica particles with $\theta = 152°$. Hydrophobic particles with θ less than this were found to be ineffective in stabilizing octane films.

Initially, film thinning is forced by withdrawing octane from the film. When the films are relatively thick it is possible to distinguish particles on the two sides of the film (the depth of field of the microscope is about 1μm); both are independently ordered monolayers. When the films are thin, such that both particle monolayers can be observed simultaneously, a spontaneous lateral movement of particles towards each other occurs giving an increase in particle concentration in the films (Fig. 17.34b). About 40ms after the image in (b) was obtained a crystalline disc bridging the film has formed (Fig. 17.34c), and grows rapidly. The particle trajectories are clearly seen during this period of film thinning (Fig. 17.34c and d). The bridging particles, although close

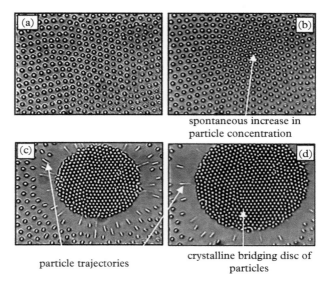

Figure 17.34 *Vertical films of octane in water stabilized by silanized silica particles (diameter 3 μm) with $\theta = 152°$. The sequence (a) to (d) shows successive stages in film thinning. Reprinted with permission from T.S. Horozov, R. Aveyard, and J.H. Clint, Langmuir, 2005, 21, 2330. Copyright 2005 American Chemical Society.*

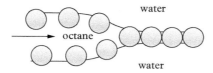

Figure 17.35 *Representation of the 'unzipping' of an octane film effected by forcing octane into the film.*

(4.2. 4.2.μm laterally between centres), are not close-packed within the disc, and the disc is in equilibrium with well-ordered dilute monolayers away from its periphery.

The crystalline films are very stable and can be closed down by pumping in octane. During this closure the disc initially shrinks without ejecting particles; the particles are forced closer together until ultimately they are close-packed. Then, further closure causes the film to 'unzip' (represented in Fig. 17.35), when particles at the edge of the film become detached from one of the film surfaces. The zipping process is reversible, but at some extent of closure the surfaces of the film spontaneously detached and a stable film consisting of separate ordered particle monolayers is restored.

....

GENERAL READING

Colloidal Particles at Liquid Interfaces (ed. B.P. Binks and T.S. Horozov). Cambridge University Press, 2006. This is a collection of learned review articles written by experts in their fields. It covers a wide range of topics including theoretical treatments and relevance in emulsions, foams and flotation systems.

....

REFERENCES

A. Amirfazlia, and A.W. Neumann, Status of the three-phase line tension: a review. *Adv. Colloid Interface Sci.*, 2004, **110** 121–141.

R. Aveyard, Can Janus particles give thermodynamically stable Pickering emulsions? *Soft Matter*, 2012, **8**, 5233–5240.

R. Aveyard, B.D. Beake, and J.H. Clint, Wettability of spherical particles at liquid surfaces. *J. Chem. Soc., Faraday Trans.*, 1996, **92**, 4271–4277.

R. Aveyard, B.P. Binks, J.H. Clint, P.D.I. Fletcher, T.S. Horozov, B. Neumann, V.N. Paunov, J. Annesley, S.W. Botchway, D. Nees, A.W. Parker, A.D. Ward, and A.N. Burgess, Measurement of long-range repulsive forces between charged particles at an oil–water interface. *Phys. Rev. Lett.*, 2002, **88**, 246102.

R. Aveyard, J.H. Clint, D. Nees, and V.N. Paunov, Size-dependent lens angles for small oil lenses on water. *Colloids Surf. A*, 1999, **146**, 95–111.

R. Aveyard, J.H. Clint, D. Nees, and V.N. Paunov, Compression and structure of monolayers of charged latex particles at air/water and octane/water interfaces. *Langmuir*, 2000a, **16**, 1969–1979.

R. Aveyard, J.H. Clint, D. Nees, and N. Quirke, Structure and collapse of particle monolayers under lateral pressure at the octane/aqueous surfactant solution interface. *Langmuir*, 2000b, **16**, 8820–8828.

B.P. Binks (Ed.), Colloidal particles at liquid interfaces. *Phys. Chem. Chem. Phys.*, 2007, **9**, 6285 *et seq.*

B.P. Binks, J.H. Clint, P.D.I. Fletcher, T.J.G. Lees, and P. Taylor, Growth of gold nanoparticle films driven by the coalescence of particle-stabilized emulsion drops. *Langmuir*, 2006, **22**, 4100–4103.

M.P. Boneva, K.D. Danov, N.C. Christov, and P.A. Kralchevsky, Attraction between particles at a liquid interface due to the interplay of gravity- and electric-field-induced interfacial deformations. *Langmuir*, 2009, **25**, 9129–9139.

D.L. Cheung and S.A.F. Bon, Stability of Janus nanoparticles at fluid interfaces. *Soft Matter*, 2009, **5**, 3969–3976.

H.-L. Cheng and S.S. Velankar, Film climbing of particle-laden interfaces. *Colloids Surf. A*, 2008, **315**, 275–284.

S. Dasgupta, T. Auth, and G. Gompper, Nano- and microparticles at fluid and biological interfaces. *J. Phys.: Condens. Matter*, 2017, **29**, 373003.

T.S. Horozov, R. Aveyard, and J.H. Clint, Particle zips: vertical emulsion films with particle monolayers at their surfaces. *Langmuir*, 2005, **21**, 2330–2341.

T.S. Horozov, R. Aveyard, J.H. Clint, and B.P. Binks, Order–disorder transition in monolayers of modified monodisperse silica particles at the octane–water interface. *Langmuir*, 2003, **19**, 2822–2829.

S. Jiang and S. Granick, Janus balance of amphiphilic colloidal particles. *J. Chem. Phys.*, 2007, **127**, 161102.

P.A. Kralchevsky and K. Nagayama, Capillary forces between colloidal particles. *Langmuir*, 1994, **10**, 23–36.

J. Mingins and A. Scheludko, Attachment of spherical particles to the surface of a pendant drop and the tension of the wetting perimeter. *J. Chem. Soc., Faraday Trans 1*, 1979, **75**, 1–6.

M.G. Nikolaides, A.R. Bausch, M.F. Hsu, A.D. Dinsmore, M.P. Brenner, C. Gay, and D.A. Weitz, Electric-field-induced capillary attraction between like-charged particles at liquid interfaces. *Nature*, 2002, **420**, 299.

T. Ondarçuhu, P. Fabre, E. Raphaël, and M. Veyssié. Specific properties of amphiphilic particles at fluid interfaces. *J. Phys. (Paris)*, 1990, **51**, 1527–1536.

P. Pieranski, Two-dimensional interfacial colloidal crystals. *Phys. Rev. Lett.*, 1980, **45**, 569.

E.J. Stancik, M. Kouhkan, and G.G. Fuller, Coalescence of particle-laden fluid interfaces. *Langmuir*, 2004, **20**, 90–94.

T.M. Ruhland, A.H. Gröschel, N. Ballard, T.S. Skelhon, A. Walther, A.H.E. Müller, and S.A.F. Bon, Influence of Janus particle shape on their interfacial behavior at liquid–liquid interfaces. *Langmuir*, 2013, **29**, 1388–1394.

18

Emulsions stabilized by solid particles

In the previous chapter it is seen that small solid particles adsorb at fluid interfaces and, in some respects at least, behave like surfactants. In the present chapter attention will be directed to the role of adsorbed particle monolayers in determining emulsion type and emulsion stability. Usually the particles studied have been much larger than (low molar mass) surfactant molecules, but colloidal particles (nanoparticles) with dimensions of the order of a few nanometres are of also of interest in the present context. A major difference between larger particles and surfactant molecules is that the former are adsorbed irreversibly. For example, the free energy change for the adsorption from water of a spherical particle, with radius 100 nm and a contact angle (through water) of 60°, to the oil/water interface (tension 0.050 N m^{-1}) is about $-10^5 kT$ (see Fig. 17.2); that of a typical surfactant molecule is expected to be about -10 to $-20kT$ (Box 17.2). Therefore, although surfactant molecules can exchange readily between an adsorbed monolayer and bulk solution, in general a particle cannot because it is trapped in a deep energy well at the fluid interface. In the present context however, this does not mean that particles adsorbed on one emulsion droplet cannot transfer to another droplet. French et al. (2016) made an elegant study that involved mixing two samples of Pickering emulsions prepared using particles with different colours. Particles can be shared between contacting droplets (just as they can bridge planar thin films as described in the previous chapter (see §17.5.3)). Particle transfer between droplets is effected when disaggregation of droplets occurs and bridging particles originating from one of the drops are left behind on the other drop.

It has long been realized that solid particles can stabilize emulsions very effectively, and indeed solid particles are present in a range of industrial emulsion systems or formulations including foodstuffs, pharmaceuticals, agrochemicals, and paints. In this chapter some basic ideas relevant to the stabilization of emulsions by particles (Box 18.1) are explored, in particular what determines the type of emulsion formed (O/W or W/O), what is the free energy of emulsion formation, and can it be negative, that is, can solid-stabilized emulsions be thermodynamically (rather than kinetically) stable?

Surfactants: In Solution, at Interfaces and in Colloidal Dispersions. Bob Aveyard. © Bob Aveyard 2019.
Published in 2019 by Oxford University Press. DOI: 10.1093/oso/9780198828600.001.0001

Box 18.1   All shapes and sizes—particles galore

In this chapter the behaviour of *model systems* is considered. However, Pickering emulsions are encountered in a range of processes and products in the commercial and industrial worlds, and a great variety of 'particles' have been used in their formulation. Unwanted Pickering emulsions can also occur, for example, in crude oil recovery where water-in-oil emulsions stabilized by mineral particles are formed. Pickering emulsions can be very stable relative to surfactant-stabilized emulsions, and conventional molecular surfactants can be undesirable in some applications (e.g. in food formulations). Important attributes of particles used for emulsion stabilization include their wettability, their size and their shape. Examples of particles used have included carbon nanotubes, magnetic nanoparticles, silica, clay, hydroxyapatite, cyclodextrins (cyclic oligosaccharides) and various natural stabilizers including starch, various proteins, chitosan (a linear polysaccharide obtained from the shells of crustaceans) and bacteria-related particles (Yang et al., 2017). Particles are often treated to adjust their wettability. Perhaps some of the most commonly encountered Pickering emulsions in everyday life are in foodstuffs (e.g. whipping cream stabilized by fat particles and ice cream by ice crystals). Biologically-based particles such as starch granules and protein microparticles are candidates for use in food formulations (Berton-Carabin and Schroën, 2015). Pickering emulsions can be used as templates for fabrication of materials. For example, the dispersed liquid phase of an emulsion can be extracted to give hollow microcapsules (*colloidosomes*) whose shells are made up of the stabilizing particles. Colloidosome size ranges typically from of the order of 1 μm to 1 mm, and various shapes are observed. Microencapsulation of active materials such as pesticides, viruses, drugs or fragrances etc. provides a route to their controlled release.

18.1   What type of emulsions do particles stabilize?

The significance of the hydrophile–lipophile balance (HLB) in systems with oil, water, and surfactant at equilibrium is covered in Chapter 14. When a surfactant, above its critical aggregation concentration (*cac*), forms either normal micelles in the aqueous phase or oil-in-water (O/W) microemulsion droplets, the (macro) emulsion formed on homogenization of the system (consisting say of equal volumes of oil and water) is of the O/W type. Conversely for systems in which surfactant aggregation occurs in the oil phase to give reverse micelles or water-in-oil (W/O) microemulsion droplets, the system forms W/O emulsions on agitation. That is, the type of emulsion formed is that in which the continuous phase contains the surfactant *aggregates* above the *cac*. It has been shown (Chapter 14) how these phenomena can be understood in terms of the curvature properties of close-packed surfactant monolayers adsorbed at the oil/water interface.

The preferred curvature of a close-packed surfactant monolayer at an oil/water interface is linked to the effective molecular shape and hence the packing factor P (see §9.5.1 and Table 9.3) of the surfactant molecules. The quantity P is defined as the ratio

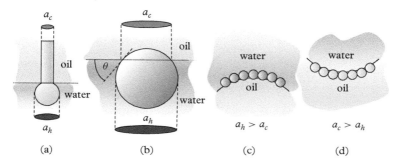

Figure 18.1 *Representation of (a) a surfactant molecule and (b) a spherical particle at a fluid interface. For the surfactant molecule the effective geometry is discussed in terms of the head and tail cross-sectional areas a_h and a_c respectively. An adsorbed spherical particle can be treated similarly; in this case the ratio of a_c to a_h is determined by the contact angle θ (measured through water) as illustrated. (c, d) Close-packed monolayers of spherical particles at curved oil/water interfaces. (c) is for hydrophilic particles ($\theta < 90°$), which stabilize oil drops in water, and (d) is for hydrophobic particles ($\theta > 90°$) which stabilize water drops in oil.*

a_c/a_h, where a_c and a_h are, respectively, the effective cross-sectional areas of the chain and head groups (Fig. 18.1(a)). A close-packed surfactant monolayer tends to curve with the larger surfactant group ('head' or 'tail') on the outside of the monolayer. For example, if $P < 1$ the chain cross-sectional area is less than that of the head group and so, for surfactant concentrations above the *cac*, normal micelles or microemulsion droplets form in the aqueous phase. A similar approach can be applied to close-packed monolayers of spherical particles, with a_c and a_h being defined as illustrated in Fig. 18.1b. The packing factor for a spherical particle is therefore related to the contact angle θ, that is, to the particle *wettability*. For hydrophilic particles ($\theta < 90°$) $P = \sin^2\theta < 1$, and for hydrophobic particles ($\theta > 90°$) $P = (1/\sin^2\theta) > 1$. For $\theta = 90°$, $P = 1$. Close-packed monolayers of spherical particles at curved oil/water interfaces are illustrated in Fig. 18.1, where (c) is for hydrophilic particles ($\theta < 90°$), which stabilize oil drops in water, and (d) is for hydrophobic particles ($\theta > 90°$) which stabilize water drops in oil.

Images are shown in Fig. 18.2 of water drops, coated with polystyrene particles (diameter 2.5 μm), in octane; the particles have carboxyl groups on their surfaces and are mutually repulsive. In the image (a) of the spherical droplet (diameter 250 μm), the distance between particle centres is 7 μm, and the particle monolayer is clearly not close-packed. The drop is resting on a plane octane/water interface coated with similar particles at a similar surface concentration. Image (b) in Fig. 18.2 shows an irregularly shaped particle-coated water drop in octane. Ramsden in 1903 observed that emulsion drops stabilized by particle adsorption often exhibited 'grotesque shapes'. The significance of this is referred to later in §18.3.2.

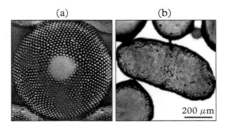

Figure 18.2 *Water drops, coated with polystyrene particles (diameter 2.5 μm), in octane. The particles carry surface carboxyl groups making them mutually repulsive. (a) Spherical drop, diameter 250 μm. The particle monolayer is not close-packed, the particle centres being separated by 7 μm. (b) An irregularly-shaped drop. Reproduced from R. Aveyard, J.H. Clint, and T.S. Horozov, Phys. Chem. Chem. Phys., 2003, 5, 2398, with permission from PCCP Owner Societies.*

18.2 Some experimental findings on solid-stabilized (Pickering) emulsions

Solid-stabilized emulsions are usually named after Pickering, who observed that particles that are more readily wetted by water than by oil stabilize O/W emulsions (Pickering, 1907). However (see e.g. Binks, 2002), Ramsden had noted some 4 years earlier that particles can form 'membranes' that envelop air bubbles and oil drops in water (Ramsden, 1903). Later, Finkle et al. (1923) elucidated the relationship that exists between solid particle wettability (θ at the oil/water interface) and the type of emulsion stabilized by the solids. The favoured emulsion type formed in systems with equal volumes of dodecane and water, stabilized by particles exhibiting a range of values of θ, is indicated in Table 18.1. As will be see, however, emulsion type also depends on the volume fractions of oil and water present.

The process of forming an O/W emulsion (by energy input) from water containing dispersed hydrophilic particles, and oil is illustrated in Fig. 18.3. For relatively low concentrations of particles, the emulsion drop sizes formed on homogenization depends on particle concentration, the drop sizes falling as the availability of particles rises. Ultimately the drop size levels off, and any excess of particles over those coating the emulsion droplets remains in the continuous water phase. These particles can, under certain conditions, form networks which render the continuous phase, and hence the emulsion, viscous (see §15.4). This can be an advantage, and it prevents creaming or sedimentation.

18.2.1 Particle distribution between phases

Particles in a system with oil and aqueous phases are usually expected to partition almost entirely in favour of the phase which preferentially wets the particles. This can be seen as follows. The particle concentrations in oil and in water respectively are designated $[p]_o$ and $[p]_w$, so the distribution ratio of particles between oil and water, K, is $([p]_o/[p]_w)$.

Table 18.1 *Relationship between emulsion type and contact angle θ in systems with equal volumes of water and dodecane*

Solid	$\theta/°$	Emulsion type	Solid	$\theta/°$	Emulsion type
barium sulfate	0	O/W	hydrophobic silica	135	W/O
calcium carbonate	43	O/W	polystyrene	152	W/O
hydrophilic silica	38	O/W	PTFE	147	W/O

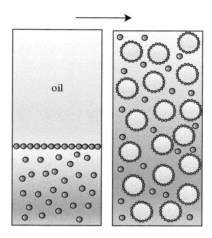

Figure 18.3 *Formation of an oil-in-water emulsion from water, containing dispersed hydrophilic particles, and oil. An excess of particles exists in the water phase over that required to coat the droplets. For clarity free particles and those adsorbed on droplets are drawn on a different scale.*

The fraction of particles in the oil phase, f_o, is therefore

$$f_o = [p]_o/\left([p]_o + [p]_w\right) = K/(1 + K) \tag{18.2.1}$$

By standard thermodynamics, and noting from Young's equation that $(\gamma_{so} - \gamma_{sw}) = \gamma_{ow} \cos \theta$, the free energy change for the transfer of a particle radius r from water to oil, $\Delta_t^{w \to o} G$, is

$$\Delta_t^{w \to o} G = -kT \ln K = 4\pi r^2 (\gamma_{so} - \gamma_{sw}) = 4\pi r^2 \gamma_{ow} \cos \theta \tag{18.2.2}$$

so that K is given as

$$K = \exp\left(-4\pi r^2 \gamma_{ow} \cos \theta / kT\right) \tag{18.2.3}$$

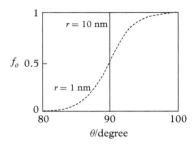

Figure 18.4 *Distribution of spherical particles between oil and water as a function of contact angle. The ordinate is the fraction of particles in the oil phase at equilibrium, and has been calculated from (18.2.1) and (18.2.3). The interfacial tension γ_{ow} has been taken as 0.010 N m^{-1} and the dashed curve is for particles with radius r = 1 nm and the full (effectively vertical) line is for r = 10 nm.*

Some values of the fraction of particles in oil as a function of contact angle, calculated by use of (18.2.1) and (18.2.3), are shown in Fig. 18.4. The oil/water interfacial tension has been taken as 0.010 N m^{-1} (appropriate say for a system of octanol and water) and particles with radii of 1 nm and 10 nm are represented. For the smaller particles the distribution changes rather gradually with θ, whereas for the larger particles the distribution is very sharp, f_o changing from 0 to 1 over a very small range of θ. The distribution becomes sharper the greater the tension and the particle size.

18.2.2   Emulsion inversion

Given the similarities between particles and surfactants in the stabilization of emulsions, it might be expected that inversion of particle-stabilized emulsions can be brought about by changing the particle wettability, either by using a range of particles with different contact angles, or by using mixtures of say two types of particle with different wettability. This is analogous to changing the system HLB in some way in the formation of surfactant-stabilized emulsions. Phase inversion can also be effected by changing the volume fractions of oil and water, using a given particle as stabilizer. If one of the phases is sufficiently conducting (e.g. dilute aqueous NaCl), emulsion type and emulsion inversion can be simply detected by measuring the emulsion conductivity, as discussed in Chapter 14.

An example of how the emulsion type can be inverted by changing the oil to water volume ratio is illustrated in Fig. 18.5. Emulsion conductivities are shown as a function of the water volume fraction, ϕ_w, in the emulsion. The particles used are hydrophobic silica particles and are initially dispersed in the oil (toluene) phase.[1] The emulsions are prepared by addition of toluene, with its dispersed particles, to water

---

[1] The silica particles have a primary (i.e. un-aggregated) diameter of between 5 and 30 nm. They are prepared from hydrophilic silica by treatment with dichlorodimethylsilane and the surface density of Si–OH groups (hydrophilic) and Si–O–Si(CH$_3$)$_2$ (hydrophobic) groups are the same at about 1 nm^{-2} (Binks and Lumsdon, 2000a).

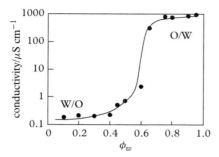

Figure 18.5 *Inversion of toluene + water emulsions containing 2 wt % hydrophobic silica particles initially dispersed in oil. Toluene phase is added sequentially to water and the emulsion conductivity measured. Conductivities are plotted against the water volume fraction in the emulsion. Redrawn from Binks and Lumsdon (2000a).*

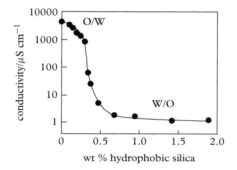

Figure 18.6 *Inversion of water + toluene emulsions by changing the average particle wettability. All emulsions contain equal volume of oil and water and the water phase contains a fixed 2 wt % hydrophilic particles; variable amounts of hydrophobic particles are present in oil as indicated on the abscissa. Redrawn from Binks and Lumsdon (2000b).*

during homogenization. As can be seen from Fig. 18.5, at low ϕ_w, the emulsions have low conductivity, that is, they consist of water drops dispersed in oil. This is the 'natural' emulsion type for stabilization by hydrophobic particles. However, when ϕ_w reaches about 0.6 the emulsions invert to O/W type and the conductivity rises. Both types of emulsion are very stable towards coalescence (particularly close to the inversion condition, unlike systems with surfactant—see §14.2.4) but have very different average drop sizes, 1 μm diameter in the case of the W/O emulsions and 100 μm for the O/W dispersions. In the O/W emulsions the oil drops cream at a greater rate than the smaller water drops sediment in W/O emulsions.

The average wettability of particles can be varied as mentioned by mixing hydrophilic and hydrophobic particles in different proportions, and emulsion inversion can be effected (at say $\phi_o = \phi_w = 0.5$) by using such particle mixtures. This is illustrated by the conductivity results shown in Fig. 18.6, again for toluene + water emulsions

Table 18.2 *Emulsion type and stability to coalescence of emulsions formed from equal volumes of toluene and water and silica particles exhibiting a range of percentages of SiOH groups remaining after silanization. Results of Binks and Lumsdon (2000c), as reported in Aveyard et al. (2003a)*

% SiOH	Emulsion type	Time	% coalescence
100	O/W	2 min	90
79	O/W	8 min	5
76	O/W	3 years	0
67	W/O	3 years	0
50	W/O	3 years	0
20	W/O	2 min	89

prepared from equal volumes of oil and water. The aqueous phases contain a fixed 2 wt % hydrophilic silica particles and the oil phases have varying amounts of hydrophobic particles from zero to 2 wt %. As seen from the changes in conductivity, emulsion inversion from O/W type, when zero or low amounts of hydrophobic particles are present, to W/O occurs for around 0.4 wt % hydrophobic particles originally in oil.

As with emulsions formed by changing oil/water volume fractions, the stability of both O/W and W/O emulsions towards coalescence is high. The stability to creaming and sedimentation however decreases approaching inversion as a result of increasing drop sizes. As mentioned, inversion can also be achieved by using a series of particles each with a different wettability (Binks and Lumsdon, 2000c). In systems with equal volumes of toluene and water with silica particles, it is found that for the extreme values of θ (both high and low), where the free energy of particle attachment to the oil/water interface is relatively small, emulsions are unstable to coalescence. At intermediate θ, however, the stability to coalescence becomes very high. Some results taken from Binks and Lumsdon (2000c) are given in Table 18.2. The silica samples were treated with dimethyldichlorosilane which reacts with the surface silanol groups (see footnote 1).

18.2.3   Multiple emulsions

A multiple emulsion is one in which the droplets of dispersed phase themselves contain smaller drops of the continuous phase of the emulsion, for example, water drops-in-oil drops-in-water (W/O/W) emulsions. They can be prepared using two surfactant types, one hydrophilic in nature which adsorbs primarily at the surface of oil drops in water, and the other hydrophobic which adsorbs mainly at the surface of water drops in oil. The emulsions are not very stable however, usually coalescing relatively quickly to give a simple emulsion. Stability can be improved by addition of solid particles as 'cosurfactants', for example, as in the addition of hydrophobically modified clay

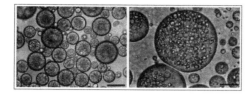

Figure 18.7 *Left: Multiple emulsion of water-in-triglyceride-in-water (scale bar 50 μm), and right: toluene-in-water-in-toluene multiple emulsion (scale bar 20 μm). Reproduced with permission from R. Aveyard, B.P. Binks, and J.H. Clint, Adv. Colloid Interface Sci., 2003, 100–102, 503.*

particles to O/W/O emulsions, the clay particles adsorbing at the outer oil/water interface preventing the coalescence of the (larger) water droplets (Sekine et al., 1999).

Since solid particles can be very effective emulsion stabilizers, and both O/W and W/O emulsions can be formed depending on particle wettability, it might be expected that multiple emulsions can be formed using solid particles alone without the need for surfactant. Further, particles have the advantage in this context that they are irreversibly adsorbed. Indeed, Binks et al. (2002) succeeded in the preparation of very stable multiple emulsions, of both W/O/W and O/W/O type, using only solid particles as stabilizers. Photomicrographs of such emulsions prepared by Binks and co-workers are given in Fig. 18.7, showing O/W/O and W/O/W stabilized by two types of silica particles differing by 25 per cent SiOH groups at the surface. The emulsions are made in two stages. For example, multiple W/O/W emulsions consisting of a triglyceride oil and water were prepared by first forming a W/triglyceride emulsion with high shear, the more hydrophobic silica particles being dispersed in triglyceride. Then, using a low shear rate this emulsion was emulsified into water containing the more hydrophilic particles.

18.3   Stability of solid-stabilized emulsions

Solid-stabilized emulsions can be extremely stable in a kinetic sense; examples are given in Table 18.2 of emulsions which show no coalescence after 3 years. However, the emulsions do not form spontaneously and energy input is required. The particles around a droplet provide a mechanical barrier to coalescence, but for thermodynamic stability (as in microemulsions—see Box 10.4, Chapter 10) the free energy of formation must be negative. Some of the factors that influence the magnitude and sign of the free energy change that accompanies the formation of emulsions stabilized by solid particles are explored below. The circumstances, if any, under which a Pickering emulsion might conceivably be thermodynamically stable are examined. It is supposed in the discussions that the emulsions (containing *spherical* droplets) are stabilized by smooth, monodisperse *spherical* homogeneous particles or by *spherical* Janus particles with areas of different wettability. Although such emulsions are unlikely to be encountered in systems of practical interest it is nonetheless useful to understand the behaviour expected for an idealized system, so that the effects of likely deviations from this model can be probed.

18.3.1 Kinetically stable Pickering emulsions

The free energy change, $\Delta_{emul}G$, accompanying the formation of a monodisperse emulsion is given by the product of the free energy change for the formation of a droplet, $\Delta_d G$, and the number, n_d, of droplets formed. Strictly account should also be taken of the change in entropy when the droplets are dispersed in the continuous phase, but this is ignored here since it is likely to be negligible except where the free energy of droplet formation is extremely small. Levine and Bowen (1991) made calculations of the free energy change accompanying the formation of particle-coated emulsion droplets and Aveyard et al. (2003b) subsequently built on this work to calculate the free energy of emulsion formation. It can be supposed hypothetically that the formation of a particle-coated emulsion droplet first involves the creation of a bare drop which is then followed by the adsorption of the spherical particles (radius r). It is assumed here the particle layer coating a drop is hexagonally close packed.[2] Bare oil/water interface still exists of course between the adsorbed particles and the area depends on the depth of particle immersion, that is, on the contact angle as shown in Fig. 18.8. The maximum fraction of oil/water area that can be removed (at $\theta = 90°$) in a hexagonally close-packed layer of spheres is $\pi/2\sqrt{3} = 0.9069$, as indicated by the horizontal dashed line in the figure. Since the adsorbed particles are partly immersed in the droplet, adsorption causes the drop to swell and so the drop radius R (up to the oil/water interface) increases.[3] The adsorption also involves the de-mixing of the particles from the phase in which they are originally dispersed, and this is accompanied by a negative entropy (positive free energy) change.

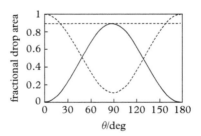

Figure 18.8 *Effect of contact angle on fraction of drop area removed by particle adsorption (full curve) and fraction of drop area remaining (dashed curve).*

---

[2] It is not possible to have perfect hexagonal packing of small spheres on the surface of a large sphere, but the errors incurred in assuming perfect packing are unimportant in the context.

[3] As an example, the surface of a drop with un-swollen radius of 10 μm contains of the order of 3.6×10^4 hexagonally close-packed particles with radius 100 nm. The radius of the particle-coated (swollen) drop up to the oil/water interface is about 10.03 μm for $\theta_p = 70°$.

It is worth noting at this stage that (kinetically) stable emulsions can be formed in which adsorbed particles on droplets are not close-packed (see Fig. 18.2a). In the calculations of $\Delta_{emul}G$ to be discussed now, however, close-packing of particle layers is assumed, rather than choosing an arbitrary coverage by particles. The effects of changing the coverage of surfaces by particles on the free energy of surface formation are however explored later in §18.3.2.

The free energy of formation of a monodisperse emulsion can be expressed

$$\Delta_{emul}G = n_d \Delta_d G = n_d \left\{ \mathcal{A}_{ow}\gamma_{ow} + n_p \left(\Delta_{ads}G - T\Delta_{ads}S \right) \right\} \qquad (18.3.1)$$

where n_p is the number of particles adsorbed onto a drop. The area of the bare, notionally swollen drop, is denoted $\mathcal{A}_{ow}$, and $\Delta_{ads}G$ and $\Delta_{ads}S$ and are, respectively, the free energy and the entropy of adsorption (de-mixing in the case of entropy) per particle. An example of the way in which $\Delta_{emul}G$ varies with contact angle is given in Fig. 18.9. The results were obtained for the formation of 1 m³ of monodisperse emulsion consisting of equal volumes of oil and water. The oil/water tension was taken to be 0.040 N m⁻¹ and the initial un-swollen drop radius is 1 μm and the particle radius is 10 nm. For θ in the range 0 to 90° the particles are expected to distribute almost entirely in favour of the water phase and above 90° to the oil (see Fig. 18.4). Correspondingly in the calculations, for $\theta < 90°$ (where the emulsion is O/W type), adsorption was taken to be from water, and for $\theta > 90°$ (W/O emulsions formed) adsorption is from the oil.[4] The emulsions contain about 1.2×10^{17} drops each carrying about 4×10^4 close-packed adsorbed particles.

It is seen in Fig. 18.9 that all values of the calculated $\Delta_{emul}G$ are positive. No variation of the assumed values of the quantities used (γ_{ow}, R, r, or packing arrangements of the

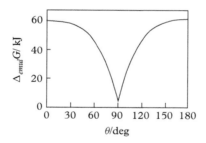

Figure 18.9 *Free energies of emulsion formation as a function of the contact angle of the particles with the oil/water interface. Free energies are for the formation of 1 m³ emulsion from equal volumes of oil and water. Drop radius is 1 μm and that of the particles is 10 nm. The oil/water interfacial tension is taken as 0.040 N m⁻¹.*

---

[4] It is possible to calculate $\Delta_{emul}G$ for the formation of say OW emulsions for $\theta > 90°$ and assume the particles are initially dispersed in water. In this case negative values of $\Delta_{emul}G$ can result, but this does not indicate the existence of stable emulsions since the lowest free energy of the system corresponds to particles being preferentially wetted by the oil phase not water. When $\Delta_{emul}G$ is calculated for adsorption from the oil phase then the right limb of the plot in Fig. 18.9 is obtained.

particles) leads to negative free energies of emulsion formation. Thus the simple model used here for Pickering emulsions indicates that they are expected to be thermodynamically unstable. As seen in Fig. 18.9 the free energy falls with increasing contact angle in the range 0–90° and rises again up to 180°. The changes in free energy result from the variation of $\Delta_{ads}G$ with θ (see Fig. 17.2) and the changes in the area of free oil/water interface between adsorbed particles on the drops (Fig. 18.8). As θ increases below 90°, $\Delta_{ads}G$, which is always negative, becomes more negative. At the same time the area of uncovered oil/water interface decreases, giving a decreasingly positive contribution to $\Delta_{emul}G$. Thus both contributions lead to a fall in the positive $\Delta_{emul}G$. However, for all θ, the positive free energy associated with free oil/water interface outweighs the negative contribution from particle adsorption.

In the above treatment no account has been taken of lateral repulsion between adsorbed particles (see §17.4.3), but is likely to be present in concentrated particle layers. If such repulsion is introduced into the treatment, the free energy of emulsion formation would become more positive since the net free energy of adsorption would become less negative. Lateral repulsion between particles will be considered further in the discussion of formation of Pickering emulsions using Janus particles (see Fig. 18.13).

18.3.2   Can thermodynamically stable Pickering emulsions exist?

It is interesting that Sacanna et al. (2007) report the formation of thermodynamically stable monodisperse emulsions from oil, water, and nanoparticles. In this case, however, the behaviour of the system was quite complex and amphiphilic ions derived from hydrolysis of the oil phase are implicated in the stabilization process (Kraft et al., 2010). Nonetheless, it encourages the investigation of ways that could, in principle at least, lead to the formation of thermodynamically stable systems, and some possibilities are now examined.

Could negative line tension lead to thermodynamic stability?

In the discussion so far the expression used for the adsorption free energy of a particle, $\Delta_a G$, has not included effects of line tension τ in the three-phase contact line around an adsorbed particle. It has been seen how line tension can be included giving the expression for the free energy of adsorption from water ($\theta < 90°$):

$$\Delta_{ads}^{w \to ow} G = \frac{2\pi r \tau}{\sin\theta}\,(1 - \cos\theta) - \pi r^2 \gamma_{ow}(1 - \cos\theta)^2 \qquad (17.2.4)$$

For $\theta > 90°$ the sign in each of the brackets is positive. If this expression, rather than (17.1.4), is used for $\Delta_{ads}G$ in (18.3.1) values of $\Delta_{emul}G$ which include effects of line tension can be obtained. It will be recalled that, unlike interfacial tension, line tension can be either positive (line tends to contract) or negative (line tends to expand). Clearly, if τ (which is an energy per unit length of contact line) is positive, then the adsorption free energy and hence $\Delta_{emul}G$ will be more positive. If however τ is negative, it is possible

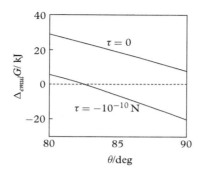

Figure 18.10 *Free energies of monodisperse emulsion formation as a function of θ. Free energies are for 1 m³ emulsion formed from equal volumes of oil and water. Drop radius is 1 μm and that of the particles is 10 nm. The oil/water interfacial tension is taken as 50 mN m⁻¹. The upper line is for τ = 0 and the lower for τ = −10⁻¹⁰ N.*

to have a negative free energy of emulsion formation (as calculated using (18.3.1)). The effects are expected to become more marked the smaller the particles because the length of contact line, relative to the various relevant surface areas, increases. Some values of $\Delta_{emul}G$ in systems where $\tau = 0$ or negative are compared in Fig. 18.10. For values of the various parameters given in the legend it is seen that introduction of a negative line tension ($\tau = -10^{-10}$ N) can render $\Delta_{emul}G$ negative, especially for contact angles not far from 90° (below 90°, as shown, or above 90°).[5] However, in general the magnitude and sign of τ is unlikely to be known for a given system and it is not possible at present to devise systems in which negative line tension might confer thermodynamic stability. It is worth remarking that *positive* line tension, if sufficiently large, can prevent the adsorption of particles, in which case of course even kinetic stability becomes impossible (§17.2.1 and Fig. 17.9).

Could the use of Janus particles confer thermodynamic stability on Pickering emulsions?

The negative free energy of adsorption of particles gives a negative contribution to $\Delta_{emul}G$. The question is, can $\Delta_a G$ be made sufficiently more negative by the use of Janus particles (§17.3) such that $\Delta_{emul}G$ also becomes negative? Expressions were obtained in §17.3 for the free energy of adsorption of Janus particles, and these can be used in conjunction with (18.3.1) to compute $\Delta_{emul}G$. Examples are taken below of O/W emulsions stabilized by hexagonally close-packed polar particles (radius $r = 100$ nm) with apolar caps. The total volume of a monodisperse emulsion, consisting of equal

---

[5] Negative $\Delta_{emul}G$ as calculated using (18.3.1) can imply that the effective interfacial tension, γ_{eff}, around emulsions drops is negative, as discussed in §18.3.2. If so, since drops are fluid the surface coverage by particles can be reduced, and γ_{eff} made less negative or zero, by drops changing their shape.

volumes of oil and water, is taken to be 1 m^3 as before and the un-swollen drop radius R is 10 μm. The oil/water interfacial tension is taken as 0.050 N m^{-1}.

The free energy of adsorption from water to the oil/water interface of a polar particle with an apolar cap, whose size is defined by the angle α as shown in Fig. 17.11, is

$$\Delta_{ads}^{w\rightarrow ow} G = 2\pi r^2 \gamma_{ow} \left\{ (1 - \cos \alpha)(\cos \theta_a - \cos \theta_p) - 0.5(1 - \cos \theta_p)^2 \right\} \qquad (17.3.1)$$

where θ_p and θ_a are the contact angles of, respectively, the polar and apolar surfaces with the oil/water interface. As explained in §17.3, (17.3.1) is used to calculate $\Delta_{ads}^{w\rightarrow ow} G$ for θ_p in the range $\alpha < \theta_p < (\theta_p)_t$; $(\theta_p)_t$ is the value of θ_p at which the Janus particle becomes equally wetted by the oil and aqueous phases. For $\theta_p > (\theta_p)_t$ the particle is preferentially wetted by the oil. Putting values of $\Delta_a^{w\rightarrow ow} G$ obtained from (17.3.1) into (18.3.1) yields the values of $\Delta_{emul}G$ presented in Fig. 18.11. Each line corresponds to a given value of α as indicated. As seen, $\Delta_{emul}G$ is negative for a range of values of θ_p and α (for a fixed θ_a).

As pointed out (§17.4.3), it is likely that particles of interest will carry some charge at the surface and repulsion can take place between adsorbed particles, mainly through the oil phase. It is this that gives rise to the highly ordered monolayers observed experimentally (see e.g. Fig. 17.18d). The repulsion will reduce the magnitude of the negative free energy of adsorption of a particle, and so the region of negative free energy of emulsion formation will become less extensive, or possibly disappear.

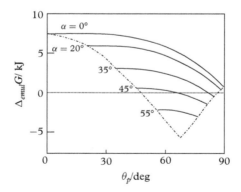

Figure 18.11 *Free energy of formation of 1 m^3 (with $\phi_o = \phi_w = 0.5$) monodisperse O/W Pickering emulsions (drop radius $R = 10$ μm) stabilized by Janus particles as a function of θ_p. Values assumed for the calculations are $\gamma_{ow} = 0.050$ N m^{-1}, $r = 100$ nm, and $\theta_a = 150°$. The curves are for the values of α indicated and are truncated at α at the low angle end and at θ_{trans} at the high end. Redrawn from Aveyard (2012).*

The average electrical energy of repulsion, U_{elec}, of a particle in a hexagonally packed monolayer of spherical particles is given by (Aveyard et al., 2003b)

$$U_{elec} = \frac{q^2}{2\pi\varepsilon_0\varepsilon_r}\left[\frac{1}{L} - \frac{1}{\sqrt{L^2 + 16\varsigma^2}} + \frac{1}{L}\ln\left\{\frac{1 + \sqrt{1 + 16\varsigma^2/L^2}}{2}\right\}\right] \qquad (18.3.2)$$

As explained in §17.4.3, the charge on a particle/oil interface is taken to be a point charge q a distance $\varsigma = r(3 + \cos\theta_p)/2$ from the lowest part of the particle/water interface as illustrated in Fig. 18.12. The particle is supposed to have a low relative permittivity ε_r, similar to that of the oil; ε_r is about 2–3 for polystyrene and 3–4 for silica, similar to that for alkanes of about 2. In (18.3.2), L is the separation between centres of neighbouring particles in the monolayer ($L = 2r$ for contacting particles in a hexagonally close-packed monolayer). The (positive) repulsive energy per particle must be added to $\Delta_{ads}G$ to give the net free energy of particle adsorption into the monolayer.

The effect on $\Delta_{emul}G$ of including electrical repulsion between particles is illustrated in Fig. 18 13: the assumed values for the various quantities are given in the figure legend.

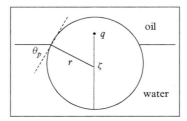

Figure 18.12 *A spherical particle with low relative permittivity resting at an oil/water interface. The charge on the particle is represented as a point charge q a distance ς from the lowest part of the particle/water interface.*

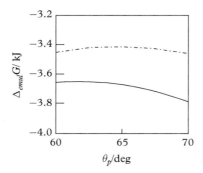

Figure 18.13 *Variation of $\Delta_{emul}G$ with θ_p for the formation of 1 m³ emulsion from equal volumes of oil and water with $\alpha = 60°$ and $\theta_a = 150°$, $\gamma_{ow} = 0.050$ N m⁻¹, r = 100 nm, and R = 10 μm. The lower curve is for $\sigma = 0$ and the upper curve for $\sigma = 800$ μC m⁻². Redrawn from Aveyard (2012).*

The surface charge density at the solid/oil interface of the adsorbed particles is taken to be 800 μC m^{-2}, which is at the upper end of anticipated values. Although lateral electrical repulsion between particles appears to have a significant effect on $\Delta_{emul}G$, the values still remain negative.

It is likely that in a nearly close-packed particle monolayer lateral repulsion considerably greater than the electrical repulsion considered exits, possibly arising from solvation (hydration) forces (§11.3.5) or solid elastic forces. The repulsive hydration energy, U_{hyd}, between *two* spherical particles can be expressed (Molina-Bolivar and Ortega-Vinuasa, 1999)

$$U_{hyd} = \pi r P_o \lambda^2 \exp(-S/\lambda) \tag{18.3.3}$$

in which λ is a decay length, P_o a hydration force constant and S the closest separation of particle surfaces. From experimental data, λ ranges from 0.2 to 1.1 nm and P_o (for hydrophilic surfaces) is in the range 10^6 to 5×10^8 N m^{-2}.

In the calculations described so far, hexagonal close-packing of spherical particles on spherical droplets has been assumed, although such packing might well be unattainable in practice. The effects of reducing the fractional coverage of droplets by particles is now explored, and in what follows it is convenient to consider the free energies, $\Delta^u G$, of forming *unit area of plane* particle-coated oil/water interface. The (in the context) small effects due to droplet surface curvature, van der Waals attractions between adsorbed particles and $\Delta_{ads}S$ (in (18.3.1)) are ignored, in which case $\Delta^u G$ can be taken to be proportional to $\Delta_{emul}G$, and is given by

$$\Delta^u G = \gamma_{ow} + \left(\Delta_{ads}G + U_{elec} + U_{hyd}/2\right) / \left(2r'^2\sqrt{3}\right)$$
$$= \gamma_{ow} + \Delta^u_{ads}G + \Delta^u_{elec}G + \Delta^u_{hyd}G \tag{18.3.4}$$

where

$$r' = L/2 = r/\sqrt{x} \tag{18.3.5}$$

and $\Delta^u_{ads}G$ is the adsorption free energy per unit area; L is the separation between centres of neighbouring particles. The number of particles per unit area at hexagonal close-packing, n_{cp}, is $1/(2r^2\sqrt{3})$, and the fractional surface coverage by particles, x, is the ratio of the number of particles in unit area to n_{cp}.

Some values of $\Delta^u G$ are shown in Fig. 18.14 as a function of surface coverage by Janus particles, together with the components $(\gamma_{ow} + \Delta^u_{ads}G)$, $\Delta^u_{elec}G$, and $\Delta^u_{hyd}G$. The hydration contribution has been calculated using $\lambda = 1.1$ nm and $P_o = 5 \times 10^8$ N m^{-2}. For $\theta_p = 60°$, for example, only 75 per cent of the particle surface is immersed in the aqueous phase and so as a crude approximation $\Delta^u_{hyd}G$ used in the figure is 0.75 that given by use of (18.3.3). The largest contributors to $\Delta^u G$ are γ_{ow} (positive) and $\Delta^u_{ads}G$ (negative). In the range of x from 0.9 to 1, the closest separation of particle surfaces, S, varies from about 11 nm to zero and in this interval the electrical contribution to $\Delta^u G$ barely changes. On the other hand, the very short-range hydration force results in

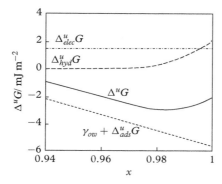

Figure 18.14 *Free energy of forming unit area of plane particle-covered oil/water interface, $\Delta^u G$, as a function of fractional surface coverage x (full line). Dashed curves are for the components of $\Delta^u G$ as indicated. Parameters assumed are: $r = 100$ nm, $\gamma_{ow} = 0.050$ N m^{-1}, $\alpha = 50°$, $\theta_a = 150°$, $\theta_p = 60°$, and $\sigma = 800$ μC m^{-2}. Parameters used in the calculation of $\Delta^u_{hyd} G$ are (see text) $\lambda = 1.1$ nm and $P_o = 5 \times 10^8$ N m^{-2}. Redrawn from Aveyard (2012).*

a value of $\Delta^u_{hyd} G$ of virtually zero (Box 18.2) until $x \approx 0.96$, when it then begins to rise exponentially with x. Using the parameters given in the legend to Fig. 18.14, the total free energy of forming unit area of interface becomes negative and exhibits a minimum of approximately -3 mJ m^{-2} at around $x = 0.99$, and is zero for $x \sim 0.923$ (not shown in the figure).

Box 18.2 How low can you get?—tension around a particle-coated droplet

It has been shown that values of $\Delta_{emul} G$ and $\Delta^u G$, calculated using (18.3.1) and (18.3.4) respectively, can be negative and this has been taken to mean that thermodynamically stable Pickering emulsions could possibly exist. The quantity $\Delta^u G$ is the effective interfacial tension, now designated γ_{eff}, of a particle coated oil/water interface; negative $\Delta^u G$ means that the tension of the interface is negative. So, what does this mean in practice?

As seen in §17.4.1, the surface coverage of a plane particle monolayer at an oil/water interface can be varied experimentally using a Langmuir trough. By analogy with the commonly employed treatment of surfactant systems (Chapter 5) the surface pressure π of the particle monolayer is written

$$\pi = \gamma_{ow} - \gamma_{eff} \tag{18.3.6}$$

where γ_{ow} is the interfacial tension of the oil/water interface in the absence of particles. It has been found experimentally that the 'collapse' of a particle monolayer under lateral compression (at $\pi = \pi_c$) consists (in the systems studied) of monolayer folding (Fig. 17.25), which involves an expansion of the interface. Onset of collapse occurs when the magnitude

Box 18.2 *Continued*

of the applied surface pressure becomes equal to the interfacial tension of the particle-free interface, that is, when γ_{eff} is zero (Fig. 17.26).

For the system represented in Fig. 18.14, $\Delta^u G = 0$ for $x \sim 0.923$, and negative for x greater than this. In real systems compression of particle layers around emulsion droplets could conceivably occur as a result of droplet coalescence, resulting in γ_{eff} becoming negative. Negative interfacial tension would result in a tendency for the droplet surface to expand which, if particles remained in place (as expected if the particle adsorption free energy is large relative to kT) would lead to non-spherical droplets, possibly similar to the one depicted in Fig. 18.2. In general however, it is observed that drops in Pickering emulsions are spherical implying the droplet interfacial tension is zero or positive. Note that the particles coating the spherical droplet in Fig. 18.2 are not close-packed.

Interestingly Wu et al. (2016) have prepared Pickering emulsions with non-spherical droplets, using a variety of oils and cyclodextrins as emulsion stabilizer. The cyclodextrin forms small crystals in situ at the oil/water interface, which act as the stabilizers. As an example of the emulsions formed, those with liquid paraffin as the oil phase had broadly spheroidal droplets. This raises the prospect of designing emulsions with required droplet shapes, although more needs to be understood about the origins of droplet morphology.

Work described above relates to emulsions stabilized by *spherical* Janus particles. Tu et al. (2013) made calculations of the free energies of Pickering emulsion formation assuming *dumbbell-shaped* Janus particles and came to the similar conclusion that such emulsions could be thermodynamically stable.

18.4   Foams, dry water, liquid marbles, and other structures

In the chapter so far only solid-stabilized emulsions have been considered, that is, dispersions of one liquid coated with particles in a second immiscible liquid. Other possibilities exist however in which gas replaces one of the liquids, and some fascinating systems can result. Liquid droplets in, for example, air as well as air bubbles in liquid can be stabilized by particles. The stability of thin liquid/liquid films in the presence of adsorbed solid particles has already been discussed in §17.5. Also, thin liquid films can be stabilized by particles *within* the films giving rise to oscillatory structural forces as discussed in §11.3.4, although the concern here is with systems with *adsorbed* particles.

Froth flotation of minerals has long been known in which air bubbles flow through an aqueous phase capturing solid particles at their surfaces (§14.10.1). These dynamic 'foams' however are not stable and break when the passage of air bubbles ceases. It is only relatively recently that very long-lived solid-stabilized foams have been prepared.

The main factors in such stabilization appear to be particle hydrophobicity, shape, size and concentration (Horozov, 2008).

Stable foams have been prepared by Algarova et al. (2004) using polymer microrods with average length 23.5 μm and diameters less than 1 μm. The contact angle of the particles (through water) at the air/water interface was approximately 80°. Foams could be formed by simply shaking about a 1 wt % aqueous dispersion of the particles and were stable for several weeks even when open to the air. The bubbles in the foams were said to be sterically stabilized by thick 'hairy' polymer layers that formed rigid protective shells around the bubbles and did not allow the formation of thin aqueous films between bubbles. That the relative hydrophobicity of the particles was important in the stabilization process was shown by adding sodium dodecylsulfate (SDS) to the foams. This made the particle more wetted by water and lowered the contact angle, which led to foam breakdown within 30 min. It is interesting that the microrods used mimic hydrophobic filamentous bacteria that cause foams in waste water treatment. The bacteria have a diameter of the order of 1 μm and lengths between 50 and 200 μm.

Although the filament-like particles are so effective in foam stabilization, highly stable aqueous foams can also been formed using spherical silica nano- and microparticles (Binks and Horozov, 2005). Foams that are stable to collapse, disproportionation and coalescence can be prepared by tailoring the particle hydrophobicity.

Particle-stabilized aqueous foams can be transformed into water-in-air powder—an intriguing kind of system referred to as 'dry water' (Binks and Murakami, 2006). Dry water consists of (say) 95 per cent water as droplets stabilized by adsorbed particles in air, and has the appearance of, and can be handled just like, a dry powder. These systems have interesting potential uses since, *inter alia*, the water droplets can take up gases such as carbon dioxide and methane (Wang et al., 2008).

Single macroscopic drops of liquid stabilized by hydrophobic solid particles, on a liquid or solid substrate, have long been known and are referred to as liquid marbles (Aussillous and Quéré, 2006). A liquid marble can be simply prepared by placing a drop of water on a pile of hydrophobic particles; the drop runs down, spontaneously picking up particles as it goes (Boys, 1959). The particles can be natural substances such as soot or lycopodium, or synthetic hydrophobic particles such as hydrophobized silica. Particle sizes are in the range 0.1–100 μm and the drop sizes are much larger, of the order of millimetres. An interesting example of naturally occurring liquid marbles is provided by aphids that roll their excrement in particles to get rid of it.

The high kinetic stability of emulsions and foams formed with solid particles is exploited in the formation of various kinds of structures (Studart et al., 2007). Porous materials can be formed by drying and sintering solid-stabilized wet foams and emulsions. Porous polymeric, metallic and ceramic systems are produced in this way, for example, aluminium foams stabilized by silicon carbide particles at high temperatures (>660°C). It is possible to tailor make macro-porous materials with open or closed cells of varying size.

....

REFERENCES

R.G. Algarova, D.S. Warhadpande, V.N. Paunov, and O.D. Velev, Foam superstabilization by polymer microrods. *Langmuir*, 2004, **20**, 10371–10374.

P. Aussillous and D. Quéré, Properties of liquid marbles. *Proc. R. Soc. A*, 2006, **462**, 973–999.

R. Aveyard, Can Janus particles give thermodynamically stable Pickering emulsions? *Soft Matter*, 2012, **8**, 5233–5240.

R. Aveyard, B.P. Binks, and J.H. Clint, Emulsions stabilised solely by colloidal particles. *Adv. Colloid Interface Sci.*, 2003a, **100–102**, 503–546.

R. Aveyard, J.H. Clint, and T.S. Horozov, Aspects of the stabilisation of emulsions by solid particles: effects of line tension and monolayer curvature energy. *Phys. Chem. Chem. Phys.*, 2003b, **5**, 2398–2409.

C.C. Berton-Carabin and K. Schroën, Pickering emulsions for food applications: background, trends, and challenges. *Annu. Rev. Food Sci. Technol.*, 2015, **6**, 12.1–12.35.

B.P. Binks, Particles as surfactants—similarities and differences. *Curr. Opin. Colloid Interface Sci.*, 2002, **7**, 21–41.

B.P. Binks and T.S. Horozov, Aqueous foams stabilized solely by silica nanoparticles. *Angew. Chem. Int. Ed.*, 2005, **44**, 3722–3725.

B.P. Binks and S.O. Lumsdon, Catastrophic phase inversion of water-in-oil emulsions stabilized by hydrophobic silica. *Langmuir*, 2000a, **16**, 2539–2547.

B.P. Binks and S.O. Lumsdon, Transitional phase inversion of solid-stabilized emulsions using particle mixtures. *Langmuir*, 2000b, **16**, 3748–3756.

B.P. Binks and S.O. Lumsdon, Influence of particle wettability on the type and stability of surfactant-free emulsions. *Langmuir*, 2000c, **16**, 8622–8631.

B.P. Binks and R. Murakami, Phase inversion of particle-stabilized materials from foams to dry water. *Nat. Mater.*, 2006, **5**, 865.

B.P. Binks, A.K.F. Dyab, and P.D.I. Fletcher, Multiple emulsions stabilised solely by nanoparticles. In: *Proceedings of the 3rd World Congress on Emulsions*, Lyon, France, 2002.

C.V. Boys, *Soap Bubbles*. Dover, New York, 1959.

P. Finkle, H.D. Draper, and J.H. Hildebrand, The theory of emulsification. *J. Am. Chem. Soc.*, 1923, **45**, 2780–2788.

D.J. French, A.T. Brown, A.B. Schofield, J. Fowler, P. Taylor, and P.S. Clegg, The secret life of Pickering emulsions: particle exchange revealed using two colours of particle. *Sci. Rep.*, 2016, **6**, 31401.

T.S. Horozov, Foams and foam films stabilised by solid particles. *Curr. Opin. Colloid Interface Sci.*, 2008, **13**, 134–140.

D.J. Kraft, B. Luijjes, J.W.J. de Folter, A.P. Philipse, and W.K. Kegel, Evolution of equilibrium Pickering emulsions: a matter of time scales. *J. Phys. Chem. B*, 2010, **114**, 12257–12263.

S. Levine and B.D. Bowen, Capillary interaction of spherical particles adsorbed on the surface of an oil/water droplet stabilized by the particles. Part I. *Colloids Surf.*, 1991, **59**, 377–386.

J.A. Molina-Bolivar and J.L. Ortega-Vinuasa, How proteins stabilize colloidal particles by means of hydration forces. *Langmuir*, 1999, **15**, 2644–2653.

S.U. Pickering, Emulsions. *J. Chem. Soc., Trans.*, 1907, **91**, 2001.

W. Ramsden, Separation of solids in the surface-layers of solutions and 'suspensions' (observations on surface-membranes, bubbles, emulsions, and mechanical coagulation). Preliminary account. *Proc. R. Soc. London*, 1903, **72**, 156–164.

S. Sacanna, W.K. Kegel, and A.P. Philipse, Thermodynamically stable Pickering emulsions. *Phys. Rev. Lett.*, 2007, **98**, 158301.

T. Sekine, K. Yoshida, F. Matsuzaki, T. Yanaki, and M. Yamaguchi, A novel method for preparing oil-in-water-in-oil type multiple emulsions using organophilic montmorillonite clay mineral. *J. Surfactants Deterg.*, 1999, **2**, 309–315.

A.R. Studart, U.T. Gonzenbach, I. Akartuna, E. Tervoort, and L.J. Gaukler, Materials from foams and emulsions stabilized by colloidal particles. *J. Mater. Chem.*, 2007, **17**, 3283–3289.

F. Tu, B.J. Park, and D. Lee, Thermodynamically stable emulsions using Janus dumbbells as colloid surfactants. *Langmuir*, 2013, **29**, 12679–12687.

W. Wang, C.L. Bray, D.J. Adams, and A.I. Cooper, Methane storage in dry water gas hydrates. *J. Am. Chem. Soc.*, 2008, **130**, 11608–11609.

L. Wu, Z. Liao, M. Liu, X. Yin, X Li, M. Wang, X. Lu, N. Lv, V Singh, Z. He, H. Li, and J. Zhang, Fabrication of non-spherical Pickering emulsion droplets by cyclodextrins mediated molecular self-assembly. *Colloids Surf. A*, 2016, **490**, 163–172.

Y. Yang, Z. Fang, X. Chen, W. Zhang, Y. Xie, Y. Chen, Z. Liu, and W. Yuan, An overview of Pickering emulsions: solid-particle materials, classification, morphology, and applications. *Front. Pharmacol.*, 2017, **8**, 287.

Themes and connections

The way in which the material in the book has been assembled into, and arranged within, chapters is to some extent arbitrary since topics can often be grouped and ordered in a variety of useful ways, depending on the interests of the reader. In part this is catered for in the text by providing cross references, and of course connections can be made by referring to the index of the book. However, in order to help the reader a little more explicitly, a series of major topics has been selected and relevant material to be found both within a chapter or section as well as throughout the book is referred under the topic headings. A student of surface chemistry or a research worker embarking on an industrial project involving the use of surfactants, or indeed someone assigned to produce a university or industrial course in surface chemistry or surfactant science, may wish to understand in a general way something about, for example, interfacial tension, capillarity, the curvature properties of surfactant monolayers or the concept of the hydrophile-lipophile balance (HLB). The material referred to below under a given topic (which is illustrative rather than exhaustive) is what might have been included in a chapter devoted to that topic. It is hoped that the presentation of some of the material in this way will help to give the reader a broader and more coherent understanding of the subject matter. Topics selected are:

- Aggregation and phase behaviour of surfactants
- Capillarity and curved liquid interfaces
- Contact angle
- Curvature properties of close-packed monolayers and some consequences
- Emulsions
- The hydrophile–lipophile balance (HLB)
- Interfacial tension
- Forces within monolayers and between surfaces
- Solid particles and particle monolayers at fluid interfaces
- Surfactant adsorption and adsorbed layers
- Wetting, spreading, and thin liquid films

Micelle formation in aqueous solution

Micelle formation occurs over a very narrow range of surfactant concentrations around the *critical micelle concentration, cmc*. Micelle formation can be treated using a *mass action model* (§9.2.1) or by supposing it is analogous to *condensation* (§9.2.2). *Standard thermodynamic quantities of micelle formation* can be obtained from *cmc* values obtained over a range of temperature using either approach.

Cmc and surfactant molecular structure

The value of the *cmc* depends on *surfactant molecular structure* (§9.3). The *driving force for micelle formation* is the dehydration of the hydrophobic surfactant tail groups (§2.2). For a given head group the *cmc* falls by roughly a factor of three for each additional methylene group in a hydrocarbon tail group. The more hydrophilic the head group the higher the *cmc* (Table 9.1).

Effects of inert electrolyte

Low concentrations of inert electrolyte affect the *cmc* and *degree of dissociation of ionic* surfactants (§9.6), and cause Winsor transitions in oil+water systems (§10.2.4). Much higher salt concentrations are required to promote Winsor transitions in the presence of *nonionic* surfactants (§10.2.4).

Surfactant packing factor

The surfactant *packing factor P* is defined as the ratio of the effective cross-sectional areas of head and tail groups and is related to the *preferred curvature* of close-packed surfactant monolayers (§9.5). Its value determines the *shape of micelles/aggregates* (§9.5) as well as the *size and type of microemulsion* drops in Winsor systems (§10.1.2). Preferred monolayer curvature and Winsor behaviour are associated with the *system HLB* and *preferred emulsion type* (§14.3).

AGGREGATION AND PHASE BEHAVIOUR OF SURFACTANTS

A molecular thermodynamic approach to micellization

A simple *molecular thermodynamic approach* in which, conceptually, the micelle is formed by condensation of chains to give the micelle core and then adding head groups to the core surface, can describe micelle formation and predict reasonable *micellar size distributions* (§9.4).

Effects of HLB variables

In systems with comparable amounts of oil and water, aggregation (to give micelles or *microemulsion droplets*) can occur in either phase depending on, for example, T, salt concentration and alkane chain length (for alkane oils) (§10.1.2). These variables are termed *HLB variables* (§14.3) and determine preferred monolayer curvature, and hence Winsor phase behaviour (§10.1.2) and favoured emulsion type (§14.3).

Mixed surfactant micelles

In systems with two (or more) surfactants present in solution *mixed micelles* can form. The *cmc* varies with the mol fractions of surfactants present (§9.7). Binary mixed micelles have been treated empirically as *regular solutions* which yields an interaction parameter β that reflects the nature of the interactions between the different surfactant head groups in the micelle.

Dynamic nature of micelles

Micelles are dynamic structures and exhibit two relaxation times, one related to monomer exchange and the other to micellar breakdown (§9.10). *Microemulsion drops* are also dynamic structures and droplets can fuse and break up and they exhibit shape fluctuations and polydispersity (§10.3).

Solubilization in micelles

Micelles in water can *solubilze* nonpolar material in their cores; amphiphilic material can also be incorporated into micelles (§9.8.1). Solubilization *changes the cmc* (§9.8.2). Large amounts of oil (water) can be solubilized in water (oil) containing surfactant (§9.8.3). Solubilization can change micelle *shape* (§9.8.3). Micelles and surfactant molecules can also associate with polymer molecules (§9.9).

Ways of expressing phase behaviour

There are various ways of *expressing phase behaviour*, depending on systems of interest. In (water + surfactant) systems over a wide concentration range, phases are often noted on diagrams with composition as abscissa and T as ordinate. Phase behaviour in 3-component systems is shown on triangular phase diagrams (§9.11; §10.4.1), and behaviour as a function of T is then illustrated in a triangular prism (§9.11; §10.4.2). Phase behaviour in systems with varying amounts of oil and water with a fixed overall amount of surfactant can be usefully expressed, as a function of temperature, in so-called Shinoda diagrams (§10.4.2).

Hemimicelles

Hemimicelles can form at solid/liquid interfaces under appropriate conditions (§7.5.3).

What is capillarity?

Capillarity (§3.1) is a manifestation of *surface or interfacial tension* (§3.2) that maintains mechanical equilibrium between two contacting fluids, with a curved interface, at different pressures.

Capillary rise and solid/fluid tensions

Capillary rise or depression is related to the relative values of solid/air and solid/liquid interfacial tensions, and hence to the *work of cohesion* of a liquid and the *work of adhesion* between liquid and solid (§3.3)

The Laplace and Kelvin equations

Liquid menisci and liquids in capillary tubes exhibit *curved interfaces*. There is a pressure drop across a curved liquid/fluid interface (*Laplace equation*, §3.4.1). The vapour pressure of a liquid depends on the curvature of its surface (*Kelvin equation*, §3.4.2).

CAPILLARITY AND CURVED LIQUID INTERFACES

See also Curvature properties of close-packed monolayers and some consequences

Capillary rise and intermolecular forces

The observation of *capillary rise/depression* played a crucial role in the understanding of *intermolecular forces* (§3.1). Measurement of capillary rise can be used to determine surface tension of pure liquids (§3.4.1).

Nucleation of liquids and capillary condensation

The *nucleation of supersaturated vapours* to form droplets, as well as *capillary condensation*, are influenced by surface tension of the liquid and the drop size or surface curvature (§3.4.2). Aspects of micelle formation are analogous to nucleation of liquids (§9.4).

Capillarity and measurement of interfacial tension

Capillarity is intimately associated with *experimental methods* used for the determination of surface and interfacial tensions in fluid/fluid systems (§3.4.1 and Appendix A3).

Ostwald ripening and bubble disproportionation

Ostwald ripening in emulsions (§14.6.4) and *bubble disproportionation* in wet foams (§14.10.2) arise from the curvature of droplet and bubble interfaces and the pressure difference on the two sides of the curved interfaces.

What is the contact angle?

The *contact angle θ* of a *liquid in contact with a solid* is the macroscopic angle that the liquid makes with the solid at the three phase contact line (§3.3; §16.3.1; §17.1).

Contact angle between a thin liquid film and the adjoining meniscus

Since the tension of a surface of a thin liquid film (in contact with its meniscus) differs from that of the bulk solution (as the result of the action of surface forces across the film), a *contact angle exists between the surface of a film and the adjoining meniscus* (§12.2.1).

Advancing and receding contact angles and contact angle hysteresis

The contact angle often referred to is the 'static' or 'equilibrium' contact angle. But the contact angle observed when a liquid is advancing over a solid surface usually differs from that observed for a receding liquid. Thus both *advancing* and *receding contact angles*, and hence *contact angle hysteresis*, can be observed (§16.3.2). Hysteresis occurs when the solid surface is chemically heterogeneous and/or rough (§16.3.5).

CONTACT ANGLE

Lens angle for a lens resting on a liquid substrate

For a small *drop of a non-spreading liquid resting on a second immiscible liquid*, the relevant angle is the *lens angle*, which depends on the various interfacial tensions in the system, as well as on gravity (§16.2.3).

Spreading coefficient and work of adhesion; the Young and Young–Dupré equations

The contact angle is related to the *spreading coefficient S* of a liquid on a solid surface (§16.3.1) and to the *work of adhesion W_A* between a solid and a liquid, via the *Young–Dupré equation*, and to the various interfacial tension in the system via *Young's equation* (§3.3).

Contact angle and surfactant adsorption

Addition of surfactant can modify the contact angle of a solution with a solid, and the angle usually changes with surfactant concentration as a result of adsorption at the various interfaces present, and the resultant changes in the interfacial tensions (§16.3.4).

Measurement of contact angle

The contact angle for a liquid in contact with a solid surface can be *measured* in a variety of ways using either *force tensiometry* or *goniometry* (observation of the profile of a sessile drop (§16.3.2).

Surface roughness and superhydrophobic surfaces

Surface texture (roughness) can give rise to *superhydrophobic surfaces* (apparently very high contact angle) (§16.3.5).

The precursor film

The contact angle *varies with time during spreading* of a drop of liquid on a solid surface, and the drop is preceded by a thin *precursor film* (§16.4.1; §16.4.2).

The surfactant packing factor *P*

The surfactant *packing factor P* is defined as the ratio of the *effective* cross-sectional areas of surfactant tail and head groups (§9.5.1). *P* depends on prevailing conditions such as *T*, salt concentration and oil molecular structure in oil + water systems (§10.2.1).

Distribution of surfactant between oil and water

In dilute surfactant systems with oil and water present (*Winsor systems*) surfactant monomers and aggregates (micelles or microemulsion droplets) *distribute between the oil and aqueous phases* independently. The phase in which aggregates appear is determined by surfactant monolayer curvature properties (i.e. by the packing factor (§10.1.1) and hence by the HLB variables (§10.1.2; §14.3).

P and the system HLB

The *system HLB* does not depend on the surfactant molecular structure alone but also on the so-called *HLB variables* (§14.3). These variables are the ones that affect effective molecular structure and hence the surfactant packing factor and Winsor transitions (§10.1.2).

P and the size and shape of surfactant aggregates

Monolayer curvature properties (hence the value of the packing factor) determine *the shape (and size) of micellar aggregates* in dilute aqueous solution (§9.5) and also (in part) the structure of *mesophases in concentrated surfactant systems* (§9.11). Values of the packing factor also determine the *size and type of microemulsion droplets* formed in Winsor systems (§10.2.1).

Variation of tension at the *cac* with HLB variables

The rate of change of the oil/water interfacial tension γ_c (attained at the *cac*) with an HLB variable (e.g. *T*, salt concentration, amount of cosurfactant present), reflects the difference in the value of the appropriate quantity associated with the flat and spontaneously curved oil/water interface (§10.2.4).

CURVATURE PROPERTIES OF CLOSE-PACKED MONOLAYERS AND SOME CONSEQUENCES

See also The hydrophile–lipophile balance (HLB) and Surfactant adsorption and adsorbed layers

---

Curvature energy of close-packed monolayers

A close-packed surfactant monolayer at an oil/water interface has a *spontaneous curvature* and energy is required to change the curvature. This *curvature energy*, per unit area, g_c is related to the *bending elastic moduli* κ and $\bar{\kappa}$ by the *Helfrich equation* (§10.2.2). The modulus κ is often of the order of kT (Table 10.2). The bending elastic moduli are related to the structures of the surfactant-rich phases in WIII systems (§10.2.3).

HLB and Bancroft's rule

The *hydrophile–lipophile balance* (HLB) is an empirical concept which assigns an *HLB number* to a surfactant based on its molecular structure. *High HLB surfactants stabilize O/W emulsions and low HLB W/O emulsions*. The HLB concept is associated with *Bancroft's rule*: the favoured emulsion type is that in which surfactant is 'soluble' in the 'continuous phase' (§14.3). A modern understanding is that the emulsion type is that in which the *surfactant aggregates are present in the continuous phase*. Aggregates may be only a small fraction of total surfactant present in the system (§10.1.1).

Ultralow oil/water interfacial tensions

The work of forming an interface arises from extension of the surface and then bending it away from its spontaneous curvature. *Ultralow oil/water interfacial tensions* attained in the Winsor III region arise since the work of extension is very low and because the spontaneous curvature is close to zero. The very low tensions in the WI and WII regions are taken to arise from bending the surfaces away from the spontaneous curvature to a flat configuration (§10.2.2).

P for spherical particles in monolayers

As for surfactant molecules, a *packing factor* can be defined for *spherical solid particles* at an oil/water interface; *P* is related to the contact angle, θ, of the particle with the oil/water interface. Hydrophilic particles ($\theta < 90°$; $P < 1$) tend to stabilize O/W emulsions, and hydrophobic particles ($\theta > 90°$; $P > 1$) tend to stabilize W/O emulsions (§18.1).

What are emulsions?

Emulsions consist of *drops of one liquid dispersed in another immiscible liquid* (dispersion medium) (§14.1). Drops are *stabilized against coalescence* by adsorption of low molar mass surfactant, polymer or, in *Pickering emulsions*, by small adsorbed solid particles (Chapter 18).

Surfactant hydrophile-lipophile balance (HLB)

HLB is an empirical concept which assigns a *number* to a *surfactant molecule* on the basis of its molecular structure. *High HLB* surfactants tend to stabilize O/W emulsions and *low HLB* W/O emulsions. The concept is associated with *Bancroft's rule (qv)*. However, it is not the surfactant structure alone that determines emulsion type. The *nature of the oil phase*, *salt concentration in the aqueous phase*, and *temperature* (so-called *HLB variables*) are also implicated. It is preferable to speak of low or high HLB *systems* rather than surfactants (§14.3).

Surface forces and emulsions

Surface forces acting across a thin emulsion film can include dispersion forces (§11.2.1), electrical double layer (repulsive) interactions (§11.2.3), steric forces (§11.3.1), forces resulting from bridging of surfaces (§11.3.2), depletion forces (§11.3.3), oscillatory structural forces, solvation forces and hydrophobic forces (§11.3.4), The DLVO theory of colloid stability considers only dispersion and electrical interactions, but the other forces are often present in emulsion systems. Unlike solid particles, emulsion droplets are deformable (Chapter 14, Box 14.2) and can adhere across a thin emulsion film; the film has a contact angle with the adjacent bulk drop surface (§14.6.2).

Preparation of emulsions

There are various procedures, including the use of *comminution* (§14.2.1) and *condensation* methods (§14.2.2), as well as the so-called *PIT method* for preparation of *nanoemulsions* (§14.2.4).

EMULSIONS

See also HLB, curvature properties of close-packed monolayers; forces between surfaces; solid particles; thin liquid films.

System HLB and packing factor *P*

These are associated with *effective surfactant molecular shape*, which is described by a surfactant *packing factor P* (the ratio of the cross-sectional areas of surfactant tail and head groups) (§9.5.1). For *Winsor I systems*, which form O/W emulsions, $P < 1$; *Winsor II* systems form W/O emulsions and $P > 1$ (§10.2.1). Winsor transitions, using *a given surfactant*, are effected by *changing an HLB parameter* (§14.3). The *spontaneous curvature (sign and value) of a close-packed surfactant* monolayer is determined by P (§10.2.1).

Packing factor of spherical particles

As for surfactant molecules, a *packing factor* can be defined for *spherical solid particles* at an oil/water interface; P is related to the contact angle, θ, of the particle with the oil/water interface. Hydrophilic particles ($\theta < 90°$; $P < 1$) tend to stabilize O/W emulsions, and hydrophobic particles ($\theta > 90°$; $P > 1$) tend to stabilize W/O emulsions (§18.1).

Which type of emulsion is formed?

The empirical *Bancroft's rule* states the *favoured emulsion type* is that in which the *surfactant is 'more soluble' in the continuous phase* (§14.3). A *modern version* is that the favoured emulsion is that in which *aggregated* surfactant resides in the continuous phase. In some systems the bulk of the surfactant may be present as *monomers in the droplet phase* (§10.1.1 and §10.1.2). Ultimately the favoured emulsion type depends on the preferred curvature of the close-packed monolayers adsorbed at drop surfaces (§14.4).

The (kinetic) stability of emulsions

Kinetic stability (§14.1) is very dependent on the stability of the thin *emulsion films* that separate droplets in close proximity (§12.2). Thin film stability depends on the way in which *surface forces* (*qv*) across a film vary with film thickness, which is conveniently discussed in terms of a *disjoining pressure isotherm* (§12.2.2).

Time evolution of emulsion structure

Emulsion structure evolves in various ways including *sedimentation (creaming), droplet aggregation and coalescence* as well as *Ostwald ripening* (§14.6). Emulsions can also be *inverted* from O/W to W/O and vice versa by various means (§14.5).

Emulsion breaking

Unwanted emulsions are '*broken*' by causing droplets to coalesce for example, by changing the HLB variables so that the conditions no longer favour the unwanted emulsion type (§14.7).

HLB number of surfactant molecule

The *hydrophile–lipophile balance* concept was originally an empirical one based on surfactant molecular structure alone (§14.3). The hydrophilicity of the head group and hydrophobicity of the surfactant tail group are expressed as numbers, H and L respectively, which are suitably combined to give an *HLB number* for the surfactant (§14.3.1). A surfactant with a high HLB number *tends to* stabilize O/W emulsions and one with a low HLB number stabilizes W/O emulsions.

Distribution of surfactant between oil and water

Surfactant monomers and micelles distribute independently between oil and water. *Aggregates* form in the phase which allows the spontaneous curvature of the surfactant layer to be attained. Monomers distribute independently of aggregates which may be predominantly in a different phase to the aggregates (§10.1.1). Monomer distribution is not obviously related to the relative solubility of surfactant in oil or water in the absence of the other phase.

System HLB

System HLB in systems with comparable volumes of oil and water together with surfactant above the *cac*, is related to the *locus of formation of the surfactant aggregates*. In Winsor I systems for example the aqueous phase contains O/W microemulsion droplets and the favoured emulsion type is also O/W; the continuous phase of the emulsion contains the aggregates (§10.1.2).

Bancroft's rule

Bancroft's rule: the favoured emulsion type given in a system with comparable volumes of oil and water present with surfactant stabilizer is that in which the surfactant is more 'soluble' in the continuous phase of the emulsion (§14.3).

**THE HYDROPHILE–
LIPOPHILE BALANCE (HLB)**

See also Curvature properties
of close-packed monolayers
and some consequences

Why does preferred monolayer curvature determine emulsion type?

Emulsion droplet surfaces have *very low curvature* relative to that of microemulsion droplets. However, the curvature properties of surfactant mono-layers determine the preferred emulsion type through the net *curvature of the growing neck formed between two coalescing droplets* (§14.4).

HLB and Bancroft's rule

The *HLB concept* (*qv*) takes no account of the prevailing conditions (e.g. salt concentration in the water phase or the temperature of the system) or the chemical structure of the oil molecules. *Bancroft's rule* makes no distinction between the way in which surfactant monomers and micelles distribute between oil and aqueous phases (§14.3).

The packing factor and HLB variables

The preferred curvature of close-packed surfactant monolayers at the oil/water interface can be discussed in terms of the *packing factor*, P, which is the ratio of the effective cross-sectional areas of the chain and head groups of the surfactant (§9.5.1). The cross-sectional areas depend upon the *HLB variables* such as aqueous phase salt concentration, T, nature of the oil phase and presence of cosurfactant or second surfactant (§10.1.2; §10.2.4; §14.3).

P and HLB variables in relation to microemulsion and emulsion type

For *Winsor I systems*, which form O/W emulsions when homogenized, the packing factor $P < 1$. *Winsor II* systems form W/O emulsions and $P > 1$. Winsor transitions, using a given surfactant, are effected by *changing an HLB parameter*. Clearly, assignment of an *HLB number* (*qv*) to the surfactant in these circumstances is not useful (§14.3).

Definition of surface tension

Definition for liquids (§3.2) and *solids* (§16.3.3). For fluid/fluid interfaces surface (interfacial) tension is the isothermal reversible work needed to form unit area of interface. For solids with non-equilibrium surfaces it can be defined as the force/unit length in the surface (§16.3.3).

Measurement of surface (interfacial) tension

Techniques for the measurement of surface and interfacial tensions of fluid interfaces are described in Appendix 3. *Changes* in the 'surface tension' of a solid caused by adsorption can be obtained from a knowledge of adsorbed amounts (§16.3.3).

Components of interfacial tensions

Notionally interfacial tensions can usefully be split into contributions arising from various interactions that operate across the interface, for example, dispersion forces and 'polar' forces (§3.3 and §16.3.3).

INTERFACIAL TENSION

See also Capillarity and curved liquid interfaces

Entry, wetting, and spreading

The thermodynamic feasibility of a liquid drop in an immiscible bulk liquid entering the surface depends on the various interfacial tensions appearing in the *classical entry coefficient, E* (§12.3.2). It is possible that entry does not occur when feasible as a result of the presence of a *thin metastable film* between drop and bulk liquid surface. The various tensions, and the film tension in the metastable film are related to the *disjoining pressure* in the film (§12.3.2).

Surface tension and intermolecular forces

Relationship to intermolecular forces and to *capillarity* (*qv*). Molecules at surface have fewer neighbours and higher energy than those in bulk liquid (§3.2).

Thin liquid films and film tension

For thin liquid films (with mutually interacting surfaces) the tensions differ from the normal interfacial tensions. The film tension is the sum of the two interfacial tensions. Interactions between the surfaces can be described in terms of the *disjoining pressure* in the film (§12.2.2).

Work of cohesion and work of adhesion

The work to split a column of liquid of unit cross section to give two units of area of liquid/vapour interface is twice the surface tension, that is, *work of cohesion*. Similarly, the *work of adhesion* of two liquids is the sum of the two surface tensions less the interfacial tension between the liquids (§3.3). The work of adhesion between solid and liquid is related to the contact angle of the liquid with the solid by the Young–Dupré equation (§3.3).

Adsorption and interfacial tension

Interfacial tension is *lowered by adsorption* and increased by desorption according to the Gibbs adsorption equation (§4.3). Tensions can become *ultralow in Winsor II systems*, where the preferred *monolayer curvature is close to zero* (§10.1.3).

Dynamic surface tension, kinetics of adsorption and surface rheological properties

Adsorption at a newly formed interface occurs over a period of time, during which the interfacial tension falls. This tension is termed the *dynamic tension* (§6.1), and is related to surface dilational rheological properties (§6.2.2) and the kinetics of adsorption (§6.3).

Lateral forces within monolayers

Lateral forces between adsorbed surfactant molecules can arise from various sources including area exclusion, electrical repulsion between ionic head groups, chain–chain attraction, etc. The existence of forces is reflected in the forms of *adsorption isotherms and surface equations of state* (§5.3.2).

Lateral electrical repulsion

Electrical repulsion between ionic surfactant head groups at liquid/fluid interfaces can be modulated by adjustment of the concentration of *inert electrolyte* in the aqueous phase (§5.4.1).

Chain–chain attractive interactions

These interactions can be important in monolayers at the air/water interface (§5.3.2) and at the solid/water interface (§7.3.2).

Forces between a surface and surfactant head and tail groups

These forces contribute to the standard free energy of adsorption (§5.3.1) of a molecule. At oil/water and air/water interfaces head groups are hydrated in water. At air/water interfaces neighbouring surfactant chains can mutually attract (§5.3.2). Chains can interact with the surface of a hydrophobic solid in contact with water (§7.3.2). Ionic head groups can interact attractively or repulsively with hydrophilic solids at solid/water interfaces (§7.5.2).

Lateral forces in mixed surfactant monolayers

Attractive and repulsive interactions can occur between *surfactant head groups in mixed surfactant layers*. These interactions can be treated using either a two-dimensional regular solution model (§5.5.3) or two-dimensional van der Waals approach (§5.5.4).

FORCES WITHIN MONO-LAYERS AND BETWEEN SURFACES

See also Thin liquid films and Surfactant adsorption and adsorbed layers

Forces acting between surfaces in proximity

These forces are termed *surface forces* (Chapter 11) and depend on separation of the surfaces. Forces include dispersion forces (§11.2.1), electrical double layer (repulsive) interactions §11.2.3), steric forces (§11.3.1), forces resulting from bridging of surfaces (§11.3.2), depletion forces (§11.3.3), oscillatory structural forces, solvation forces and hydrophobic forces (§11.3.4).

Measurement of surface forces

Surface forces can be *measured directly* using the 'surface force apparatus' (SFA) and atomic force microscopy (AFM) (§11.5). The measurement of *disjoining pressures* in thin planar liquid films is discussed in §12.2.2.

DLVO theory of colloid stability

The way in which surface forces vary with separation of the surfaces of particles in dispersions determines whether or not the dispersions are (kinetically) stable. The *DLVO theory of colloid stability* considers the net force arising from attractive van der Waals forces and repulsive electrical double layer forces as a function of particle separation (§11.2).

Surface forces and stability of particulate dispersions and emulsions

A wider range of forces than considered in the DLVO theory is possible and can be exploited in the preparation of *stable solid particulate dispersions* (Chapter 13). A similar range of forces can act between emulsion droplets, but droplets, unlike solid particles can be deformable (§14.1) and can coalesce (§14.6.3).

Surface forces in thin liquid films

In *thin liquid films the two film interfaces interact* and the film tension differs from twice the corresponding macroscopic interfacial tensions (§12.1). An equilibrium symmetrical thin liquid film has a finite contact angle with its adjacent meniscus that is simply related to the *interaction energy across unit area of film* (§12.2.1).

Disjoining pressure and surface forces

Interactions across thin liquid films are conveniently discussed in terms of the *disjoining pressure* Π_D that the interactions generate normal to the film surfaces; the pressure varies with film thickness h, that is, with surface separation. A *disjoining pressure isotherm* ($\Pi_D(h)$ vs h) has a similar shape to a plot of the corresponding interaction energy curve $V(h)$ vs h (§12.2.2).

Adsorption of solid particles; adsorption free energy

Small solid particles *adsorb strongly at liquid/fluid interfaces*. The *adsorption free energy* has contributions from changes in area of solid/liquid interfaces and from the removal, on adsorption, of liquid/fluid interface (§17.1). For *very small particles* the existence of a *line tension* τ at the three phase contact line around an adsorbed particle can in principle affect the adsorption free energy significantly (§17.2). The free energy is more positive if τ is positive, and more negative if τ is negative.

The packing factor and preferred monolayer curvature

Close-packed monolayers of, say, spherical particles at an oil/water interface have a *preferred curvature*. An analogue of the *packing factor P* for surfactant molecules (§9.5.1) can be defined for a spherical particle in terms of the contact angle, θ, it makes with the oil/water interface (§18.1). The preferred curvature of a close-packed particle monolayer of spherical particles is zero for $\theta = 90°$ (oil and water phases wet the particle equally) and $P = \sin^2 \theta$ is unity.

Observation of particle monolayers on the Langmuir trough

If particles are large enough (say 1μm diameter for spherical particles) *monolayers of particles* at fluid interfaces can be *visualized microscopically*. The monolayers can be compressed or expanded using a *Langmuir trough* technique (§17.4.1). When particles are *close-packed*, monolayers can *fold* rather than eject particles (§17.4.4).

SOLID PARTICLES AND PARTICLE MONOLAYERS AT FLUID INTERFACES

Monolayer structure and lateral forces between adsorbed particles

The *structure in particle monolayers* (§17.4.2) is a reflection of the lateral interactions between the particles (§ 17.4.3). Highly *organized monolayers* can be formed as a result of lateral repulsion between particles *through the oil phase*. Lateral forces between two adsorbed particles at an oil/water interface can be measured directly using an optical tweezer technique (§17.4.3). Monolayer behaviour can be discussed in terms of *surface equations of state*, just as for surfactant monolayers (§5.3.2).

Thin liquid films stabilized by particles

Particles are able to *stabilize thin liquid films* (§17.5.1). Particle-stabilized films can be observed microscopically if the particles are sufficiently large (§17.5.2). Observed film properties depend on whether the films are O/W/O or W/O/W and if the particles are hydrophilic or hydrophobic (§17.5.3).

Contact angle, distribution of particles between oil and water, and the stabilization of emulsions

The *distribution of spherical particles between oil and water* is sharp for particles with radius 10 nm and larger (§18.2.1): for $\theta < 90°$ particles reside virtually all in the aqueous phase and for $\theta > 90°$ in oil. In systems with roughly equal volumes of water and oil, O/W emulsions are stabilized by particles with $\theta < 90°$ and W/O emulsions are formed if $\theta > 90°$ (§18.2.2). The situation is broadly analogous to that described by Bancroft's rule for surfactant-stabilized emulsions (§14.3).

Spherical Janus particles

Spherical *Janus particles* have areas of different wettability, say *a polar and an apolar cap*. Such particles are amphiphilic like surfactant molecules (§17.3.1). The introduction of an apolar (polar) cap onto a polar (apolar) particle increases the magnitude of the (negative) adsorption free energy for adsorption to oil/water interfaces (§17.3.2).

Why surfactants adsorb

The driving force for adsorption from aqueous solution to air/water or oil/water interfaces is *chain dehydration* (§2.2). At solid/fluid (water) interfaces various interactions are possible depending on the solid surface and surfactant molecular structure (§7.3).

Effect of adsorption on interfacial tension

Adsorption lowers surface/interfacial tension, according to the *Gibbs adsorption equation* (§4.3). Above, but close to the *cmc* of a surfactant solution the tension remains constant (§9.1) and the monolayer is *close-packed*. Adsorption in Winsor systems gives rise to *ultralow oil/water interfacial tensions* (§10.1.3). *Surface pressure*, π, of an adsorbed layer is equal to the lowering of the tension caused by adsorption (§4.4).

Dynamic surface tension and adsorption kinetics

Adsorption at a freshly formed interface is time-dependent and gives rise to *dynamic surface tensions* (§6.1). Various factors influence the *kinetics of adsorption* (§6.3).

Marangoni flow and stabilization of soap films

When part of a liquid surface is expanded rapidly surfactant flows from the neighbouring surface to even out the surface concentration. The flowing surfactant carries with it underlying liquid. This is termed *Marangoni flow*, and is important for the *stabilization of thin liquid films* (e.g. soap films) (§12.4), such as those in foams (§14.8).

SURFACTANT ADSORPTION AND ADSORBED LAYERS
See also Curvature properties of close-packed monolayers and some consequences, and HLB

The adsorption isotherm and standard free energy of adsorption

The *adsorption isotherm* equation describes the variation of surface concentration of surfactant with its bulk concentration (§5.3.2). The proportionality constant contains the *standard free energy of adsorption* $\Delta_{ads}\mu^\circ$ (§5.3.1). The disposition of adsorbed surfactant molecules at interfaces can be deduced from the way in which $\Delta_{ads}\mu^\circ$ varies with chain length in an homologous series of surfactants.

The surfactant packing factor, monolayer curvature and HLB variables

The relative values of head and tail group cross sectional areas of a surfactant in a close-packed monolayer often mean the monolayer *tends to bend*. The ratio of areas defines a *packing factor P* (§9.5.1), which depends not only on surfactant molecular structure but also on a range of so-called *HLB factors* (§10.2; §14.3). *Microemulsion drop size* depends on the tendency of the monolayers coating the droplets to bend (§10.2.2).

Surfactants, surface forces and dispersion stability

Surface forces between monolayer-covered interfaces in close proximity are largely governed by the monolayers rather than the underlying bulk phases(§11.2.1).Adsorbed surfactant layers around droplets and solid particles are often responsible for the kinetic stability of a dispersion (Chapters 11, 13, and 14).

Surface rheological behaviour

Fluid adsorbed layers offer resistance to shear and have *shear rheological properties* (§6.2.1). Adsorbed layers also exhibit surface *dilational rheological properties* (§6.2.2). Insoluble films are purely elastic whereas soluble films are viscoelastic, the dilational viscoelastic modulus being frequency dependent for an oscillatory change in surface area.

Surface equations of state

The variation of surface pressure with area per molecule is described by a *surface equation of state*. At fluid/fluid interfaces a monolayer can be treated either as a (mobile) two-dimensional gas or two-dimensional solution (§5.3.2). Monolayers at solid/fluid interfaces are often treated as being localized (§7.4). In either case lateral interactions can occur between adsorbed molecules (§5.3.2; §7.3.2).

Physical methods for monolayer structure

There is a variety of modern physical methods available for the direct probing of surfactant monolayer structure (Chapter 8).

Aggregation at interfaces and hemimicelles

Surfactant aggregation at interfaces is due to lateral forces between adsorbed molecules and can give rise to either two-dimensional condensed phases (§5.3.2; §7.5.1) (seen e.g. using BAM (§8.5)) or hemimicelles (which can be observed using AFM (§7.5.3).

Complete and incomplete wetting and partial non-wetting

For *incomplete wetting* of a solid by a liquid the contact angle $\theta < 90°$; for a *partially non-wetting* liquid $\theta > 90°$. *Complete wetting* ($\theta = 0°$) to give a *duplex film* is commonly observed but *complete non-wetting* ($\theta = 180°$) is rare (§3.3). Liquid droplets on *rough or crenelated surfaces* can however give very high contact angles (§16.3.5).

Disjoining pressure and the Hamaker constant

For wetting in a system where the only surface interactions arise from London van der Waals interactions (e.g. alkanes on water in air) the disjoining pressure is related to the effective Hamaker constant for the system and hence to the wetting behaviour (§16.2.1 and §16.2.2).

WETTING, SPREADING, AND THIN LIQUID FILMS
See also Contact angle

Thin liquid films and surface forces

A *thin liquid film* is one in which *surface forces* act between the two surfaces, and the tension of the two interfaces differ from the tension of bulk solution (§12.2.1). Films can be *symmetrical* (§12.2) or *asymmetrical* (§12.3). Surface forces in thin liquid films are conveniently considered in terms of disjoining pressure isotherms (§12.2.2).

Film drainage and rupture

Drainage and rupture of soap films is inhibited by the monolayers of adsorbed surfactant. The concentration gradient of adsorbed surfactant as a vertical film drains arrests movement of the interfaces and inhibits drainage (§12.4). If a thin liquid film is subjected to shock such that there is a local stretching and thinning, the *film elasticity* associated with *Marangoni effect* heals the weak spot in the film (§6.2.2; §12.4).

Drop entry into a liquid-fluid interface

The entry of a submerged liquid droplet into a liquid/fluid interface in an equilibrium system is thermodynamically feasible if the *classical entry coefficient E* (defined in terms of the various interfacial tensions) exceeds 0. A stable thin film formed between the drop and the interface may however prevent entry in practice (§12.3.2; Box 14.7; §16.2.1).

Dynamic spreading and the precursor film

The dynamic spreading of a small drop of liquid on a smooth horizontal solid surface is preceded by a *precursor thin film* formed by flow of liquid from the drop. The drop effectively spreads over the precursor film (§16.4.1). The presence of surfactant has a considerable effect on spreading dynamics. So-called *superspreading* of surfactant solutions can occur on hydrophobic surfaces. On hydrophilic surfaces *stick-slip motion* at the spreading edge of drops can occur as can the formation of *dendrites* (§16.4.2).

Equilibrium film thickness: disjoining pressure isotherm and capillary pressure

An *equilibrium thickness* of a thin liquid film is attained when the *capillary pressure* is equal to the *disjoining pressure* on a limb of the disjoining pressure isotherm with a negative slope. Thin films can have two equilibrium thicknesses; for soap films these correspond to the *common black film*, which contains liquid, and the *Newton black film* which is essentially a bilayer of surfactant molecules with hydrated head groups (§12.2.2).

Duplex and thin liquid films and the spreading coefficient

Spreading of a liquid on another liquid or solid can give a thick (*duplex*) film or a drop or lens of liquid which can be in equilibrium with *a thin liquid film* (with which it has a finite contact angle) or an *adsorbed layer* which exerts a surface pressure (§12.3.3). For spreading to give a duplex film the *classical spreading coefficient, S,* is negative (§16.2.1).

Thin liquid films and emulsion and foam stability and breakdown

The stability of both emulsions (§14.1) and foams (§14.8) is very dependent on the stability of, respectively, the thin *emulsion films* that separate droplets in close proximity, and *foam films*. A (dry) foam is effectively an array of soap films connected by a network of Plateau borders. *Destabilization (breaking) of emulsions and foams* often involves causing the rupture of the emulsion or foam films (§14.7; §14.11).

Index

Bold numbers refer to major sections or chapters and *italic* numbers to subsections; subsequent page numbers are not indexed in these cases.